AF781836

Table of Contents[1]

Similarity, Proof, and Trigonometry

[1]Each lesson is ONE day, and ONE day is considered a 45-minute period.

Geometry • Module 2

Similarity, Proof, and Trigonometry

OVERVIEW

Just as rigid motions are used to define congruence in Module 1, so dilations are added to define similarity in Module 2.

To be able to define similarity, there must be a definition of similarity transformations and, consequently, a definition for dilations. Students are introduced to the progression of terms beginning with scale drawings, which they first study in Grade 7 (Module 1 Topic D), but in a more observational capacity than in high school Geometry: Students determine the scale factor between a figure and a scale drawing or predict the lengths of a scale drawing, provided a figure and a scale factor. In Topic A, students begin with a review of scale drawings in Lesson 1, followed by two lessons on how to systematically create scale drawings. The study of scale drawings, specifically the way they are constructed under the ratio and parallel methods, gives us the language to examine dilations. The comparison of why both construction methods (MP.7) result in the same image leads to two theorems: the triangle side splitter theorem and the dilation theorem. Note that while dilations are defined in Lesson 2, it is the dilation theorem in Lesson 5 that begins to tell us how dilations behave (**G-SRT.A.1**, **G-SRT.A.4**).

Topic B establishes a firm understanding of how dilations behave. Students prove that a dilation maps a line to itself or to a parallel line and, furthermore, dilations map segments to segments, lines to lines, rays to rays, circles to circles, and an angle to an angle of equal measure. The lessons on proving these properties, Lessons 7–9, require students to build arguments based on the structure of the figure in question and a handful of related facts that can be applied to the situation (e.g., the triangle side splitter theorem is called on frequently to prove that dilations map segments to segments and lines to lines) (MP.3, MP.7). Students apply their understanding of dilations to divide a line segment into equal pieces and explore and compare dilations from different centers.

In Topic C, students learn what a similarity transformation is and why, provided the right circumstances, both rectilinear and curvilinear figures can be classified as similar (**G-SRT.A.2**). After discussing similarity in general, the scope narrows, and students study criteria for determining when two triangles are similar (**G-SRT.A.3**). Part of studying triangle similarity criteria (Lessons 15 and 17) includes understanding side length ratios for similar triangles, which begins to establish the foundation for trigonometry (**G-SRT.B.5**). The final two lessons demonstrate the usefulness of similarity by examining how two ancient Greek mathematicians managed to measure the circumference of the earth and the distance to the moon, respectively (**G-MG.A.1**).

In Topic D, students are laying the foundation to studying trigonometry by focusing on similarity between right triangles in particular (the importance of the values of corresponding length ratios between similar triangles is particularly apparent in Lessons 16, 21, and 25). Students discover that a right triangle can be divided into two similar sub-triangles (MP.2) to prove the Pythagorean theorem (**G-SRT.B.4**). Two lessons are spent studying the algebra of radicals that is useful for solving for sides of a right triangle and computing trigonometric ratios.

An introduction to trigonometry, specifically right triangle trigonometry and the values of side length ratios within right triangles, is provided in Topic E by defining the sine, cosine, and tangent ratios and using them to find missing side lengths of a right triangle (**G-SRT.B.6**). This is in contrast to studying trigonometry in the context of functions, as is done in Algebra II of this curriculum. Students explore the relationships between sine, cosine, and tangent using complementary angles and the Pythagorean theorem (**G-SRT.B.7**, **G-SRT.B.8**). Students discover the link between how to calculate the area of a non-right triangle through algebra versus trigonometry. Topic E continues with a study of the laws of sines and cosines to apply them to solve for missing side lengths of an acute triangle (**G-SRT.D.10**, **G-SRT.D.11**). Topic E closes with Lesson 34, which introduces students to the functions $\arcsin$, $\arccos$, and $\arctan$, which are formally taught as inverse functions in Algebra II. Students use what they know about the trigonometric functions sine, cosine, and tangent to make sense of $\arcsin$, $\arccos$, and $\arctan$. Students use these new functions to determine the unknown measures of angles of a right triangle.

Throughout the module, students are presented with opportunities to apply geometric concepts in modeling situations. Students use geometric shapes to describe objects (**G-MG.A.1**) and apply geometric methods to solve design problems where physical constraints and cost issues arise (**G-MG.A.3**).

Focus Standards

Understand similarity in terms of similarity transformations.

G-SRT.A.1 Verify experimentally the properties of dilations given by a center and a scale factor:

a. A dilation takes a line not passing through the center of the dilation to a parallel line, and leaves a line passing through the center unchanged.

b. The dilation of a line segment is longer or shorter in the ratio given by the scale factor.

G-SRT.A.2 Given two figures, use the definition of similarity in terms of similarity transformations to decide if they are similar; explain using similarity transformations the meaning of similarity for triangles as the equality of all corresponding pairs of angles and the proportionality of all corresponding pairs of sides.

G-SRT.A.3 Use the properties of similarity transformations to establish the AA criterion for two triangles to be similar.

Prove theorems involving similarity.

G-SRT.B.4 Prove theorems about triangles. *Theorems include: a line parallel to one side of a triangle divides the other two proportionally, and conversely; the Pythagorean Theorem proved using triangle similarity.*

G-SRT.B.5 Use congruence and similarity criteria for triangles to solve problems and to prove relationships in geometric figures.

Define trigonometric ratios and solve problems involving right triangles.

G-SRT.C.6 Understand that by similarity, side ratios in right triangles are properties of the angles in the triangle, leading to definitions of trigonometric ratios for acute angles.

G-SRT.C.7 Explain and use the relationship between the sine and cosine of complementary angles.

G-SRT.C.8 Use trigonometric ratios and the Pythagorean Theorem to solve right triangles in applied problems.★

Apply geometric concepts in modeling situations.

G-MG.A.1 Use geometric shapes, their measures, and their properties to describe objects (e.g., modeling a tree trunk or a human torso as a cylinder).★

G-MG.A.3 Apply geometric methods to solve design problems (e.g., designing an object or structure to satisfy physical constraints or minimize cost; working with typographic grid systems based on ratios).★

Extension Standards

Apply trigonometry to general triangles.

G-SRT.D.9 (+) Derive the formula $A = 1/2\ ab\ \sin(C)$ for the area of a triangle by drawing an auxiliary line from a vertex perpendicular to the opposite side.

G-SRT.D.10 (+) Prove the Laws of Sines and Cosines and use them to solve problems.

G-SRT.D.11 (+) Understand and apply the Law of Sines and the Law of Cosines to find unknown measurements in right and non-right triangles (e.g., surveying problems, resultant forces).

Foundational Standards

Draw, construct, and describe geometrical figures and describe the relationships between them.

7.G.A.1 Solve problems involving scale drawings of geometric figures, including computing actual lengths and areas from a scale drawing and reproducing a scale drawing at a different scale.

Understand congruence and similarity using physical models, transparencies, or geometry software.

8.G.A.3 Describe the effect of dilations, translations, rotations, and reflections on two-dimensional figures using coordinates.

8.G.A.4 Understand that a two-dimensional figure is similar to another if the second can be obtained from the first by a sequence of rotations, reflections, translations, and dilations; given two similar two-dimensional figures, describe a sequence that exhibits the similarity between them.

8.G.A.5 Use informal arguments to establish facts about the angle sum and exterior angle of triangles, about the angles created when parallel lines are cut by a transversal, and the angle-angle criterion for similarity of triangles. *For example, arrange three copies of the same triangle so that the sum of the three angles appears to form a line, and give an argument in terms of transversals why this is so.*

Focus Standards for Mathematical Practice

MP.3 **Construct viable arguments and critique the reasoning of others**. Critical to this module is the need for dilations in order to define similarity. In order to understand dilations fully, the proofs in Lessons 4 and 5 to establish the triangle side splitter and the dilation theorems require students to build arguments based on definitions and previously established results. This is also apparent in Lessons 7, 8, and 9, when the properties of dilations are being proven. Though there are only a handful of facts students must point to in order to create arguments, how students reason with these facts determine if their arguments actually establish the properties. It is essential to communicate effectively and purposefully.

MP.7 **Look for and make use of structure**. Much of the reasoning in Module 2 centers around the interaction between figures and dilations. It is unsurprising, then, that students must pay careful attention to an existing structure and how it changes under a dilation, for example, why it is that dilating the key points of a figure by the ratio method results in the dilation of the segments that join them. The math practice also ties into the underlying idea of trigonometry: how to relate the values of corresponding ratio lengths between similar right triangles and how the value of a trigonometric ratio hinges on a given acute angle within a right triangle.

Terminology

New or Recently Introduced Terms

- **Cosine** (Let θ be the angle measure of an acute angle of the right triangle. The *cosine of θ of a right triangle* is the value of the ratio of the length of the adjacent side (denoted *adj*) to the length of the hypotenuse (denoted *hyp*). As a formula, $\cos\theta = \text{adj}/\text{hyp}$.)

- **Dilation** (For $r > 0$, a *dilation with center* C *and scale factor* r is a transformation $D_{C,r}$ of the plane defined as follows:
  1. For the center C, $D_{C,r}(C) = C$, and
  2. For any other point P, $D_{C,r}(P)$ is the point Q on $\overrightarrow{CP}$ so that $CQ = r \cdot CP$.)
- **Sides of a Right Triangle** (The *hypotenuse* of a right triangle is the side opposite the right angle; the other two sides of the right triangle are called the *legs*. Let θ be the angle measure of an acute angle of the right triangle. The *opposite side* is the leg opposite that angle. The *adjacent side* is the leg that is contained in one of the two rays of that angle (the hypotenuse is contained in the other ray of the angle).)
- **Similar** (Two figures in a plane are *similar* if there exists a similarity transformation taking one figure onto the other figure. A congruence is a similarity with scale factor 1. It can be shown that a similarity with scale factor 1 is a congruence.)
- **Similarity Transformation** (A *similarity transformation* (or *similarity*) is a composition of a finite number of dilations or basic rigid motions. The *scale factor* of a similarity transformation is the product of the scale factors of the dilations in the composition; if there are no dilations in the composition, the scale factor is defined to be 1. A similarity is an example of a transformation.)
- **Sine** (Let θ be the angle measure of an acute angle of the right triangle. The *sine of* θ *of a right triangle* is the value of the ratio of the length of the opposite side (denoted *opp*) to the length of the hypotenuse (denoted *hyp*). As a formula, $\sin\theta = \text{opp}/\text{hyp}$.)
- **Tangent** (Let θ be the angle measure of an acute angle of the right triangle. The *tangent of* θ *of a right triangle* is the value of the ratio of the length of the opposite side (denoted *opp*) to the length of the adjacent side (denoted *adj*). As a formula, $\tan\theta = \text{opp}/\text{adj}$.)

  Note that in Algebra II, sine, cosine, and tangent are thought of as functions whose domains are subsets of the real numbers; they are not considered as values of ratios. Thus, in Algebra II, the values of these functions for a given θ are notated as $\sin(\theta)$, $\cos(\theta)$, and $\tan(\theta)$ using function notation (i.e., parentheses are included).

Familiar Terms and Symbols[2]

- Composition
- Dilation
- Pythagorean theorem
- Rigid motions
- Scale drawing
- Scale factor
- Slope

[2]These are terms and symbols students have seen previously.

Suggested Tools and Representations

- Compass and straightedge

Assessment Summary

Assessment Type	Administered	Format	Standards Addressed
Mid-Module Assessment Task	After Topic C	Constructed response with rubric	G-SRT.A.1, G-SRT.A.2, G-SRT.A.3, G-SRT.B.4, G-SRT.B.5, G-MG.A.1, G-MG.A.3
End-of-Module Assessment Task	After Topic E	Constructed response with rubric	G-SRT.B.4, G-SRT.B.5, G-SRT.C.6, G-SRT.C.7, G-SRT.C.8

Topic A

Scale Drawings

G-SRT.A.1, G-SRT.B.4, G-MG.A.3

Focus Standards:	G-SRT.A.1	Verify experimentally the properties of dilations given by a center and a scale factor: a. A dilation takes a line not passing through the center of the dilation to a parallel line, and leaves a line passing through the center unchanged. b. The dilation of a line segment is longer or shorter in the ratio given by the scale factor.
	G-SRT.B.4	Prove theorems about triangles. *Theorems include: a line parallel to one side of a triangle divides the other two proportionally, and conversely; the Pythagorean Theorem proved using triangle similarity.*
	G-MG.A.3	Apply geometric methods to solve design problems (e.g., designing an object or structure to satisfy physical constraints or minimize cost; working with typographic grid systems based on ratios).★
Instructional Days:	5	
Lesson 1:	Scale Drawings (P)[1]	
Lesson 2:	Making Scale Drawings Using the Ratio Method (P)	
Lesson 3:	Making Scale Drawings Using the Parallel Method (P)	
Lesson 4:	Comparing the Ratio Method with the Parallel Method (S)	
Lesson 5:	Scale Factors (S)	

Students embark on Topic A with a brief review of scale drawings and scale factor, which they last studied in Grades 7 and 8. In Lesson 1, students recall the properties of a well-scaled drawing and practice creating scale drawings using basic construction techniques. Lessons 2 and 3 explore systematic techniques for creating scale drawings. With the ratio method, students dilate key points of a figure according to the scale factor to produce a scale drawing (**G-SRT.A.1**). Note that exercises within Lesson 2 where students apply the ratio method to solve design problems relate to the modeling standard **G-MG.A.3**. With the parallel method, students construct sides parallel to corresponding sides of the original figure to create a scale drawing. Lesson 4 is an examination of these two methods, with the goal of understanding why the methods produce identical drawings. The outcome of this comparison is the triangle side splitter theorem, which states that a segment splits two sides of a triangle proportionally if and only if it is parallel to the third side (**G-SRT.B.4**).

[1]Lesson Structure Key: **P**-Problem Set Lesson, **M**-Modeling Cycle Lesson, **E**-Exploration Lesson, **S**-Socratic Lesson

This theorem is then used in Lesson 5 to establish the dilation theorem: A dilation from a center O maps a segment PQ to a segment $P'Q'$ so that $P'Q' = r \cdot PQ$; additionally, if O is not contained in $\overleftrightarrow{PQ}$ and $r \neq 1$, then $\overleftrightarrow{PQ} || \overleftrightarrow{P'Q'}$.

As opposed to work done in Grade 8 on dilations, where students observed how dilations behaved and experimentally verified properties of dilations by examples, high school Geometry is anchored in explaining why these properties are true by reasoned argument. Grade 8 content focused on *what* was going on, while high school Geometry content focuses on explaining *why* it occurs. This is particularly true in Lessons 4 and 5, where students rigorously explain their explorations of dilations using the ratio and parallel methods to build arguments that establish the triangle side splitter and dilation theorems (MP.3).

EUREKA MATH™

Lesson 1: Scale Drawings

Student Outcomes

- Students review properties of scale drawings and are able to create them.

Lesson Notes

Lesson 1 reviews the properties of a scale drawing before studying the relationship between dilations and scale drawings in Lessons 2 and 3. Students focus on scaling triangles using construction tools and skills learned in Module 1. The lesson begins by exploring how to scale images using common electronics. After students work on scaling triangles given various pieces of initial information, the lesson is tied together by showing how triangle scaling can be used in programming a phone to scale a complex image.

Note that students first studied scale drawings in Grade 7 (Module 1 Lessons 16–22). Teachers may need to modify the exercises to include the use of graph paper, patty paper, and geometry software (e.g., freely available GeoGebra) to make the ideas accessible.

Classwork

Opening (2 minutes)

- A common feature on cell phones and tablets is the ability to scale, that is, to enlarge or reduce an image by putting a thumb and index finger to the screen and making a pinching (to reduce) or spreading movement (to enlarge) as shown in the diagram below.

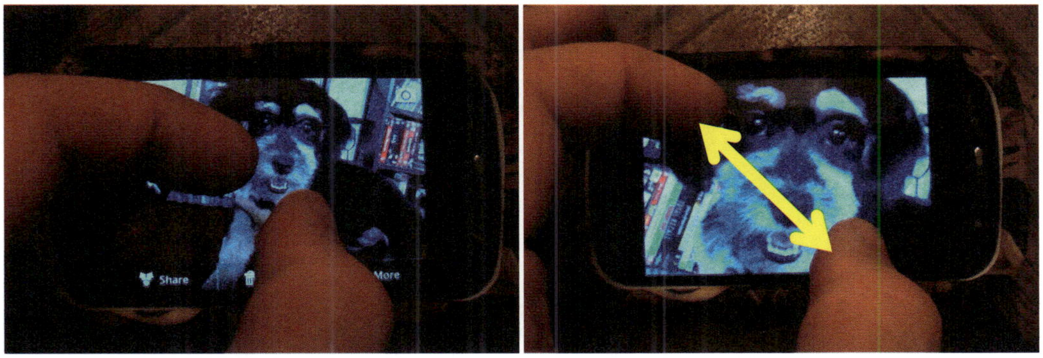

- Notice that as the fingers move outward on the screen (shown on the right), the image of the puppy is enlarged on the screen.
- How did the code for this feature get written? What general steps must the code dictate?
- Today we review a concept that is key to tackling these questions.

Opening Exercise (2 minutes)

Opening Exercise

Above is a picture of a bicycle. Which of the images below appears to be a well-scaled image of the original? Why?

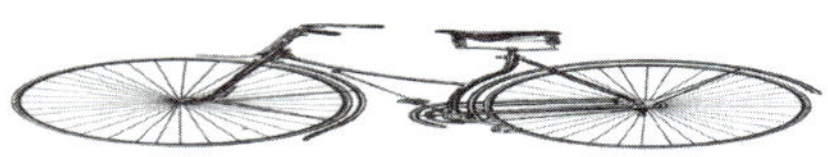  

Only the third image appears to be a well-scaled image since the image is in proportion to the original.

As mentioned in the Lesson Notes, students have seen scale drawings in Grades 4 and 7. The Opening Exercise is kept brief to reintroduce the idea of what it means to be well scaled without going into great depth yet.

After the Opening Exercise, refer to the Opening, and re-pose the initial intent. Several questions are provided to help illustrate the pursuit of the lesson. The expectation is not to answer these questions now but to keep them in mind as the lesson progresses.

- How did the code to scale images get written? What kinds of instructions guide the scaling process? What steps take an original figure to a scale drawing? How is the program written to make sure that images are well scaled and not distorted?

To help students answer these questions, the problem is simplified by examining the process of scaling a simpler figure: a triangle. After tackling this simpler problem, it is possible to revisit the more complex images.

Example 1 (8 minutes)

Example 1 provides students with a triangle and a scale factor. Students use a compass and straightedge to create the scale drawing of the triangle.

> **Example 1**
>
> **Use construction tools to create a scale drawing of $\triangle ABC$ with a scale factor of $r = 2$.**

Scaffolding:

- One way to facilitate Example 1 is to use graph paper, with $\triangle ABC$ located at $A(-3,-2)$, $B(5,-2)$, and $C(-3,6)$.
- An alternative challenge is to use a scale of $r = \frac{1}{2}$.

Before students begin creating the scale drawing, review the following.

- Recall that the scale factor r is the ratio of any length in a scale drawing relative to its corresponding length in the original figure. A scale factor $r > 1$ results in an enlargement of the original figure. A scale factor of $0 < r < 1$ results in a reduction of the original figure.

For further background information, refer to Grade 7 (Module 1 Lesson 17).

- Since we know that a scale drawing can be created without concern for location, a scale drawing can be done in two ways: (1) by drawing it so that one vertex coincides with its corresponding vertex, leaving some overlap between the original triangle and the scale drawing and (2) by drawing it completely independent of the original triangle. Two copies of the original triangle have been provided for you.

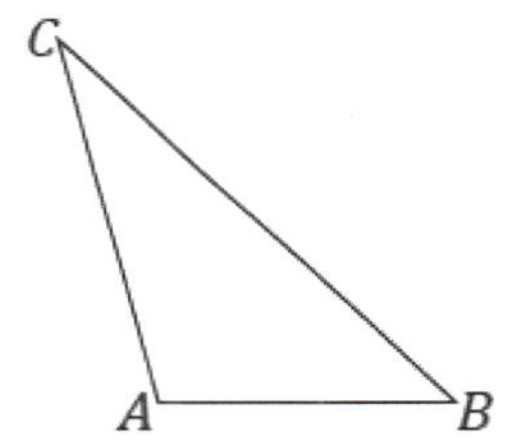

MP.5

***Solution 1:** Draw $\overrightarrow{AB}$. To determine B', adjust the compass to the length of AB. Then reposition the compass so that the point is at B, and mark off the length of $\overline{AB}$; label the intersection with $\overrightarrow{AB}$ as B'. C' is determined in a similar manner. Join B' to C'.*

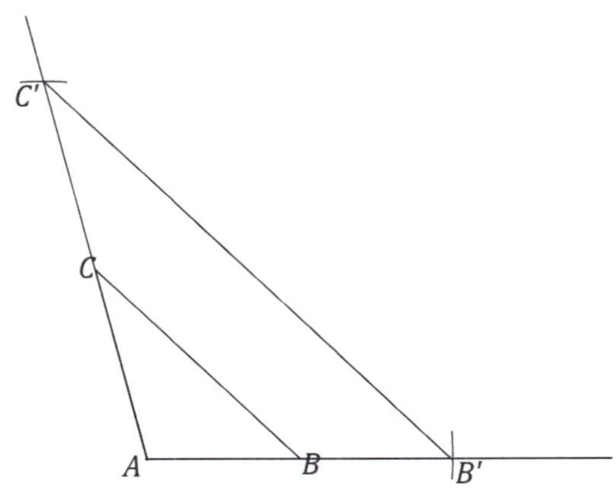

MP.5

Solution 2: *Draw a segment that will be longer than double the length of $\overrightarrow{AB}$. Label one end as A'. Adjust the compass to the length of $\overline{AB}$, and mark off two consecutive such lengths along the segment, and label the endpoint as B'. Copy $\angle A$. Determine C' along the $\overrightarrow{AC}$ in the same way as B'. Join B' to C'.*

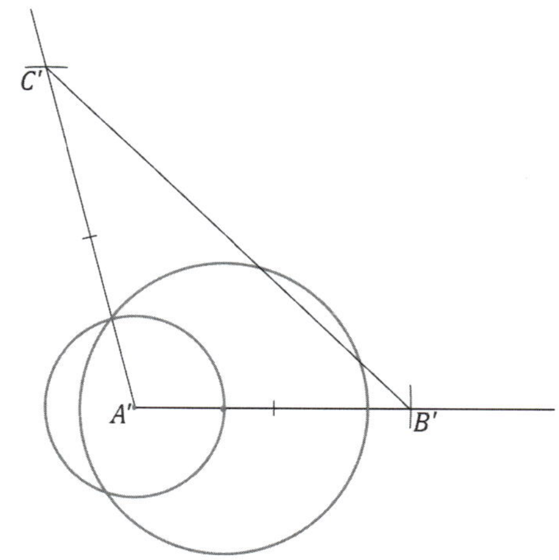

- Why do both solutions yield congruent triangles? Both triangles begin with the same criteria: Two pairs of sides that are equal in length and a pair of included angles that are equal in measurement. By SAS, a unique triangle is determined. Since $\triangle ABC$ was scaled by a factor of 2 in each case, the method in which we scale does not change the outcome; that is, we have a triangle with the same dimensions whether we position it on top of or independent of the original triangle.
- Regardless of which solution method you used, measure the length of $\overline{BC}$ and $\overline{B'C'}$. What do you notice?
  - $\overline{B'C'}$ is twice the length of $\overline{BC}$.
- Now measure the angles $\angle B$, $\angle C$, $\angle B'$, and $\angle C'$. What do you notice?
  - *The measures of $\angle B$ and $\angle B'$ are the same, as are $\angle C$ and $\angle C'$.*

Discussion (3 minutes)

- What are the properties of a well-scaled drawing of a figure?
  - *A well-scaled drawing of a figure is one where corresponding angles are equal in measure, and corresponding lengths are all in the same proportion.*
- What is the term for the constant of proportionality by which all lengths are scaled in a well-scaled drawing?
  - *The scale factor is the constant of proportionality.*
- If somewhere else on your paper you created the scale drawing in Example 1 but oriented it at a different angle, would the drawing still be a *scale* drawing?
  - *Yes. The orientation of the drawing does not change the fact that it is a scale drawing; the properties of a scale drawing concern only lengths and relative angles.*

Reinforce this by considering the steps of Solution 2 in Example 1. The initial segment can be drawn anywhere, and the steps following can be completed as is. Ensure that students understand and rehearse the term *orientation,* and record a student-friendly definition.

- Return to the three images of the original bicycle. Why are the first two images classified as not well-scaled?
  - *The corresponding angles are not equal in measurement, and the corresponding lengths are not in constant proportion.*

Exercise 1 (4 minutes)

Students scale a triangle by a factor of $r = 3$ in Exercise 1. Either of the above solution methods is acceptable. As students work on Exercise 1, take time to circulate and check for understanding. Note that teachers may choose to provide graph paper and have students create scale drawings on it.

Exercise 1

1. **Use construction tools to create a scale drawing of $\triangle DEF$ with a scale factor of $r = 3$. What properties does your scale drawing share with the original figure? Explain how you know.**

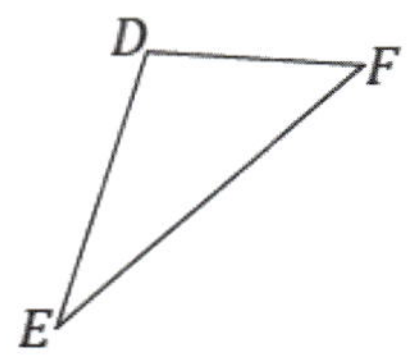

By measurement, I can see that each side is three times the length of the corresponding side of the original figure and that all three angles are equal in measurement to the three corresponding angles in the original figure.

Make sure students understand that any of these diagrams are acceptable solutions.

A solution where $\overline{ED'}$ and $\overline{EF'}$ are drawn first.

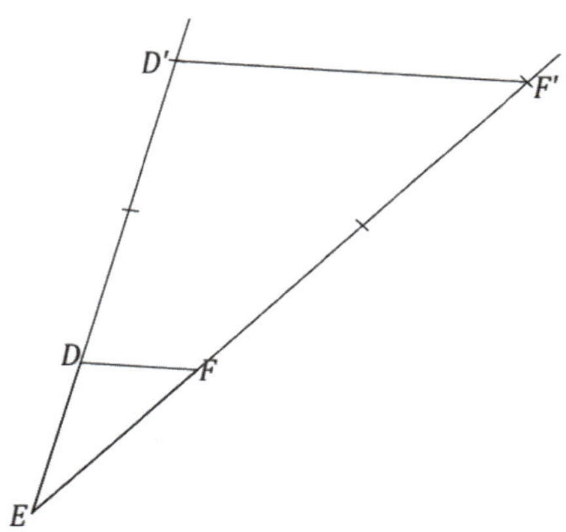

A solution where $\overline{FE'}$ and $\overline{FD'}$ are drawn first.

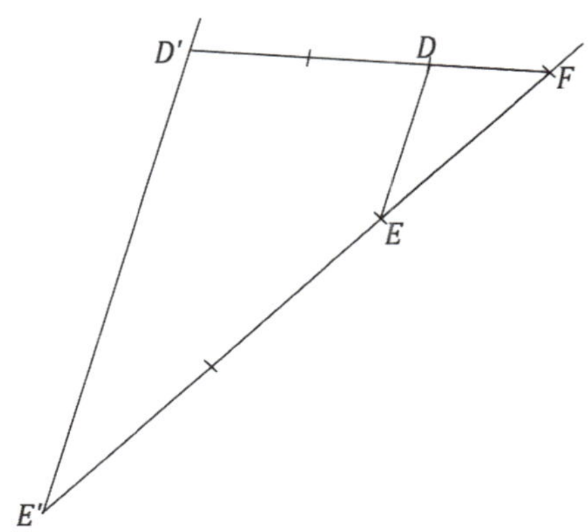

A solution where $\overline{DE'}$ and $\overline{DF'}$ are drawn first.

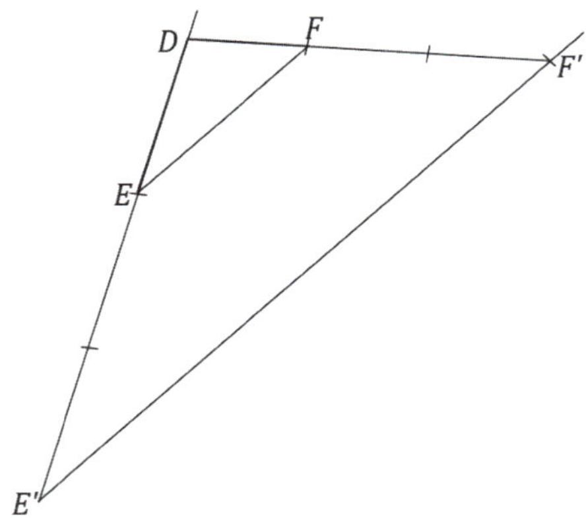

Example 2 (4 minutes)

Example 2 provides a triangle and a scale factor of $r = \frac{1}{2}$. Students use a compass to locate the midpoint.

Example 2

Use construction tools to create a scale drawing of $\triangle XYZ$ with a scale factor of $r = \frac{1}{2}$.

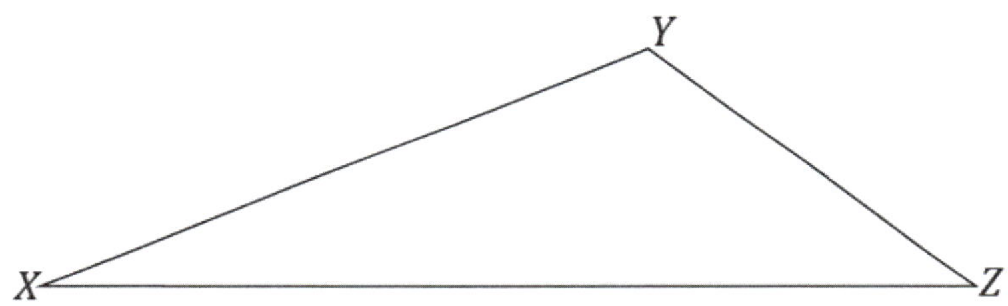

- Which construction technique have we learned that can be used in this question that was not used in the previous two problems?
  - *We can use the construction to determine the perpendicular bisector to locate the midpoint of two sides of $\triangle XYZ$.*

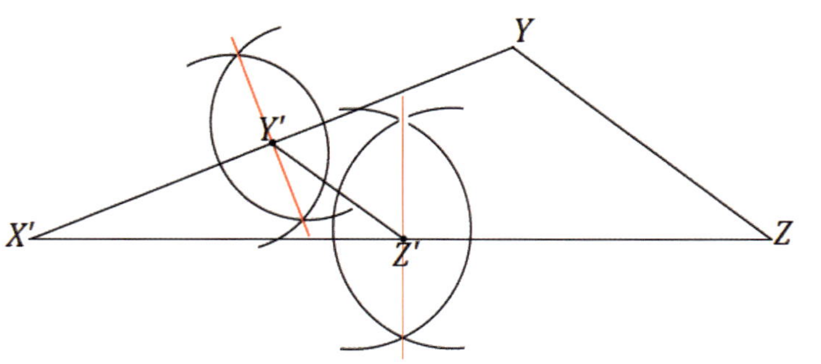

Scaffolding:
For students struggling with constructions, consider having them measure the lengths of two sides and then determine the midpoints.

EUREKA MATH™

As the solutions to Exercise 1 showed, the constructions can be done on other sides of the triangle (i.e., the perpendicular bisectors of $\overline{YZ}$ and $\overline{XZ}$ are acceptable places to start.)

Exercises 2–4 (13 minutes)

Have students complete Exercise 2 and, if time allows, go on to Exercises 3 and 4. In Exercise 2, using a scale factor of $r = \frac{1}{4}$ is a natural progression following the use of a scale factor of $r = \frac{1}{2}$ in Example 1. Prompt students to consider how $\frac{1}{4}$ relates to $\frac{1}{2}$. They should recognize that the steps of the construction in Exercise 2 are similar to those in Example 1.

Exercises 2–4

2. Use construction tools to create a scale drawing of $\triangle PQR$ with a scale factor of $r = \frac{1}{4}$. What properties do the scale drawing and the original figure share? Explain how you know.

> *Scaffolding:*
> For students who are ready for a challenge, consider asking them to use a scale factor of $r = \frac{3}{4}$.

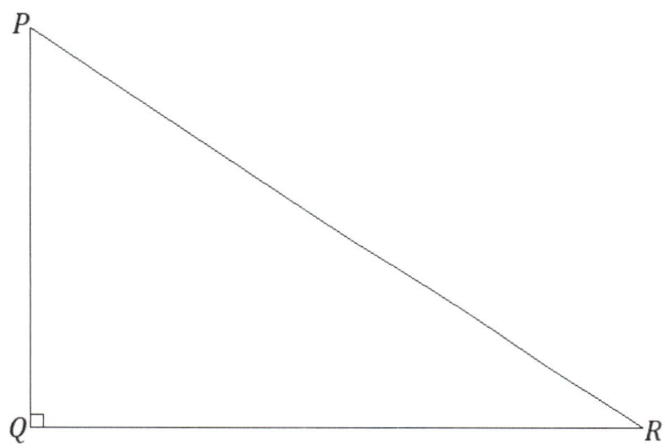

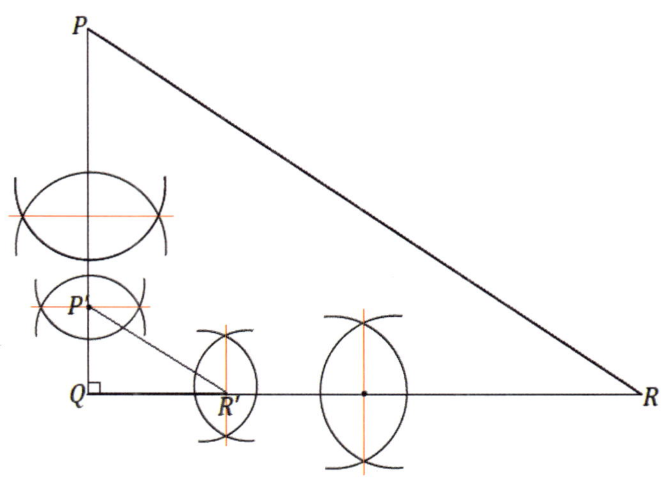

By measurement, I can see that all three sides are each one-quarter the lengths of the corresponding sides of the original figure, and all three angles are equal in measurement to the three corresponding angles in the original figure.

3. **Triangle EFG is provided below, and one angle of scale drawing $\triangle E'F'G'$ is also provided. Use construction tools to complete the scale drawing so that the scale factor is $r = 3$. What properties do the scale drawing and the original figure share? Explain how you know.**

Extend either ray from G'. Use the compass to mark off a length equal to $3EG$ on one ray and a length equal to $3FG$ on the other. Label the ends of the two lengths E' and F', respectively. Join E' to F'.

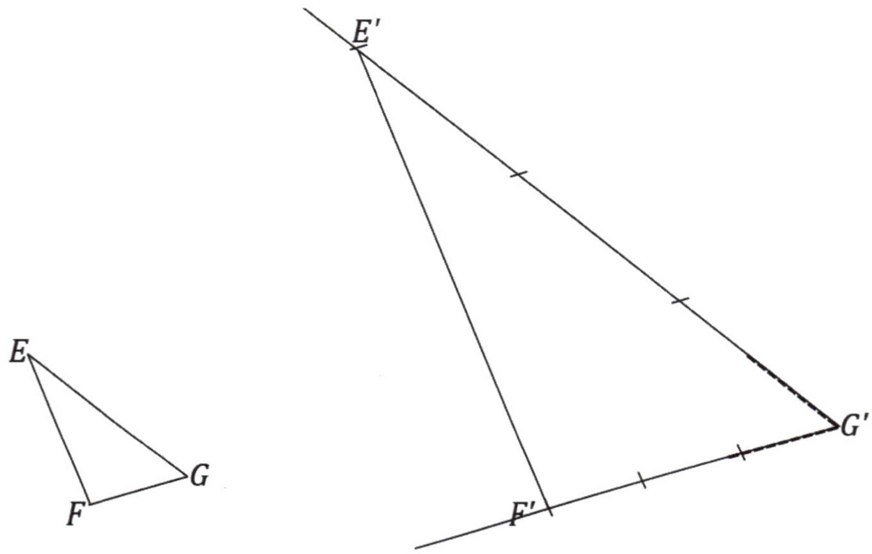

By measurement, I can see that each side is three times the length of the corresponding side of the original figure and that all three angles are equal in measurement to the three corresponding angles in the original figure.

4. **Triangle ABC is provided below, and one side of scale drawing $\triangle A'B'C'$ is also provided. Use construction tools to complete the scale drawing and determine the scale factor.**

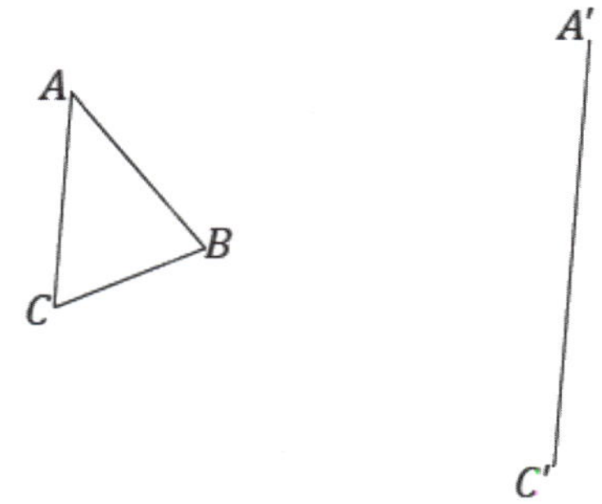

***One possible solution:** We can copy $\angle A$ and $\angle C$ at points A' and C' so that the new rays intersect as shown and call the intersection point B'. By measuring, we can see that $A'C' = 2AC$, $A'B' = 2AB$, and $B'C' = 2BC$. We already know that $m\angle A' = m\angle A$ and $m\angle C' = m\angle C$. By the triangle sum theorem, $m\angle B' = m\angle B$.*

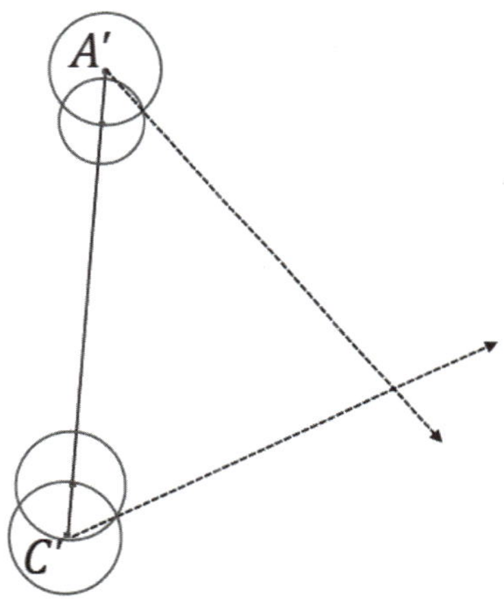

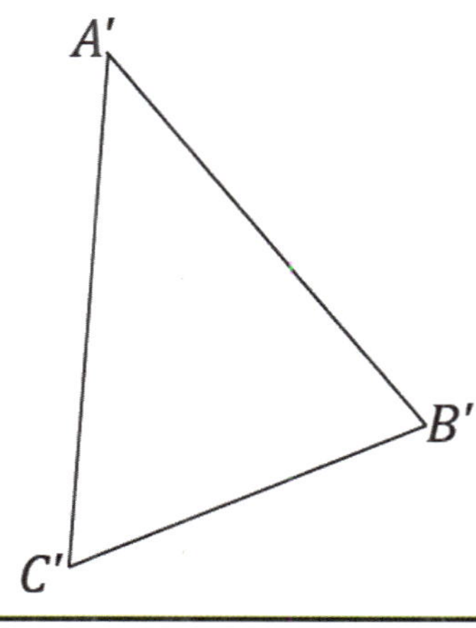

ffolding:

- If students struggle with constructing an angle of equal measure, consider allowing them to use a protractor to measure angles A and C, and draw angles at A' and C', respectively, with equal measures. This alleviates time constraints; however, know that constructing an angle is a necessary skill to be addressed in remediation.
- Use patty paper or geometry software to allow students to focus on the concept development.

Discussion (3 minutes)

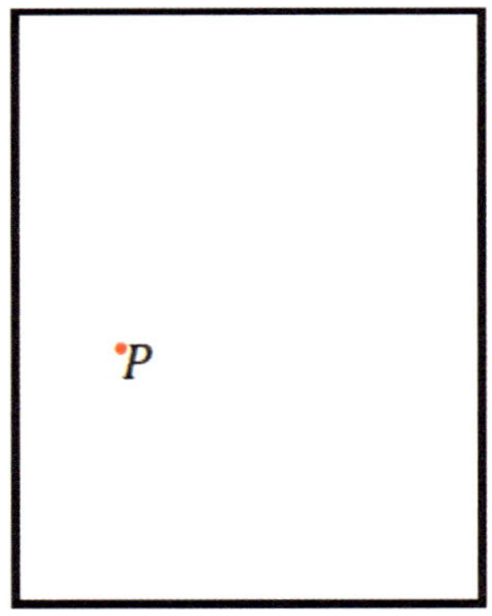

- In the last several exercises, we constructed or completed scale drawings of triangles, provided various pieces of information. We now return to the question that began the lesson.
- What does the work we did with scaled triangles have to do with understanding the code that is written to tell a phone or a computer how to enlarge or reduce an image? Here is one possible way.
- Consider the following figure, which represents an image or perhaps a photo. A single point P is highlighted in the image, which can easily be imagined to be one of many points of the image (e.g., it could be just a single point of the bicycle in the Opening Exercise).
- If we know by how much we want to enlarge or reduce the image (i.e., the scale factor), we can use what we know about scaling triangles to locate where this point ends up in the scale drawing of the image.
- For example, if we were to place our thumb and index finger on the points T and I, respectively, and make the zoom-in movement so that our thumb and index finger end up at T' and I', respectively, the scale factor that corresponds to the zoom-in motion dictates where P' is located.
- Therefore, we can generalize how the code to enlarge or reduce an image on a phone or a tablet is written. For every point P in an image, a triangle can be formed with vertices P, T, and I. Since we are able to scale triangles, we are then also able to scale entire images, point by point. In fact, we can use this process not just for the code to program electronics but also to scale a complex image by hand if we so wished.

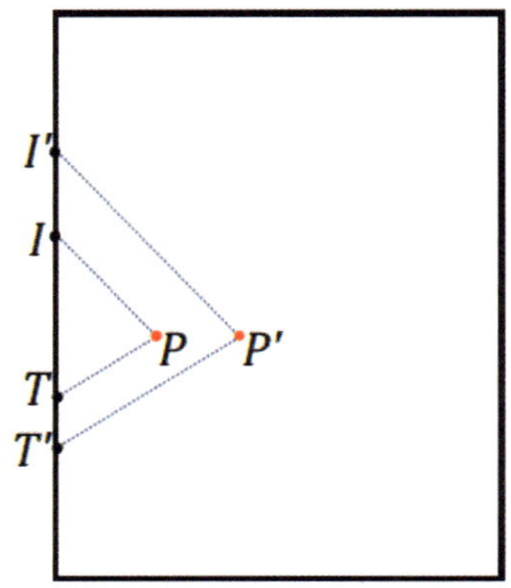

Closing (1 minute)

- What are the key properties of a scale drawing relative to its original figure?
  - *There are two properties of a scale drawing of a figure: Corresponding angles are equal in measurement, and corresponding lengths are proportional in measurement.*
- If we were to take any of the scale drawings in our examples and place them in a different location or rotate them on our paper, would it change the fact that the drawing is still a scale drawing?
  - *No, the properties of a scale drawing have to do with lengths and relative angles, not location or orientation.*
- Provided a triangle and a scale factor or a triangle and one piece of the scale drawing of the triangle, it is possible to create a complete scale drawing of the triangle using a compass and straightedge. No matter which method is used to create the scale drawing, we rely on triangle congruence criteria to ensure that a unique triangle is determined.

Exit Ticket (5 minutes)

Name ______________________________ Date ______________

Lesson 1: Scale Drawings

Exit Ticket

Triangle ABC is provided below, and one side of scale drawing $\triangle A'B'C'$ is also provided. Use construction tools to complete the scale drawing and determine the scale factor. What properties do the scale drawing and the original figure share? Explain how you know.

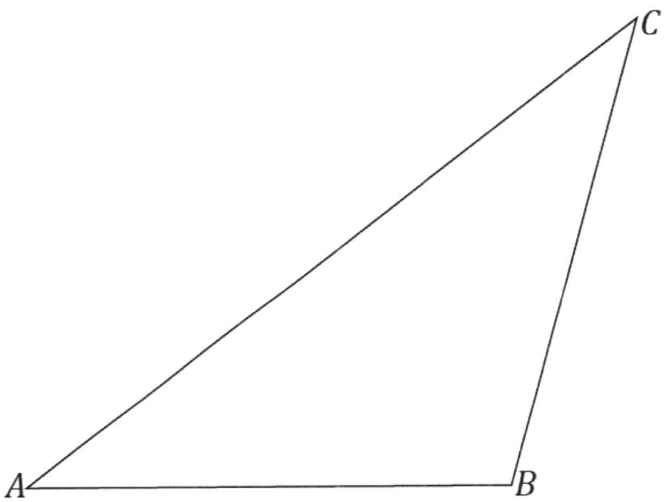

Exit Ticket Sample Solutions

Triangle ABC is provided below, and one side of scale drawing $\triangle A'B'C'$ is also provided. Use construction tools to complete the scale drawing and determine the scale factor. What properties do the scale drawing and the original figure share? Explain how you know.

One possible solution: Since the scale drawing will clearly be a reduction, use the compass to mark the number of lengths equal to the length of $\overline{A'B'}$ along $\overline{AB}$. Once the length of $\overline{A'C'}$ is determined to be $\frac{1}{2}$ the length of $\overline{AB}$, use the compass to find a length that is half the length of $\overline{AB}$ and half the length of $\overline{BC}$. Construct circles with radii of lengths $\frac{1}{2}AC$ and $\frac{1}{2}BC$ from the A' and B', respectively, to determine the location of C', which is at the intersection of the two circles.

By measurement, I can see that each side is $\frac{1}{2}$ the length of the corresponding side of the original figure and that all three angles are equal in measurement to the three corresponding angles in the original figure.

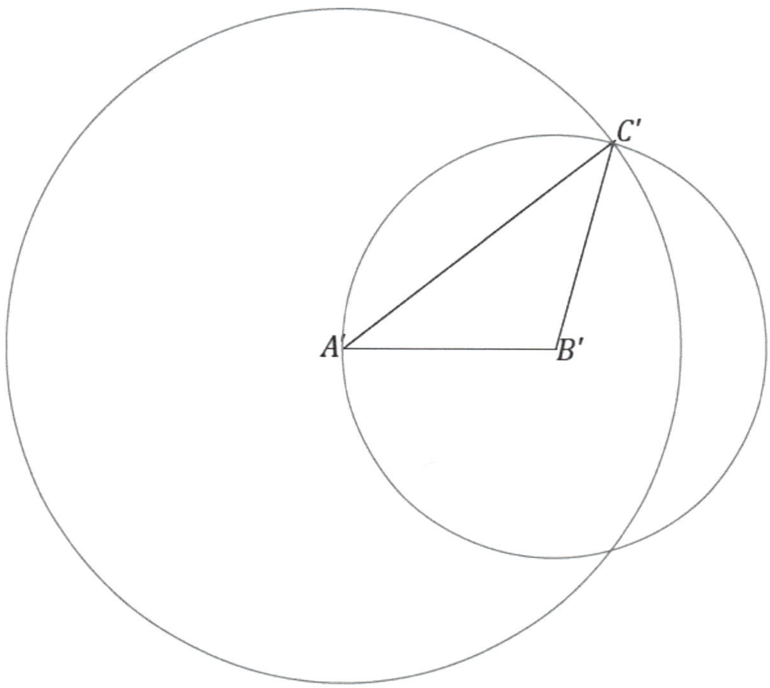

Problem Set Sample Solutions

1. **Use construction tools to create a scale drawing of $\triangle ABC$ with a scale factor of $r = 3$.**

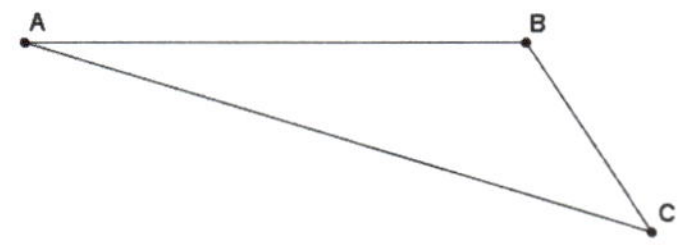

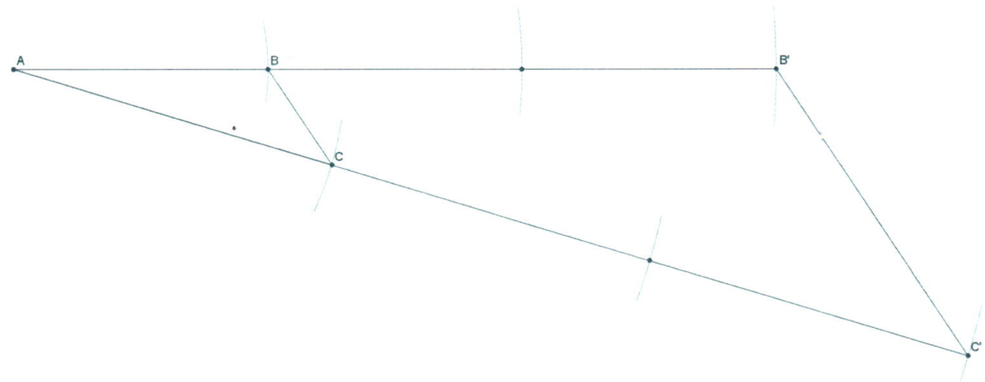

2. **Use construction tools to create a scale drawing of $\triangle ABC$ with a scale factor of $r = \frac{1}{2}$.**

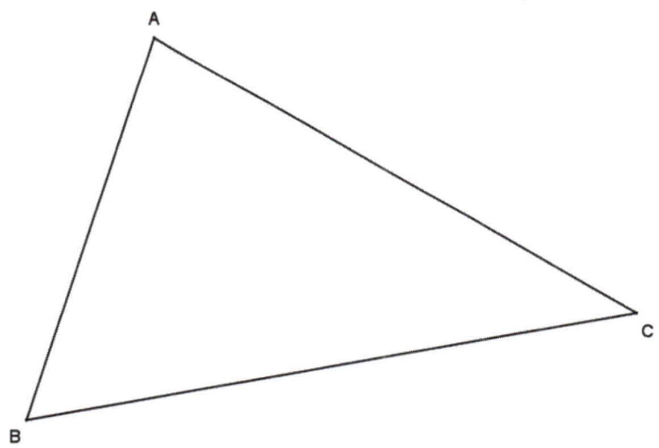

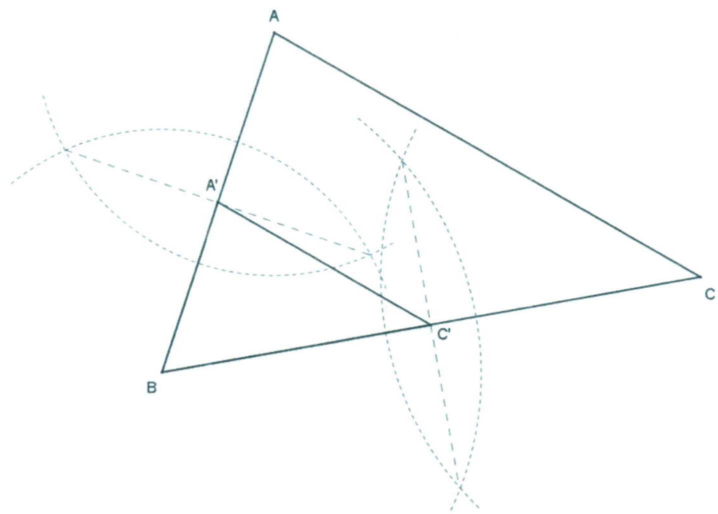

3. Triangle EFG is provided below, and one angle of scale drawing $\triangle E'F'G'$ is also provided. Use construction tools to complete a scale drawing so that the scale factor is $r = 2$.

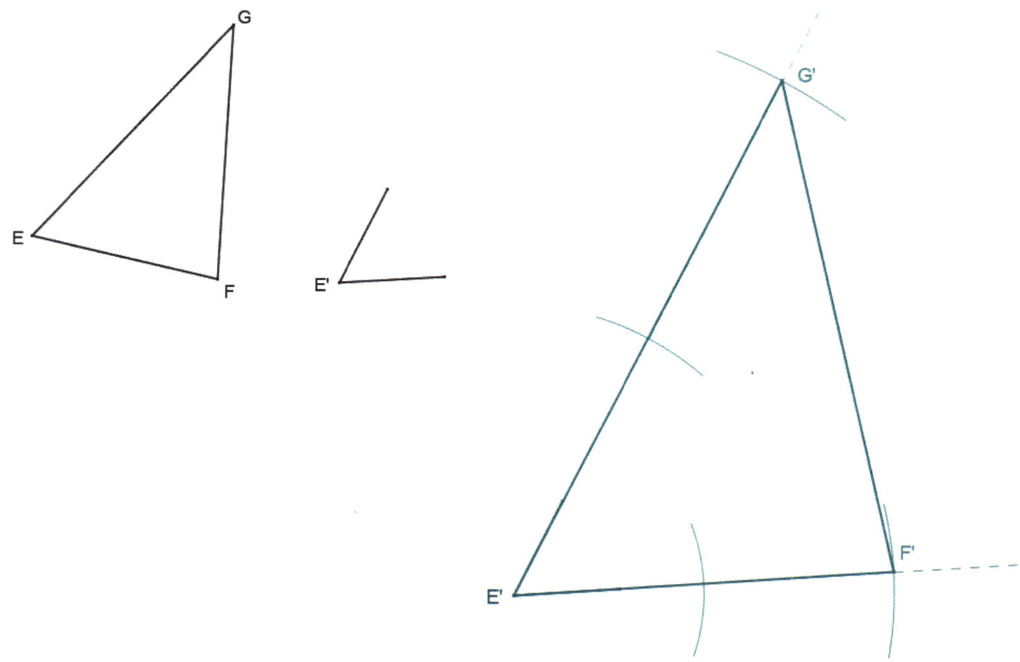

4. Triangle MTC is provided below, and one angle of scale drawing $\triangle M'T'C'$ is also provided. Use construction tools to complete a scale drawing so that the scale factor is $r = \frac{1}{4}$.

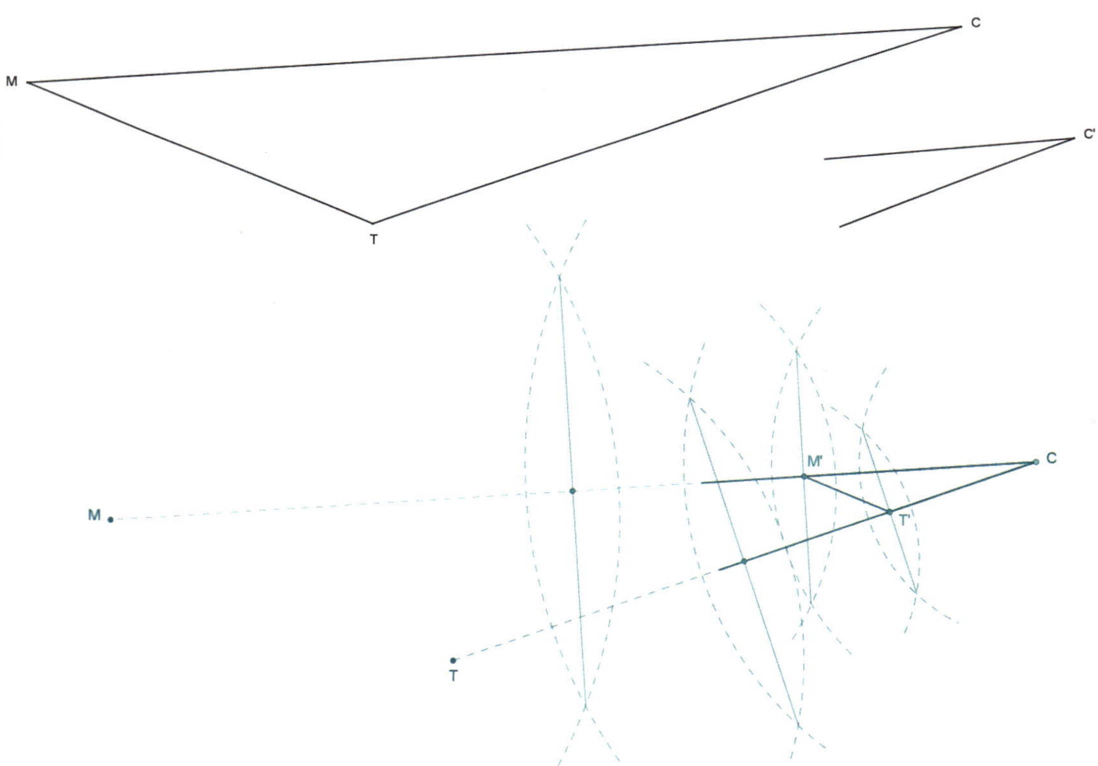

EUREKA MATH™

5. Triangle ABC is provided below, and one side of scale drawing $\triangle A'B'C'$ is also provided. Use construction tools to complete the scale drawing and determine the scale factor.

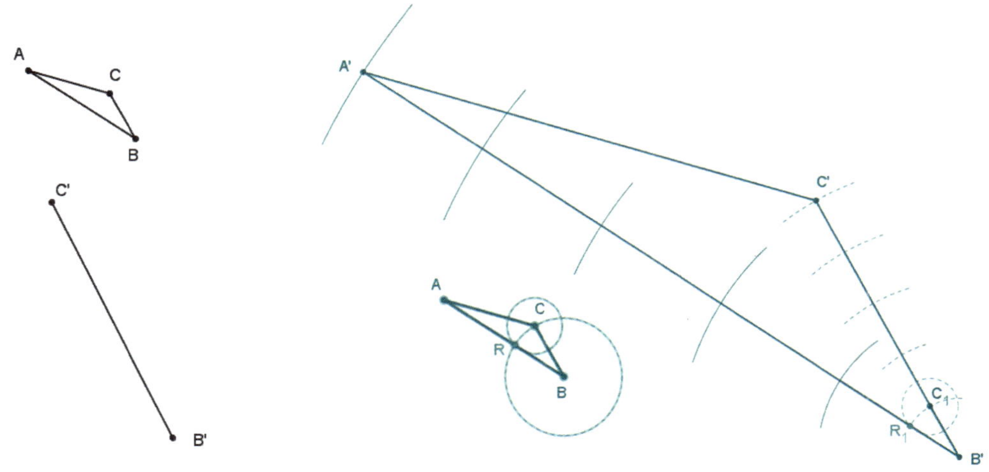

The ratio of $B'C' : BC$ is $5:1$, so the scale factor is 5.

6. Triangle XYZ is provided below, and one side of scale drawing $\triangle X'Y'Z'$ is also provided. Use construction tools to complete the scale drawing and determine the scale factor.

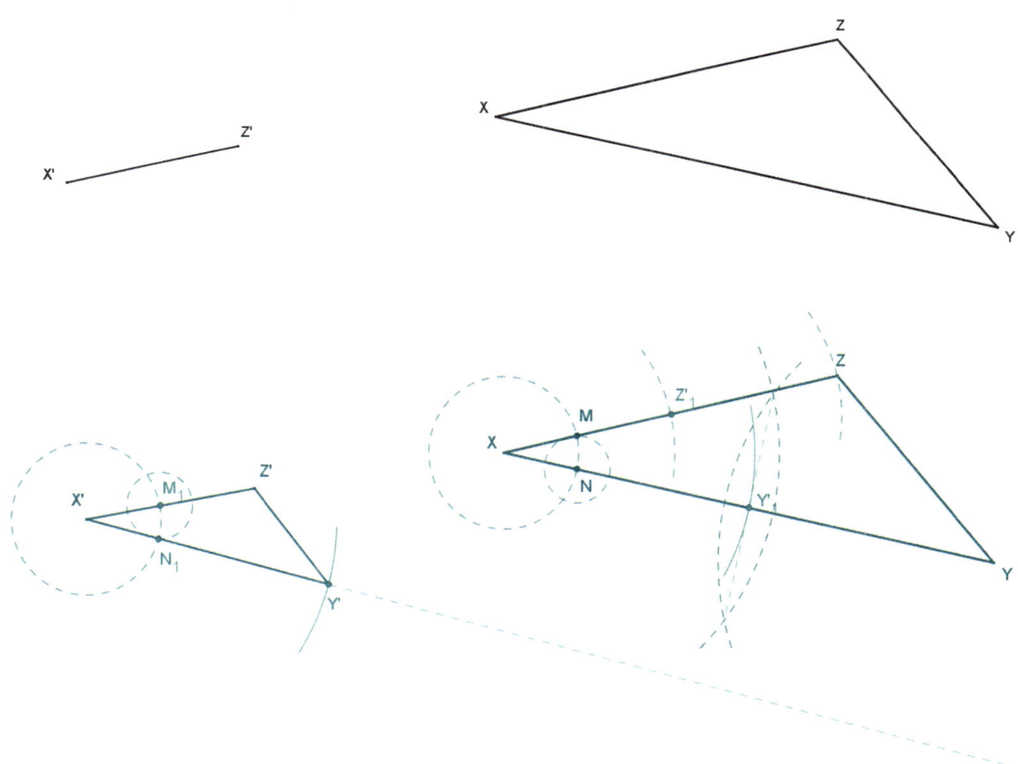

The ratio of $X'Z' : XZ$ is $1:2$, so the scale factor is $\frac{1}{2}$.

7. Quadrilateral $GHIJ$ is a scale drawing of quadrilateral $ABCD$ with scale factor r. Describe each of the following statements as always true, sometimes true, or never true, and justify your answer.

a. $AB = GH$

Sometimes true, but only if $r = 1$.

b. $m\angle ABC = m\angle GHI$

Always true because $\angle GHI$ corresponds to $\angle ABC$ in the original drawing, and angle measures are preserved in scale drawings.

c. $\frac{AB}{GH} = \frac{BC}{HI}$

Always true because distances in a scale drawing are equal to their corresponding distances in the original drawing times the scale factor r, so $\frac{AB}{GH} = \frac{AB}{r(AB)} = \frac{1}{r}$ and $\frac{BC}{HI} = \frac{BC}{r(BC)} = \frac{1}{r}$.

d. $\text{Perimeter }(GHIJ) = r \cdot \text{Perimeter}(ABCD)$

Always true because the distances in a scale drawing are equal to their corresponding distances in the original drawing times the scale factor r, so

$\text{Perimeter}(GHIJ) = GH + HI + IJ + JG$

$\text{Perimeter}(GHIJ) = r(AB) + r(BC) + r(CD) + r(DA)$

$\text{Perimeter}(GHIJ) = r(AB + BC + CD + DA)$

$\text{Perimeter}(GHIJ) = r \cdot \text{Perimeter}(ABCD)$.

e. $\text{Area }(GHIJ) = r \cdot \text{Area}(ABCD)$ where $r \neq 1$

Never true because the area of a scale drawing is related to the area of the original drawing by the factor r^2. The scale factor $r > 0$ and $r \neq 1$, so $r \neq r^2$.

f. $r < 0$

Never true in a scale drawing because any distance in the scale drawing would be negative as a result of the scale factor and, thus, cannot be drawn since distance must always be positive.

Lesson 2: Making Scale Drawings Using the Ratio Method

Student Outcomes

- Students create scale drawings of polygonal figures by the ratio method.
- Given a figure and a scale drawing from the ratio method, students answer questions about the scale factor and the center.

Lesson Notes

In Lesson 1, students created scale drawings in any manner they wanted, as long as the scale drawings met the criteria of well-scaled drawings. Lesson 2 introduces students to a systematic way of creating a scale drawing: the ratio method, which relies on dilations. Students dilate the vertices of the provided figure and verify that the resulting image is in fact a scale drawing of the original. It is important to note that we approach the ratio method as a method that strictly dilates the vertices. After some practice with the ratio method, students dilate a few other points of the polygonal figure and notice that they lie on the scale drawing. They may speculate that the dilation of the entire figure is the scale drawing, but this fact is not generalized in Lesson 2.

Note that students need rulers, protractors, and calculators for this lesson.

Classwork

Opening Exercise (2 minutes)

> **Opening Exercise**
>
> **Based on what you recall from Grade 8, describe what a *dilation* is.**
>
> ***Student responses will vary; students may say that a dilation results in a reduction or an enlargement of the original figure or that corresponding side lengths are proportional in length, and corresponding angles are equal in measure. The objective is to prime them for an in-depth conversation about dilations; take one or two responses and move on.***

Scaffolding:

Providing an example of a dilation (such as in the image below) may help students recall details about dilations.

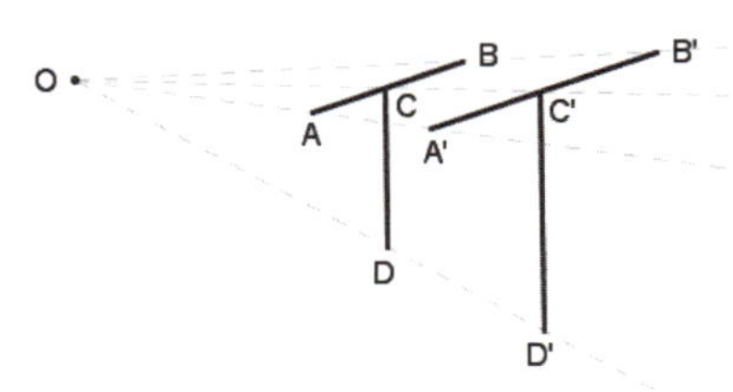

Discussion (5 minutes)

- In Lesson 1, we reviewed the properties of a scale drawing and created scale drawings of triangles using construction tools. We observed that as long as our scale drawings had angles equal in measure to the corresponding angles of the original figure and lengths in constant proportion to the corresponding lengths of the original figure, the location and orientation of our scale drawing did not concern us.
- In Lesson 2, we use a systematic process of creating a scale drawing called the ratio method. The ratio method dilates the vertices of the provided polygonal figure. The details that we recalled in the Opening Exercise are characteristics that are consistent with scale drawings too. We will verify that the resulting image created by dilating these key points is in fact a scale drawing.

- Recall the definition of a dilation:

Definition	**Characteristics**
For $r > 0$, a dilation with center O and scale factor r is a transformation $D_{O,r}$ of the plane defined as follows: For the center O, $D_{O,r}(O) = O$, and For any other point P, $D_{O,r}(P)$ is the point P on the ray $\overrightarrow{OP}$ so that $\|OP'\| = r \cdot \|OP\|$.	- Preserves angles - Names a center and a scale factor
Dilation	
Examples	**Non-Examples**
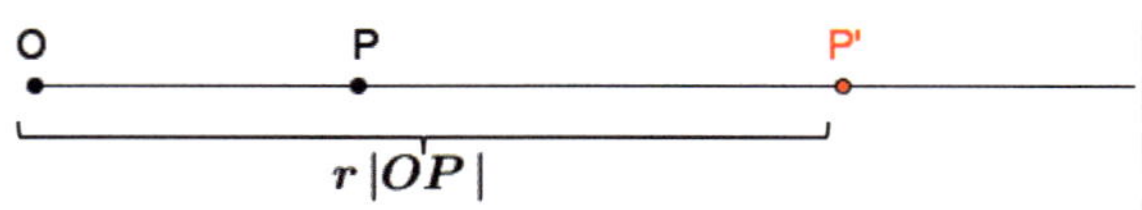	- Rigid motions such as translations, rotations, reflections

Note that students last studied dilations in Grade 8 Module 3. At that time, the notation used was not the capital letter D, but the full word dilation. Students have since studied rigid motion notation in Geometry Module 1 and should be familiar with the style of notation presented here.

> *Scaffolding:*
> - Consider displaying a poster with the definition and notation of dilation.
> - Ask students to draw some examples to demonstrate understanding, such as, "Draw a segment OP on your paper. Dilate the segment using $D_{O,3}$."

- A dilation is a rule (a function) that moves points in the plane a specific distance along the ray that originates from a center O. What determines the distance a given point moves?
    - *The location of the scaled point is determined by the scale factor and the distance of the original point from the center.*
- What can we tell about the scale factor of a dilation that pulls any point that is different from the center toward the center O?
    - *We know that the scale factor for a dilation where a point is pulled toward the center must be* $0 < r < 1$.
- What can we tell about the scale factor of a dilation that pushes all points, except the center, away from the center O?
    - *The scale factor for a dilation where a point is pushed away from the center must be* $r > 1$.
- A point, different from the center, that is unchanged in its location after a dilation must have a scale factor of $r = 1$.
- Scale factor is always a positive value, as we use it when working with distance. If we were to use negative values for scale factor, we would be considering distance as a negative value, which does not make sense. Hence, scale factor is always positive.

Example 1 (8 minutes)

Examples 1–2 demonstrate how to create a scale drawing using the ratio method. In this example, the ratio method is used to dilate the vertices of a polygonal figure about center O, by a scale factor of $r = \frac{1}{2}$.

- To use the ratio method to scale any figure, we must have a scale factor and center in order to dilate the vertices of a polygonal figure.
- In the steps below, we have a figure with center O and a scale factor of $r = \frac{1}{2}$. What effect should we expect this scale factor to have on the image of the figure?
  - *Since the scale factor is a value less than one (but greater than zero), the image should be a reduction of the original figure. Specifically, each corresponding length should be half of the original length.*

Example 1

Create a scale drawing of the figure below using the ratio method about center O and scale factor $r = \frac{1}{2}$.

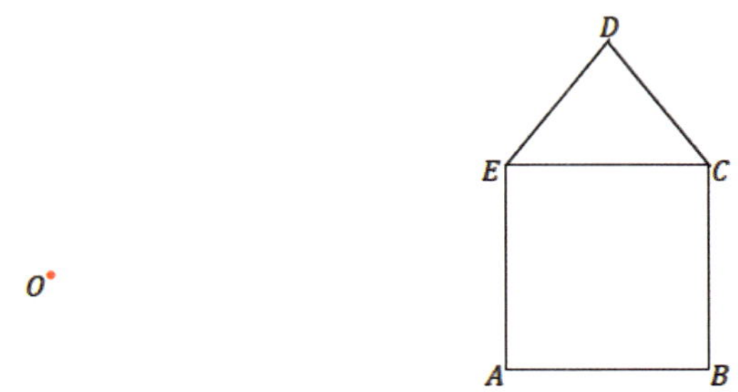

Step 1. Draw a ray beginning at O through each vertex of the figure.

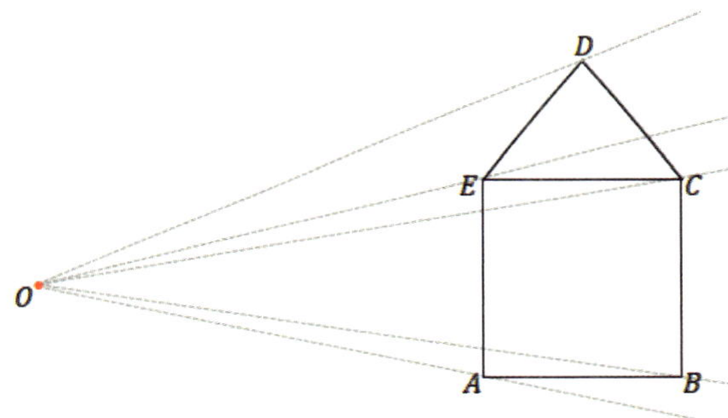

Step 2. Dilate each vertex along the appropriate ray by scale factor $r = \frac{1}{2}$. Use the ruler to find the midpoint between O and D and then each of the other vertices. Label each respective midpoint with prime notation (e.g., D').

- Why are we locating the midpoint between O and D?
  - *The scale factor tells us that the distance of the scaled point should be half the distance from O to D, which is the midpoint of $\overline{OD}$.*

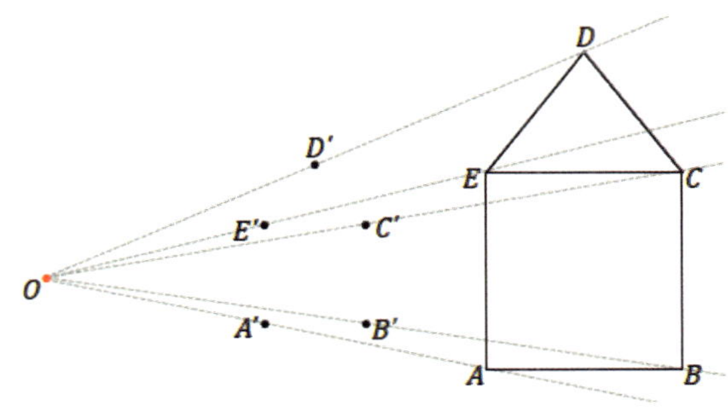

Step 3. Join vertices in the way they are joined in the original figure (e.g., segment $A'B'$ corresponds to segment AB).

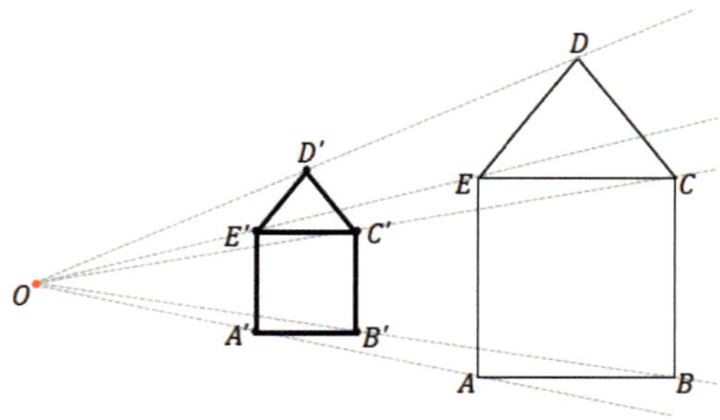

MP.5

- Does $A'B'C'D'E'$ look like a scale drawing? How can we verify whether $A'B'C'D'E'$ is really a scale drawing?
  - *Yes. We can measure each segment of the original and the scale drawing; the segments of $A'B'C'D'E'$ appear to be half as long as their corresponding counterparts in $ABCDE$, and all corresponding angles appear to be equal in measurement; the image is a reduction of the original figure.*
- It is important to notice that the scale factor for the scale drawing is the same as the scale factor for the dilation.

Students may notice that in the triangle formed by the center and the endpoints of any segment on the original figure, the dilated segment forms the mid-segment of the triangle.

Have students measure and confirm that the length of each segment in the scale drawing is half the length of each segment in the original drawing and that the measurements of all corresponding angles are equal. The quadrilateral $ABCD$ is a square, and all four angles are $90°$ in measurement. The measurement of $\angle D = 80°$, and the measurements of $\angle C$ and $\angle E$ are both $50°$. The measurements of the side lengths are not provided because they differ from the images that appear in print form.

Scaffolding:
Teachers may want to consider using patty paper as an alternate means to measuring angles with a protractor in the interest of time.

Exercise 1 (5 minutes)

Scaffolding:
In preparing for this lesson, consider whether the class has time for each example and exercise. If time is short, consider moving from Example 1 to Example 2.

Exercise 1

1. **Create a scale drawing of the figure below using the ratio method about center O and scale factor $r = \frac{3}{4}$. Verify that the resulting figure is in fact a scale drawing by showing that corresponding side lengths are in constant proportion and the corresponding angles are equal in measurement.**

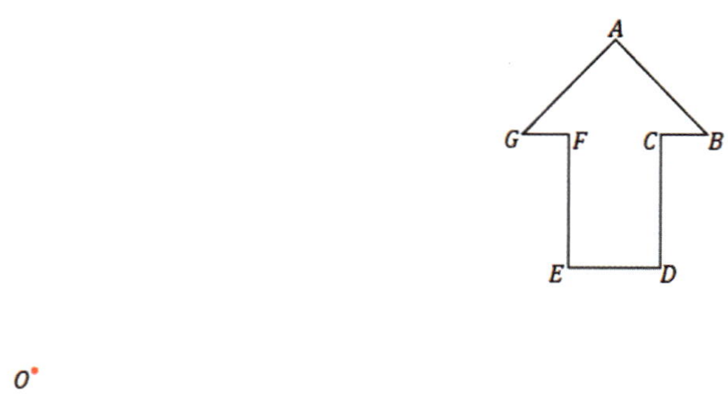

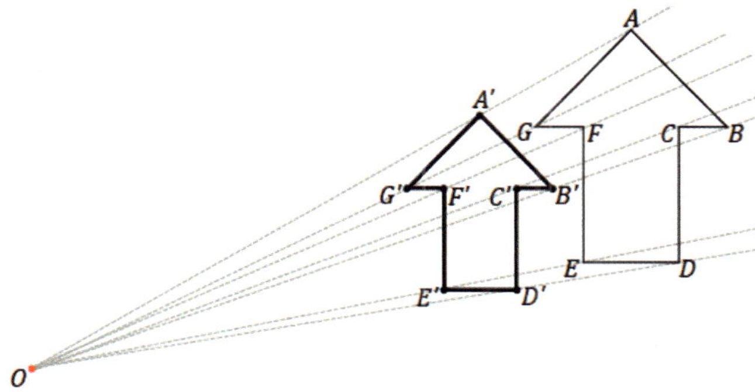

Verification of the reduced figure should show that the length of each segment in the scale drawing is $\frac{3}{4}$ the length of each segment in the original figure (e.g., $A'G' = \frac{3}{4}(AG)$). The angle measurements are as follows (when an angle could be interpreted as having two possible measurements, the smaller one was selected in all cases): $m\angle A = 90°$, $m\angle B = 45°$, $m\angle C = 90°$, $m\angle D = 90°$, $m\angle E = 90°$, $m\angle F = 90°$, and $m\angle G = 45°$.

Example 2 (7 minutes)

Example 2

a. **Create a scale drawing of the figure below using the ratio method about center O and scale factor $r = 3$.**

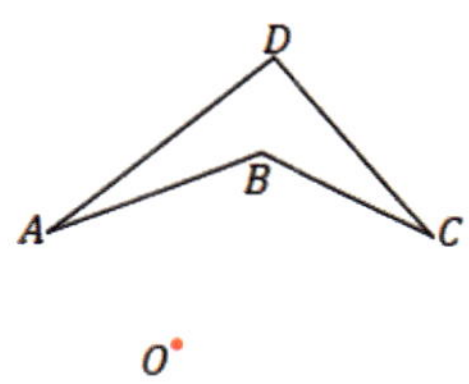

Step 1. Draw a ray beginning at O through each vertex of the figure.

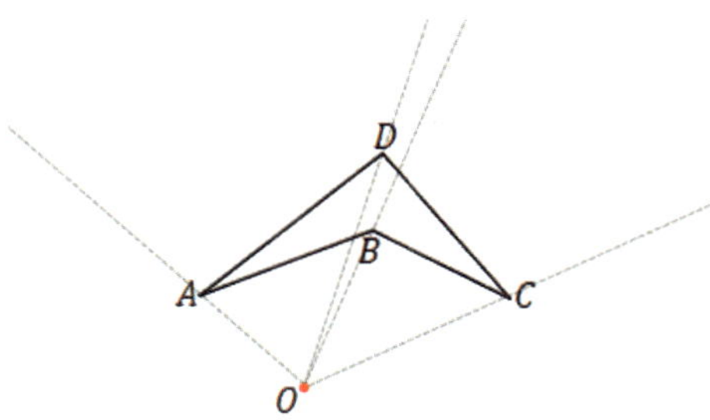

Step 2. Use your ruler to determine the location of A' on $\overrightarrow{OA}$; A' should be three times as far from O as A. Determine the locations of B' and C' in the same way along the respective rays.

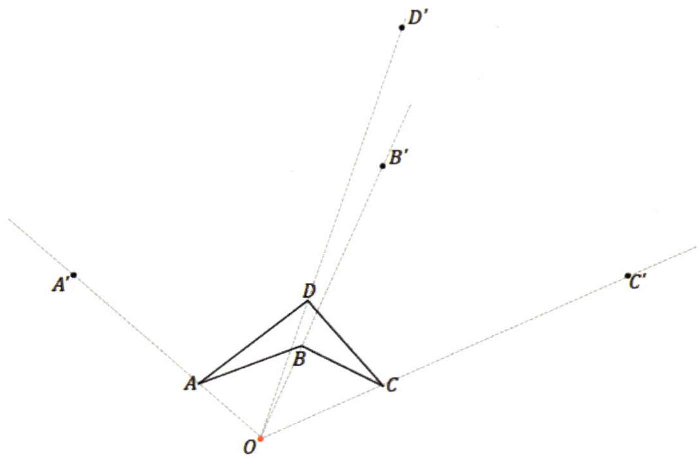

Step 3. Draw the corresponding line segments (e.g., segment $A'B'$ corresponds to segment AB).

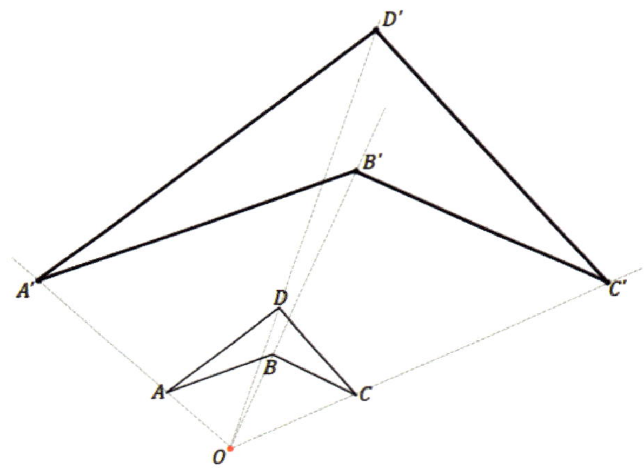

- Does $A'B'C'D'$ look like a scale drawing of $ABCD$?
  - *Yes*
- How can we verify whether $A'B'C'D'$ is really a scale drawing of $ABCD$?
  - *We can measure each segment of the original and the scale drawing; the segments of $A'B'C'D'$ should be three times as long as their corresponding counterparts in $ABCD$, and all corresponding angles should be equal in measurement; the image is an enlargement of the original figure.*

Have students measure and confirm that the length of each segment in the scale drawing is three times the length of each segment in the original drawing and that the measurements of all corresponding angles are equal. The measurements of the angles in the figure are as follows: $m\angle A = 17°$, $m\angle B = 134°$ (the smaller of the two possible options of measuring the angle was selected, but either will do), $m\angle C = 22°$, $m\angle D = 23°$. Again, the measurements of the side lengths are not provided because they differ from the images that appear in print form.

b. **Locate a point X so that it lies between endpoints A and B on segment AB of the original figure in part (a). Use the ratio method to locate X' on the scale drawing in part (a).**

Sample response:

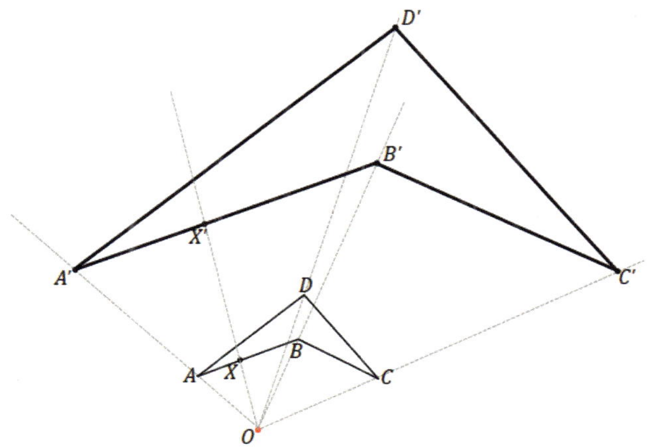

- Consider that everyone in class could have chosen a different location for X between points A and B. What does the result of part (b) imply?
  - *The result of part (b) implies that all the points between AB are dilated to corresponding points between points A' and B'.*
- It is tempting to draw the conclusion that the dilation of the vertices is the same as the dilation of each segment onto corresponding segments in the scale drawing. Even though this appears to be the case here, we wait until later lessons to definitively show whether this is actually the case.

c. **Imagine a dilation of the same figure as in parts (a) and (b). What if the ray from the center passed through two distinct points, such as B and D? What does that imply about the locations of B' and D'?**

Both B' and D' also lie on the same ray.

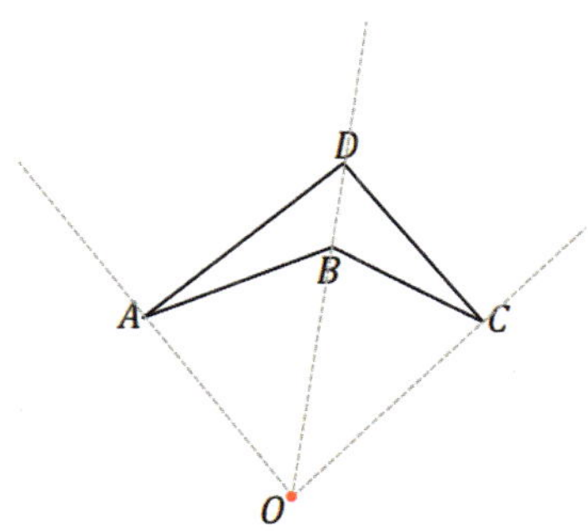

Exercises 2–6 (11 minutes)

Exercises 2–6

2. $\triangle A'B'C'$ is a scale drawing of $\triangle ABC$ drawn by using the ratio method. Use your ruler to determine the location of the center O used for the scale drawing.

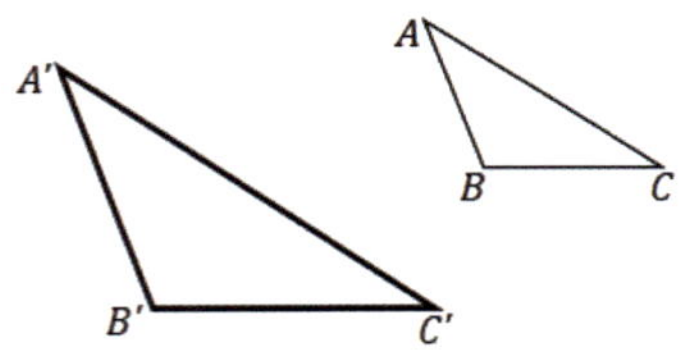

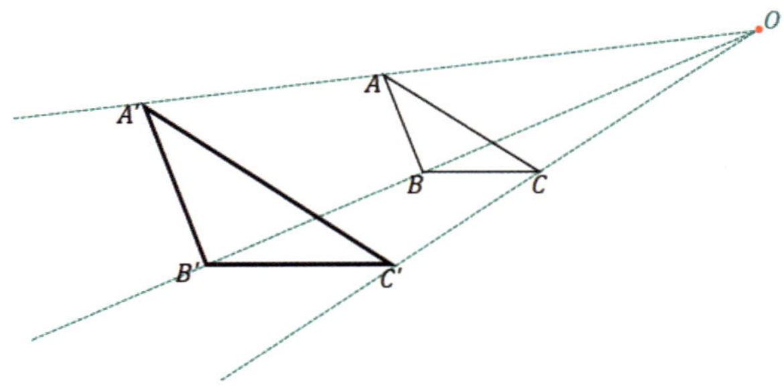

3. Use the figure below with center O and a scale factor of $r = \frac{5}{2}$ to create a scale drawing. Verify that the resulting figure is in fact a scale drawing by showing that corresponding side lengths are in constant proportion and that the corresponding angles are equal in measurement.

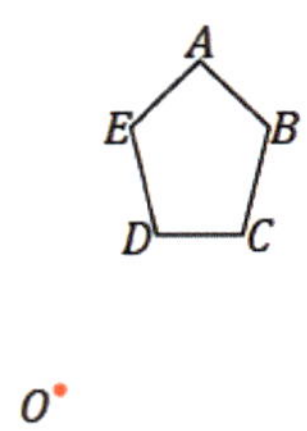

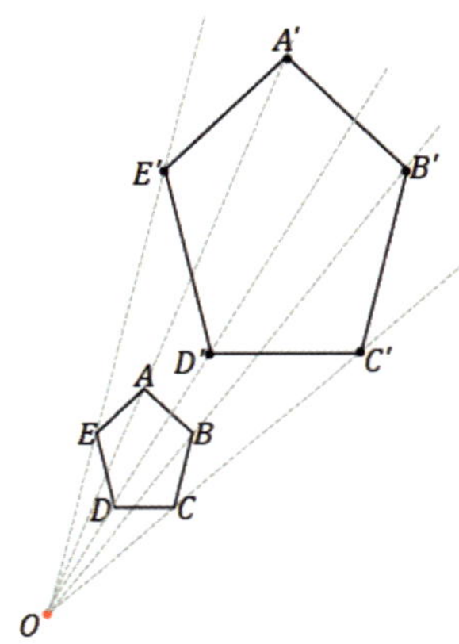

Verification of the enlarged figure should show that the length of each segment in the scale drawing is 2.5 times the length of each segment in the original figure (e.g., $A'B' = 2.5(AB)$). The angle measurements are $m\angle A = 94°$, $m\angle B = 118°$, $m\angle C = 105°$, $m\angle D = 105°$, and $m\angle E = 118°$.

4. Summarize the steps to create a scale drawing by the ratio method. Be sure to describe all necessary parameters to use the ratio method.

To use the ratio method to create a scale drawing, the problem must provide a polygonal figure, a center O, and a scale factor. To begin the ratio method, draw a ray that originates at O and passes through each vertex of the figure. We are dilating each vertex along its respective ray. Measure the distance between O and a vertex and multiply it by the scale factor. The resulting value is the distance away from O at which the scaled point will be located. Once all the vertices are dilated, they should be joined in the same way as they are joined in the original figure.

5. A clothing company wants to print the face of the Statue of Liberty on a T-shirt. The length of the face from the top of the forehead to the chin is 17 feet, and the width of the face is 10 feet. Given that a medium-sized T-shirt has a length of 29 inches and a width of 20 inches, what dimensions of the face are needed to produce a scaled version that will fit on the T-shirt?

   a. What shape would you use to model the face of the statue?

      Answers may vary. Students may say triangle, rectangle, or circle.

   b. Knowing that the maximum width of the T-shirt is 20 inches, what scale factor is needed to make the width of the face fit on the shirt?

      Answers may vary. Sample response shown below.

$$\frac{20}{120} = \frac{1}{6}$$

      The width of the face on the T-shirt will need to be scaled to $\frac{1}{6}$ the size of the statue's face.

   c. What scale factor should be used to scale the length of the face? Explain.

      Answers may vary. Students should respond that the scale factor identified in part (b) should be used for the length.

      To keep the length of the face proportional to the width, a scale factor of $\frac{1}{6}$ should be used.

   d. Using the scale factor identified in part (c), what is the scaled length of the face? Will it fit on the shirt?

      Answers may vary.

$$\frac{1}{6}(204) = 34$$

      The scaled length of the face would be 34 inches. The length of the shirt is only 29 inches, so the face will not fit on the shirt.

   e. Identify the scale factor you would use to ensure that the face of the statue was in proportion and would fit on the T-shirt. Identify the dimensions of the face that will be printed on the shirt.

      Answers may vary. Scaling by a factor of $\frac{1}{7}$ produces dimensions that are still too large to fit on the shirt. The largest scale factor that could be used is $\frac{1}{8}$, producing a scaled width of 15 inches and a scaled length of 25.5 inches.

   f. The T-shirt company wants the width of the face to be no smaller than 10 inches. What scale factors could be used to create a scaled version of the face that meets this requirement?

      Scale factors of $\frac{1}{8}, \frac{1}{9}, \frac{1}{10}, \frac{1}{11}$, or $\frac{1}{12}$ could be used to ensure the width of the face is no smaller than 10 inches.

   g. If it costs the company $\$0.005$ for each square inch of print on a shirt, what is the maximum and minimum costs for printing the face of the Statue of Liberty on one T-shirt?

      The largest scaled face would have dimensions 15×25.5, meaning the print would cost approximately $\$1.91$ per shirt. The smallest scaled face would have dimensions 10×17, meaning the print would cost $\$0.85$ per shirt.

6. Create your own scale drawing using the ratio method. In the space below:
   a. Draw an original figure.
   b. Locate and label a center of dilation O.
   c. Choose a scale factor r.
   d. Describe your dilation using appropriate notation.
   e. Complete a scale drawing using the ratio method.

   Show all measurements and calculations to confirm that the new figure is a scale drawing. The work here will be your answer key.

   Next, trace your original figure onto a fresh piece of paper. Trade the traced figure with a partner. Provide your partner with the dilation information. Each partner should complete the other's scale drawing. When finished, check all work for accuracy against your answer key.

   Answers will vary. Encourage students to check each other's work and to discover the reason for any discrepancies found between the author's answers and the partner's answers.

MP.3

Scaffolding:

Figures can be made as simple or as complex as desired—a triangle involves fewer segments to keep track of than a figure such as the arrow in Exercise 1. Students should work with a manageable figure in the allotted time frame.

Closing (2 minutes)

Ask students to summarize the key points of the lesson. Additionally, consider asking students the following questions independently in writing, to a partner, or to the whole class.

- To create a scale drawing using the ratio method, each vertex of the original figure is dilated about the center O by scale factor r. Once all the vertices are dilated, they are joined to each other in the same way as in the original figure.
- The scale factor tells us whether the scale drawing is being enlarged ($r > 1$) or reduced ($0 < r < 1$).
- How can it be confirmed that what is drawn by the ratio method is in fact a scale drawing?
   - *By measuring the side lengths of the original figure and the scale drawing, we can establish whether the corresponding sides are in constant proportion. We can also measure corresponding angles and determine whether they are equal in measure. If the side lengths are in constant proportion, and the corresponding angle measurements are equal, the new figure is in fact a scale drawing of the original.*
- It is important to note that though we have dilated the vertices of the figures for the ratio method, we do not definitively know if each segment is dilated to the corresponding segment in the scale drawing. This remains to be seen. We cannot be sure of this even if the scale drawing is confirmed to be a well-scaled drawing. We learn how to determine this in the next few lessons.

Exit Ticket (5 minutes)

Name ______________________________ Date ______________

Lesson 2: Making Scale Drawings Using the Ratio Method

Exit Ticket

One of the following images shows a well-scaled drawing of $\triangle ABC$ done by the ratio method; the other image is not a well-scaled drawing. Use your ruler and protractor to make the necessary measurements and show the calculations that determine which is a scale drawing and which is not.

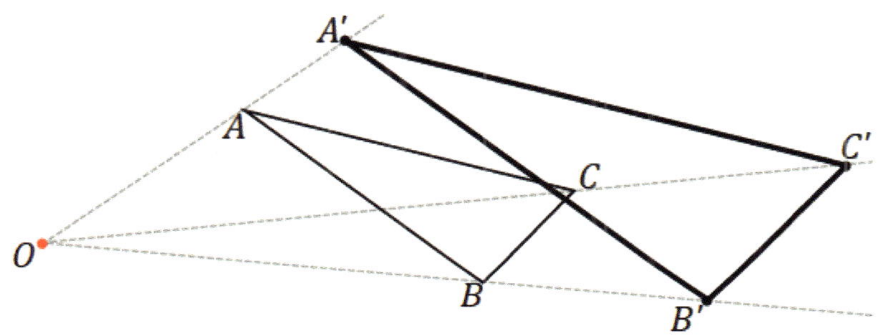

Figure 1

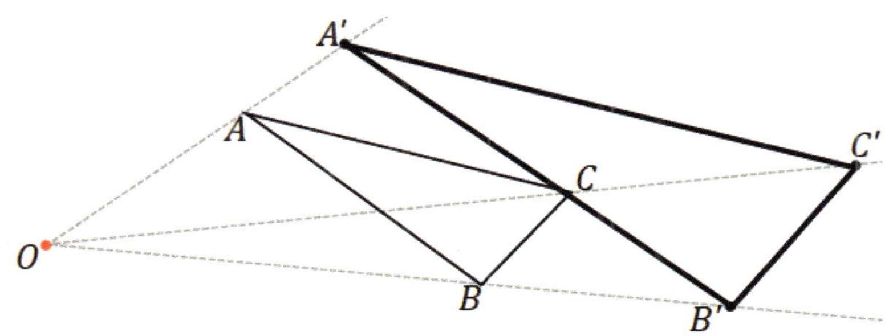

Figure 2

Exit Ticket Sample Solutions

One of the following images shows a well-scaled drawing of $\triangle ABC$ done by the ratio method; the other image is not a well-scaled drawing. Use your ruler and protractor to make the necessary measurements and show the calculations that determine which is a scale drawing and which is not.

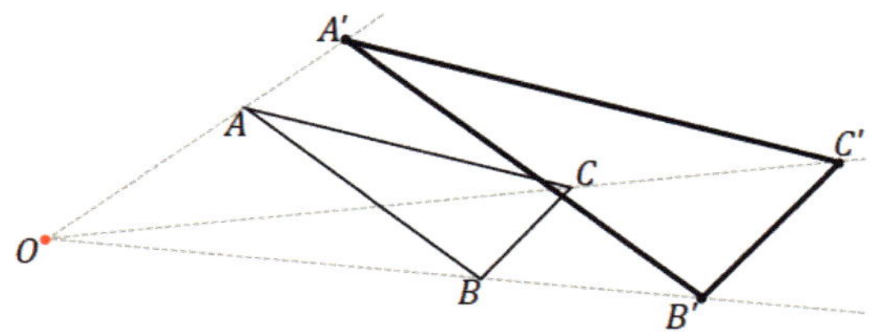

Figure 1

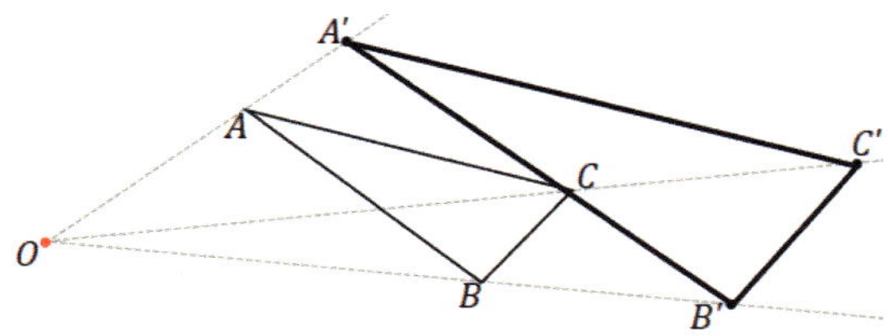

Figure 2

Figure 1 shows the true scale drawing.

$\triangle ABC$ angle measurements: $m\angle A = 22°$, $m\angle B = 100°$, $m\angle C = 58°$, which are the same for $\triangle A'B'C'$ in Figure 1. The value of the ratios of $OA'{:}\,OA$, $OB'{:}\,OB$, and $OC'{:}\,OC$ are the same.

$\triangle A'B'C'$ in Figure 2 has angle measurements $m\angle A' = 20°$, $m\angle B' = 99°$, $m\angle C' = 61°$, and the value of the ratios of $OA'{:}\,OA$, $OB'{:}\,OB$, and $OC'{:}\,OC$ are not the same.

Problem Set Sample Solutions

Considering the significant construction needed for the Problem Set questions, teachers may feel that a maximum of three questions is sufficient for a homework assignment. It is up to the teacher to assign what is appropriate for the class.

1. **Use the ratio method to create a scale drawing about center O with a scale factor of $r = \frac{1}{4}$. Use a ruler and protractor to verify that the resulting figure is in fact a scale drawing by showing that corresponding side lengths are in constant proportion and the corresponding angles are equal in measurement.**

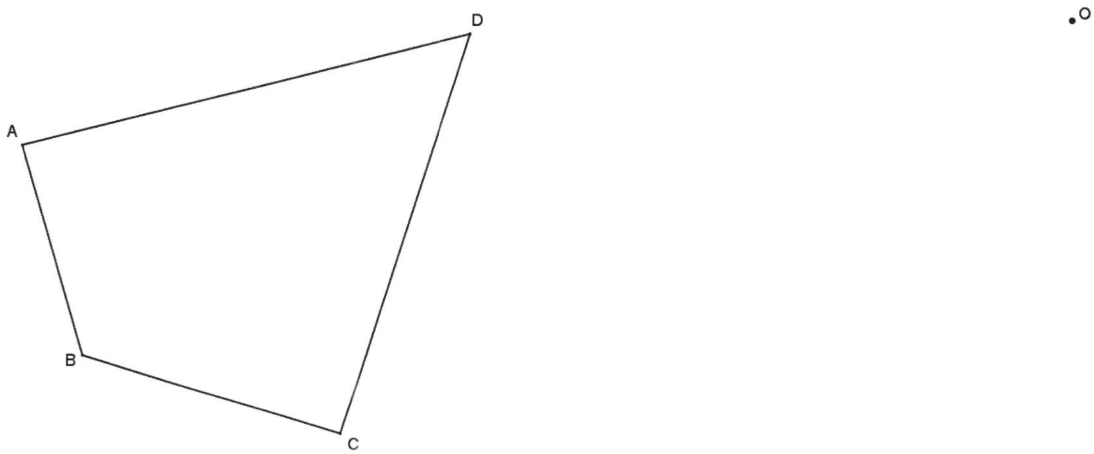

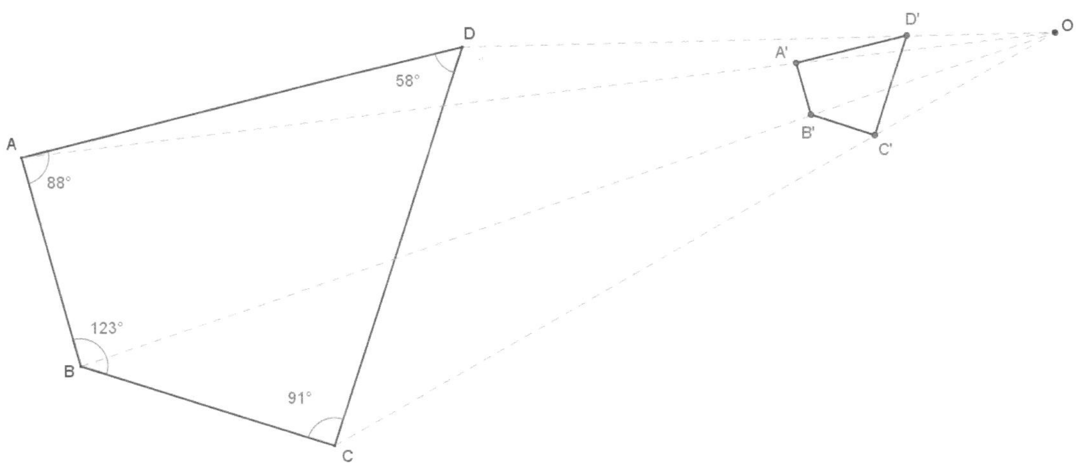

The measurements in the figure are $m\angle A = 88°$, $m\angle B = 123°$, $m\angle C = 91°$, and $m\angle D = 58°$. All side-length measurements of the scale drawing should be in the constant ratio of $1:4$.

2. Use the ratio method to create a scale drawing about center O with a scale factor of $r = 2$. Verify that the resulting figure is in fact a scale drawing by showing that corresponding side lengths are in constant proportion and that the corresponding angles are equal in measurement.

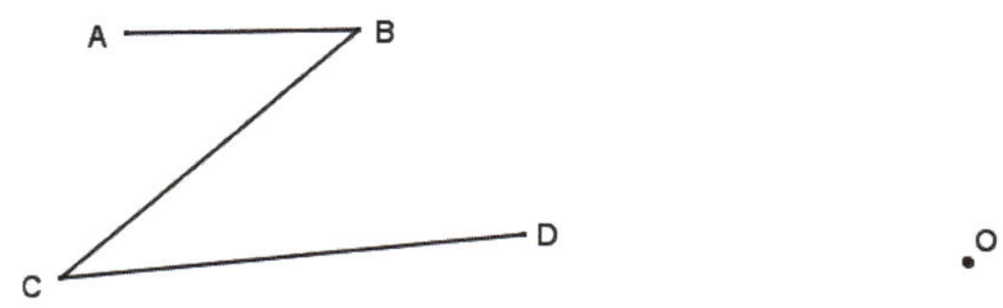

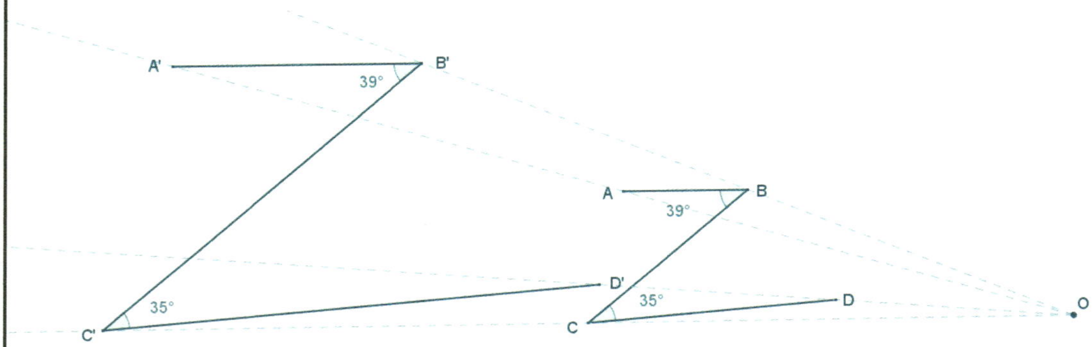

The measurements in the figure are $m\angle B = 39°$ and $m\angle C = 35°$. All side-length measurements of the scale drawing should be in the constant ratio of $2:1$.

3. Use the ratio method to create two scale drawings: $D_{O,2}$ and $D_{P,2}$. Label the scale drawing with respect to center O as $\triangle A'B'C'$ and the scale drawing with respect to center P as $\triangle A''B''C''$.

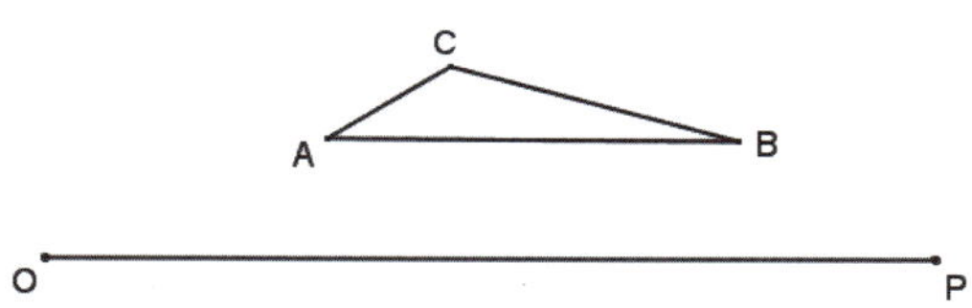

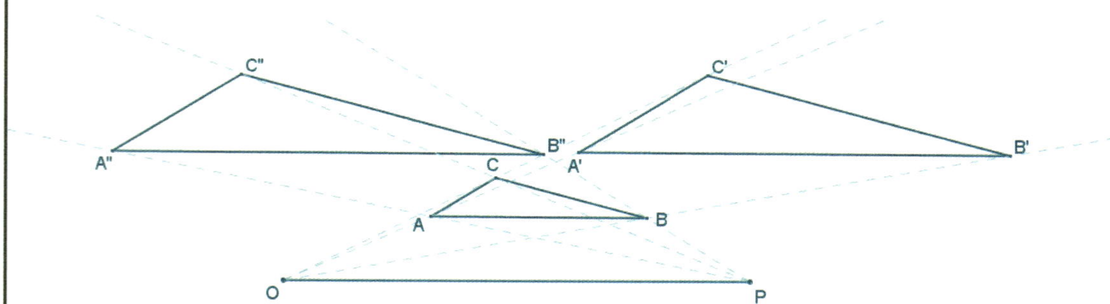

What do you notice about the two scale drawings?

They are both congruent since each was drawn with the same scale factor.

What rigid motion can be used to map $\triangle A'B'C'$ onto $\triangle A''B''C''$?

Answers may vary. For example, a translation by vector $\overrightarrow{A'A''}$ is acceptable.

4. **Sara found a drawing of a triangle that appears to be a scale drawing. Much of the drawing has faded, but she can see the drawing and construction lines in the diagram below. If we assume the ratio method was used to construct $\triangle A'B'C'$ as a scale model of $\triangle ABC$, can you find the center O, the scale factor r, and locate $\triangle ABC$?**

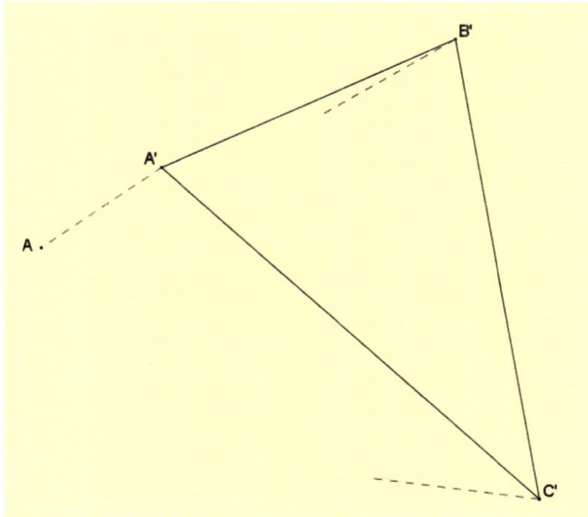

Extend ray $A'A$ and the partial ray drawn from either B' or C'. The point where they intersect is center O.

$\frac{OA'}{OA} = \frac{3}{2}$; the scale factor is $\frac{3}{2}$. Locate B $\frac{2}{3}$ of the distance from O to B' and C $\frac{2}{3}$ of the way from O to C'. Connect the vertices to show original $\triangle ABC$.

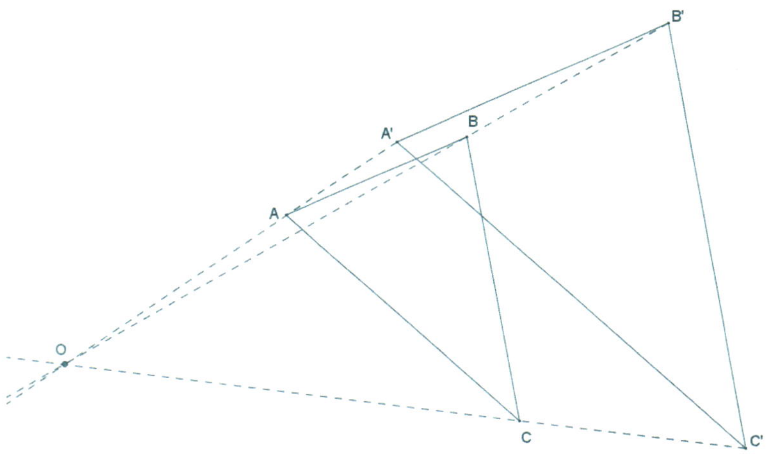

5. Quadrilateral $A'''B'''C'''D'''$ is one of a sequence of three scale drawings of quadrilateral $ABCD$ that were all constructed using the ratio method from center O. Find the center O, each scale drawing in the sequence, and the scale factor for each scale drawing. The other scale drawings are quadrilaterals $A'B'C'D'$ and $A''B''C''D''$.

Note to the teacher: One option is to simplify this diagram by joining vertices D''' and B''', forming a triangle.

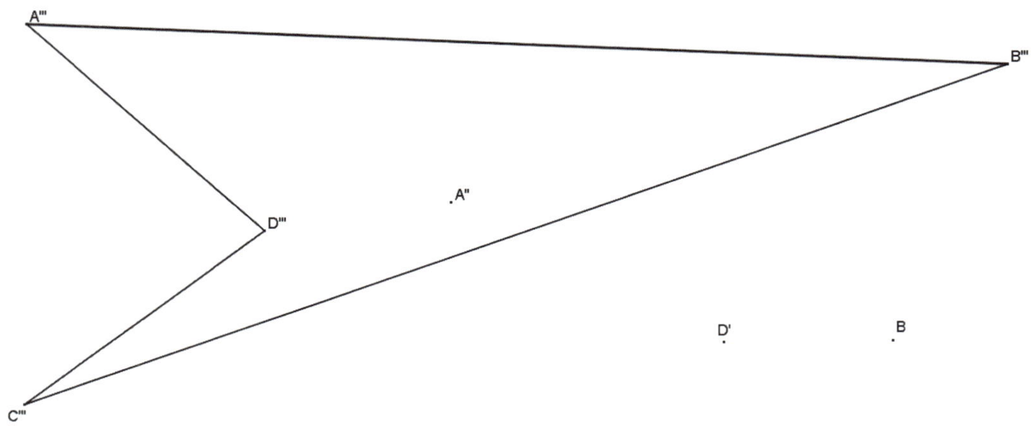

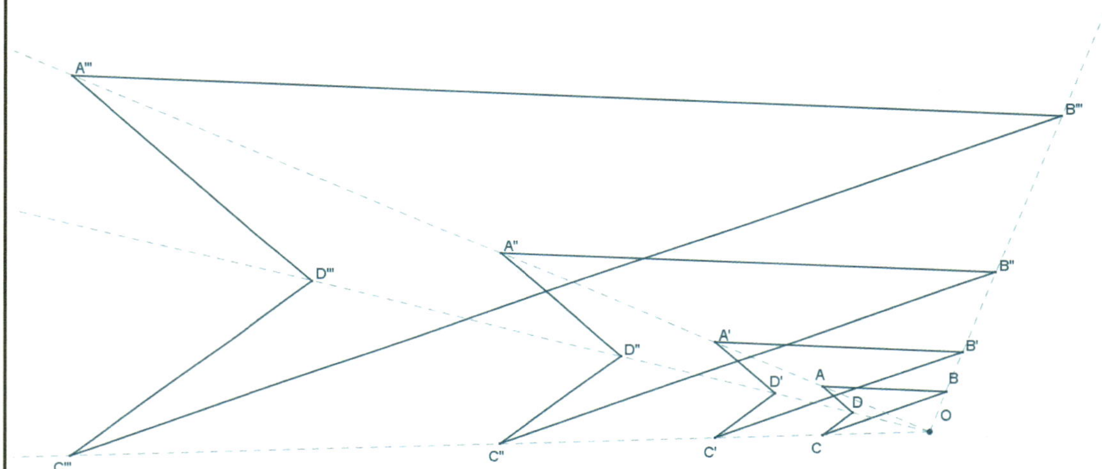

Each scale drawing is created from the same center point, so the corresponding vertices of the scale drawings should align with the center O. Draw any two of $\overline{A'''A''}$, $\overline{D'''D'}$, or $\overline{B'''B}$ to find center O at their intersection.

The ratio of $OA'':OA$ is $4:1$, so the scale factor of figure $A''B''C''D''$ is 4.

The ratio of $OD':OD$ is $2:1$, so the scale factor of figure $A'B'C'D'$ is 2.

The ratio of $OB''':OB$ is $8:1$, so the scale factor of figure $A'''B'''C'''D'''$ is 8.

6. Maggie has a rectangle drawn in the corner of an $8\frac{1}{2}$-inch by 11-inch sheet of printer paper as shown in the diagram. To cut out the rectangle, Maggie must make two cuts. She wants to scale the rectangle so that she can cut it out using only one cut with a paper cutter.

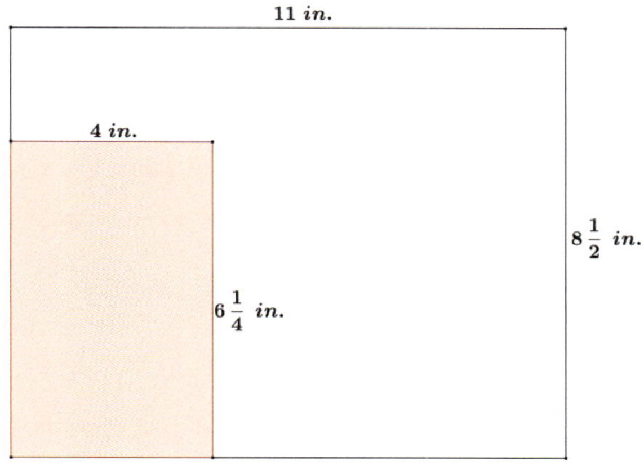

a. What are the dimensions of Maggie's scaled rectangle, and what is its scale factor from the original rectangle?

If the rectangle is scaled from the corner of the paper at which it currently sits, the maximum height of the rectangle will be $8\frac{1}{2}$ inches.

$$k = \frac{8\frac{1}{2}}{6\frac{1}{4}} = \frac{34}{25}$$

The scale factor to the enlarged rectangle is $\frac{34}{25}$.

$$y = \frac{34}{25}(4)$$

$$y = \frac{136}{25} = 5.44$$

Using the scale factor, the width of the scaled rectangle is 5.44 inches.

b. After making the cut for the scaled rectangle, is there enough material left to cut another identical rectangle? If so, what is the area of scrap per sheet of paper?

The total width of the sheet of paper is 11 inches, which is more than $2(5.44) = 10.88$ inches, so yes, there is enough material to get two identical rectangles from one sheet of paper. The resulting scrap strip measures $8\frac{1}{2}$ inches by 0.12 inches, giving a scrap area of $1.02\ \text{in}^2$ per sheet of paper.

Lesson 3: Making Scale Drawings Using the Parallel Method

Student Outcomes

- Students create scale drawings of polygonal figures by the parallel method.
- Students explain why angles are preserved in scale drawings created by the parallel method using the theorem of parallel lines cut by a transversal.

Lesson Notes

In Lesson 3, students learn the parallel method as yet another way of creating scale drawings. The lesson focuses on constructing parallel lines with the use of setsquares, although parallel lines can also be constructed using a compass and straightedge; setsquares reduce the time needed for the construction. Straightedges, compasses, and setsquares are needed for this lesson (setsquares can be made in class; refer to Grade 7 Module 6 Lesson 7). Rulers are allowed in one instance in the lesson (Example 1, part (b)), but the question can completed without them as long as compasses (or other devices for maintaining a fixed distance) are available.

Classwork

Opening Exercise (2 minutes)

The purpose of this Opening Exercise is to get students thinking about how parallel lines can be used to create a dilation. Accept all ideas, and use responses to segue to Example 1.

MP.2

Opening Exercise

Dani dilated $\triangle ABC$ from center O, resulting in $\triangle A'B'C'$. She says that she completed the drawing using parallel lines. How could she have done this? Explain.

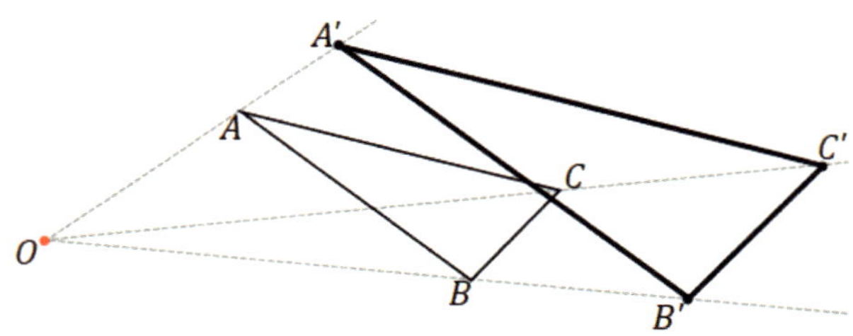

Accept reasonable suggestions from students and proceed to Example 1.

Example 1 (6 minutes)

Example 1 is intended to remind students how to create a parallel line with the use of a setsquare. Provide students with compasses, or allow the use of measurement for part (b).

Scaffolding:
For further practice with setsquares, see Grade 7 Module 6 Lesson 7. If students are familiar with using a setsquare, they may skip to Example 2.

Example 1

a. **Use a ruler and setsquare to draw a line through C parallel to $\overline{AB}$. What ensures that the line drawn is parallel to $\overline{AB}$?**

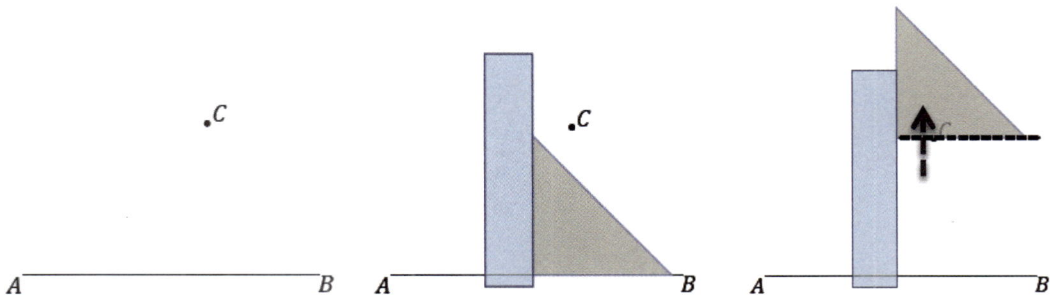

Since the setsquare is in the shape of a right triangle, we are certain that the legs of the setsquare are perpendicular. Then, when one leg is aligned with $\overline{AB}$, and the other leg is flush against the ruler and can slide along the ruler, the $90°$ angle between the horizontal leg and the ruler remains fixed; in effect, there are corresponding angles that are equal in measure. The two lines must be parallel.

b. **Use a ruler and setsquare to draw a parallelogram $ABCD$ around $\overline{AB}$ and point C.**

Step 1

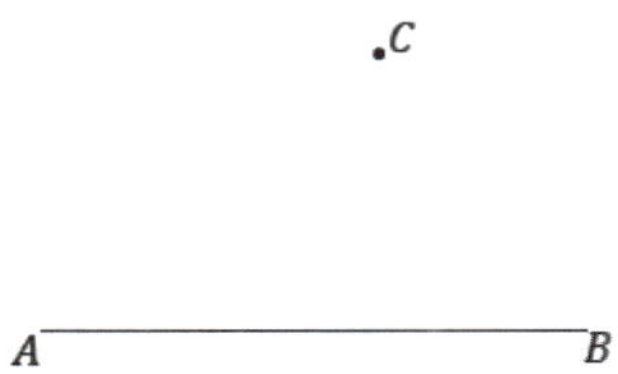

Step 2

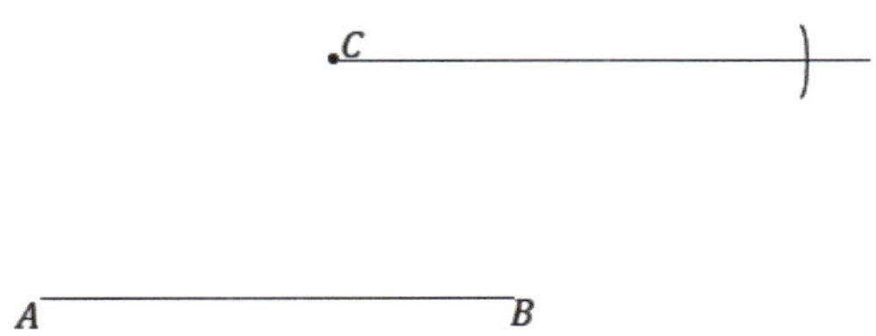

Step 3

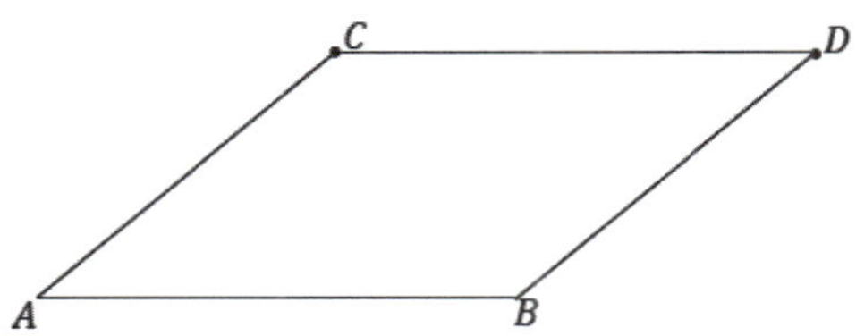

Example 2 (10 minutes)

Example 1 demonstrates how to create a scale drawing using the parallel method.

- The basic parameters and initial steps to the parallel method are like those of the initial steps to the ratio method; a ray must be drawn from the center through all vertices, and one corresponding vertex of the scale drawing must be determined using the scale factor and ruler. However, as suggested by the name of the method, the following steps require a setsquare to draw a segment parallel to each side of the figure.

Example 2

Use the figure below with center O and a scale factor of $r = 2$ and the following steps to create a scale drawing using the parallel method.

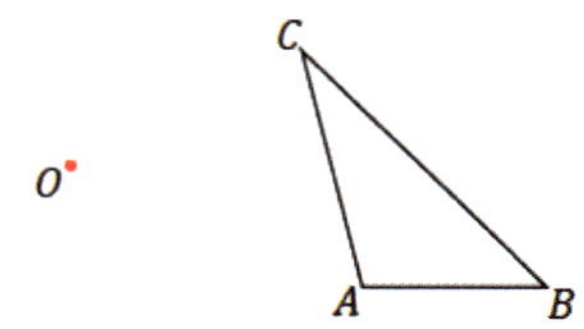

Step 1. Draw a ray beginning at O through each vertex of the figure.

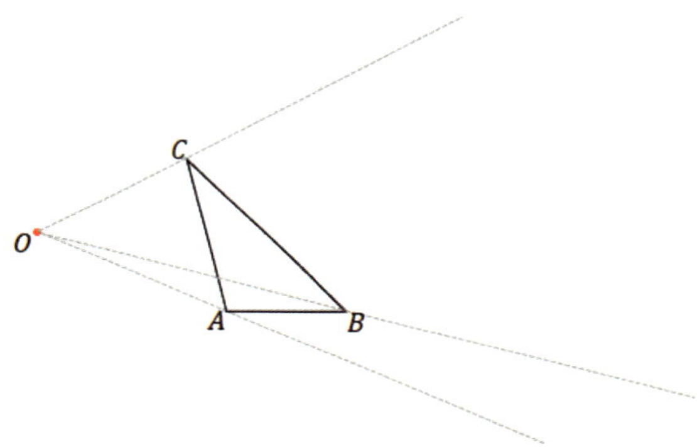

Step 2. Select one vertex of the scale drawing to locate; we have selected A'. Locate A' on $\overrightarrow{OA}$ so that $OA' = 2OA$.

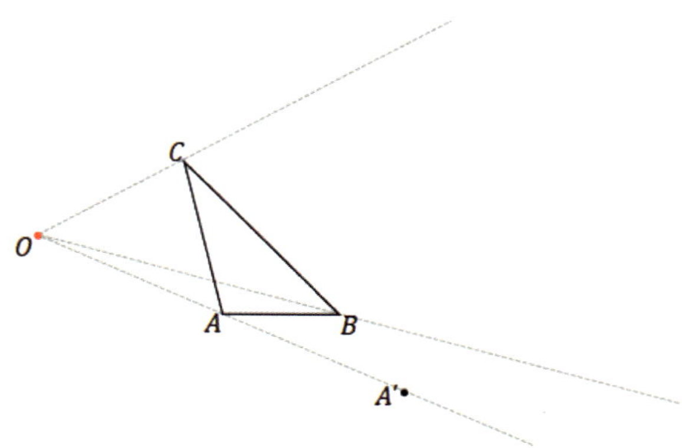

Step 3. Align the setsquare and ruler as in the image below; one leg of the setsquare should line up with side $\overline{AB}$, and the perpendicular leg should be flush against the ruler.

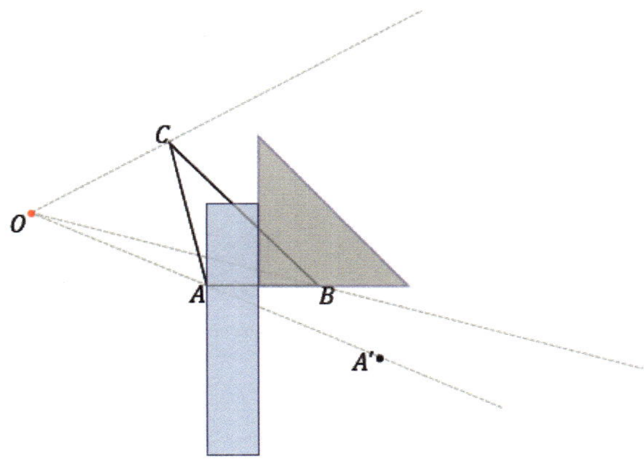

Scaffolding:

These steps should be modeled for all students, but depending on student needs, these images may need to be made larger, or teachers may need to model the steps more explicitly.

Step 4. Slide the setsquare along the ruler until the edge of the setsquare passes through A'. Then, along the perpendicular leg of the setsquare, draw the segment through A' that is parallel to $\overline{AB}$ until it intersects with $\overrightarrow{OB}$, and label this point B'.

Ask students to summarize how they created the scale drawing and why they think this method works.

It may happen that it is not possible to draw the entire parallel segment $A'B'$ due to the position of the setsquare and the location of B'. Alert students that this may happen and that they should simply pick up the setsquare (or ruler) and complete the segment once it has been started.

It may even happen that the setsquare is not long enough to meet point A'. In such a case, a ruler can be placed flush against the other leg of the setsquare, and then the setsquare can be removed and a line drawn through A'.

In a similar vein, if any of the rays is not long enough, extend it so that the intersection between the parallel segment and that ray is visible.

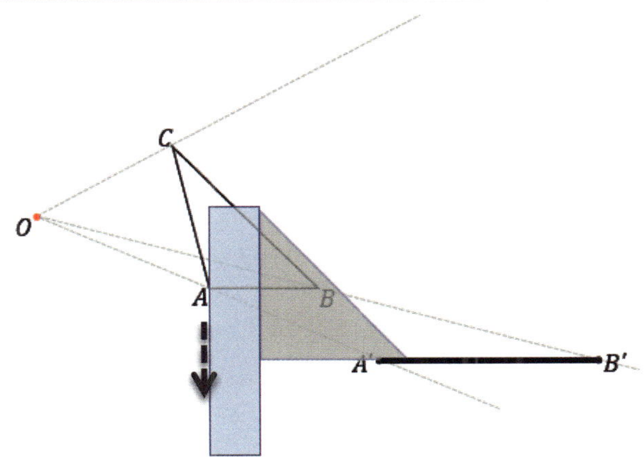

MP.1

Step 5. **Continue to create parallel segments to determine each successive vertex point. In this particular case, the setsquare has been aligned with $\overline{AC}$. This is done because, in trying to create a parallel segment from $\overline{BC}$, the parallel segment was not reaching B'. This could be remedied with a larger setsquare and longer ruler, but it is easily avoided by working on the segment parallel to $\overline{AC}$ instead.**

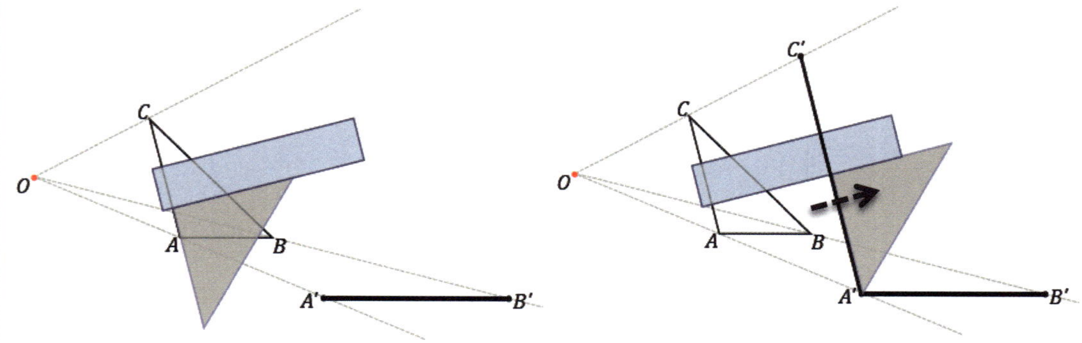

Step 6. **Use your ruler to join the final two unconnected vertices.**

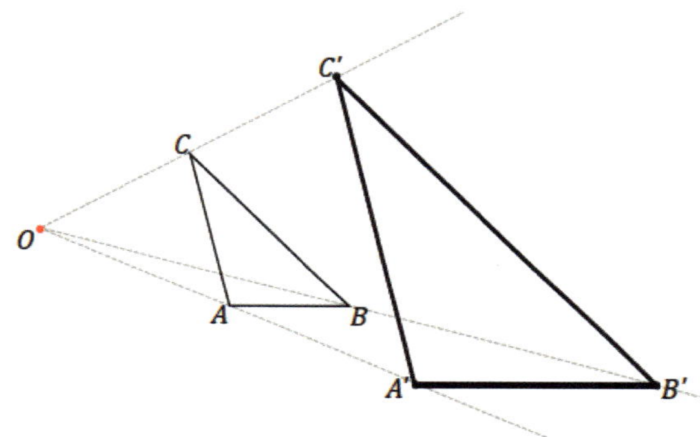

Have students show that $\triangle A'B'C'$ is a scale drawing; measure and confirm that the length of each segment in the scale drawing is twice the length of each segment in the original drawing and that the measurements of all corresponding angles are equal. $\triangle ABC$ angle measurements are $m\angle A = 104°$, $m\angle B = 44°$, and $m\angle C = 32°$. The measurements of the side lengths are not provided because they differ from the images that appear in print form.

> *Scaffolding:*
>
> Patty paper may facilitate measurements in the Examples and Exercises, but students should be prepared to use measuring tools in an Exit Ticket or Assessment.

- We want to note here that though we began with a scale factor of $r = 2$ for the dilation, we consider the resulting scale factor of the scale drawing created by the parallel method separately. As we can see by trying the dilations out, the scale factor for the dilation and the scale factor for the scale drawing by the parallel method are in fact one and the same.
- There is a concrete reason why this is, but we do not go into the explanation of why the parallel method actually yields a scale drawing with the same scale factor as the dilation until later lessons.

Exercises 1–2 (10 minutes)

Exercise 1 differs from Example 1 in one way: in Example 1, students had to locate the initial point A', whereas in Exercise 1, students are provided with the location of the initial point but not told explicitly how far from the center the point is. Teachers should use their discretion to decide if students are ready for the slightly altered situation or whether they need to retry a problem as in Example 1.

Exercises

1. **With a ruler and setsquare, use the parallel method to create a scale drawing of $WXYZ$ by the parallel method. W' has already been located for you. Determine the scale factor of the scale drawing. Verify that the resulting figure is in fact a scale drawing by showing that corresponding side lengths are in constant proportion and that corresponding angles are equal in measurement.**

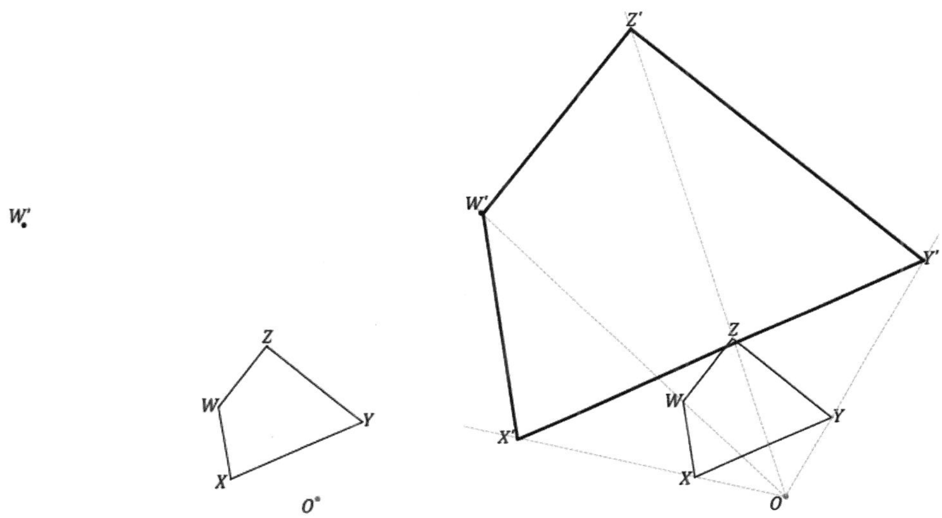

The scale factor is 3. Verification of the enlarged figure should show that the length of each segment in the scale drawing is three times the length of each segment in the original figure (e.g., $W'X' = 3(WX)$). The angle measurements are $m\angle W = 132°$, $m\angle X = 76°$, $m\angle Y = 61°$, and $m\angle Z = 91°$.

2. With a ruler and setsquare, use the parallel method to create a scale drawing of $DEFG$ about center O with scale factor $r = \frac{1}{2}$. Verify that the resulting figure is in fact a scale drawing by showing that corresponding side lengths are in constant proportion and that the corresponding angles are equal in measurement.

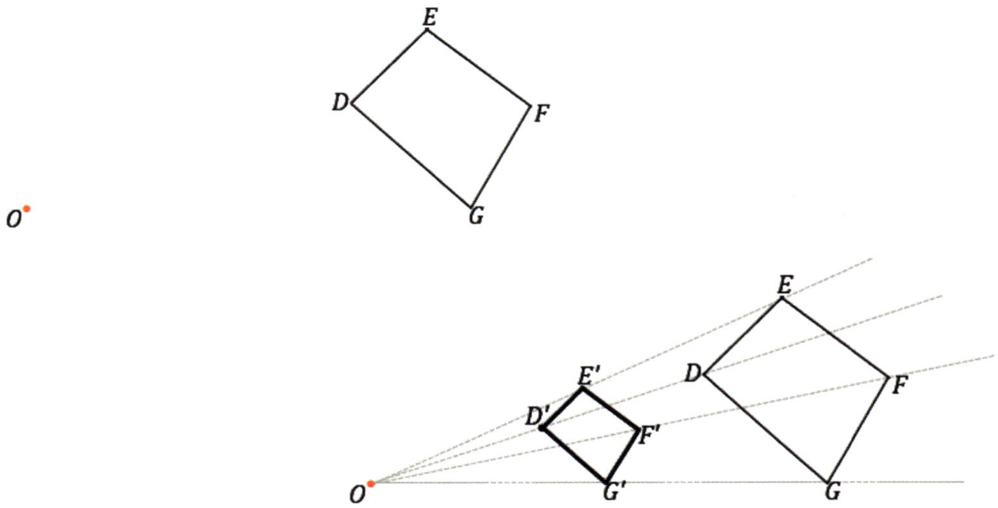

Verification of the reduced figure should show that the length of each segment in the scale drawing is one half the length of each segment in the original figure, (e.g., $D'E' = \frac{1}{2}(DE)$). The angle measurements are $m\angle D = 85°$, $m\angle E = 99°$, $m\angle F = 97°$, and $m\angle G = 79°$.

Discussion (5 minutes)

- So far we have verified that corresponding angles between figures and their scale drawings are equal in measurement by actually measuring each pair of angles. Instead of measuring each time, we can recall what we know about parallel lines cut by transversals to verify that corresponding angles are in fact equal in measurement. How would you explain this? Mark the following figure as needed to help explain.
  - *If a transversal intersects two parallel lines, then corresponding angles are equal in measurement. Since we have constructed corresponding segments to be parallel, we are certain that $\overline{AC} \parallel \overline{A'C'}$ and $\overline{AB} \parallel \overline{A'B'}$. We make use of the corresponding angles fact twice to show that corresponding angles $\angle A$ and $\angle A'$ are equal in measurement. A similar argument shows the other angles are equal in measurement.*

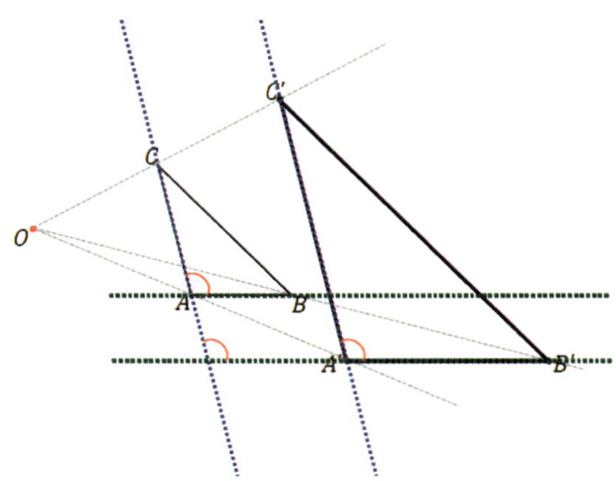

Exercise 3 (5 minutes)

The center O lies within the figure in Exercise 3. Ask students if they think this affects the resulting scale drawing, and allow them to confer with a neighbor.

3. With a ruler and setsquare, use the parallel method to create a scale drawing of pentagon $PQRST$ about center O with scale factor $r = \frac{5}{2}$. Verify that the resulting figure is in fact a scale drawing by showing that corresponding side lengths are in constant proportion and that corresponding angles are equal in measurement.

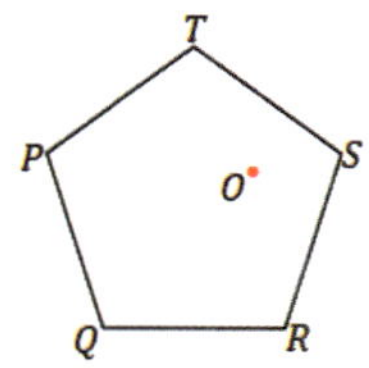

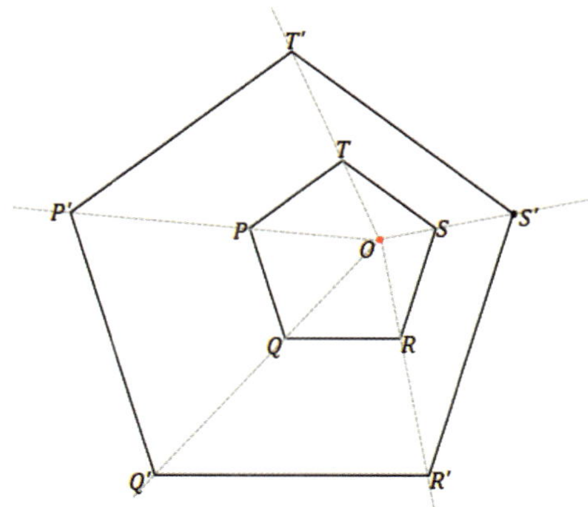

Verification of the enlarged figure should show that the length of each segment in the scale drawing is two-and-a-half times the length of each segment in the original figure (e.g., $P'T' = \frac{5}{2}(PT)$). Each of the angles has a measurement of $108°$.

Closing (2 minutes)

Ask students to summarize the key points of the lesson. Additionally, consider having them answer the following questions independently in writing, to a partner, or to the whole class.

- How are dilations and scale drawings related?
  - *Dilations can be used to create scale drawings by the ratio method or the parallel method.*
- To create a scale drawing using the ratio method, a center, a figure, and a scale factor must be provided. Then the dilated vertices can either be measured or located using a compass. To use the parallel method, a center, a figure, and a scale factor or one provided vertex of the dilated figure must be provided. Then use a setsquare to help construct sides parallel to the sides of the original figure, beginning with the side that passes through the provided dilated vertex.

Exit Ticket (5 minutes)

Name ______________________________ Date ______________

Lesson 3: Making Scale Drawings Using the Parallel Method

Exit Ticket

With a ruler and setsquare, use the parallel method to create a scale drawing of quadrilateral $ABCD$ about center O with scale factor $r = \frac{3}{4}$. Verify that the resulting figure is in fact a scale drawing by showing that corresponding side lengths are in constant proportion and that the corresponding angles are equal in measurement.

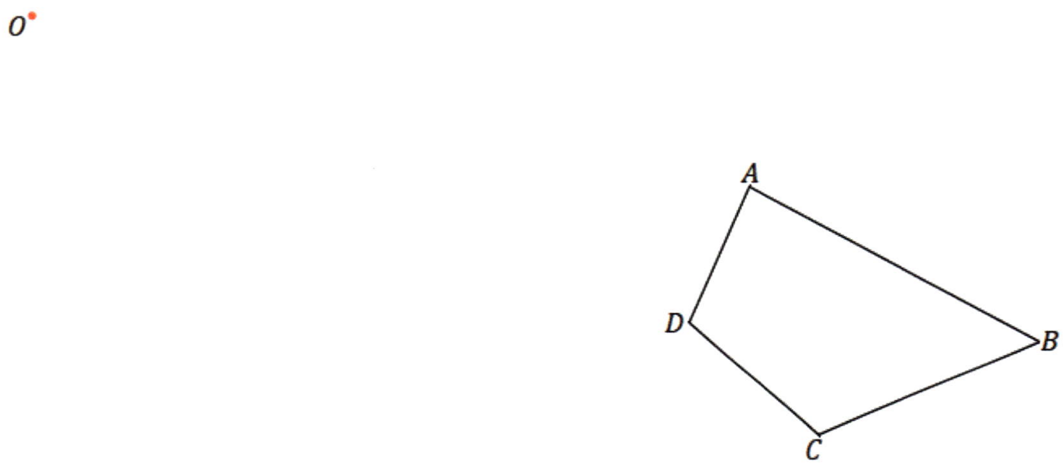

What kind of error in the parallel method might prevent us from having parallel, corresponding sides?

Exit Ticket Sample Solutions

With a ruler and setsquare, use the parallel method to create a scale drawing of quadrilateral $ABCD$ about center O with scale factor $r = \frac{3}{4}$. Verify that the resulting figure is in fact a scale drawing by showing that corresponding side lengths are in constant proportion and that the corresponding angles are equal in measurement.

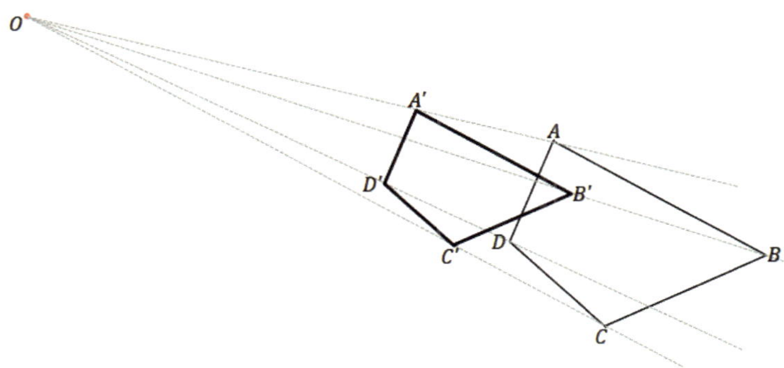

The measurements of the angles in the figure are $m\angle A = 86°$, $m\angle B = 51°$, $m\angle C = 115°$, and $m\angle D = 108°$. All side-length measurements of the scale drawing should be in the constant ratio of $3:4$.

What kind of error in the parallel method might prevent us from having parallel, corresponding sides?

If the setsquare is not aligned with the segment of the figure, you will not create a parallel segment. Also, if the setsquare is not perfectly flush with the ruler, it will not be possible to create a segment parallel to the segment of the figure.

Problem Set Sample Solutions

1. With a ruler and setsquare, use the parallel method to create a scale drawing of the figure about center O. One vertex of the scale drawing has been provided for you.

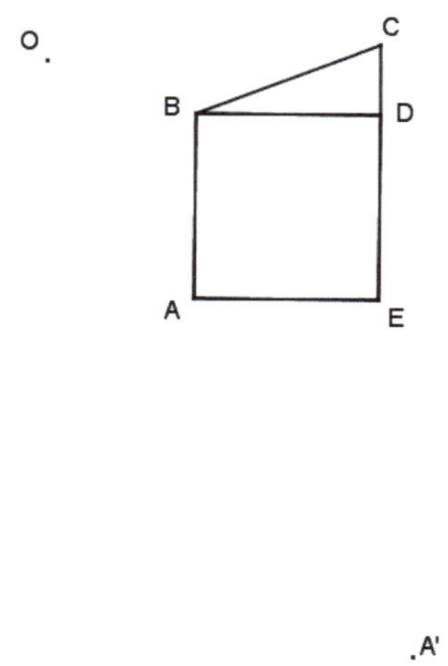

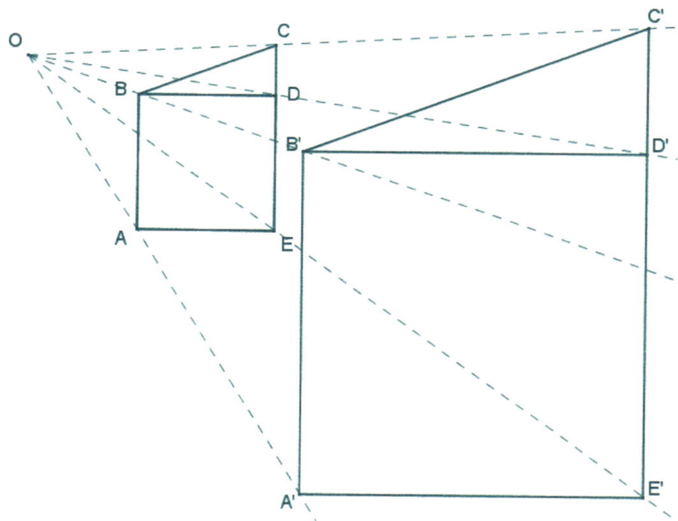

Determine the scale factor. Verify that the resulting figure is in fact a scale drawing by showing that corresponding side lengths are in constant proportion and that the corresponding angles are equal in measurement.

The scale factor is $r = \frac{5}{2}$. *The measurements of the angles in the figure are* $m\angle A = m\angle B = m\angle D = m\angle E = 90°$, $m\angle DBC = 20°$, $m\angle CDB = 90°$, *and* $m\angle C = 70°$. *All side-length measurements of the scale drawing are in the constant ratio of* $5:2$.

2. **With a ruler and setsquare, use the parallel method to create a scale drawing of the figure about center O and scale factor $r = \frac{1}{3}$. Verify that the resulting figure is in fact a scale drawing by showing that corresponding side lengths are in constant proportion and the corresponding angles are equal in measurement.**

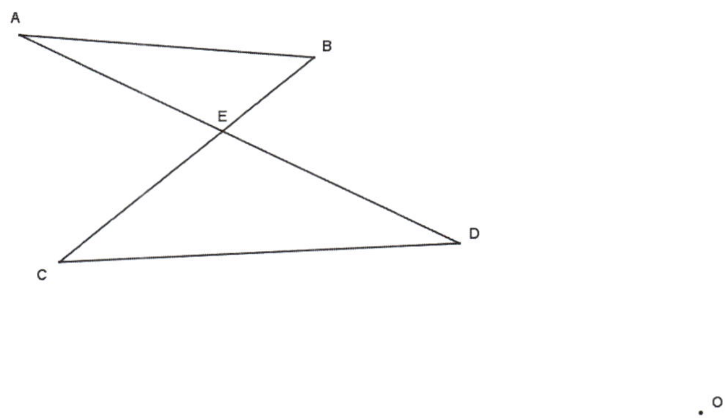

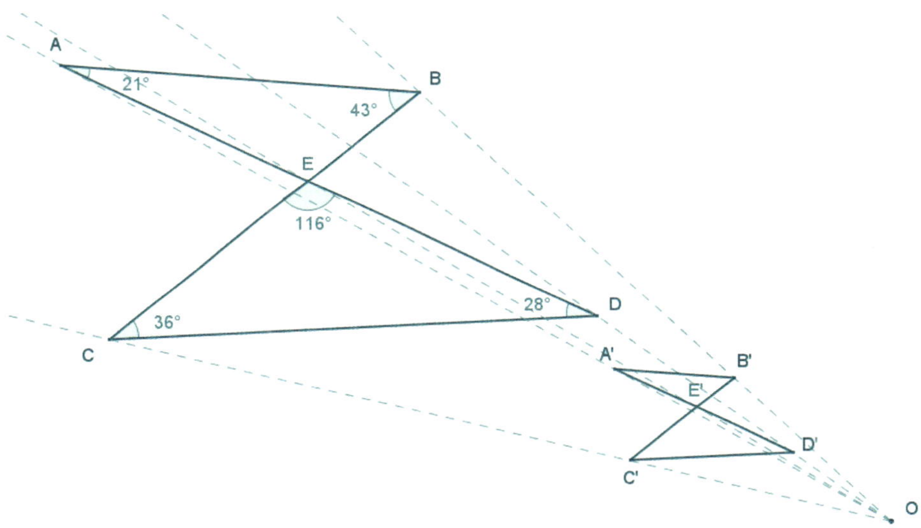

The measurements of the angles in the figure are $m\angle A = 21°$, $m\angle B = 43°$, $m\angle C = 36°$, $m\angle D = 28°$, and $m\angle CED = m\angle BEA = 116°$. All side-length measurements of the scale drawing are in the constant ratio of $1:3$.

3. With a ruler and setsquare, use the parallel method to create the following scale drawings about center O: (1) first use a scale factor of 2 to create $\triangle A'B'C'$, and (2) then, with respect to $\triangle A'B'C'$, use a scale factor of $\frac{2}{3}$ to create scale drawing $\triangle A''B''C''$. Calculate the scale factor for $\triangle A''B''C''$ as a scale drawing of $\triangle ABC$. Use angle and side length measurements and the appropriate proportions to verify your answer.

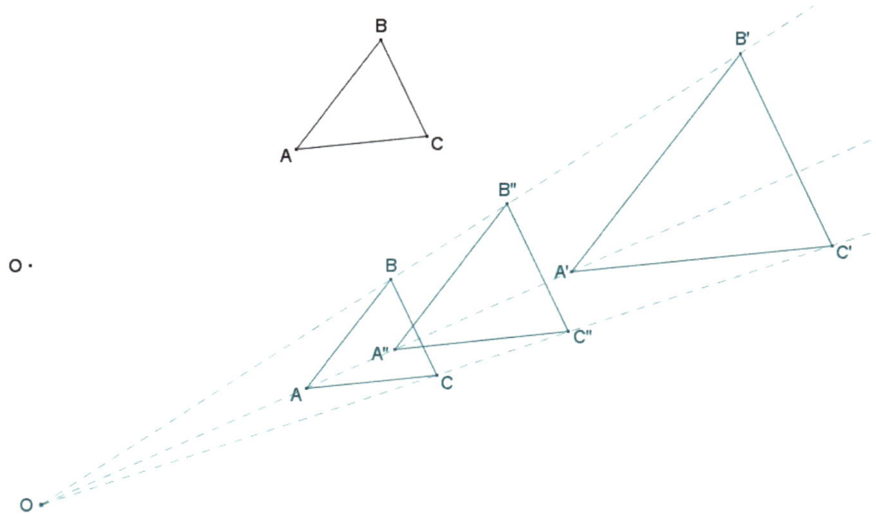

The scale factor of $\triangle A'B'C'$ relative to $\triangle ABC$ is 2. The measurements of the angles in the figure are $m\angle A \approx 46°$, $m\angle B \approx 64°$, and $m\angle C \approx 70°$. The scale factor from $\triangle ABC$ to $\triangle A''B''C''$ is $\frac{2}{3} \cdot 2 = \frac{4}{3}$. Solutions should show the side-length proportions between $\triangle A''B''C''$ and $\triangle ABC$: $\frac{A''B''}{AB} = \frac{B''C''}{BC} = \frac{C''A''}{CA} = \frac{4}{3}$.

4. Follow the direction in each part below to create three scale drawings of $\triangle ABC$ using the parallel method.

   a. With the center at vertex A, make a scale drawing of $\triangle ABC$ with a scale factor of $\frac{3}{2}$.

   b. With the center at vertex B, make a scale drawing of $\triangle ABC$ with a scale factor of $\frac{3}{2}$.

   c. With the center at vertex C, make a scale drawing of $\triangle ABC$ with a scale factor of $\frac{3}{2}$.

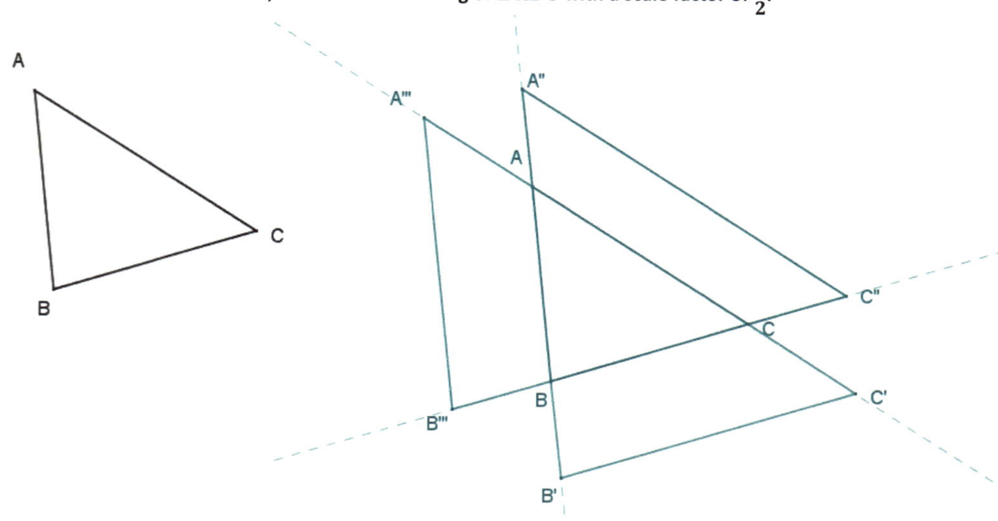

EUREKA MATH™

d. **What conclusions can be drawn about all three scale drawings from parts (a)–(c)?**

The three scale drawings are congruent.

$AB' = A''B = \frac{3}{2}AB$	***Dilation using scale factor*** $\frac{3}{2}$.
$\angle A'' \cong \angle BAC$	***Corresponding*** $\angle$***'s formed by parallel lines are congruent.***
$\angle B' \cong \angle ABC$	***Corresponding*** $\angle$***'s formed by parallel lines are congruent.***
$\triangle AB'C' \cong \triangle A''BC''$	***ASA***

A similar argument can be used to show $\triangle AB'C' \cong \triangle A'''B'''C$***, and by transitivity of congruence, all three scale drawings are congruent to each other.***

5. **Use the parallel method to make a scale drawing of the line segments in the following figure using the given** W'**, the image of vertex** W**, from center** O**. Determine the scale factor.**

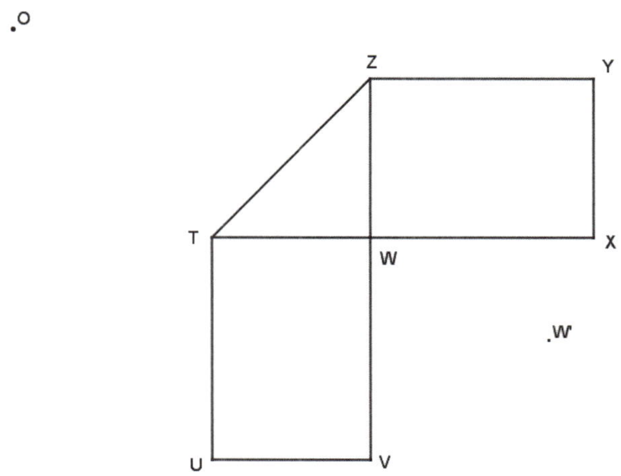

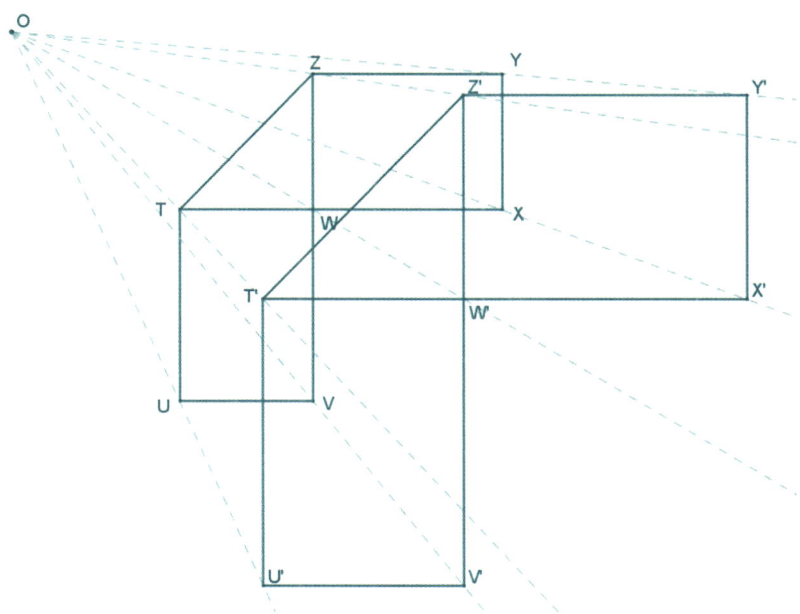

The ratio of $OW' : OW$ ***is*** $3:2$***, so the scale factor is*** $\frac{3}{2}$***. All corresponding lengths are in the ratio of*** $3:2$***.***

Use your diagram from Problem 1 to answer this question.

6. If we switch perspective and consider the original drawing $ABCDE$ to be a scale drawing of the constructed image $A'B'C'D'E'$, what would the scale factor be?

   If the original figure were the scale drawing, and the scale drawing were the original figure, the scale factor would be $\frac{2}{5}$.

Lesson 4: Comparing the Ratio Method with the Parallel Method

Student Outcomes

- Students understand that the ratio and parallel methods produce the same scale drawing and understand the proof of this fact.
- Students relate the equivalence of the ratio and parallel methods to the *triangle side splitter theorem*: A line segment splits two sides of a triangle proportionally if and only if it is parallel to the third side.

Lesson Notes

This lesson challenges students to understand the reasoning that connects the ratio and parallel methods that have been used in the last two lessons for producing a scale drawing. The Opening Exercises are important to the discussions that follow and are two important ideas in their own right. The first key idea is that two triangles with the same base that have vertices on a line parallel to the base are equal in area. The second key idea is that two triangles with different bases but equal altitudes have a ratio of areas that is equal to the ratio of their bases. Following the Opening Exercises, students and the teacher show that the ratio method and parallel method are equivalent. The concluding discussion shows how that work relates to the triangle side splitter theorem.

Classwork

Today, our goal is to show that the parallel method and the ratio method are equivalent; that is, given a figure in the plane and a scale factor $r > 0$, the scale drawing produced by the parallel method is congruent to the scale drawing produced by the ratio method. We start with two easy exercises about the areas of two triangles whose bases lie on the same line, which helps show that the two methods are equivalent.

Opening Exercise (10 minutes)

Students need the formula for the area of a triangle. The first exercise is a famous proposition of Euclid's (Proposition 37 of Book 1). Consider asking students to go online after class to read how Euclid proves the proposition.

Give students two minutes to work on the first exercise in groups, and walk around the room answering questions and helping students draw pictures. After two minutes, go through the proof on the board, answering questions about the parallelogram along the way. Repeat the process for Exercise 2.

Do not look for pristine proofs from students on these exercises, merely for confirmation that they understand the statements. For example, in Exercise 1, they should understand that two triangles between two parallel lines with the same base must have the same area. These two exercises help avoid the quagmire of drawing altitudes and calculating areas in the proofs that follow; these exercises help students to simply recognize when two triangles have the same area or recognize when the ratio of the bases of two triangles is the same as the ratio of their areas.

It is useful to leave the statements of Opening Exercises 1 and 2 on the board throughout the lesson to refer back to them.

Opening Exercise

a. **Suppose two triangles, $\triangle ABC$ and $\triangle ABD$, share the same base $\overline{AB}$ such that points C and D lie on a line parallel to $\overleftrightarrow{AB}$. Show that their areas are equal, that is, $\text{Area}(\triangle ABC) = \text{Area}(\triangle ABD)$. (Hint: Why are the altitudes of each triangle equal in length?)**

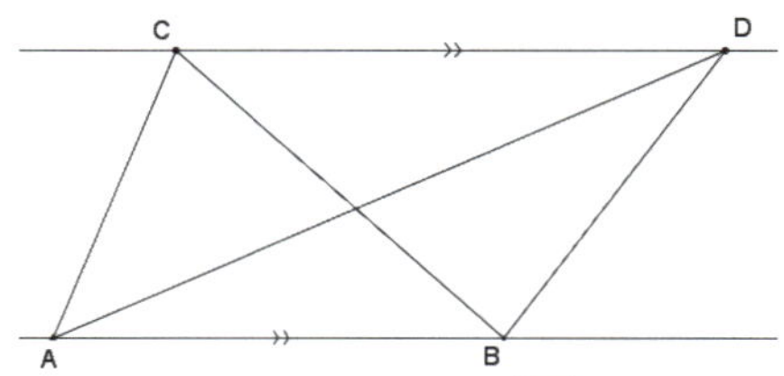

Draw a perpendicular line to $\overleftrightarrow{AB}$ through C, and label the intersection of both lines C'. Then $\overline{CC'}$ is an altitude for $\triangle ABC$. Do the same for $\triangle ABD$ to get an altitude $\overline{DD'}$.

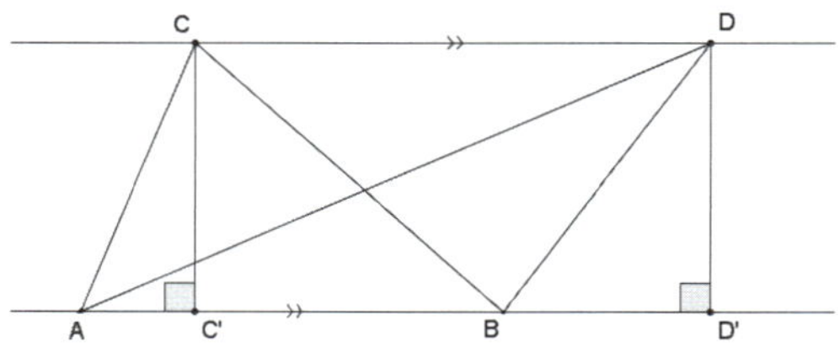

Quadrilateral $CC'D'D$ is a parallelogram and, therefore, $CC' = DD'$, both of which follow from the properties of parallelograms. Since $\overline{CC'}$ and $\overline{DD'}$ are altitudes of the triangles, we get by the area formula for triangles,

$$\text{Area}(\triangle ABC) = \frac{1}{2}AB \cdot CC' = \frac{1}{2}AB \cdot DD' = \text{Area}(\triangle ABD).$$

MP.6 Draw the first picture below while reading through Opening Exercise 2 with the class. Ask questions that check for understanding, like, "Are the points A, B, and B' collinear? Why?" and "Does it matter if B and B' are on the same side of A on the line?"

b. **Suppose two triangles have different-length bases, $\overline{AB}$ and $\overline{AB'}$, that lie on the same line. Furthermore, suppose they both have the same vertex C opposite these bases. Show that the value of the ratio of their areas is equal to the value of the ratio of the lengths of their bases, that is,**

$$\frac{\text{Area}(\triangle ABC)}{\text{Area}(\triangle AB'C)} = \frac{AB}{AB'}.$$

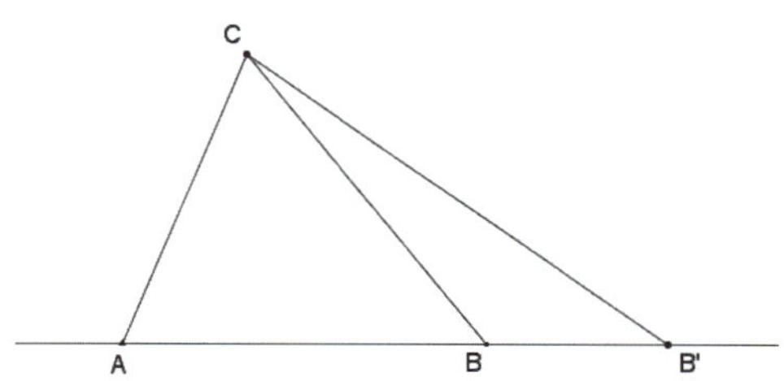

Draw a perpendicular line to $\overleftrightarrow{AB}$ through C, and label the intersection of both lines C'.

Then $\overline{CC'}$ is an altitude for both triangles.

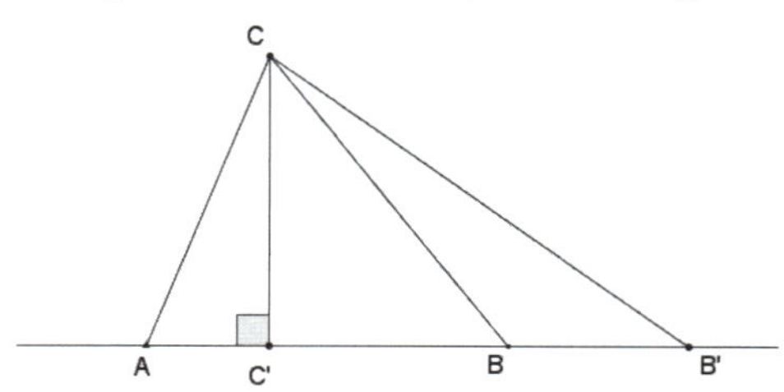

By the area formula for triangles,

$$\frac{\text{Area}(\triangle ABC)}{\text{Area}(\triangle AB'C)} = \frac{\frac{1}{2}AB \cdot CC'}{\frac{1}{2}AB' \cdot CC'} = \frac{AB}{AB'}.$$

Ask students to summarize to a neighbor the two results from the Opening Exercises. Use this as an opportunity to check for understanding.

Discussion (20 minutes)

The next two theorems generate the so-called triangle side splitter theorem, which is the most important result of this module. (In standard **G-SRT.B.4**, it is stated as, "A line parallel to one side of a triangle divides the other two proportionally, and conversely.") The triangle side splitter theorem is used many times in the next few lessons to understand dilations and similarity. Note that using the AA similarity criterion to prove the triangle side splitter theorem is *circular*: The triangle side splitter theorem is the reason why a dilation takes a line to a parallel line and an angle to another angle of equal measure, which are both needed to prove the AA similarity criterion. Thus, the triangle side splitter theorem must be proven in a way that does not invoke these two ideas. (Note that in Grade 8, we assumed the triangle side splitter theorem and its immediate consequences and, in doing so, glossed over many of the issues we need to deal with in this course.)

Even though the following two proofs are probably the simplest known proofs of these theorems (and should be easy to understand), they rely on subtle tricks that students should not be expected to discover on their own. Brilliant mathematicians constructed these careful arguments over 2,300 years ago. However, that does not mean that this part of the lesson is a lecture. Go through the proofs with students at a relaxed pace, ask them questions to check for understanding, and have them articulate the reasons for each step. If done well, the class can take joy in the clever arguments presented here!

Discussion

To show that the parallel and ratio methods are equivalent, we need only look at one of the simplest versions of a scale drawing: scaling segments. First, we need to show that the scale drawing of a segment generated by the parallel method is the same segment that the ratio method would have generated and vice versa. That is,

The parallel method $\Longrightarrow$ The ratio method,

and

The ratio method $\Longrightarrow$ The parallel method.

MP.1 & MP.3

Ask students why scaling a segment is sufficient for showing equivalence of both methods for scaling polygonal figures. That is, if we wanted to show that both methods would produce the same polygonal figure, why is it enough to only show that both methods would produce the same segment?

Students should respond that polygonal figures are composed of segments. If we can show that both methods produce the same segment, then it makes sense that both methods would work for all segments that comprise the polygonal figure.

The first implication above can be stated as the following theorem:

PARALLEL $\Longrightarrow$ RATIO THEOREM: Given $\overline{AB}$ and point O not on $\overleftrightarrow{AB}$, construct a scale drawing of $\overline{AB}$ with scale factor $r > 0$ using the parallel method: Let $A' = D_{O,r}(A)$, and ℓ be the line parallel to $\overleftrightarrow{AB}$ that passes through A'. Let B' be the point where $\overrightarrow{OB}$ intersects ℓ. Then B' is the same point found by the ratio method, that is, $B' = D_{O,r}(B)$.

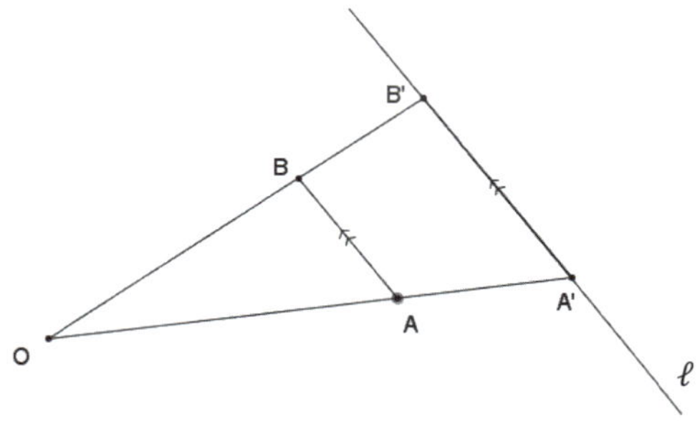

Discuss the statement of the theorem with students. Ask open-ended questions that lead students through the following points:

- $\overline{A'B'}$ is the scale drawing of $\overline{AB}$ using the parallel method. Why?
- $A' = D_{O,r}(A)$ is already the first step in the ratio method. The difference between the parallel method and the ratio method is that B' is found using the parallel method by intersecting the parallel line ℓ with $\overrightarrow{OB}$, while in the ratio method, the point is found by dilating point B at center O by scale factor r to get $D_{O,r}(B)$. We need to show that these are the same point, that is, that $B' = D_{O,r}(B)$. Since both points lie on $\overrightarrow{OB}$, this can be done by showing that $OB' = r \cdot OB$.

There is one subtlety with the above theorem to consider discussing with students. In it, we assumed—asserted really—that ℓ and $\overrightarrow{OB}$ intersect. They do, but is it clear that they do? (Pictures can be deceiving!) First, suppose that ℓ did not intersect $\overleftrightarrow{OB}$, and then by definition, ℓ and $\overleftrightarrow{OB}$ are parallel. Since ℓ is also parallel to $\overleftrightarrow{AB}$, then $\overleftrightarrow{OB}$ and $\overleftrightarrow{AB}$ are parallel (by parallel transitivity from Module 1), which is clearly a contradiction since both contain the point B. Hence, it must be that ℓ intersects $\overleftrightarrow{OB}$. But where does it intersect? Does it intersect $\overrightarrow{OB}$ or the opposite ray from O? There are two cases to consider. Case 1: If A' and O are in opposite half-planes of $\overleftrightarrow{AB}$ (i.e., as in the picture above when $r > 0$), then ℓ is contained completely in the half-plane that contains A' by the plane separation axiom. Thus, ℓ cannot intersect $\overrightarrow{BO}$, which means it must intersect $\overrightarrow{OB}$. Case 2: Now suppose that A' and O are in the same half-plane of $\overleftrightarrow{AB}$ (when $0 < r < 1$), and consider the two half-planes of ℓ. The points B and O must lie in the opposite half-planes of ℓ. (Why? Hint: What fact would be contradicted if they were in the same half-plane?) Thus, by the plane separation axiom, the line intersects $\overline{OB}$, and thus ℓ intersects $\overrightarrow{OB}$.

There is a set of theorems that revolve around when two lines intersect each other as in the paragraph above. That set falls under the general heading of "crossbar theorems." Explore these theorems with students by looking the theorems up on the web. The theorem above is written in a way that asserts that ℓ and $\overrightarrow{OB}$ intersect, and so covers up these intersection issues in a factually correct way that helps avoid unnecessarily pedantic crossbar discussions in the future.

PROOF: We prove the case when $r > 1$; the case when $0 < r < 1$ is the same but with a different picture. Construct $\overline{BA'}$ and $\overline{AB'}$ to form two triangles $\triangle BAB'$ and $\triangle BAA'$, labeled as T_1 and T_2, respectively, in the picture below.

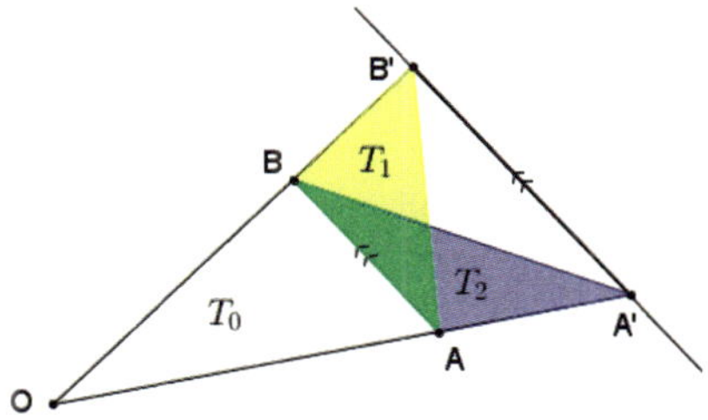

The areas of these two triangles are equal,

$$\text{Area}(T_1) = \text{Area}(T_2),$$

by Exercise 1. Why? Label $\triangle OAB$ by T_0. Then $\text{Area}(\triangle OA'B) = \text{Area}(\triangle OB'A)$ because areas add:

$$\begin{aligned}\text{Area}(\triangle OA'B) &= \text{Area}(T_0) + \text{Area}(T_2)\\ &= \text{Area}(T_0) + \text{Area}(T_1)\\ &= \text{Area}(\triangle OB'A).\end{aligned}$$

Next, we apply Exercise 2 to two sets of triangles: (1) T_0 and $\triangle OA'B$ and (2) T_0 and $\triangle OB'A$.

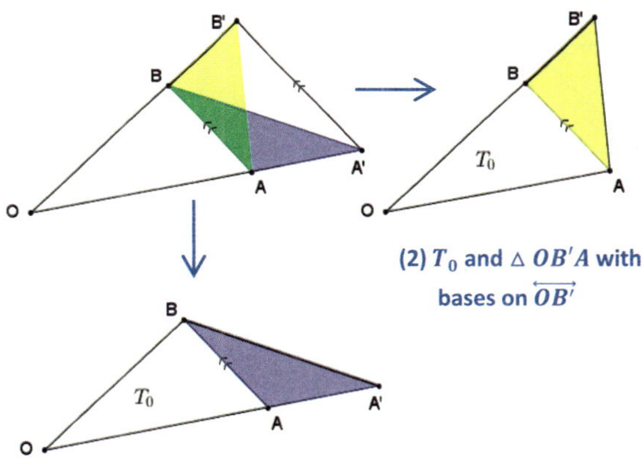

(2) T_0 and $\triangle OB'A$ with bases on $\overleftrightarrow{OB'}$

(1) T_0 and $\triangle OA'B$ with bases on $\overleftrightarrow{OA'}$

Therefore,

$$\frac{\text{Area}(\triangle OA'B)}{\text{Area}(T_0)} = \frac{OA'}{OA}, \text{ and}$$

$$\frac{\text{Area}(\triangle OB'A)}{\text{Area}(T_0)} = \frac{OB'}{OB}.$$

Since Area($\triangle\ OA'B$) = Area($\triangle\ OB'A$), we can equate the fractions: $\frac{OA'}{OA} = \frac{OB'}{OB}$. Since r is the scale factor used in dilating $\overline{OA}$ to $\overline{OA'}$, we know that $\frac{OA'}{OA} = r$; therefore, $\frac{OB'}{OB} = r$, or $OB' = r \cdot OB$. This last equality implies that B' is the dilation of B from O by scale factor r, which is what we wanted to prove.

Next, we prove the reverse implication to show that both methods are equivalent to each other.

MP.3 Ask students why showing that "the ratio method implies the parallel method" establishes equivalence. Why isn't the first implication "good enough"? (Because we do not know yet that a scale drawing produced by the ratio method would be the same scale drawing produced by the parallel method—the first implication does help us conclude that.)

This theorem is easier to prove than the previous one. In fact, the previous theorem can be used to quickly prove this one!

RATIO $\Longrightarrow$ PARALLEL THEOREM: Given $\overline{AB}$ and point O not on the $\overleftrightarrow{AB}$, construct a scale drawing $\overline{A'B'}$ of $\overline{AB}$ with scale factor $r > 0$ using the ratio method (i.e., find $A' = D_{O,r}(A)$ and $B' = D_{O,r}(B)$, and draw $\overline{A'B'}$). Then B' is the same as the point found using the parallel method.

PROOF: Since both the ratio method and the parallel method start with the same first step of setting $A' = D_{O,r}(A)$, the only difference between the two methods is in how the second point is found. If we use the parallel method, we construct the line ℓ parallel to $\overleftrightarrow{AB}$ that passes through A' and label the point where ℓ intersects $\overrightarrow{OB}$ by C. Then B' is the same as the point found using the parallel method if we can show that $C = B'$.

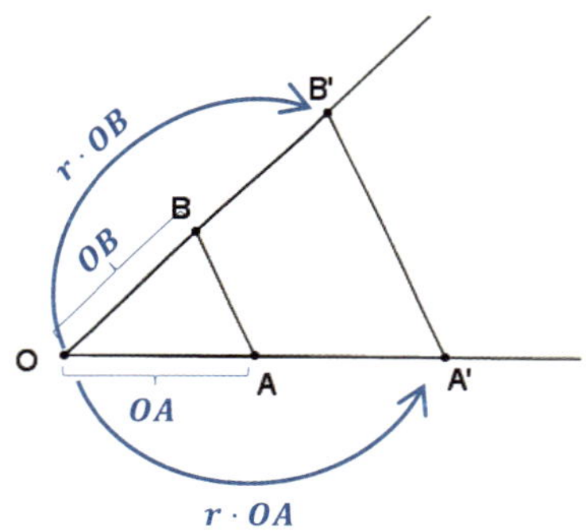

The Ratio Method

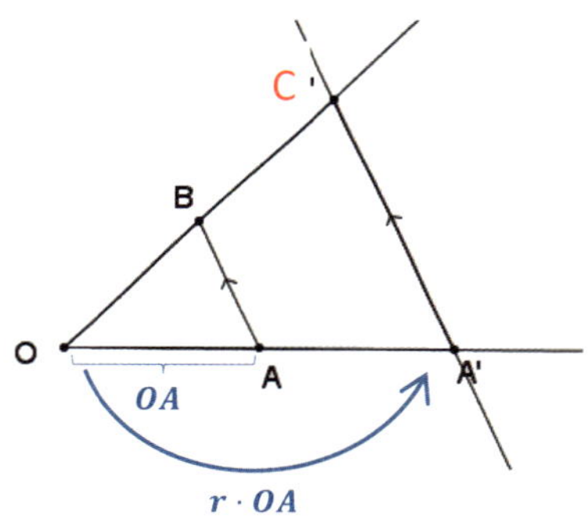

The Parallel Method

By the parallel $\Longrightarrow$ ratio theorem, we know that $C = D_{O,r}(B)$, that is, that C is the point on $\overrightarrow{OB}$ such that $OC = r \cdot OB$. But B' is also the point on $\overrightarrow{OB}$ such that $OB' = r \cdot OB$. Hence, they must be the same point.

Discussion (8 minutes)

The fact that the ratio and parallel methods are equivalent is often stated as the triangle side splitter theorem. To understand the triangle side splitter theorem, we need a definition:

SIDE SPLITTER: A line segment CD is said to *split the sides of* $\triangle OAB$ *proportionally* if C is a point on $\overline{OA}$, D is a point on $\overline{OB}$, and $\frac{OA}{OC} = \frac{OB}{OD}$ (or equivalently, $\frac{OC}{OA} = \frac{OD}{OB}$). We call line segment CD a *side splitter.*

TRIANGLE SIDE SPLITTER THEOREM: A line segment splits two sides of a triangle proportionally if and only if it is parallel to the third side.

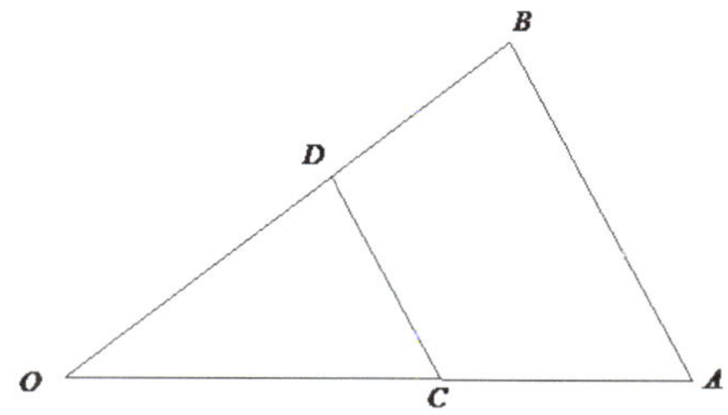

Provide students with time to read and make sense of the theorem. Students should be able to state that a line segment that splits two sides of a triangle is called a side splitter. If the sides of a triangle are split proportionally, then the line segment that split the sides must be parallel to the third side of the triangle. Conversely, if a segment that intersects two sides of a triangle is parallel to the third side of a triangle, then that segment is a side splitter.

Ask students to rephrase the statement of the theorem for a triangle $OA'B'$ and a segment AB (i.e., the terminology used in the theorems above). It should look like this:

Restatement of the triangle side splitter theorem:

In $\triangle OA'B'$, $\overline{AB}$ splits the sides proportionally (i.e., $\frac{OA'}{OA} = \frac{OB'}{OB}$) if and only if $\overline{A'B'} \parallel \overline{AB}$.

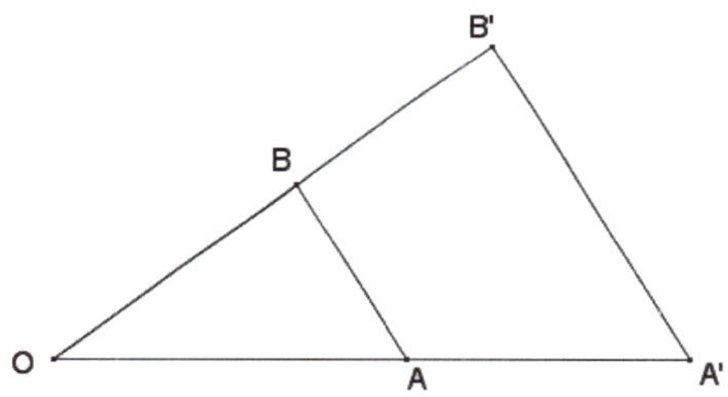

- Ask students to relate the restatement of the triangle side splitter theorem to the two theorems above. In order for students to do this, they need to translate the statement into one about dilations. Begin with the implication that $\overline{AB}$ splits the sides proportionally.
- What does $\frac{OA'}{OA} = \frac{OB'}{OB}$ mean in terms of dilations?
  - *This means that there is a dilation with scale factor $r = \frac{OA'}{OA}$ such that $D_{O,r}(A) = A'$ and $D_{O,r}(B) = B'$.*
- Which method (parallel or ratio) does the statement "$\overline{AB}$ splits the sides proportionally" correspond to?
  - *The ratio method*
- What does the ratio $\Longrightarrow$ parallel theorem imply about B'?
  - *This implies that B' can be found by constructing a line ℓ parallel to $\overline{AB}$ through A' and intersecting that line with $\overrightarrow{OB}$.*
- Since $\overline{AB} \parallel \ell$, what does that imply about $\overline{A'B'}$ and $\overline{AB}$?
  - *The two segments are also parallel as in the triangle side splitter theorem.*

- Now, suppose that $\overline{A'B'} \parallel \overline{AB}$ as in the picture below. Which method (parallel or ratio) does this statement correspond to?

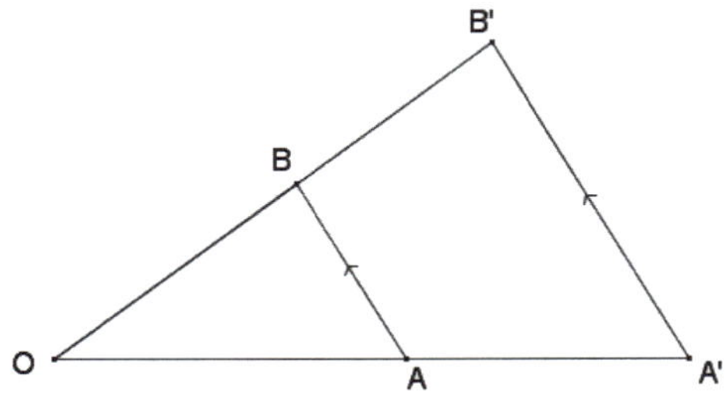

  - *This statement corresponds to the parallel method because in the parallel method, only the endpoint A of line segment AB is dilated from center O by scale factor r to get point A'. To draw $\overline{A'B'}$, a line is drawn through A' that is parallel to $\overline{AB}$, and B' is the intersection of that line and $\overrightarrow{OB}$.*
- What does the parallel $\Longrightarrow$ ratio theorem imply about the point B'?
  - *This implies that $D_{O,r}(B) = B'$ (i.e., $OB' = r \cdot OB$).*
- What does $OB' = r \cdot OB$ and $OA' = r \cdot OA$ imply about $\overline{AB}$?
  - *$\overline{AB}$ splits the sides of $\triangle OA'B$.*

Closing (2 minutes)

Ask students to summarize the main points of the lesson. Students may respond in writing, to a partner, or to the whole class.

- **THE TRIANGLE SIDE SPLITTER THEOREM**: A line segment splits two sides of a triangle proportionally if and only if it is parallel to the third side.
- Prior to this lesson we have used the ratio method and the parallel method separately to produce a scale drawing. The triangle side splitter theorem is a way of saying that we can use either method because both produce the same scale drawing.

Consider asking students to compare and contrast the two methods in their own words as a way of explaining how the triangle side splitter theorem captures the mathematics of *why* each method produces the same scale drawing.

> Lesson Summary
>
> **THE TRIANGLE SIDE SPLITTER THEOREM: A line segment splits two sides of a triangle proportionally if and only if it is parallel to the third side.**

Exit Ticket (5 minutes)

Name ______________________________ Date ______________

Lesson 4: Comparing the Ratio Method with the Parallel Method

Exit Ticket

In the diagram, $\overline{XY} \parallel \overline{AC}$. Use the diagram to answer the following:

1. If $BX = 4$, $BA = 5$, and $BY = 6$, what is BC?

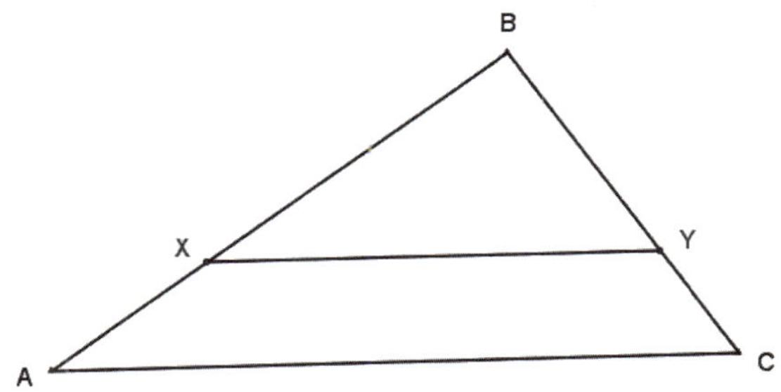

Not drawn to scale

2. If $BX = 9$, $BA = 15$, and $BY = 15$, what is YC?

Exit Ticket Sample Solutions

In the diagram, $\overline{XY} \parallel \overline{AC}$. Use the diagram to answer the following:

1. If $BX = 4$, $BA = 5$, and $BY = 6$, what is BC?

   $BC = 7.5$

2. If $BX = 9$, $BA = 15$, and $BY = 15$, what is YC?

   $YC = 10$

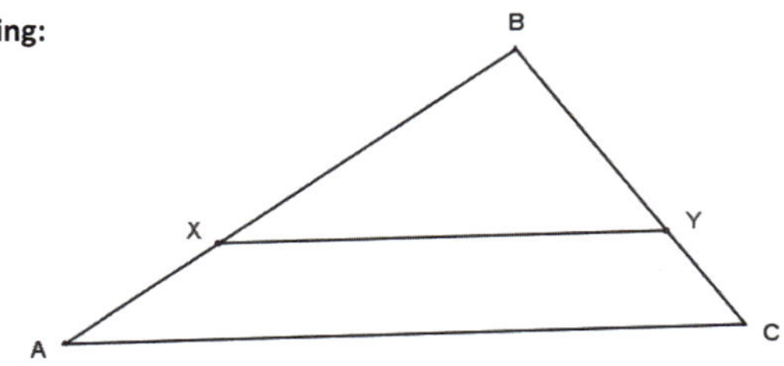

Not drawn to scale

Problem Set Sample Solutions

1. Use the diagram to answer each part below.

   a. Measure the segments in the figure below to verify that the proportion is true.

   $$\frac{OA'}{OA} = \frac{OB'}{OB}$$

   Actual measurements may vary due to copying, but students should state that the proportion is true.

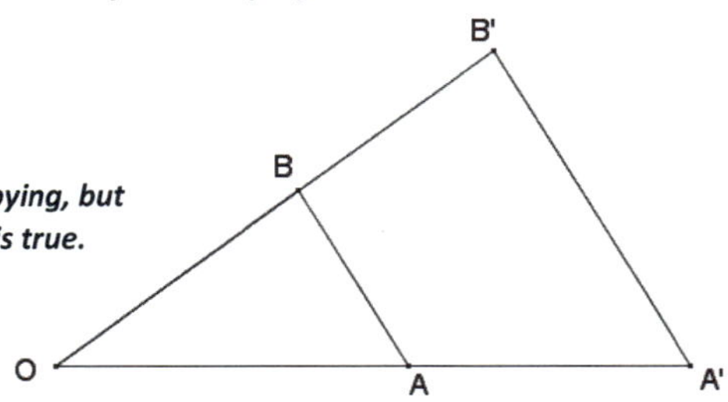

   b. Is the proportion $\frac{OA}{OA'} = \frac{OB}{OB'}$ also true? Explain algebraically.

   The proportion is also true because the reciprocals of equivalent ratios are also equivalent.

   c. Is the proportion $\frac{AA'}{OA'} = \frac{BB'}{OB'}$ also true? Explain algebraically.

   True. $OA' = OA + AA'$, and $OB' = OB + BB'$. So, using the equivalent ratios in part (a):

   $$\frac{OA'}{OA} = \frac{OB'}{OB}$$
   $$\frac{OA + AA'}{OA} = \frac{OB + BB'}{OB}$$
   $$\frac{OA}{OA} + \frac{AA'}{OA} = \frac{OB}{OB} + \frac{BB'}{OB}$$
   $$1 + \frac{AA'}{OA} = 1 + \frac{BB'}{OB}$$
   $$\frac{AA'}{OA} = \frac{BB'}{OB}.$$

2. Given the diagram below, $AB = 30$, line ℓ is parallel to $\overline{AB}$, and the distance from $\overline{AB}$ to ℓ is 25. Locate point C on line ℓ such that $\triangle ABC$ has the greatest area. Defend your answer.

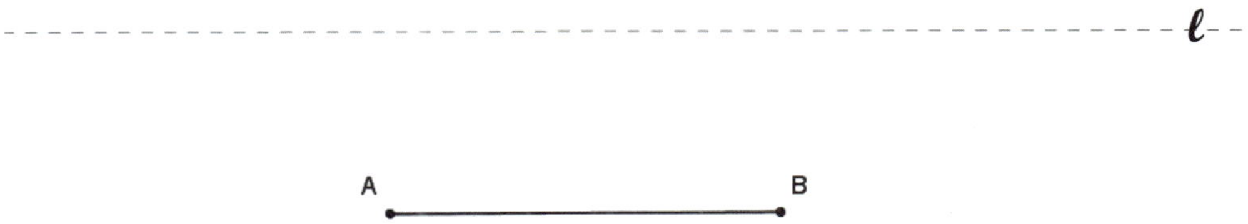

The distance between two parallel lines is constant and in this case is 25 units. $\overline{AB}$ serves as the base of all possible triangles ABC. The area of a triangle is one-half the product of its base and its height. No matter where point C is located on line ℓ, triangle ABC has a base of $AB = 30$ and a height (distance between the parallel lines) of 25. All possible triangles will therefore have an area of 375 units².

3. Given $\triangle XYZ$, $\overline{XY}$ and $\overline{YZ}$ are partitioned into equal-length segments by the endpoints of the dashed segments as shown. What can be concluded about the diagram?

The dashed lines joining the endpoints of the equal length segments are parallel to $\overline{XZ}$ by the triangle side splitter theorem.

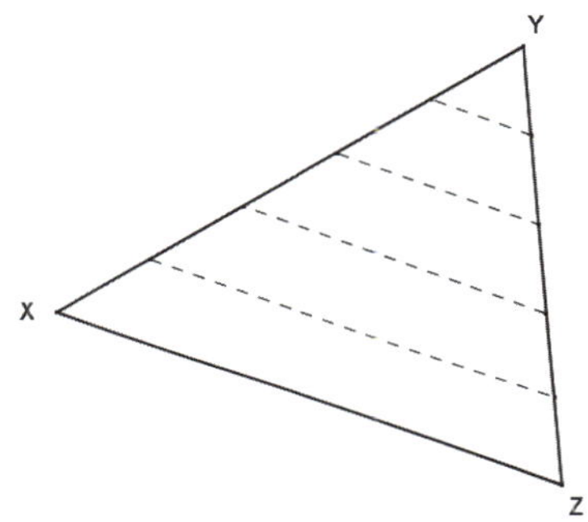

4. Given the diagram, $AC = 12$, $AB = 6$, $BE = 4$, $m\angle ACB = x°$, and $m\angle D = x°$, find CD.

Since $\angle ACB$ and $\angle D$ are corresponding angles and are both $x°$, it follows that $\overline{BC} \parallel \overline{ED}$. By the triangle side splitter theorem, $\overline{BC}$ is a proportional side splitter so $\frac{AC}{CD} = \frac{AB}{BE}$.

$$\frac{12}{CD} = \frac{6}{4}$$

$$CD = 8$$

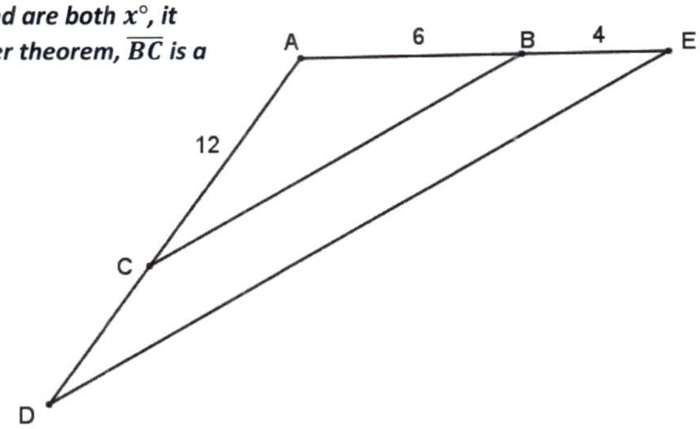

5. **What conclusions can be drawn from the diagram shown to the right? Explain.**

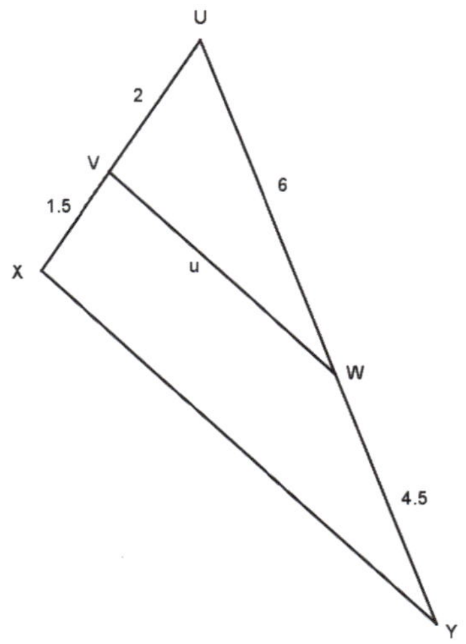

Since $\frac{3.5}{2} = \frac{7}{4}$ *and* $\frac{10.5}{6} = \frac{7}{4}$, $\frac{UX}{UV} = \frac{UY}{UW}$, *so the side splitter* $\overline{VW}$ *is a proportional side splitter.*

This provides several conclusions:

1. *The side splitter* $\overline{VW}$ *is parallel to the third side of the triangle by the triangle side splitter theorem.*
2. $\angle Y \cong \angle UWV$ *and* $\angle X \cong \angle UVW$ *because corresponding angles formed by parallel lines cut by a transversal are congruent.*
3. $\triangle UXY$ *is a scale drawing of* $\triangle UVW$ *with a scale factor of* $\frac{7}{4}$.
4. $XY = \frac{7}{4}u$ *because corresponding lengths in scale drawings are proportional.*

6. **Parallelogram $PQRS$ is shown. Two triangles are formed by a diagonal within the parallelogram. Identify those triangles, and explain why they are guaranteed to have the same areas.**

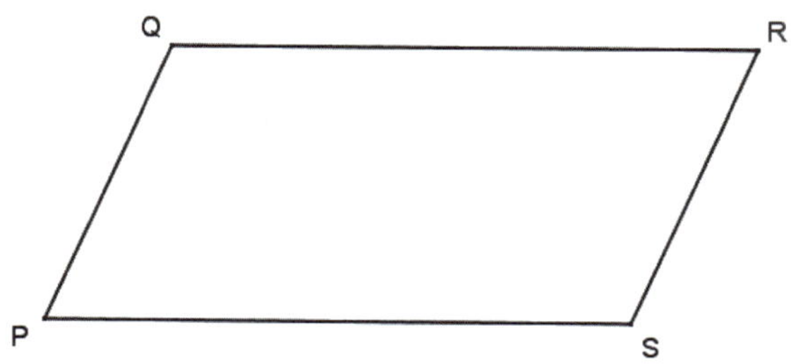

Opposite sides of a parallelogram are parallel and have the same length, so $QR = PS$, *and the distance between* $\overline{QR}$ *and* $\overline{PS}$ *is a constant,* h. *Diagonal* $\overline{PR}$ *forms* $\triangle PQR$ *and* $\triangle PSR$ *that have the same base length and the same height and, therefore, the same area.*

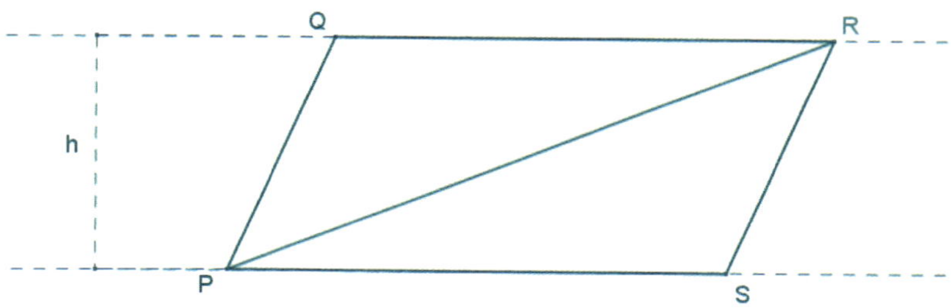

Diagonal $\overline{QS}$ *forms* $\triangle PQS$ *and* $\triangle RSQ$ *that have the same base length and the same height and, therefore, the same area.*

7. In the diagram to the right, $HI = 36$ and $GJ = 42$. If the ratio of the areas of the triangles is $\frac{\text{Area } \triangle GHI}{\text{Area } \triangle JHI} = \frac{5}{9}$, find JH, GH, GI, and JI.

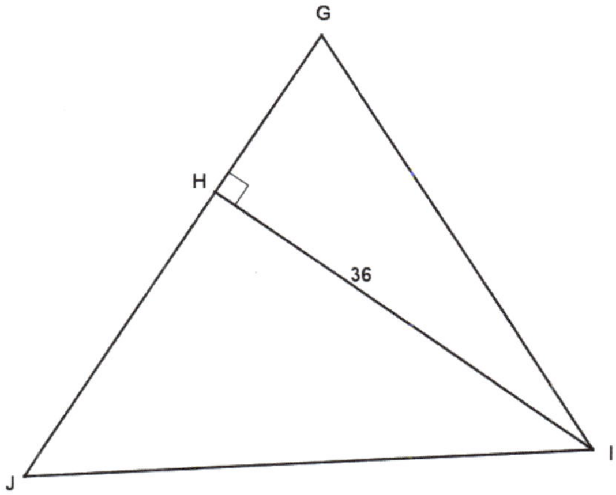

$\overline{HI}$ is the altitude of both triangles, so the bases of the triangles will be in the ratio of the areas of the triangles. $\overline{GJ}$ is composed of $\overline{JH}$ and $\overline{GH}$, so $GJ = 42 = JH + GH$.

$$\frac{GH}{JH} = \frac{5}{9}$$

$$\frac{GH}{42 - GH} = \frac{5}{9}$$

$$9(GH) = 5(42 - GH)$$

$$9(GH) = 210 - 5(GH)$$

$$14(GH) = 210$$

$GH = 15,\ JH = 27$

By the Pythagorean theorem, $JI = 45$ and $GI = 39$.

Lesson 5: Scale Factors

Student Outcomes

- Students prove the dilation theorem: If a dilation with center O and scale factor r sends point P to P' and Q to Q', then $|P'Q'| = r|PQ|$. Furthermore, if $r \neq 1$ and O, P, and Q are the vertices of a triangle, then $\overleftrightarrow{PQ} \parallel \overleftrightarrow{P'Q'}$.
- Students use the dilation theorem to show that the scale drawings constructed using the ratio and parallel methods have a scale factor that is the same as the scale factor for the dilation.

Lesson Notes

In the previous lesson, students learned about the triangle side splitter theorem, which is now used to prove the dilation theorem. In Grade 8, students learned about the fundamental theorem of similarity (FTS), which contains the concepts that are in the dilation theorem presented in this lesson. It is called the dilation theorem at this point in the module because students have not yet entered into the formal study of similarity. Some students may recall FTS from Grade 8 as they enter into the discussion following the Opening Exercise. Their prior knowledge of this topic will strengthen as they prove the dilation theorem.

Classwork

Opening Exercise (5 minutes)

Have students participate in a Quick Write. A Quick Write is an exercise where students write as much as they know about a particular topic without concern for correct grammar or spelling. The purpose of a Quick Write is for students to bring to the forefront all of the information they can recall about a particular topic. Show the prompt below, and then allow students to write for two minutes. Give students one minute to share Quick Writes with a partner. Then select students to share their thoughts with the class.

Scaffolding:

An alternate exercise is described on the following page. There are visual aids to help students make sense of the prompt. Student responses should be the same for those who need the visual as for those who do not.

Opening Exercise

Quick Write: Describe how a figure is transformed under a dilation with a scale factor $= 1$, $r > 1$, and $0 < r < 1$.

Sample student responses should include the following points:

A dilation with a scale factor of $r = 1$ produces an image that is congruent to the original figure.

A dilation with a scale factor of $r > 1$ produces an image that is larger in size than the original, but the angles of the figure are unchanged, and the lengths of the larger figure are proportional with the original figure.

A dilation with a scale factor of $0 < r < 1$ produces an image that is smaller in size than the original, but the angles of the figure are unchanged, and the lengths of the smaller figure are proportional with the original figure.

As an alternative exercise, consider showing students the following three diagrams and asking them to describe how each of the figures has been transformed with respect to the scale factor of dilation. State that in each of the figures, $\triangle ABC$ has been dilated from center O by some scale factor to produce the image $\triangle A'B'C'$. Some groups of students may benefit from seeing only one figure at a time and having a partner discussion prior to sharing their thoughts with the whole class. The sample responses above apply to this alternative exercise.

Figure 1.

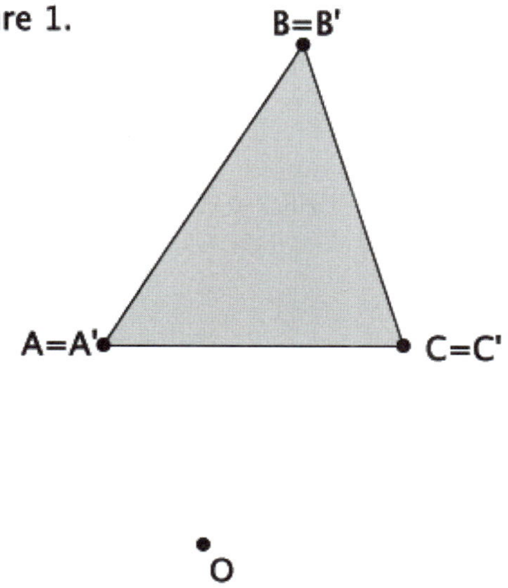

Figure 2.

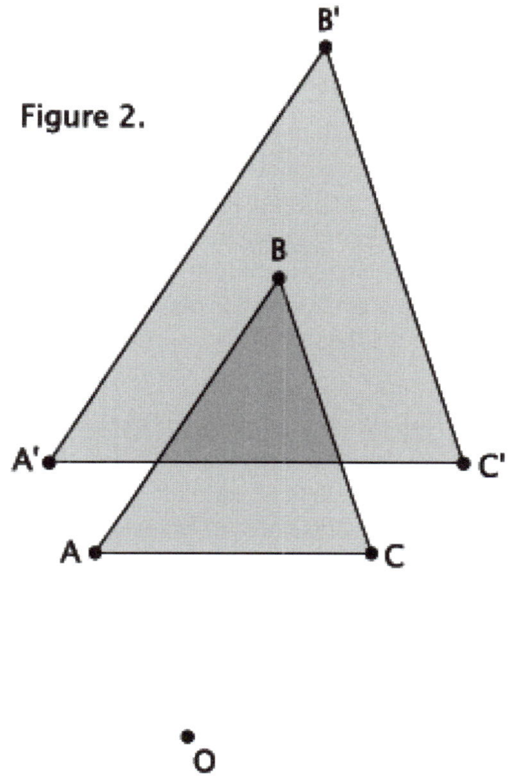

Figure 3.

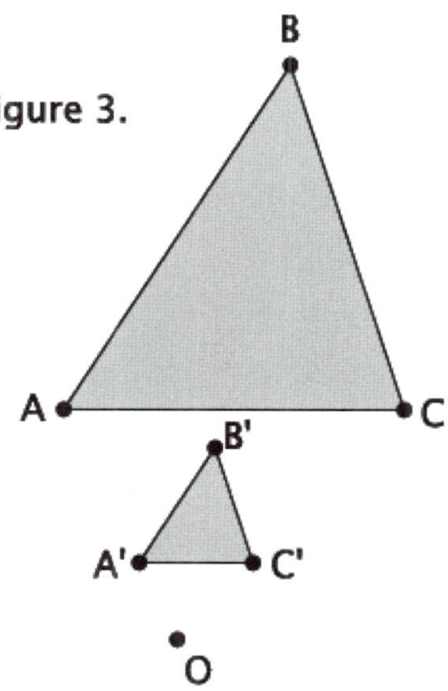

Discussion (20 minutes)

State the dilation theorem for students, and then allow them time to process what the theorem means.

> **Discussion**
>
> **DILATION THEOREM: If a dilation with center O and scale factor r sends point P to P' and Q to Q', then $|P'Q'| = r|PQ|$. Furthermore, if $r \neq 1$ and O, P, and Q are the vertices of a triangle, then $\overleftrightarrow{PQ} || \overleftrightarrow{P'Q'}$.**

- Describe in your own words what the dilation theorem states.

Provide students time to discuss the meaning of the theorem with a partner, and then select students to share their thoughts with the class. Shown below is a sample student response. Consider scripting responses that students might give. Be sure to elicit facts (1) and (2) from students to ensure their understanding of the theorem. Some students may comment on the lengths of segments OP, OP', OQ, and OQ'. These comments should be acknowledged, but make it clear that the dilation theorem focuses on the parts of a diagram that were not dilated; that is, the points P and Q were dilated, so there is an expectation of what that means and where the images of those points will end up. The focus now with respect to the theorem is on the segments PQ and $P'Q'$ and what happens to them when the points P and Q are dilated.

  - *The dilation theorem states two things: (1) If two points, P and Q, are dilated from the same center using the same scale factor, then the segment formed when you connect the dilated points P' and Q' is exactly the length of $\overline{PQ}$ multiplied by the scale factor, and (2) the lines containing the segments $P'Q'$ and PQ are parallel or equal.*
  - *For example, if points P and Q are dilated from center O by a scale factor of $r = \frac{3}{2}$, then the lines containing the segments $P'Q'$ and PQ are parallel, and $P'Q' = \frac{3}{2} \cdot PQ$, as shown below.*

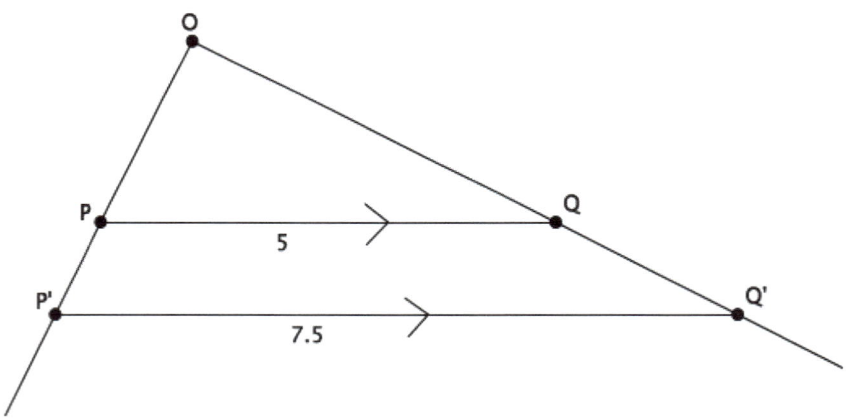

- The dilation theorem is an important theorem. To prove that the theorem is true, we use the triangle side splitter theorem (from the previous lesson) several times. When we use it, we consider different pairs of parallel lines.

- Consider the dilation theorem for $r = 1$. What impact does a scale factor of $r = 1$ have on a figure and its image? (Encourage students to use what they wrote during the Quick Write activity from the Opening Exercise.)
    - *When the scale factor is $r = 1$, then the figure remains unchanged. The scale factor being equal to 1 means that each point is taken to itself, that is, a congruence.*
- Consider the dilation theorem with the scenario of O, P, and Q not being the vertices of a triangle but in fact collinear points, as shown below. What impact does that have on the figure and its image?

    - *If the points O, P, and Q are collinear, then the dilated points P' and Q' remain on the line, and center O does not move at all.*
- A dilation of collinear points from a center with scale factor r can be used to construct the number line.

To clarify the statement "can be used to construct the number line," consider showing the series of diagrams below. Begin with the center of dilation at zero and the point x being 1 on the number line, and consider dilations of the point x from the center at zero. The successive lines below show how the point x moves when the scale factor of dilation is increased from $r = 2$ to $r = 4$.

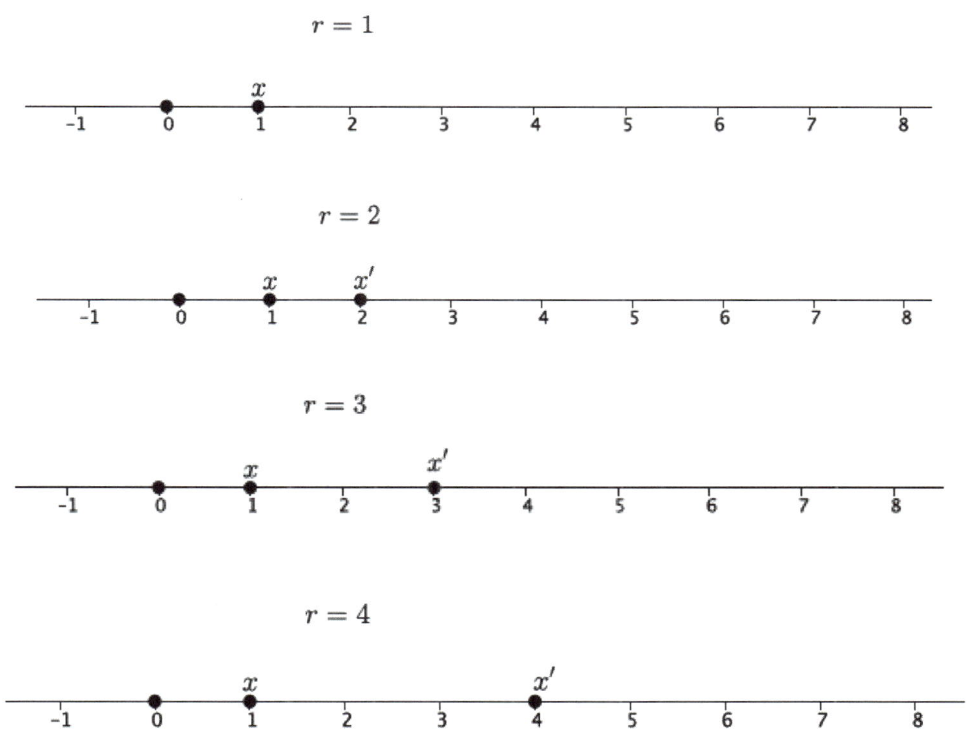

MP.8

- How is the location of the dilated point x' related to r and x?
    - *The location of the dilated point was exactly rx.*

- We could continue to dilate the point x by different scale factors to get all of the positive numbers on the number line. We can construct each fraction on the number line by dilating x by that fraction. A rotation of $180°$ around the center at zero of the whole numbers and fractions is a way to construct any rational number. If we considered rational and irrational scale factors, then we get all of the real numbers, that is, the entire real number line.

Note that we do not state that the rational and real number lines can be achieved using a negative scale factor of a dilation. Applying a negative scale factor has the same effect as a $180°$ rotation (or a reflection across zero) composed with a dilation whose scale factor is the absolute value of the negative scale factor. We do not complicate the issue by introducing a new formalism "negative dilation" into the lesson and, more broadly, the module. Therefore, all dilations have a positive scale factor.

MP.3

- Assume that points P and Q are numbers x and y on a number line, and point O, the center of dilation, is the number zero. The points P' and Q' correspond to the numbers rx and ry. Explain why the distance between rx and ry is r times the distance between x and y. Use a diagram below if necessary.

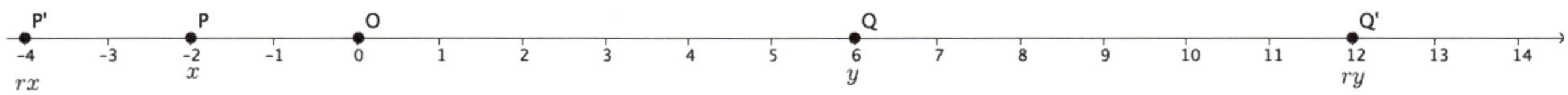

Provide students time to discuss with a partner. Then select students to share with the class. A sample student response is shown below.

- *The distance between points on the number line is the absolute value of their difference. For example, the distance between -2 and 6 is $|-2-6| = 8$ units. Then the distance between rx and ry is $|rx - ry|$. By the distributive property, $|rx - ry| = |r(x-y)|$, and since r must be positive, $|r(x-y)| = r|x-y|$. In terms of the numeric example, if $x = -2$, $y = 6$, and r is the scale factor of dilation, then*

$$|r(-2) - r(6)| = |r(-2-6)| = r|-2-6|.$$

Therefore, the distance between rx and ry is exactly r times the distance between x and y.

> *Scaffolding:*
> Students may need to see one or more numerical examples similar to the one in the sample response to arrive at the general statement.

The remaining part of this discussion uses the triangle side splitter theorem to prove the dilation theorem. The goal is to use knowledge about dilation, the triangle side splitter theorem, and the properties of parallelograms to explain why $|P'Q'| = r|PQ|$ and $\overleftrightarrow{PQ} || \overleftrightarrow{P'Q'}$. Show each step of the proof, and ask students to provide reasoning for the steps independently, with a partner, or in small groups.

> **Now consider the dilation theorem when O, P, and Q are the vertices of $\triangle OPQ$. Since P' and Q' come from a dilation with scale factor r and center O, we have $\frac{OP'}{OP} = \frac{OQ'}{OQ} = r$.**
>
> **There are two cases that arise; recall what you wrote in your Quick Write. We must consider the case when $r > 1$ and when $0 < r < 1$. Let's begin with the latter.**

Dilation Theorem Proof, Case 1

Statements	Reasons/Explanations
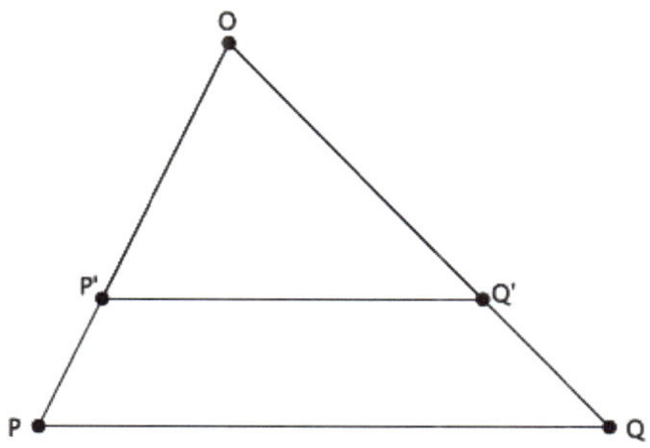	
1. A dilation with center O and scale factor r sends point P to P' and Q to Q'.	*1. Given*
2. $\frac{OP'}{OP} = \frac{OQ'}{OQ} = r$	*2. By definition of dilation: Corresponding lengths are proportional, and the ratio of the corresponding lengths are equal to the scale factor $0 < r < 1$.*
3. $\overleftrightarrow{PQ} \parallel \overleftrightarrow{P'Q'}$	*3. By the triangle side splitter theorem*
4. A dilation with center P and scale factor $\frac{PP'}{PO}$ sends point O to P' and point Q to R. Draw $\overline{P'R}$.	*4. By definition of dilation*
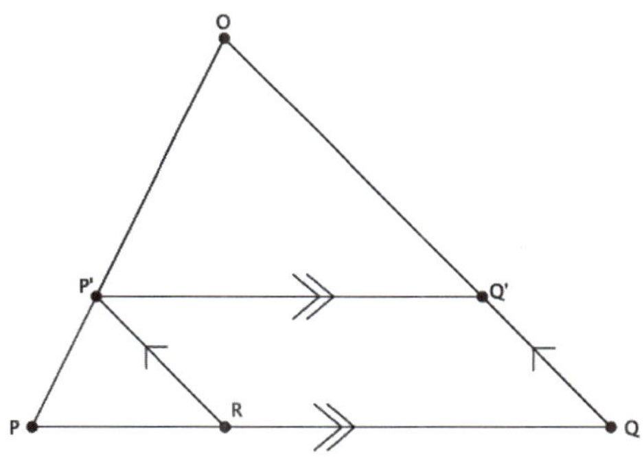	
5. $\overline{P'R} \parallel \overline{OQ'}$	*5. By the triangle side splitter theorem*
6. $RP'Q'Q$ is a parallelogram.	*6. By definition of parallelogram*

Consider asking students to state what kind of figure is formed by $RP'Q'Q$ and stating what they know about the properties of parallelograms before continuing with the next part of the discussion. Some students may need to see a parallelogram as an isolated figure as opposed to a figure within the triangles.

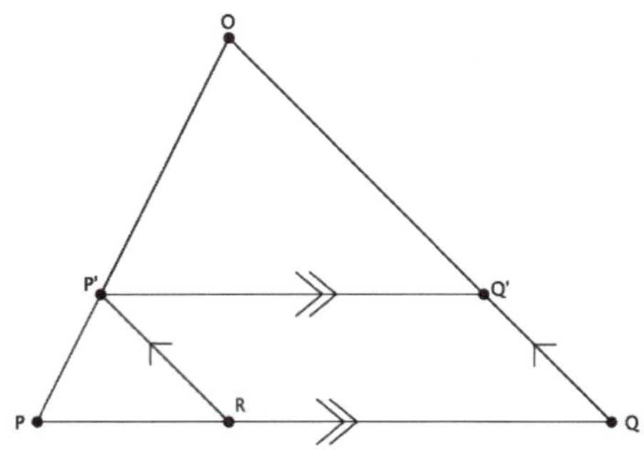

7. $RQ = P'Q'$	7. ***Opposite sides of parallelogram $RP'Q'Q$ are equal in length.***
8. $\frac{RQ}{PQ} = \frac{P'O}{PO}$	8. ***By the triangle side splitter theorem, $\overline{P'R}$ splits the sides of $\triangle POQ$ proportionally.***
9. $\frac{RQ}{PQ} = r$	9. ***By substitution (line 2)***
10. $RQ = r \cdot PQ$	10. ***Multiplication property of equality***
11. $P'Q' = r \cdot PQ$	11. ***By substitution (line 7)***

- This concludes the proof of Case 1 (when the scale factor of dilation is $0 < r < 1$) of the dilation theorem because we have shown that the dilation with center O and scale factor $0 < r < 1$ sends point P to P' and Q to Q', then $|P'Q'| = r|PQ|$, and since O, P, and Q are the vertices of a triangle, then $\overleftrightarrow{PQ} || \overleftrightarrow{P'Q'}$.

Exercises (10 minutes)

Students complete Exercise 1 independently or in pairs. Upon completion of Exercise 1, select students to share their proofs with the class. Then have students complete Exercises 2–4. Review the work for Exercises 2–4. Note that Exercise 4 is revisited in the Closing of the lesson as well.

Exercises

1. **Prove Case 2: If O, P, and Q are the vertices of a triangle and $r > 1$, show that (a) $\overleftrightarrow{PQ} \parallel \overleftrightarrow{P'Q'}$ and (b) $P'Q' = rPQ$. Use the diagram below when writing your proof.**

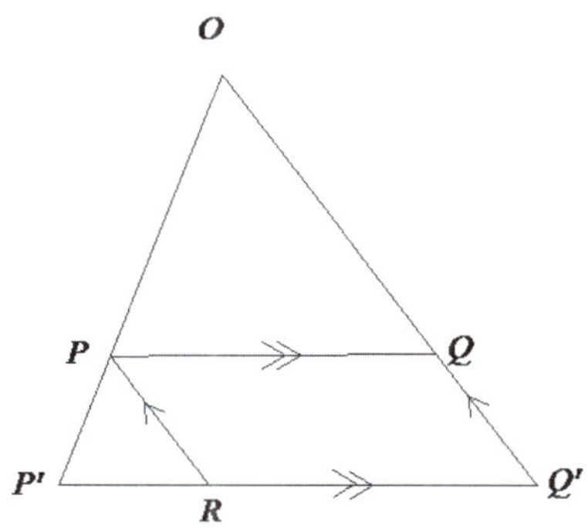

Case 2: If $r > 1$, $\overline{PQ}$ proportionally splits the sides $\overline{OP'}$ and $\overline{OQ'}$ of $\Delta OP'Q'$. By the triangle side splitter theorem, $\overleftrightarrow{P'Q'} \parallel \overleftrightarrow{PQ}$. Next, we construct a line segment $\overline{PR}$ that splits the sides of $\triangle OP'Q'$ and is parallel to side $\overline{OQ'}$. By the triangle side splitter theorem, $\overline{PR}$ splits the sides proportionally and so $\frac{P'Q'}{RQ'} = \frac{OP'}{OP} = r$. Now $RQ' = PQ$ because the lengths of opposite sides of parallelogram $RPQQ'$ are equal. So $\frac{P'Q'}{PQ} = r$ and $|P'Q'| = r|PQ|$.

> *Scaffolding:*
> - Consider having students compare and contrast this exercise with the proof they just finished. They should notice that the only difference is the scale factor of dilation.
> - It may be necessary to ask students the following questions to guide their thinking:
>
>   In the last proof, we constructed $\overline{P'R}$ so that it split the sides $\overline{PO}$ and $\overline{PQ}$ of $\triangle OPQ$ and is parallel to side $\overline{OQ}$. How does that differ for this figure?
>
>   If we apply the triangle side splitter theorem, what do we know about PQ and $P'Q'$?
>
>   What properties of a parallelogram will be useful in proving the dilation theorem?

2.

a. **Produce a scale drawing of $\triangle LMN$ using either the ratio or parallel method with point M as the center and a scale factor of $\frac{3}{2}$.**

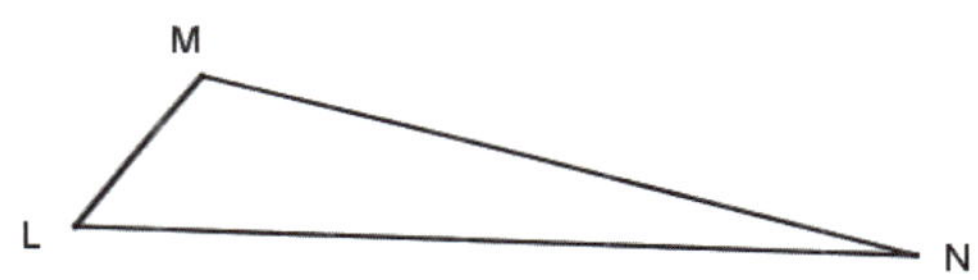

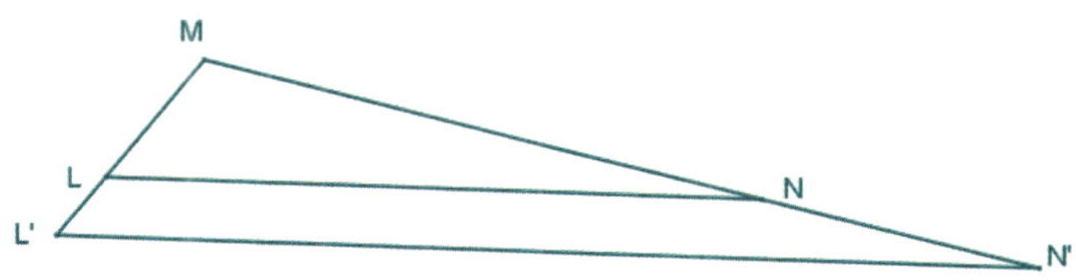

b. **Use the dilation theorem to predict the length of $\overline{L'N'}$, and then measure its length directly using a ruler.**

Lengths of the line segments can vary due to copy production. By the dilation theorem, it should be predicted that $L'N' = \frac{3}{2}LN$. This is confirmed using direct measurement.

c. **Does the dilation theorem appear to hold true?**

The dilation theorem does hold true.

3. Produce a scale drawing of $\triangle XYZ$ with point X as the center and a scale factor of $\frac{1}{4}$. Use the dilation theorem to predict $Y'Z'$, and then measure its length directly using a ruler. Does the dilation theorem appear to hold true?

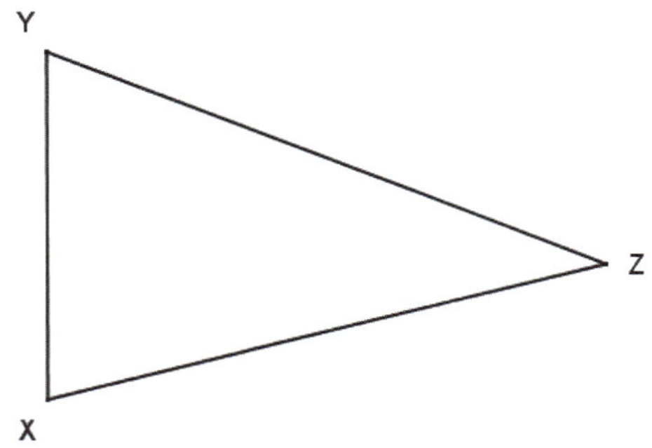

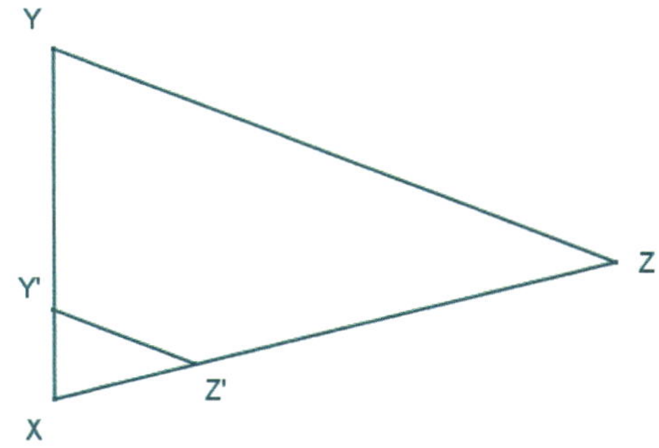

Lengths of the line segments can vary due to copy production. By the dilation theorem, it should be predicted that $Y'Z' = \frac{1}{4}YZ$. This is confirmed using direct measurement.

4. Given the diagram below, determine if $\triangle DEF$ is a scale drawing of $\triangle DGH$. Explain why or why not.

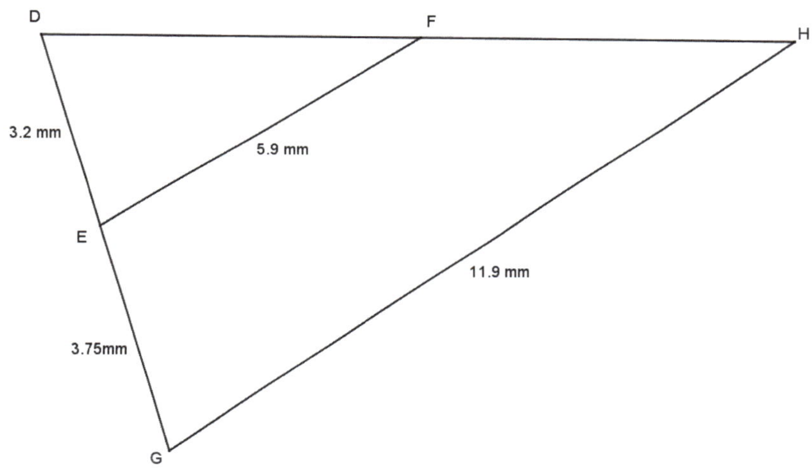

No. If $\triangle DEF$ was a scale drawing of $\triangle DGH$, then $\frac{DE}{DG} = \frac{EF}{GH}$ by the dilation theorem.

$$\frac{DE}{DG} = \frac{3.2}{6.95} \approx 0.46$$

$$\frac{EF}{GH} = \frac{5.9}{11.9} \approx 0.50$$

The ratios of the corresponding sides are not equivalent, so the drawing is not a scale drawing.

Closing (5 minutes)

Revisit Exercise 4 above. Ask students whether or not $\overline{EF}$ is parallel to $\overline{GH}$. Students should respond that because the side lengths were not in proportion that $\triangle DEF$ is not a scale drawing of $\triangle DGH$, and we would not expect the lines containing $\overline{EF}$ and $\overline{GH}$ to be parallel. Next, ask students to verbally complete the following in order to informally assess their understanding of the dilation theorem and its proof.

- Restate the dilation theorem in your own words.
- Explain how the triangle side splitter theorem was used to prove the dilation theorem.

If time permits, ask students the following question.

- We discussed how dilation was used to produce a number line. What other everyday objects may have been created in this manner?
  - *Any everyday object that is divided into equal parts can be produced using dilation, for example, a ruler or thermometer.*

Exit Ticket (5 minutes)

Name ______________________________ Date ______________

Lesson 5: Scale Factors

Exit Ticket

1. Two different points R and Y are dilated from S with a scale factor of $\frac{3}{4}$, and $RY = 15$. Use the dilation theorem to describe two facts that are known about $\overline{R'Y'}$.

2. Which diagram(s) below represents the information given in Problem 1? Explain your answer(s).

a.

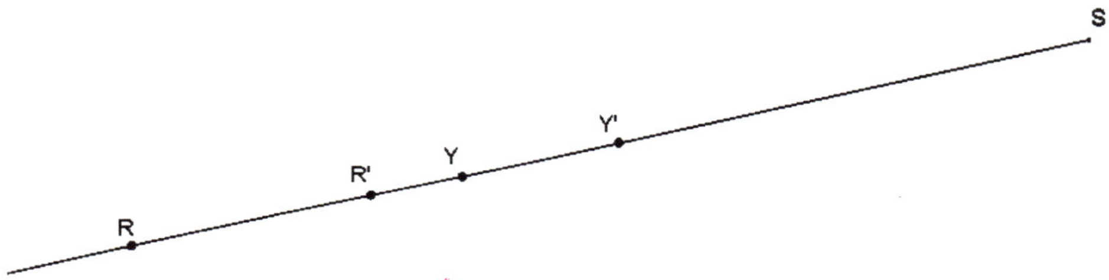

b.

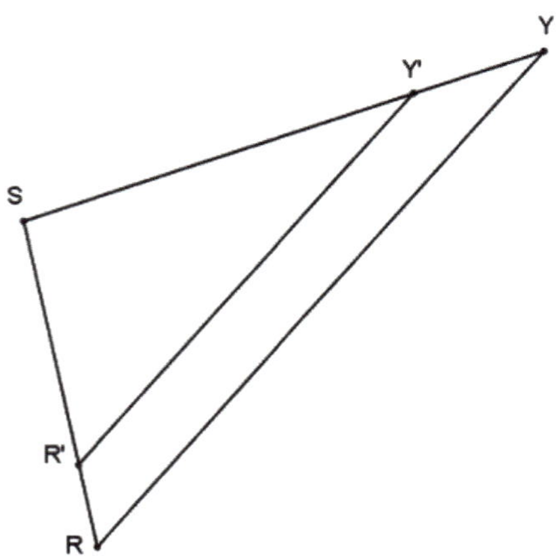

Exit Ticket Sample Solutions

1. Two different points R and Y are dilated from S with a scale factor of $\frac{3}{4}$, and $RY = 15$. Use the dilation theorem to describe two facts that are known about $\overline{R'Y'}$.

   By the dilation theorem, $R'Y' = \frac{3}{4}RY$, *so* $R'Y' = \frac{3}{4}(15) = 11.25$, *and* $\overline{R'Y'} \parallel \overline{RY}$ *or* $\overleftrightarrow{R'Y'} = \overleftrightarrow{RY}$.

2. Which diagram(s) below could represent your conclusions in Problem 1? Explain your answer(s).

   a.

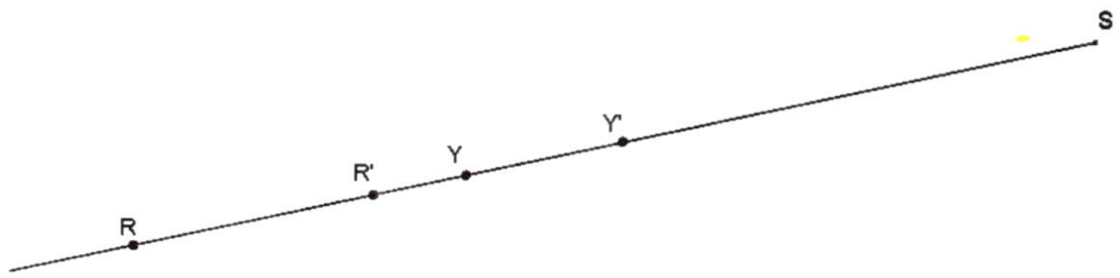

Diagram (a) can be a dilation with scale factor $\frac{3}{4}$ *since* $R'S$ *and* $Y'S$ *appear to be* $\frac{3}{4}$ *of the distances* RS *and* YS*, respectively. However, because both line segments lie on the same line,* $\overline{R'Y'} = \overline{RY}$.

   b.

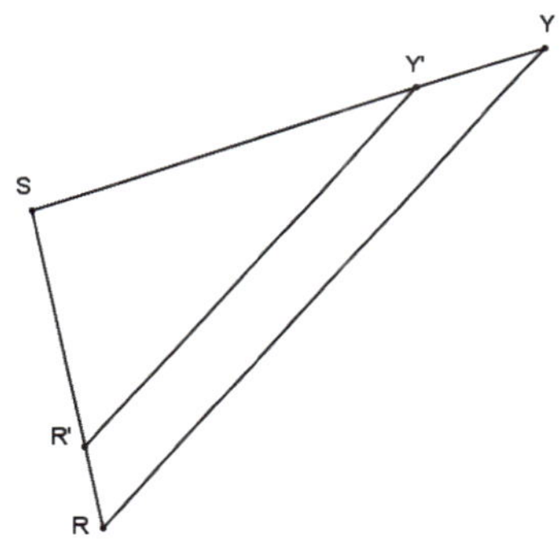

Diagram (b) could be a dilation with scale factor $\frac{3}{4}$ *since* $R'S$ *and* $Y'S$ *appear to be* $\frac{3}{4}$ *of the distances* RS *and* YS*, respectively. Because* S, Y, *and* R *are vertices of a triangle,* $\overline{R'Y'} \parallel \overline{RY}$.

Problem Set Sample Solutions

1. $\triangle AB'C'$ is a dilation of $\triangle ABC$ from vertex A, and $CC' = 2$. Use the given information in each part and the diagram to find $B'C'$.

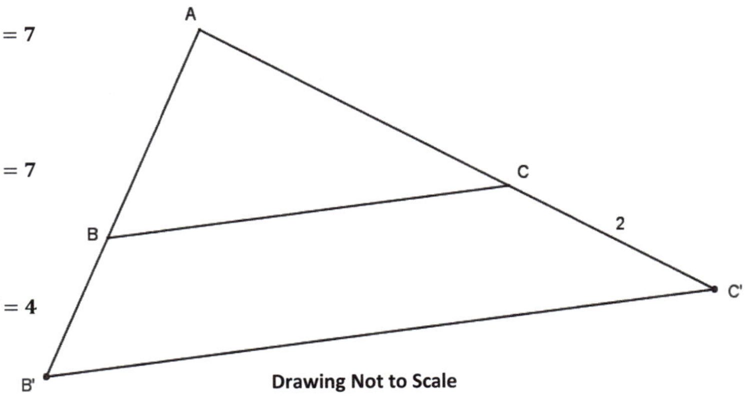

Drawing Not to Scale

   a. $AB = 9$, $AC = 4$, and $BC = 7$

      $B'C' = 10\frac{1}{2}$

   b. $AB = 4$, $AC = 9$, and $BC = 7$

      $B'C' = 8\frac{5}{9}$

   c. $AB = 7$, $AC = 9$, and $BC = 4$

      $B'C' = 4\frac{8}{9}$

   d. $AB = 7$, $AC = 4$, and $BC = 9$

      $B'C' = 18$

   e. $AB = 4$, $AC = 7$, and $BC = 9$

      $B'C' = 11\frac{4}{7}$

   f. $AB = 9$, $AC = 7$, and $BC = 4$

      $B'C' = 5\frac{1}{7}$

2. Given the diagram, $\angle CAB \cong \angle CFE$. Find AB.

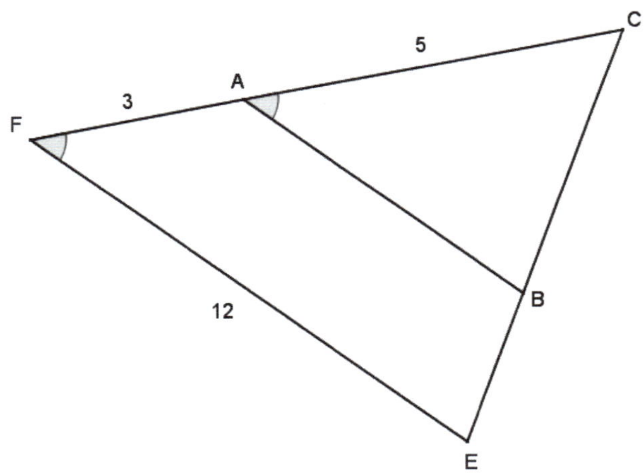

   $\overline{FE} \parallel \overline{AB}$ because when two lines are cut by a transversal, such that the corresponding angles are congruent, then the lines are parallel.

   $\triangle CFE$ is a scale drawing of $\triangle CAB$ by the parallel method.

   F is the image of A after a dilation from C with a scale factor of $\frac{8}{5}$.

   $FE = \frac{8}{5}(AB)$ by the dilation theorem.

$$12 = \frac{8}{5}(AB)$$
$$\frac{5}{8}(12) = \frac{5}{8}\left(\frac{8}{5}\right)(AB)$$
$$\frac{15}{2} = 1AB$$
$$7.5 = AB$$

3. **Use the diagram to answer each part below.**

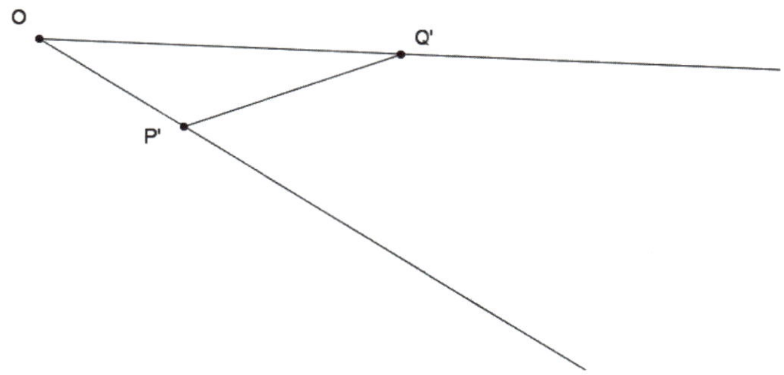

a. **$\triangle OP'Q'$ is the dilated image of $\triangle OPQ$ from point O with a scale factor of $r > 1$. Draw a possible $\overline{PQ}$.**

Placement of the segment will vary; however, by the dilation theorem, $\overline{PQ}$ must be drawn parallel to $\overline{P'Q'}$, and because scale factor $r > 1$, point P must be between O and P', and point Q must be between O and Q'.

b. **$\triangle OP''Q''$ is the dilated image of $\triangle OPQ$ from point O with a scale factor $k > r$. Draw a possible $\overline{P''Q''}$.**

Placement of the segment will vary; however, by the dilation theorem, $\overline{P''Q''}$ must be drawn parallel to $\overline{PQ}$, and because scale factor $k > r$, point P'' must be placed such that P' is between P and P'' and Q'' placed such that Q' is between Q and Q''.

Possible solutions:

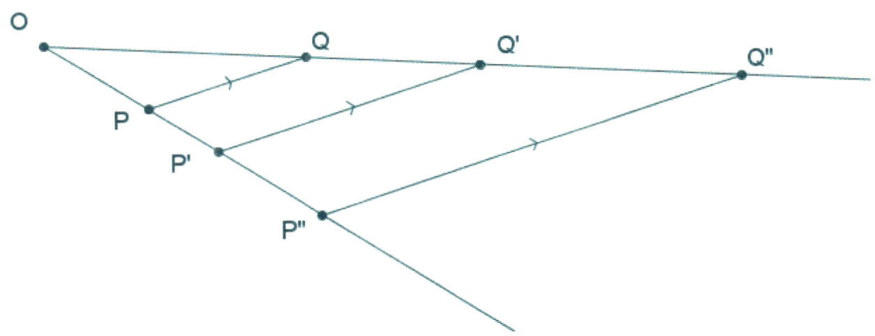

4. Given the diagram to the right, $\overline{RS} \parallel \overline{PQ}$, $\text{Area}\,(\triangle\, RST) = 15$ units², and $\text{Area}(\triangle\, OSR) = 21$ units², find RS.

ΔRST and ΔOSR have the same altitude, so the lengths of their bases are proportional to their areas.

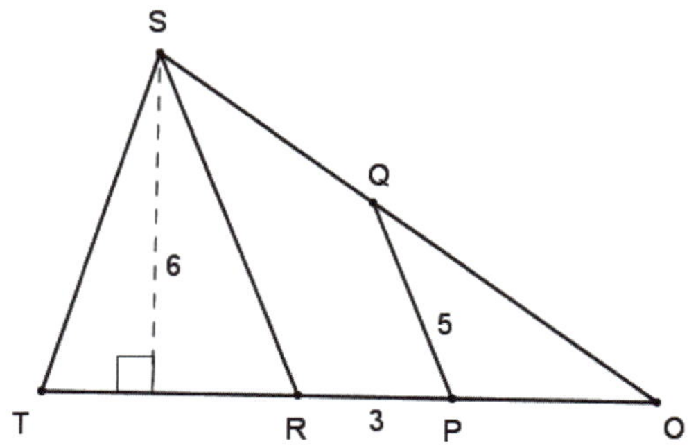

$$\text{Area}(\triangle\, RST) = \frac{1}{2}(RT)6 = 15$$
$$RT = 5$$

Since the triangles have the same height 6, the ratio of $\text{Area}(\triangle\, RST)$: $\text{Area}(\triangle\, OSR)$ is equal to the ratio RT: RO.

$$\frac{15}{21} = \frac{5}{OR}$$
$$OR = 7$$
$$OR = OP + PR$$
$$7 = OP + 3$$
$$OP = 4$$

If $\overline{RS} \parallel \overline{PQ}$, then $\triangle\, OSR$ is a scale drawing of $\triangle\, OQP$ from O with a scale factor of $\frac{7}{4}$. Therefore, R is the image of P, and S is the image of Q under a dilation from point O. By the dilation theorem:

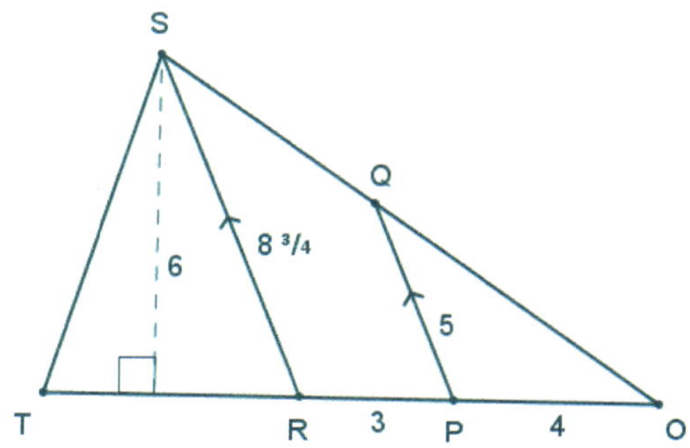

$$RS = \frac{7}{4}(5)$$
$$RS = \frac{35}{4}$$
$$RS = 8\frac{3}{4}$$

The length of segment RS is $8\frac{3}{4}$ units.

5. Using the information given in the diagram and $UX = 9$, find Z on $\overline{XU}$ such that $\overline{YZ}$ is an altitude. Then, find YZ and XZ.

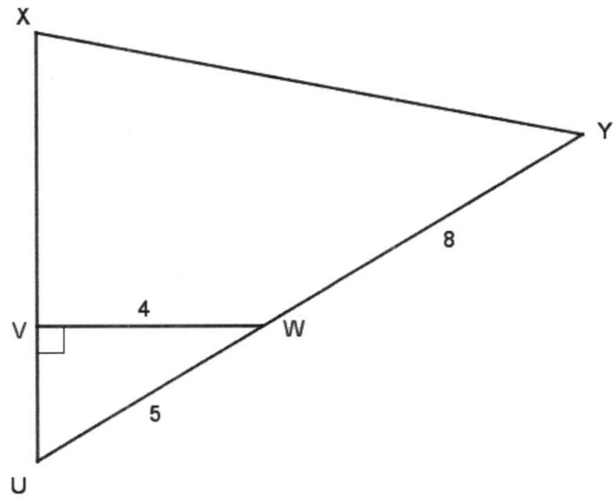

Altitude $\overline{YZ}$ is perpendicular to $\overline{UX}$; thus, $m\angle YZU = 90°$. Since $m\angle WVU = 90°$, then it follows that $\overline{YZ} \parallel \overline{WV}$ by corresponding $\angle$'s converse. So, $\triangle UYZ$ is a scale drawing of $\triangle UWV$ by the parallel method. Therefore, Z is the image of V, and Y is the image of W under a dilation from U with a scale factor of $\frac{13}{5}$.

By the dilation theorem:

$$YZ = \frac{13}{5}WV$$
$$YZ = \frac{13}{5}(4)$$
$$YZ = \frac{52}{5}$$
$$YZ = 10\frac{2}{5}$$

By the Pythagorean theorem:

$$UV^2 + VW^2 = UW^2$$
$$UV^2 + (4^2) = (5^2)$$
$$UV^2 + 16 = 25$$
$$UV^2 = 9$$
$$UV = 3$$

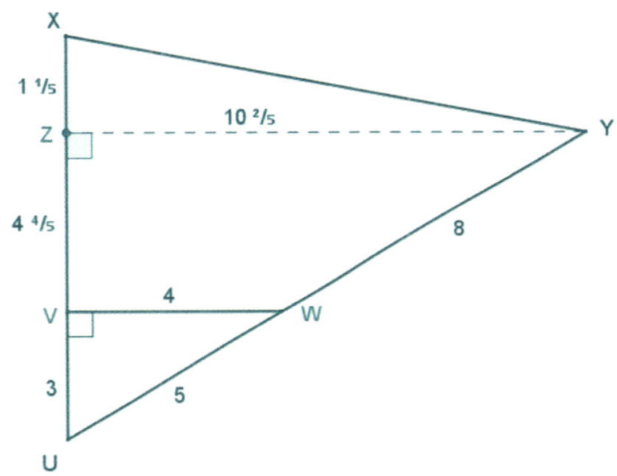

Since Z is the image of V under the same dilation:

$$UZ = \frac{13}{5}UV$$
$$UZ = \frac{13}{5}(3)$$
$$UZ = \frac{39}{5}$$
$$UZ = 7\frac{4}{5}$$

By addition:

$$XZ + UZ = UX$$
$$XZ + 7\frac{4}{5} = 9$$
$$XZ = 1\frac{1}{5}$$

The length of $\overline{YZ}$ is $10\frac{2}{5}$ units, and the length of $\overline{XZ}$ is $1\frac{1}{5}$ units.

Topic B

Dilations

G-SRT.A.1, G-SRT.B.4

Focus Standards:	G-SRT.A.1	Verify experimentally the properties of dilations given by a center and a scale factor: a. A dilation takes a line not passing through the center of the dilation to a parallel line, and leaves a line passing through the center unchanged. b. The dilation of a line segment is longer or shorter in the ratio given by the scale factor.
	G-SRT.B.4	Prove theorems about triangles. *Theorems include: a line parallel to one side of a triangle divides the other two proportionally, and conversely; the Pythagorean Theorem proved using triangle similarity.*
Instructional Days:	6	
Lesson 6:	Dilations as Transformations of the Plane (S)[1]	
Lesson 7:	How Do Dilations Map Segments? (P)	
Lesson 8:	How Do Dilations Map Lines, Rays, and Circles? (S)	
Lesson 9:	How Do Dilations Map Angles? (E)	
Lesson 10:	Dividing the King's Foot into 12 Equal Pieces (E)	
Lesson 11:	Dilations from Different Centers (E)	

Topic B is an in-depth study of the properties of dilations. Though students applied dilations in Topic A, their use in the ratio and parallel methods was to establish relationships that were consequences of applying a dilation, not directly about the dilation itself. In Topic B, students explore observed properties of dilations (Grade 8 Module 3) and reason why these properties are true. This reasoning is possible because of what students have studied regarding scale drawings and the triangle side splitter and dilation theorems. With these theorems, it is possible to establish why dilations map segments to segments, lines to lines, etc. Some of the arguments involve an examination of several sub-cases; it is in these instances of thorough examination that students must truly make sense of problems and persevere in solving them (MP.1).

In Lesson 6, students revisit the study of rigid motions and contrast the behavior of the rigid motions to that of a dilation. Students confirm why the properties of dilations are true in Lessons 7–9. Students repeatedly encounter **G.SRT.A.1a** and **b** in these lessons and build arguments with the help of the ratio and parallel

[1]Lesson Structure Key: **P**-Problem Set Lesson, **M**-Modeling Cycle Lesson, **E**-Exploration Lesson, **S**-Socratic Lesson

methods (**G.SRT.B.4**). In Lesson 10, students study how dilations can be used to divide a segment into equal divisions. Finally, in Lesson 11, students observe how the images of dilations of a given figure by the same scale factor are related, as well as the effect of a composition of dilations on the scale factor of the composition.

Lesson 6: Dilations as Transformations of the Plane

Student Outcomes

- Students review the properties of basic rigid motions.
- Students understand the properties of dilations and that a dilation is also an example of a transformation of the plane.

Lesson Notes

Topic A plunged right into the *use* of dilations to create scale drawings and arguments to prove the triangle side splitter theorem and dilation theorem. Topic B examines dilations in detail. In Grade 8 (Module 3), students discovered properties of dilations, such as that the dilation of a line maps onto another line or that the dilation of an angle maps onto another angle. We now examine how dilations differ from the other transformations and use reasoned arguments to confirm the properties of dilations observed in Grade 8.

Topic B begins with a review of the rigid motions studied in Module 1 (Lessons 12–16).

Classwork

Discussion (7 minutes)

The goal of Lesson 6 is to study dilations as transformations of the plane. Begin with a review of what a transformation is and the category of transformations studied in Module 1. The following questions can be asked as part of a whole-group discussion, or consider asking for them to be written to let students express their thoughts on paper before discussing them aloud.

> *Scaffolding:*
>
> Use Module 1 Lesson 17 Exercises 1–5 to provide students with good visuals and review the learned transformations and the conclusion that they are distance-preserving.

- Recall that our recent study of translations, reflections, and rotations was a study of transformations. With a partner, discuss what a transformation of the plane means.

Allow students time to discuss before sharing responses.

  - *A transformation of the plane is a rule that assigns each point in the plane to a unique point. We use function notation to denote transformations, that is, F denotes the transformation of a point, P, and is written as $F(P)$. Thus, a transformation moves point P to a unique point $F(P)$.*

- When we refer to the image of P by F, what does this refer to?
  - *The point $F(P)$ is called the image of P, or P'.*
  - *Recall that every point P' in the plane is the image of some point P, that is, $F(P) = P'$.*
- In Module 1, we studied a set of transformations that we described as being "rigid". What does the term rigid refer to?
  - *The transformations in Module 1—translations, reflections, and rotations—are all transformations of the plane that are rigid motions, or they are distance-preserving. A transformation is distance-preserving if, given two points P and Q, the distance between these points is the same as the distance between the images of these points, that is, the distance between $F(P)$ and $F(Q)$.*

As we know, there are a few implied properties of any rigid transformation:

a. Rigid motions map a line to a line, a ray to a ray, a segment to a segment, and an angle to an angle.

b. Rigid motions preserve lengths of segments.

c. Rigid motions preserve the measures of angles.

Exercises 1–6 (12 minutes)

It is at the teacher's discretion to assign only some or all of Exercises 1–6. Completion of all six exercises likely requires more than the allotted time.

Exercises 1–6

1. **Find the center and the angle of the rotation that takes $\overline{AB}$ to $\overline{A'B'}$. Find the image P' of point P under this rotation.**

	Step 1. Determine the location of center O as the intersection of the perpendicular bisectors of $\overline{AA'}$ and $\overline{BB'}$.
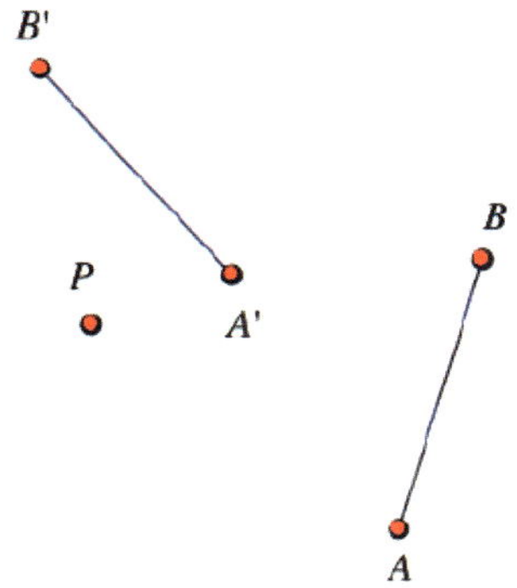	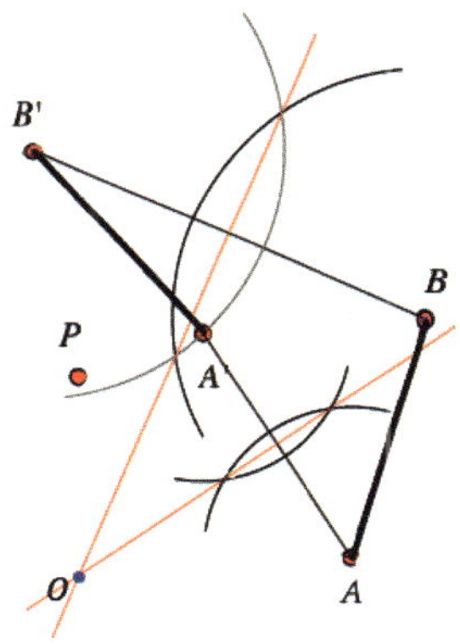
Step 2. Determine the angle of rotation by joining A to O and O to A'; the angle of rotation is $58°$ counterclockwise.	***Step 3. Rotate P by $58°$. P should remain a fixed distance away from O.***
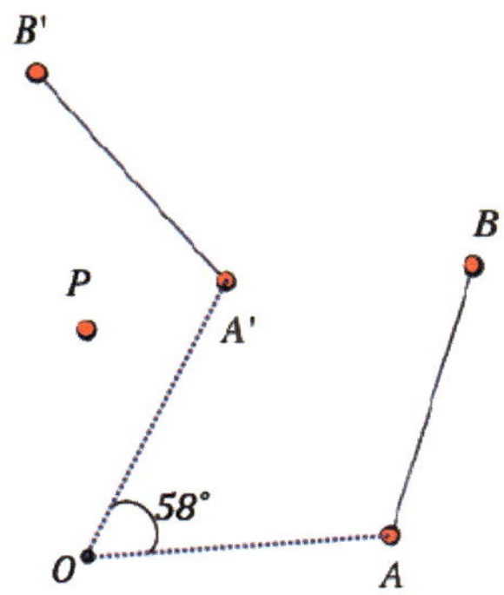	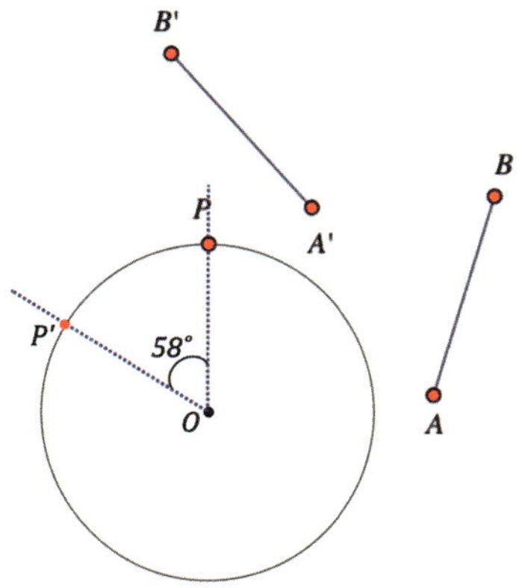

2. **In the diagram below, $\triangle B'C'D'$ is the image of $\triangle BCD$ after a rotation about a point A. What are the coordinates of A, and what is the degree measure of the rotation?**

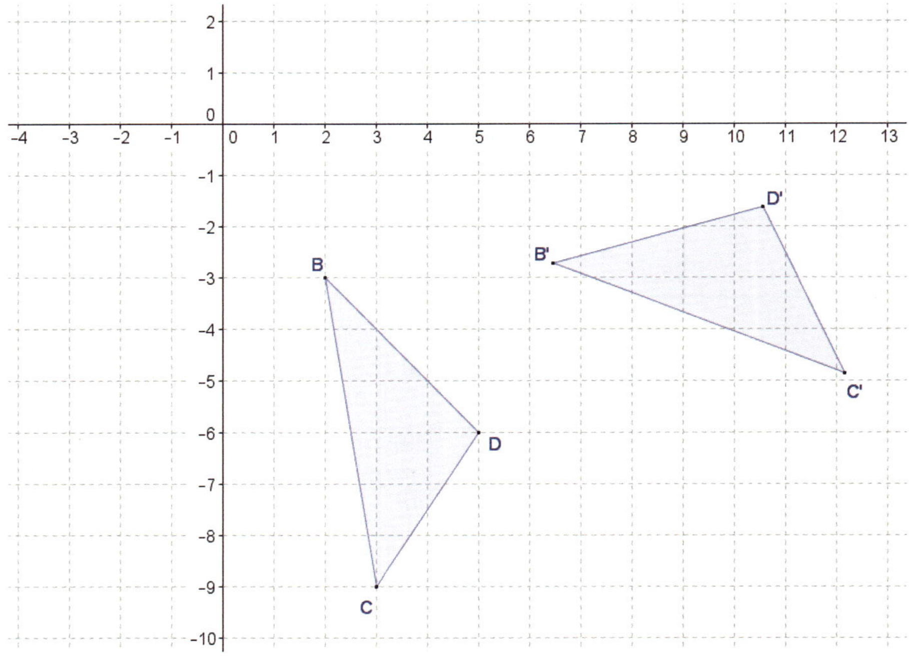

By constructing the perpendicular bisector of each segment joining a point and its image, I found the center of dilation A to be $A(4, 1)$. Using a protractor, the angle of rotation from $\triangle BCD$ to $\triangle B'C'D'$ about point $A(4, 1)$ is $60°$.

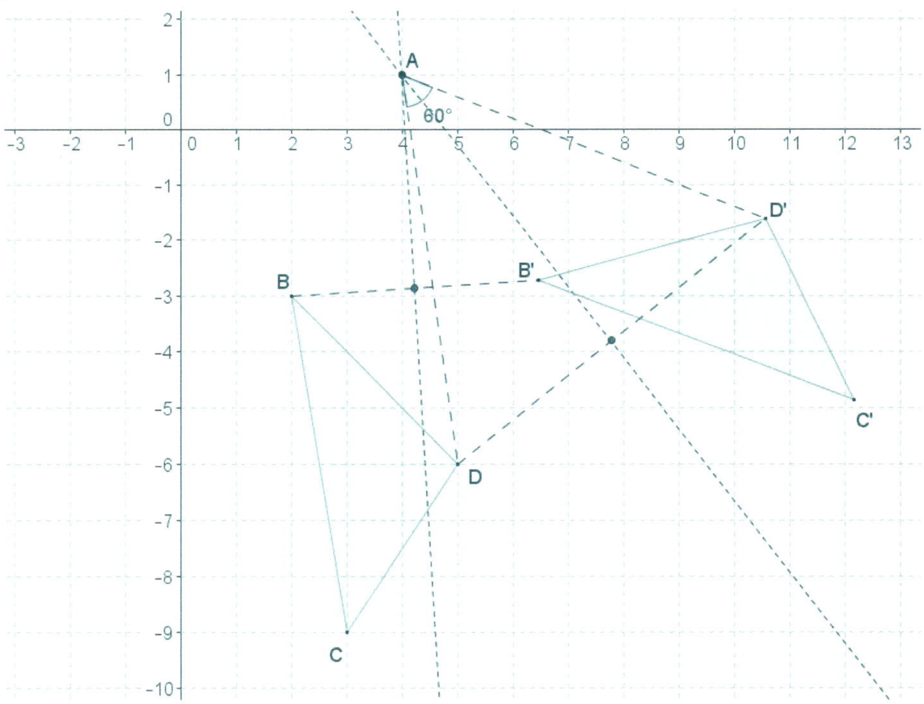

3. **Find the line of reflection for the reflection that takes point A to point A'. Find the image P' under this reflection.**

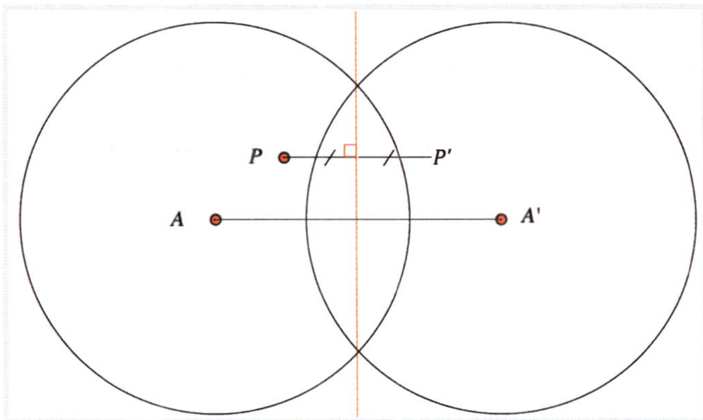

4. **Quinn tells you that the vertices of the image of quadrilateral $CDEF$ reflected over the line representing the equation $y = -\frac{3}{2}x + 2$ are the following: $C'(-2, 3)$, $D'(0, 0)$, $E'(-3, -3)$, and $F'(-3, 3)$. Do you agree or disagree with Quinn? Explain.**

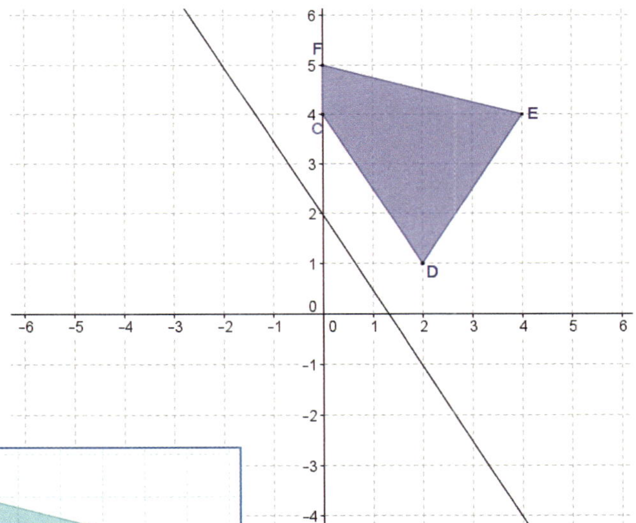

I disagree because under a reflection, an image point lies along a line through the pre-image point that is perpendicular to the line of reflection. The line representing the equation $y = \frac{2}{3}x + 4$ includes C and is perpendicular to the line of reflection; however, it does not include the point $(-2, 3)$. Similar reasoning can be used to show that Quinn's coordinates for D', E', and F' are not the images of D, E, and F, respectively, under a reflection over $y = -\frac{3}{2}x + 2$.

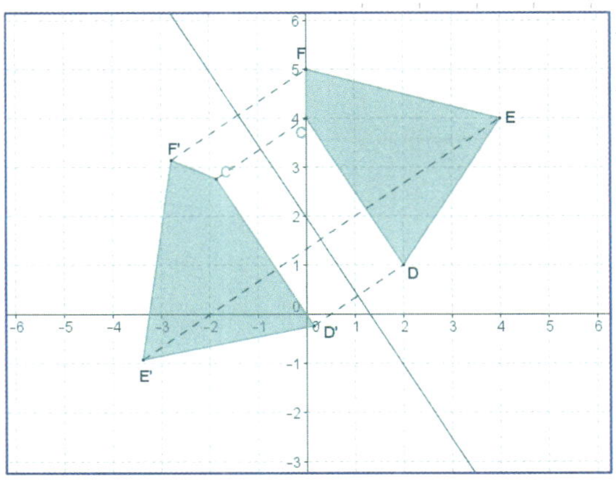

5. A translation takes A to A'. Find the image P' and pre-image P'' of point P under this translation. Find a vector that describes the translation.

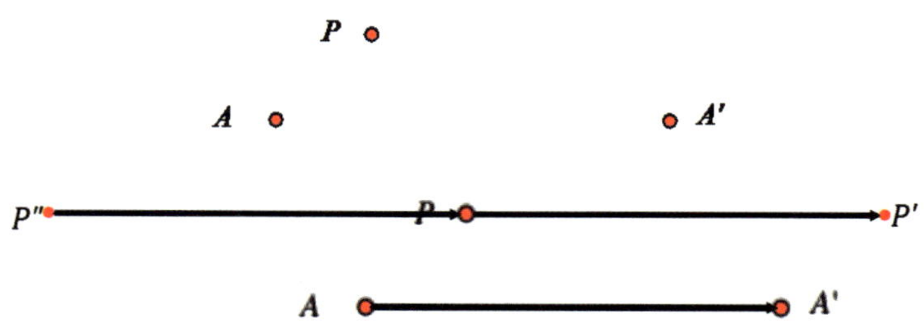

6. The point C' is the image of point C under a translation of the plane along a vector.

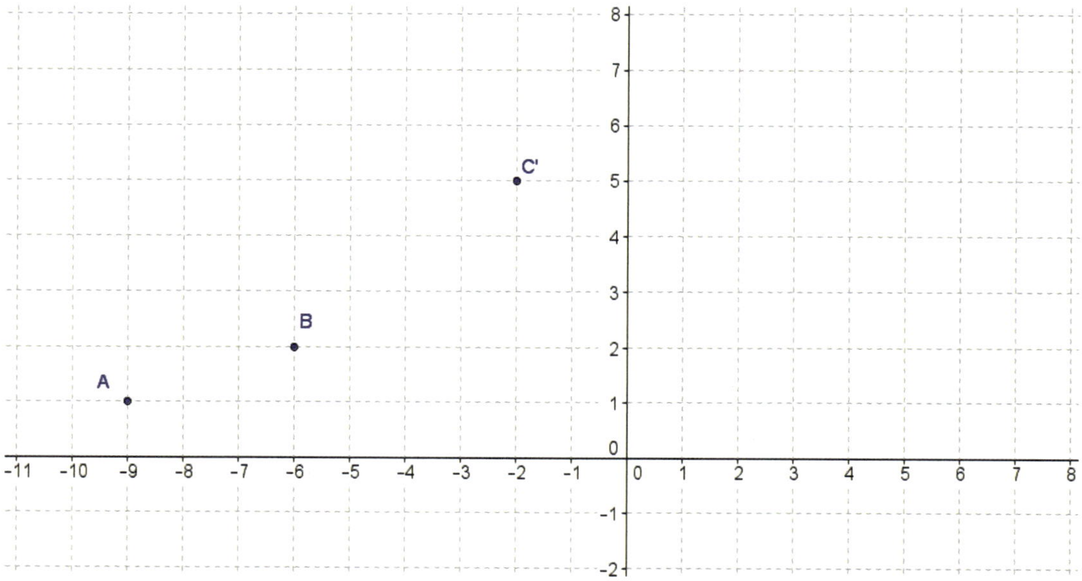

a. Find the coordinates of C if the vector used for the translation is $\overrightarrow{BA}$.

$C(1, 6)$

b. Find the coordinates of C if the vector used for the translation is $\overrightarrow{AB}$.

$C(-5, 4)$

Discussion (7 minutes)

Lead a discussion that highlights how dilations are like rigid motions and how they are different from them.

- In this module, we have used dilations to create scale drawings and to establish the triangle side splitter theorem and the dilation theorem. We pause here to inspect how dilations as a class of transformations are like rigid transformations and how they are different.
- What do dilations have in common with translations, reflections, and rotations?
  - *All of these transformations meet the basic criteria of a transformation in the sense that each follows a rule assignment where any point P is assigned to a unique point $F(P)$ in the plane. Dilations and rotations both require a center in order to define the function rule.*
- What distinguishes dilations from translations, reflections, and rotations?
  - *Dilations are not a distance-preserving transformation like the rigid transformations. For every point P, other than the center, the point is assigned to $D_{O,r}(P)$, which is the point P on $\overrightarrow{OP}$ so that the distance from $D_{O,r}(P)$ to O is $r \cdot OP$. The fact that distances are scaled means the transformation is not distance preserving.*
- From our work in Grade 8, we have seen that dilations, just like the rigid motions, do in fact map segments to segments, lines to lines, and rays to rays, but we only confirmed this experimentally, and in the next several lessons, we create formal arguments to prove why these properties hold true for dilations.
- One last feature that dilations share with the rigid motions is the existence of an inverse dilation, just as inverses exist for the rigid transformations. What this means is that composition of the dilation and its inverse takes every point in the plane to itself.
- Consider a $90°$ clockwise rotation about a center O: $R_{O,90}(P)$. The inverse rotation would be a $90°$ counter-clockwise rotation to bring the image of point P back to itself: $R_{O,-90}\left(R_{O,90}(P)\right) = R_{O,0}(P) = P$.

MP.7

- What would an inverse dilation rely on to bring the image of a dilated point P' back to P?
  - *If we were dilating a point P by a factor of 2, the image would be written as $P' = D_{O,2}(P)$. In this case, P' is pushed away from the center O by a factor of two so that it is two times as far away from O. To bring it back to itself, we need to halve the distance or, in other words, scale by a factor of $\frac{1}{2}$, which is the reciprocal of the original scale factor: $D_{O,\frac{1}{2}}\left(D_{O,2}(P)\right) = D_{O,1}(P) = P$. Therefore, an inverse dilation relies on the original center O but requires a scale factor that is the reciprocal (or multiplicative inverse) of the original scale factor.*

Exercises 7–9 (12 minutes)

Exercises 7–9

7. A dilation with center O and scale factor r takes A to A' and B to B'. Find the center O, and estimate the scale factor r.

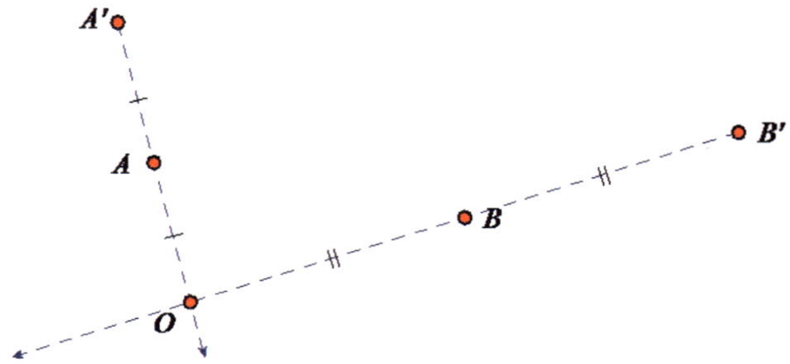

The estimated scale factor is $r \approx 2$.

8. Given a center O, scale factor r, and points A and B, find the points $A' = D_{O,r}(A)$ and $B' = D_{O,r}(B)$. Compare length AB with length $A'B'$ by division; in other words, find $\frac{A'B'}{AB}$. How does this number compare to r?

$r = 3$

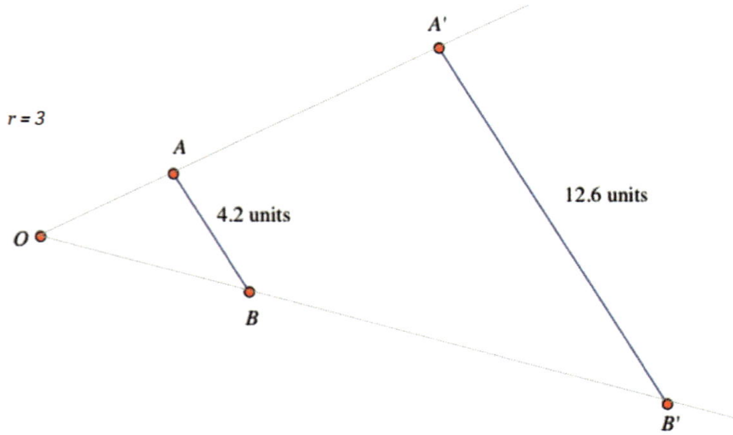

$$\frac{A'B'}{AB} = \frac{12.6}{4.2} = 3 = r$$

9. Given a center O, scale factor r, and points A, B, and C, find the points $A' = D_{O,r}(A)$, $B' = D_{O,r}(B)$, and $C' = D_{O,r}(C)$. Compare $m\angle ABC$ with $m\angle A'B'C'$. What do you find?

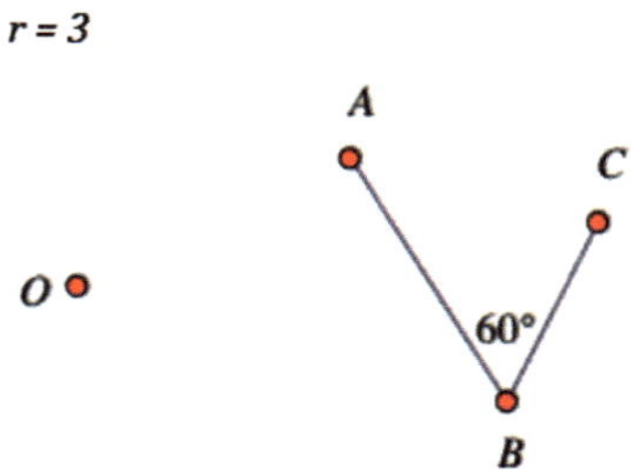

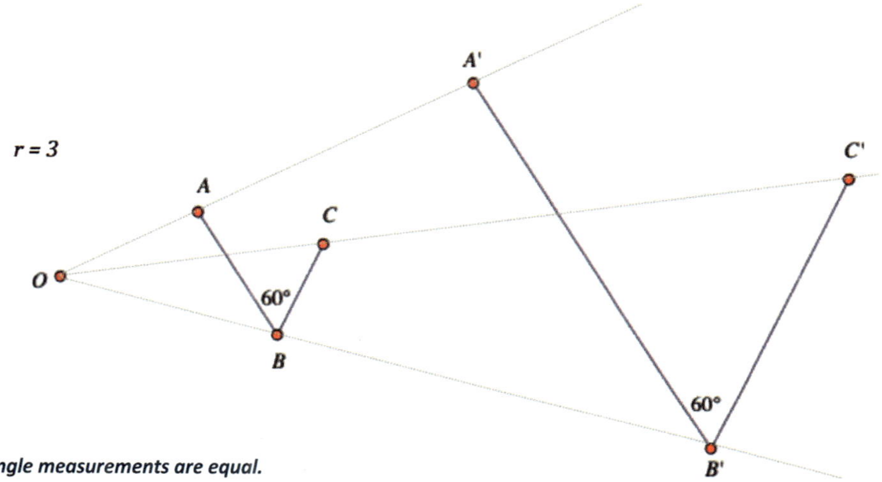

The angle measurements are equal.

Closing (2 minutes)

- We have studied two major classes of transformations: those that are distance-preserving (translations, reflections, rotations) and those that are not (dilations).
- Like rigid motions, dilations involve a rule assignment for each point in the plane and also have inverse functions that return each dilated point back to itself.
- Though we have experimentally verified that dilations share properties similar to those of rigid motions, for example, the property that lines map to lines, we have yet to establish these properties formally.

Lesson Summary

- **There are two major classes of transformations: those that are distance preserving (translations, reflections, rotations) and those that are not (dilations).**
- **Like rigid motions, dilations involve a rule assignment for each point in the plane and also have inverse functions that return each dilated point back to itself.**

Exit Ticket (5 minutes)

Name ____ Date ____________________

Lesson 6: Dilations as Transformations of the Plane

Exit Ticket

1. Which transformations of the plane are distance-preserving transformations? Provide an example of what this property means.

2. Which transformations of the plane preserve angle measure? Provide one example of what this property means.

3. Which transformation is not considered a rigid motion and why?

Exit Ticket Sample Solutions

1. **Which transformations of the plane are distance-preserving transformations? Provide an example of what this property means.**

   Rotations, translations, and reflections are distance-preserving transformations of the plane because for any two different points A and B in the plane, if F is a rotation, translation, or reflection that maps $A \to F(A)$ and $B \to F(B)$, $AB = F(A)F(B)$.

2. **Which transformations of the plane preserve angle measure? Provide one example of what this property means.**

   Rotations, translations, reflections, and dilations all preserve angle measure. If lines $\overleftrightarrow{AB}$ and $\overleftrightarrow{BC}$ are coplanar and intersect at B to form $\angle ABC$ with measure $n°$, $\overleftrightarrow{F(A)F(B)}$ (or $\overleftrightarrow{A'B'}$) and $\overleftrightarrow{F(B)F(C)}$ (or $\overleftrightarrow{B'C'}$) intersect at $F(B)$ to form $\angle F(A)F(B)F(C)$ (or $\angle A'B'C'$) that also has measure $n°$.

3. **Which transformation is not considered a rigid motion and why?**

   A dilation is not considered a rigid motion because it does not preserve the distance between points. Under a dilation $D_{O,r}$ where $r \neq 1$, $D_{O,r}(A) = A'$ and $D_{O,r}(B) = B'$, $A'B' = r(AB)$, which means that $\overline{A'B'}$ must have a length greater or less than $\overline{AB}$.

Problem Set Sample Solutions

1. **In the diagram below, A' is the image of A under a single transformation of the plane. Use the given diagrams to show your solutions to parts (a)–(d).**

   a. **Describe the translation that maps $A \to A'$, and then use the translation to locate P', the image of P.**

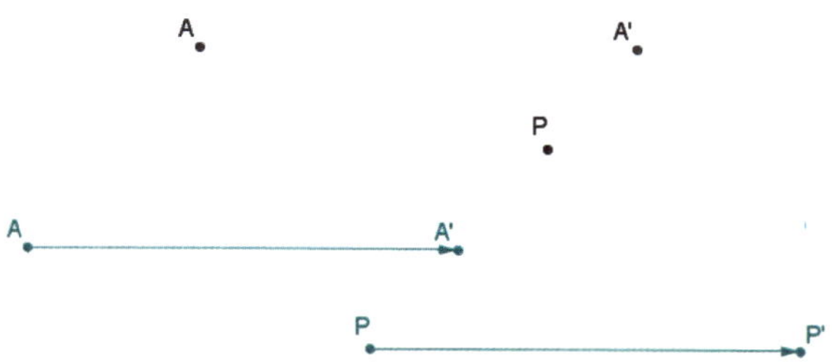

   The translation that maps $A \to A'$ is along horizontal vector $\overrightarrow{AA'}$

   b. **Describe the reflection that maps $A \to A'$, and then use the reflection to locate P', the image of P.**

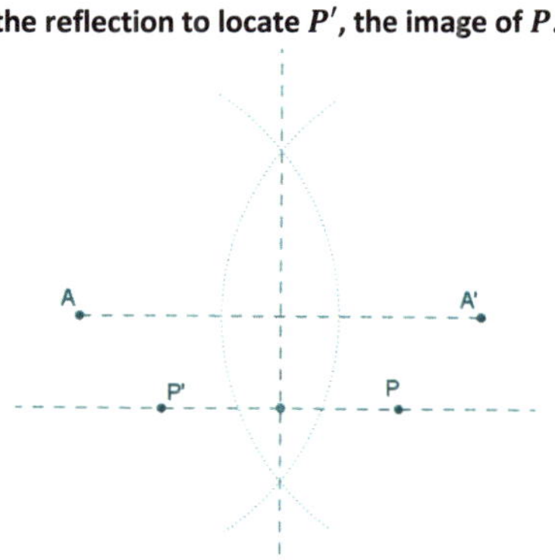

   The reflection that maps $A \to A'$ is across the perpendicular bisector of $\overline{AA'}$.

c. **Describe a rotation that maps $A \rightarrow A'$, and then use your rotation to locate P', the image of P.**

There are many possible correct answers to this part. The center of rotation C must be on the perpendicular bisector of $\overline{AA'}$ and the radius $CA \geq \frac{1}{2}AA'$.

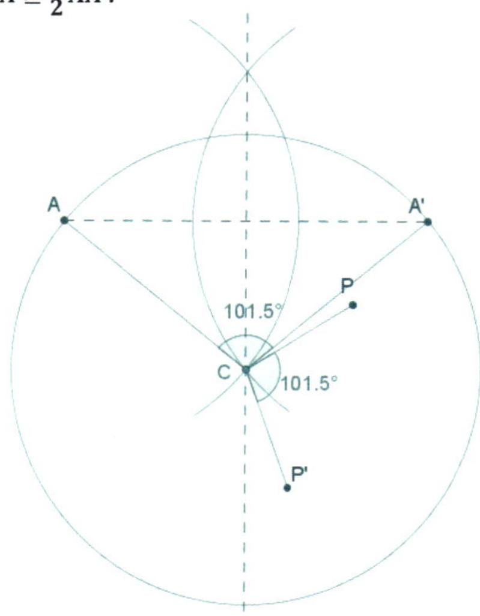

d. **Describe a dilation that maps $A \rightarrow A'$, and then use your dilation to locate P', the image of P.**

There are many possible correct answers to this part. The center of dilation must be on $\overleftrightarrow{AA'}$. If the scale factor chosen is $r > 1$, then A must be between O and A'. If the scale factor chosen is $r < 1$, then A' must be between A and O, and $OP' = r(OP)$. The sample shown below uses a scale factor $r = 2$.

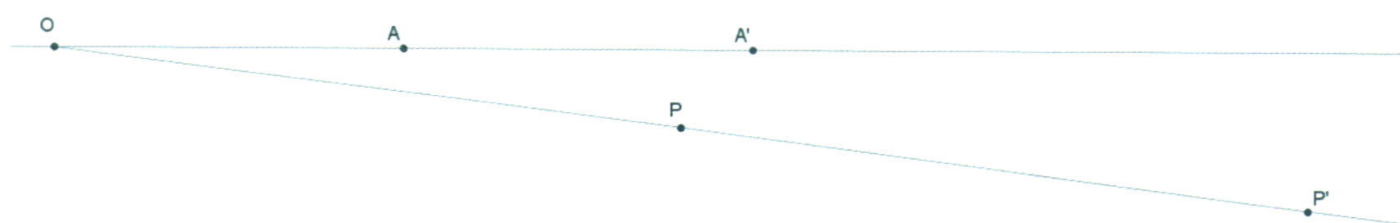

2. On the diagram below, O is a center of dilation, and $\overleftrightarrow{AD}$ is a line not through O. Choose two points B and C on $\overleftrightarrow{AD}$ between A and D.

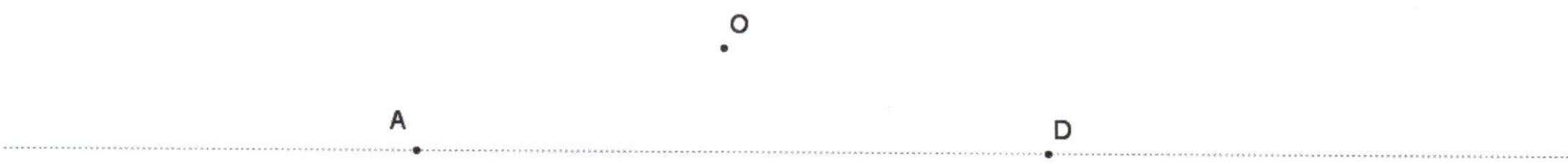

a. Dilate A, B, C, and D from O using scale factor $r = \frac{1}{2}$. Label the images A', B', C', and D', respectively.

b. Dilate A, B, C, and D from O using scale factor $r = 2$. Label the images A'', B'', C'', and D'', respectively.

c. Dilate A, B, C, and D from O using scale factor $r = 3$. Label the images A''', B''', C''', and D''', respectively.

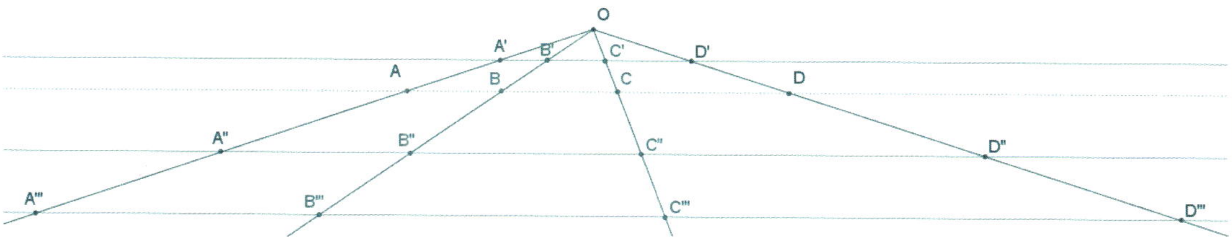

d. Draw a conclusion about the effect of a dilation on a line segment based on the diagram that you drew. Explain.

Conclusion: Dilations map line segments to line segments.

3. Write the inverse transformation for each of the following so that the composition of the transformation with its inverse maps a point to itself on the plane.

a. $T_{\overrightarrow{AB}}$

The inverse of a translation along the vector $\overrightarrow{AB}$ would be a translation along the vector $\overrightarrow{BA}$ since this vector has the same magnitude but opposite direction. This translation maps any image point to its pre-image.

b. $r_{\overleftrightarrow{AB}}$

The inverse of a reflection over line AB is the same reflection. The points P and $r_{\overleftrightarrow{AB}}(P)$ are symmetric about $\overleftrightarrow{AB}$, so repeating the reflection takes a point back to itself.

c. $R_{C,45}$

The inverse of a $45°$ rotation about a point C would be a rotation about the same point C of $-45°$, the opposite rotational direction.

d. $D_{O,r}$

The inverse of a dilation with center O and scale factor r would be a dilation from center O with a scale factor of $\frac{1}{r}$. Point A in the plane is distance OA from the center of dilation O, and its image A' would, therefore, be at a distance $r(OA)$ from O. A dilation of A' with scale factor $\frac{1}{r}$ would map the A' to a point that is a distance $\frac{1}{r}(OA') = \frac{1}{r}\left(r(OA)\right) = 1(OA) = OA$. By the definition of a dilation, points and their images lie on the same ray that originates at the center of the dilation. There is only one point on that ray at a distance OA from O, which is A.

To the teacher: Problem 4 reviews the application of dilation on the coordinate plane that was studied in depth in Grade 8.

4. **Given $U(1,3)$, $V(-4,-4)$, and $W(-3,6)$ on the coordinate plane, perform a dilation of $\triangle UVW$ from center $O(0,0)$ with a scale factor of $\frac{3}{2}$. Determine the coordinates of images of points U, V, and W, and describe how the coordinates of the image points are related to the coordinates of the pre-image points.**

Under the given dilation, U, V, and W map to U', V', and W', respectively. $U'\left(\frac{3}{2},\frac{9}{2}\right)$, $V'(-6,-6)$, and $W'\left(-\frac{9}{2},9\right)$. For each point (X,Y) on the coordinate plane, its image point is $\left(\frac{3}{2}X,\frac{3}{2}Y\right)$ under the dilation from the origin with scale factor $\frac{3}{2}$.

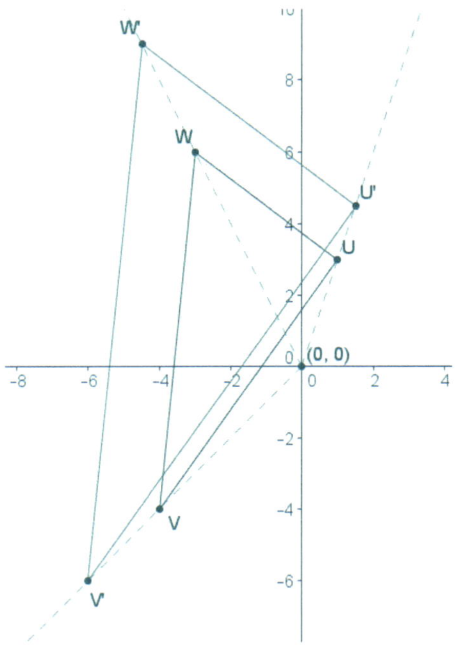

5. **Points B, C, D, E, F, and G are dilated images of A from center O with scale factors 2, 3, 4, 5, 6, and 7, respectively. Are points Y, X, W, V, U, T, and S all dilated images of Z under the same respective scale factors? Explain why or why not.**

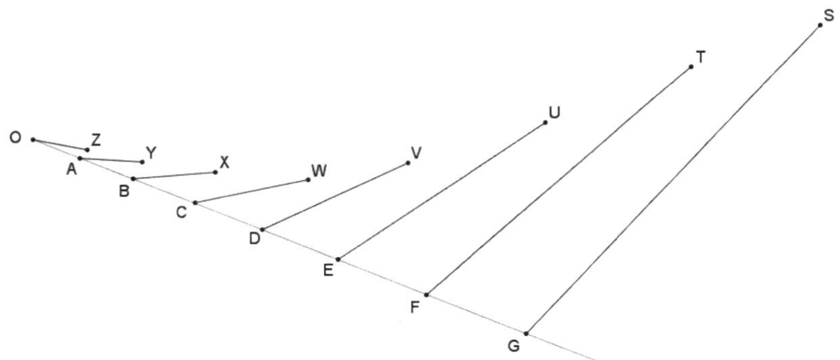

If points Y, X, W, V, U, T, and S were dilated images of Z, the images would all be collinear with O and Z; however, the points are not all on a line, so they cannot all be images of point Z from center O. We also know that dilations preserve angle measures, and it is clear that each segment meets $\overrightarrow{OA}$ at a different angle.

EUREKA MATH™

6. **Find the center and scale factor that takes A to A' and B to B', if a dilation exists.**

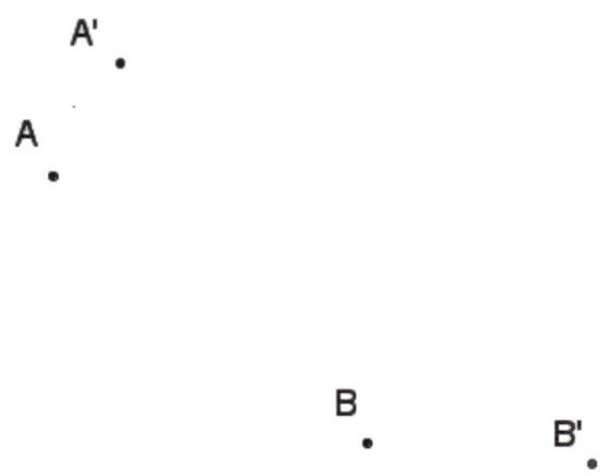

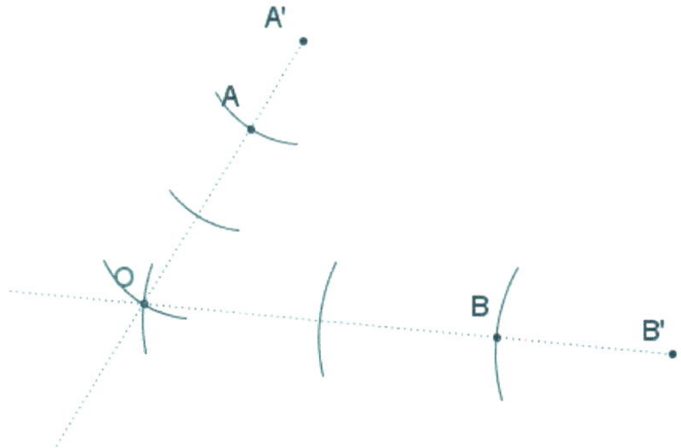

The center of the dilation is O, and the scale factor is $\frac{3}{2}$.

7. **Find the center and scale factor that takes A to A' and B to B', if a dilation exists.**

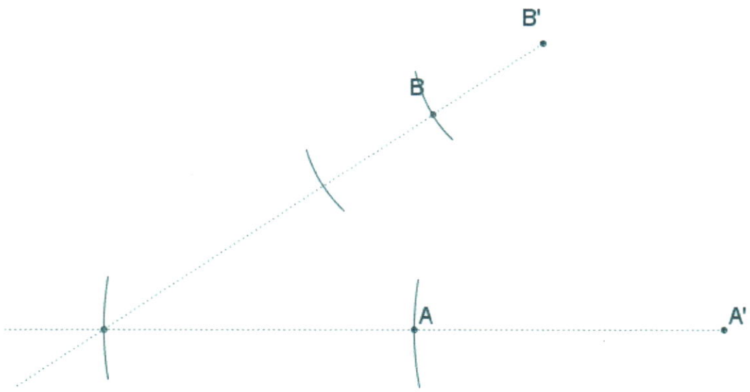

After drawing $\overrightarrow{B'B}$ and $\overrightarrow{A'A}$, the rays converge at a supposed center; however, the corresponding distances are not proportional since $\frac{OA'}{OA} = 2$ and $\frac{OB'}{OB} \neq 2$. Therefore, a dilation does not exist that maps $A \to A'$ and $B \to B'$.

It also could be shown that $\overline{AB}$ and $\overline{A'B'}$ are not parallel; therefore, the lengths are not proportional by the triangle side splitter theorem, and there is no dilation.

Lesson 7: How Do Dilations Map Segments?

Student Outcomes

- Students prove that dilation $D_{O,r}$ maps a line segment PQ to a line segment $P'Q'$, sending the endpoints to the endpoints so that $P'Q' = rPQ$. If the center O lies in line PQ or $r = 1$, then $\overleftrightarrow{PQ} = \overleftrightarrow{P'Q'}$. If the center O does not lie in line PQ and $r \neq 1$, then $\overleftrightarrow{PQ} \parallel \overleftrightarrow{P'Q'}$.
- Students prove that if $\overline{PQ}$ and $\overline{RS}$ are line segments in the plane of different lengths, then there is a dilation that maps one to the other if and only if $\overleftrightarrow{PQ} = \overleftrightarrow{RS}$ or $\overleftrightarrow{PQ} \parallel \overleftrightarrow{RS}$.

Lesson Notes

In Grade 8, students informally showed that a dilation maps a segment to a segment on the coordinate plane. The lesson includes an opening discussion that reminds students of this fact. Next, students must consider how to prove that dilations map segments to segments when the segment is not tied to the coordinate plane. We again call upon our knowledge of the triangle side splitter theorem to show that a dilation maps a segment to a segment. The goal of the lesson is for students to understand the effect that dilation has on segments, specifically that a dilation maps a segment to a segment so that its length is r times the original.

To complete the lesson in one class period, it may be necessary to skip the opening discussion and Example 4 and focus primarily on the work in Examples 1–3. Another option is to extend the lesson to two days so that all examples and exercises can be given the time required to develop student understanding of the dilation of segments.

Classwork

Opening Exercise (2 minutes)

MP.3

Opening Exercise

a. **Is a dilated segment still a segment? If the segment is transformed under a dilation, explain how.**

Accept any reasonable answer. The goal is for students to recognize that a segment dilates to a segment that is r times the length of the original.

Scaffolding:

Use a segment in the coordinate plane with endpoint $P(-4,1)$ and endpoint $Q(3,2)$. Show that a dilation from a center at the origin maps $\overline{PQ}$ to $\overline{P'Q'}$.

MP.3

b. Dilate the segment PQ by a scale factor of 2 from center O.

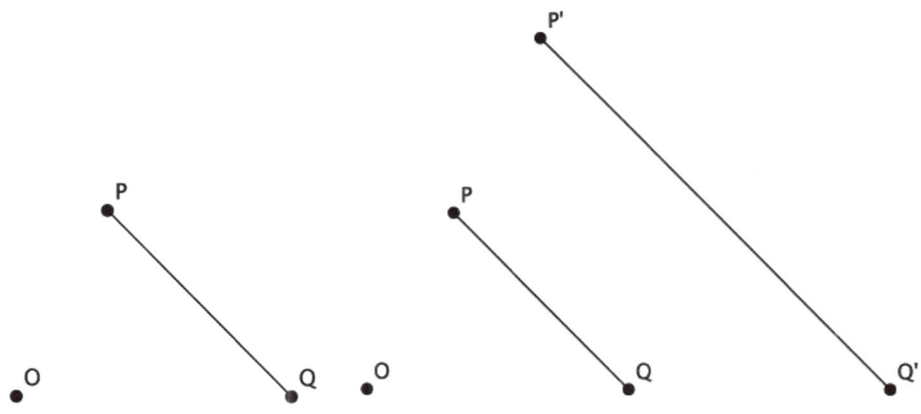

i. Is the dilated segment $P'Q'$ a segment?

Yes, the dilation of segment PQ produces a segment $P'Q'$.

ii. How, if at all, has the segment PQ been transformed?

Segment $P'Q'$ is twice the length of segment PQ. The segment has increased in length according to the scale factor of dilation.

Opening (5 minutes)

In Grade 8, students learned the multiplicative effect that dilation has on points in the coordinate plane when the center of dilation is the origin. Specifically, students learned that when the point located at (x, y) was dilated from the origin by scale factor r, the dilated point was located at (rx, ry). Review this fact with students, and then remind them how to informally verify that a dilation maps a segment to a segment using the diagram below. As time permits, review what students learned in Grade 8, and then spend the remaining time on the question in the second bullet point.

- Let $\overline{AB}$ be a segment on the coordinate plane with endpoints at $(-2, 1)$ and $(1, -2)$. If we dilate the segment from the origin by a scale factor $r = 4$, another segment is produced.
- What do we expect the coordinates of the endpoints to be?
  - *Based on what we know about the multiplicative effect dilation has on coordinates, we expect the coordinates of the endpoints of $\overline{A'B'}$ to be $(-8, 4)$ and $(4, -8)$.*

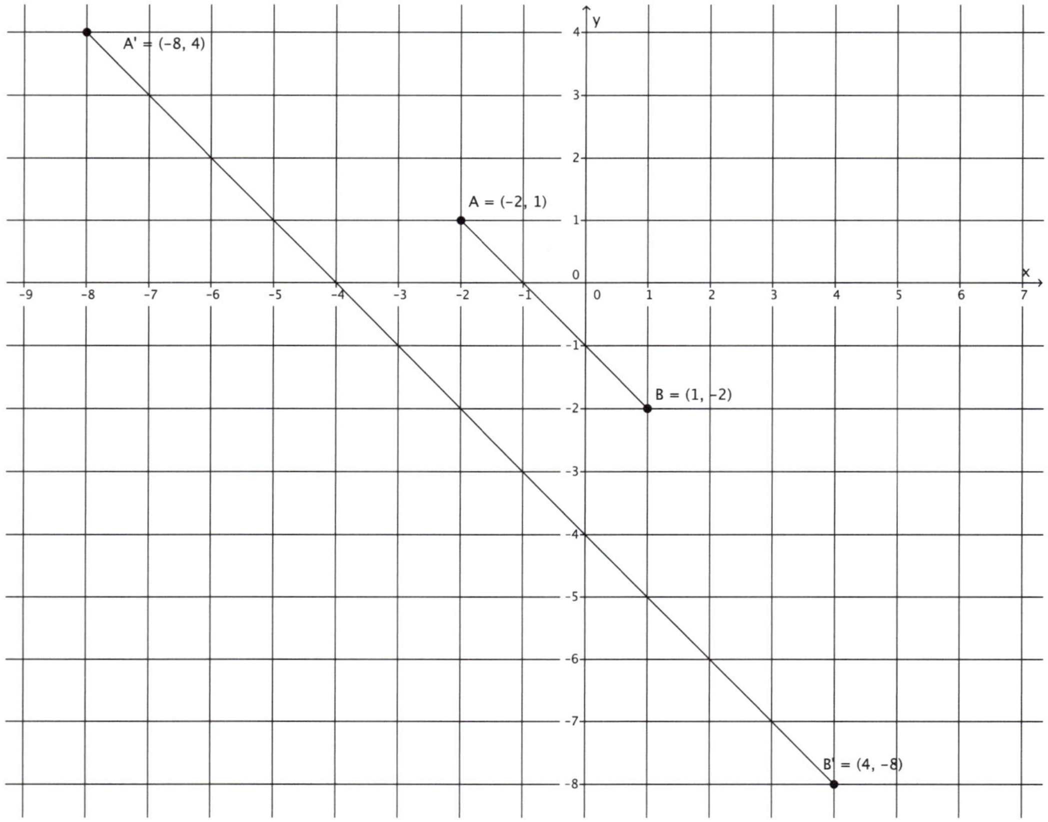

- The question becomes, how can we be sure that the dilation maps the points between A and B to the points between A' and B'? We have already shown that the endpoints move to where we expect them to, but what about the points in between? Perhaps the dilation maps the endpoints the way we expect, but all other points form an arc of a circle or some other curve that connects the endpoints. Can you think of a way we can verify that *all* of the points of segment AB map to images that lie on segment $A'B'$?
  - *We can verify other points that belong to segment AB using the same method as the endpoints. For example, points $(-1,0)$ and $(0,-1)$ of segment AB should map to points $(-4,0)$ and $(0,-4)$. Using the x- and y-axes as our rays from the origin of dilation, we can clearly see that the points and their dilated images lie on the correct ray and more importantly lie on segment $A'B'$.*
- Our next challenge is to show that dilations map segments to segments when they are not on the coordinate plane. We will prove the *preliminary dilation theorem for segments:* A dilation maps a line segment to a line segment sending the endpoints to the endpoints.

Example 1 (2 minutes)

> **Example 1**
>
> **Case 1. Consider the case where the scale factor of dilation is $r = 1$. Does a dilation from center O map segment PQ to a segment $P'Q'$? Explain.**
>
> ***A scale factor of $r = 1$ means that the segment and its image are equal. The dilation does not enlarge or shrink the image of the figure; it remains unchanged. Therefore, when the scale factor of dilation is $r = 1$, then the dilation maps the segment to itself.***

Example 2 (3 minutes)

> **Example 2**
>
> **Case 2. Consider the case where a line PQ contains the center of the dilation. Does a dilation from the center with scale factor $r \neq 1$ map the segment PQ to a segment $P'Q'$? Explain.**

At this point, students should be clear that a dilation of scale factor $r \neq 1$ changes the length of the segment. If necessary, explain to students that a scale factor of $r \neq 1$ simply means that the figure changes in length. The goal in this example and, more broadly, this lesson is to show that a dilated segment is still a segment not some other figure. The focus of the discussion should be on showing that the dilated figure is in fact a segment, not necessarily the length of the segment.

> ***Yes. The dilation sends points P and Q to points P' and Q'. Since the points P and Q are collinear with the center O, then both P' and Q' are also collinear with the center O. The dilation also takes all of the points between P and Q to all of the points between P' and Q' (again because their images must fall on the rays OP and OQ). Therefore, the dilation maps $\overline{PQ}$ to $\overline{P'Q'}$.***

Example 3 (12 minutes)

> **Example 3**
>
> **Case 3. Consider the case where $\overleftrightarrow{PQ}$ does not contain the center O of the dilation, and the scale factor r of the dilation is not equal to 1; then, we have the situation where the key points O, P, and Q form $\triangle OPQ$. The scale factor not being equal to 1 means that we must consider scale factors such that $0 < r < 1$ and $r > 1$. However, the proofs for each are similar, so we focus on the case when $0 < r < 1$.**

> *Scaffolding:*
>
> For some groups of students, it may be necessary for them to perform a dilation where the scale factor of dilation is $0 < r < 1$ so they have recent experience allowing them to better answer the second question of Example 3.

When we dilate points P and Q from center O by scale factor $0 < r < 1$, as shown, what do we know about points P' and Q'?

We know P' lies on ray OP with $OP' = r \cdot OP$, and Q' lies on ray OQ with $OQ' = r \cdot OQ$. So, $\frac{OP'}{OP} = \frac{OQ'}{OQ} = r$. The line segment $P'Q'$ splits the sides of $\triangle OPQ$ proportionally.

By the triangle side splitter theorem, we know that the lines containing $\overline{PQ}$ and $\overline{P'Q'}$ are parallel.

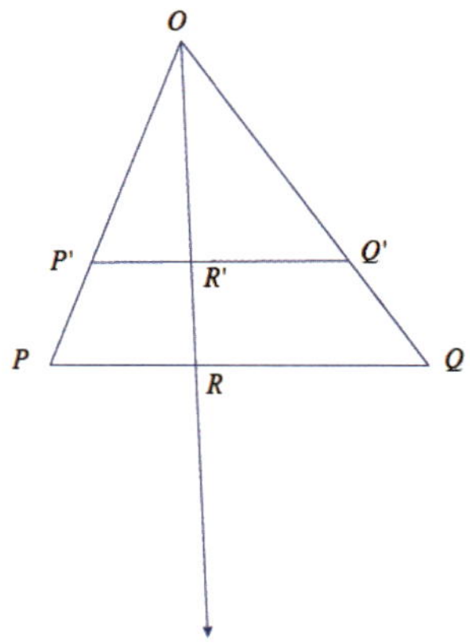

We use the fact that the line segment $P'Q'$ splits the sides of $\triangle OPQ$ proportionally and that the lines containing $\overline{PQ}$ and $\overline{P'Q'}$ are parallel to prove that a dilation maps segments to segments. Because we already know what happens when points P and Q are dilated, consider another point R that is on the segment PQ. After dilating R from center O by scale factor r to get the point R', does R' lie on the segment $P'Q'$?

Consider giving students time to discuss in small groups Marwa's proof shown on the next page on how to prove that a dilation maps collinear points to collinear points. Also consider providing students time to make sense of it and paraphrase a presentation of the proof to a partner or the class. Consider also providing the statements for the proof and asking students to provide the reasoning for each step independently, with a partner, or in small groups.

The proof below relies heavily upon the parallel postulate: Two lines are constructed, $\overleftrightarrow{P'Q'}$ and $\overleftrightarrow{P'R'}$, both of which are parallel to $\overleftrightarrow{PQ}$. Since both lines are parallel to $\overleftrightarrow{PQ}$ and contain the point P', they must be the same line by the parallel postulate. Thus, the point R' lies on the line $P'Q'$.

Note: This proof on the next page is only part of the reasoning needed to show that dilations map segments to segments. The full proof also requires showing that points between P and Q are mapped to points between P' and Q', and that this mapping is onto (that for every point on the line segment $P'Q'$, there exists a point on segment PQ that gets sent to it). See the discussion following Marwa's proof for more details on these steps.

<table>
<tr><td colspan="2">Marwa's proof of the statement: Let O be a point not on $\overleftrightarrow{PQ}$ and $D_{O,r}$ be a dilation with center O and scale factor $0 < r < 1$ that sends point P to P' and Q to Q'. If R is another point that lies on $\overleftrightarrow{PQ}$, then $D_{O,r}(R)$ is a point that lies on $\overleftrightarrow{P'Q'}$.</td></tr>
<tr><th>Statement</th><th>Reason</th></tr>
<tr><td>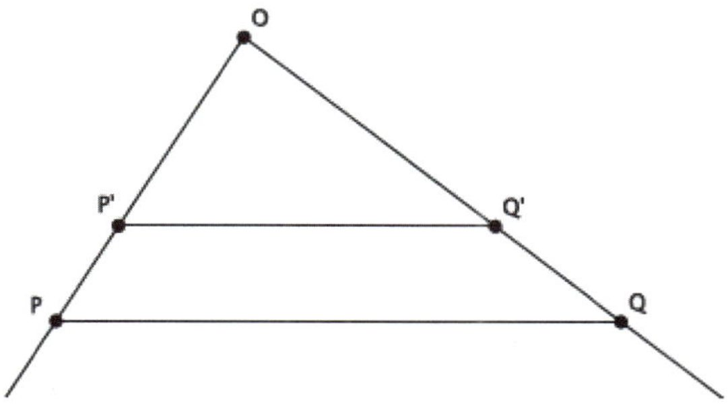
1. A dilation $D_{O,r}$ with center O and scale factor r sends point P to P' and Q to Q'.</td><td>1. Given</td></tr>
<tr><td>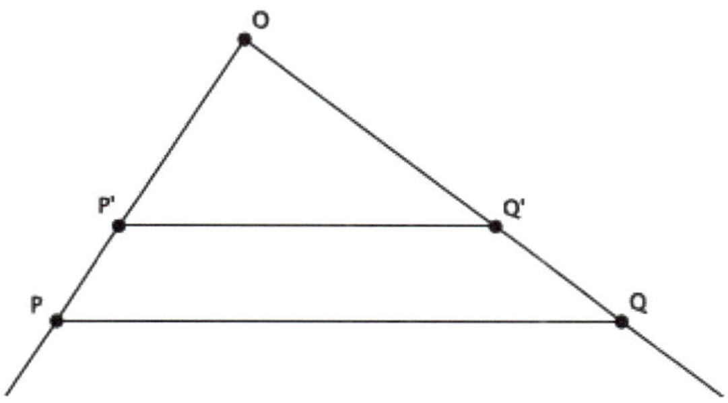
2. $\frac{OP'}{OP} = \frac{OQ'}{OQ} = r$
3. $\overleftrightarrow{PQ} \| \overleftrightarrow{P'Q'}$</td><td>2. By definition of dilation
3. By the triangle side splitter theorem</td></tr>
<tr><td>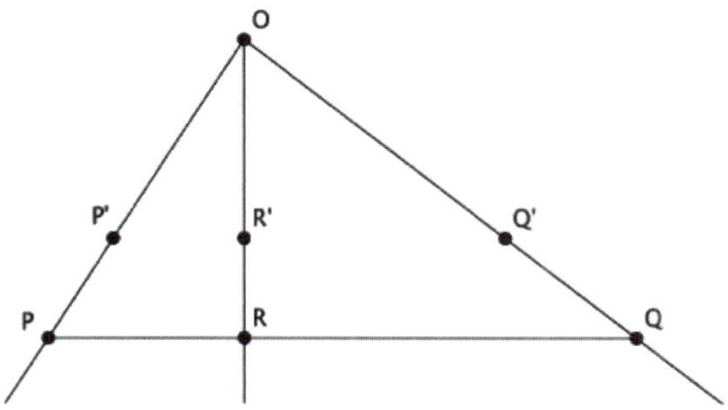
4. Let R be a point on segment PQ between P and Q. Let R' be the dilation of R by $D_{O,r}$.
5. $\frac{OP'}{OP} = \frac{OR'}{OR} = r$</td><td>4. Given
5. By definition of dilation</td></tr>
</table>

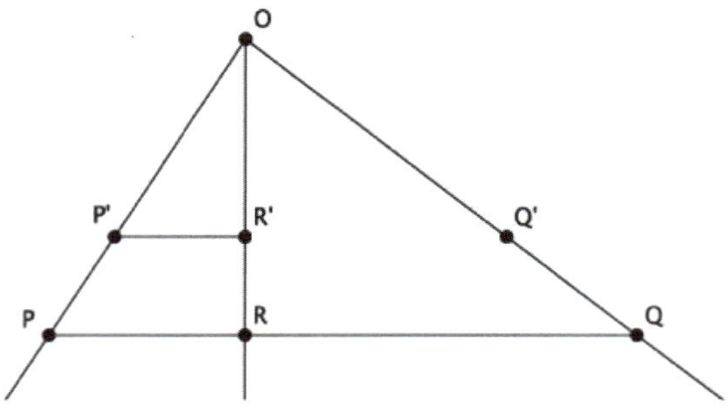 6. $\overleftrightarrow{P'R'} \parallel \overleftrightarrow{PQ}$	6. By the triangle side splitter theorem as applied to $\triangle\ OPR$ and the fact that $\overleftrightarrow{PQ}$ and $\overleftrightarrow{PR}$ are the same line
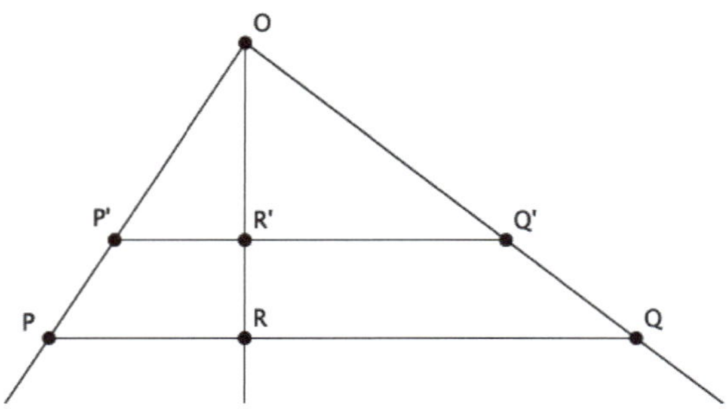 7. $\overleftrightarrow{P'R'} = \overleftrightarrow{P'Q'}$	7. By the parallel postulate: Through a given external point P', there is at most one line parallel to a given line PQ.
8. The point R' lies on $\overleftrightarrow{P'Q'}$.	8. By Step 7

There are still two subtle steps that need to be proved to show that dilations map segments to segments when the center does not lie on the line through the segment. The teacher may decide whether to show how to prove these two steps or just claim that the steps can be shown to be true. Regardless, the steps should be briefly discussed as part of what is needed to complete the proof of the full statement that dilations map segments to segments.

The first additional step that needs to be shown is that points on $\overline{PQ}$ are sent to points on $\overline{P'Q'}$, that is, if R is between P and Q, then R' is between P' and Q'. To prove this, we first write out what it means for R to be between P and Q: P, R, and Q are different points on the same line such that $PR + RQ = PQ$. By the dilation theorem (Lesson 6), $P'R' = r \cdot PR$, $R'Q' = r \cdot RQ$, and $P'Q' = r \cdot PQ$. Therefore,

$$P'R' + R'Q' = r \cdot PR + r \cdot RQ = r(PR + RQ) = r \cdot PQ = P'Q'$$

Hence, $P'R' + R'Q' = R'Q'$, and therefore R' is between P' and Q'.

The second additional step is to show that the dilation is an onto mapping, that is, for every point R' that lies on $\overline{P'Q'}$, there is a point R that lies on $\overline{PQ}$ that is mapped to R' under the dilation. To prove, we use the inverse dilation at center O with scale factor $\frac{1}{r}$ to get the point R and then follow the proof above to show that R lies on $\overline{PQ}$.

> **Putting together the preliminary dilation theorem for segments with the dilation theorem, we get**
>
> **DILATION THEOREM FOR SEGMENTS: A dilation $D_{O,r}$ maps a line segment PQ to a line segment $P'Q'$ sending the endpoints to the endpoints so that $P'Q' = rPQ$. Whenever the center O does not lie in line PQ and $r \neq 1$, we conclude $\overleftrightarrow{PQ} \parallel \overleftrightarrow{P'Q'}$. Whenever the center O lies in $\overleftrightarrow{PQ}$ or if $r = 1$, we conclude $\overleftrightarrow{PQ} = \overleftrightarrow{P'Q'}$.**
>
> **As an aside, observe that a dilation maps parallel line segments to parallel line segments. Further, a dilation maps a directed line segment to a directed line segment that points in the same direction.**

If time permits, have students verify these observations with their own scale drawings. It is not imperative that students do this activity, but this idea is used in the next lesson. At this point, students just need to observe this fact or at least be made aware of it, as it is discussed in Lesson 8.

Example 4 (7 minutes)

> **Example 4**
>
> **Now look at the converse of the dilation theorem for segments: If $\overline{PQ}$ and $\overline{RS}$ are line segments of different lengths in the plane, then there is a dilation that maps one to the other if and only if $\overleftrightarrow{PQ} = \overleftrightarrow{RS}$ or $\overleftrightarrow{PQ} \parallel \overleftrightarrow{RS}$.**
>
> **Based on Examples 2 and 3, we already know that a dilation maps a segment PQ to another line segment, say $\overline{RS}$, so that $\overleftrightarrow{PQ} = \overleftrightarrow{RS}$ (Example 2) or $\overleftrightarrow{PQ} \parallel \overleftrightarrow{RS}$ (Example 3). If $\overleftrightarrow{PQ} \parallel \overleftrightarrow{RS}$, then, because $\overline{PQ}$ and $\overline{RS}$ are different lengths in the plane, they are the bases of a trapezoid, as shown.**

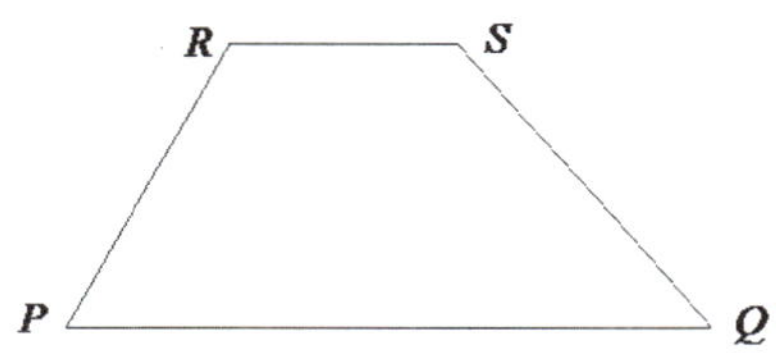

> **Since $\overline{PQ}$ and $\overline{RS}$ are segments of different lengths, then the non-base sides of the trapezoid are not parallel, and the lines containing them meet at a point O as shown.**

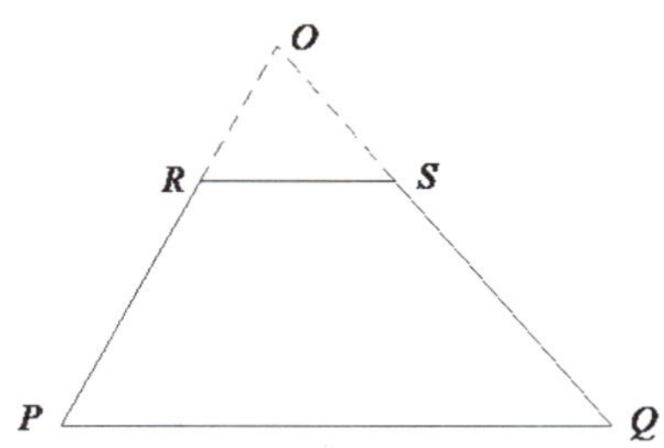

> **Recall that we want to show that a dilation maps $\overline{PQ}$ to $\overline{RS}$. Explain how to show it.**

Provide students time to discuss this in partners or small groups.

The triangle formed with vertex O, $\triangle OPQ$, has PQ as its base. Since $\overleftrightarrow{PQ} \parallel \overleftrightarrow{RS}$, then the segment RS splits the sides of the triangle proportionally by the triangle side splitter theorem. Since we know the proportional side splitters of a triangle are the result of a dilation, then we know there is a dilation from center O by scale factor r that maps points P and Q to points R and S, respectively. Thus, a dilation maps $\overline{PQ}$ to $\overline{RS}$.

The case when the segments $\overline{PQ}$ and $\overline{RS}$ are such that $\overleftrightarrow{PQ} = \overleftrightarrow{RS}$ is left as an exercise.

Exercises 1–2 (8 minutes)

Students complete Exercises 1–2 in pairs. Students may need support to complete Exercise 2. A hint is shown below that can be shared with students if necessary.

Exercises 1–2

In the following exercises, you will consider the case where the segment and its dilated image belong to the same line, that is, when $\overline{PQ}$ and $\overline{RS}$ are such that $\overleftrightarrow{PQ} = \overleftrightarrow{RS}$.

1. **Consider points P, Q, R, and S on a line, where $P = R$, as shown below. Show there is a dilation that maps $\overline{PQ}$ to $\overline{RS}$. Where is the center of the dilation?**

If we assume there is a dilation that maps $\overline{PQ}$ to $\overline{RS}$, with a scale factor so that $r = \frac{RS}{PQ}$, then the center of the dilation must coincide with endpoints P and R because, by definition of dilation, the center maps to itself. Since points P and R coincide, it must mean that the center O is such that $O = P = R$. Since the other endpoint Q of PQ lies on $\overleftrightarrow{PQ}$, the dilated image of Q must also lie on the line (draw the ray from the center through point Q, and the ray coincides with $\overleftrightarrow{PQ}$). Since the dilated image of Q must lie on the line, and point S is to the right of Q, then a dilation from center O with scale factor $r > 1$ maps $\overline{PQ}$ to $\overline{RS}$.

2. **Consider points P, Q, R, and S on a line as shown below where $\overline{PQ} \neq \overline{RS}$. Show there is a dilation that maps $\overline{PQ}$ to $\overline{RS}$. Where is the center of the dilation?**

Students may need some support to complete Exercise 2. Give them enough time to struggle with how to complete the exercise and, if necessary, provide them with the following hint: Construct perpendicular line segments $\overline{PQ'}$ and $\overline{RS'}$ as shown so that $PQ' = PQ$ and $RS' = RS$.

EUREKA MATH

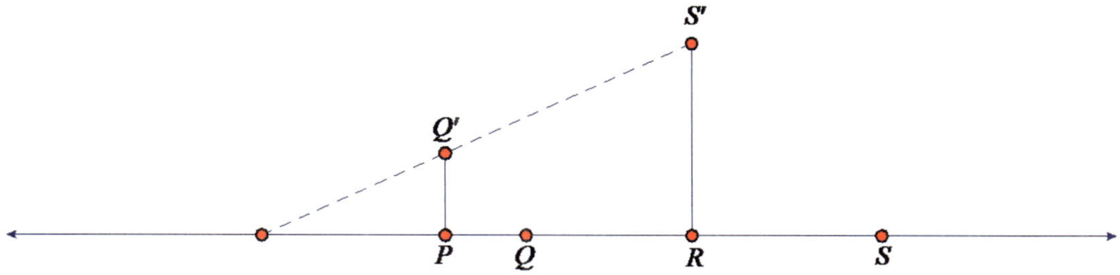

Construct perpendicular line segments of lengths PQ and RS through points P and R respectively. Note the endpoints of the perpendicular segments as Q' and S'. Draw an auxiliary line through points Q' and S' that intersects with $\overleftrightarrow{PQ}$. The intersection of the two lines is the center of dilation O. Since $\overleftrightarrow{PQ'} \| \overleftrightarrow{RS'}$ (perpendicular lines), by the triangle side splitter theorem, the segment PQ' splits the triangle $\triangle ORS'$ proportionally, so by the dilation theorem, $\frac{RS'}{PQ'} = r$. This ratio implies that $RS' = r \cdot PQ'$. By construction, $PQ' = PQ$ and $RS' = RS$. Therefore, $RS = r \cdot PQ$. By definition of dilation, since $RS = r \cdot PQ$, there is a dilation from center O with scale factor r that maps $\overline{PQ}$ to $\overline{RS}$.

Closing (3 minutes)

Revisit the Opening Exercises. Have students explain their thinking about segments and dilated segments. Were their initial thoughts correct? Did they change their thinking after going through the examples? What made them change their minds? Then ask students to paraphrase the proofs as stated in the bullet point below.

- Paraphrase the proofs that show how dilations map segments to segments when $r = 1$ and $r \neq 1$ and when the points are collinear compared to the vertices of a triangle.

Either accept or correct students' responses accordingly. Recall that the goal of the lesson is to make sense of what happens when segments are dilated. When the segment is dilated by a scale factor of $r = 1$, then the segment and its image would be the same length. When the points P and Q are on a line containing the center, then the dilated points P' and Q' are also collinear with the center producing an image of the segment that is a segment. When the points P and Q are not collinear with the center, and the segment is dilated by a scale factor of $r \neq 1$, then the point P' lies on the ray OP with $OP' = r \cdot OP$, *and* Q'*lies on* $\overrightarrow{OQ}$ *with* $Q' = r \cdot OQ$. *Then* $\frac{OP'}{OP} = \frac{OQ'}{OQ} = r$. $\overline{P'Q'}$ splits the sides of $\triangle OPQ$ proportionally, and by the triangle side splitter theorem, the lines containing $\overline{PQ}$ *and* $\overline{P'Q'}$ are parallel.

> Lesson Summary
>
> - **When a segment is dilated by a scale factor of $r = 1$, then the segment and its image would be the same length.**
> - **When the points P and Q are on a line containing the center, then the dilated points P' and Q' are also collinear with the center producing an image of the segment that is a segment.**
> - **When the points P and Q are not collinear with the center and the segment is dilated by a scale factor of $r \neq 1$, then the point P' lies on the ray OP' with $OP' = r \cdot OP$, *and* Q' *lies on ray* OQ with $OQ' = r \cdot OQ$.**

Exit Ticket (5 minutes)

Name ____ Date ________________

Lesson 7: How Do Dilations Map Segments?

Exit Ticket

1. Given the dilation $D_{O,\frac{3}{2}}$, a line segment PQ, and that O is not on $\overleftrightarrow{PQ}$, what can we conclude about the image of $\overline{PQ}$?

2. Given figures A and B below, $\overline{BA} \parallel \overline{DC}$, $\overline{UV} \parallel \overline{XY}$, and $\overline{UV} \cong \overline{XY}$, determine which figure has a dilation mapping the parallel line segments, and locate the center of dilation O. For one of the figures, a dilation does not exist. Explain why.

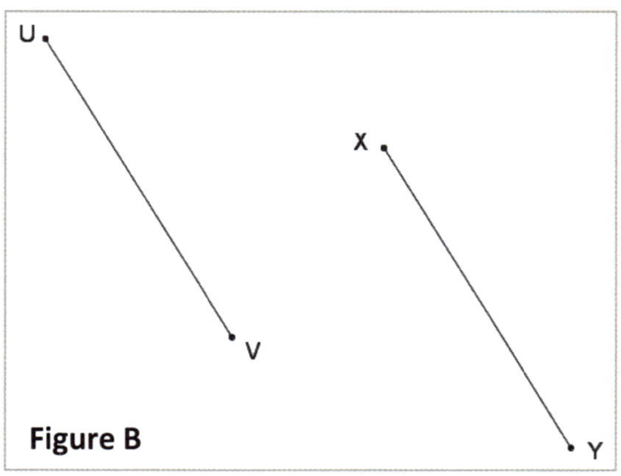

Exit Ticket Sample Solutions

1. Given the dilation $D_{O,\frac{3}{2}}$, a line segment PQ, and that O is not on $\overleftrightarrow{PQ}$, what can we conclude about the image of $\overline{PQ}$?

   Since P and Q are not in line with O, $\overline{P'Q'}$ is parallel to $\overline{PQ}$, and $P'Q' = \frac{3}{2}(PQ)$.

2. Given figures A and B below, $\overline{BA} \parallel \overline{DC}$, $\overline{UV} \parallel \overline{XY}$, and $\overline{UV} \cong \overline{XY}$, determine which figure has a dilation mapping the parallel line segments, and locate the center of dilation O. For one of the figures, a dilation does not exist. Explain why.

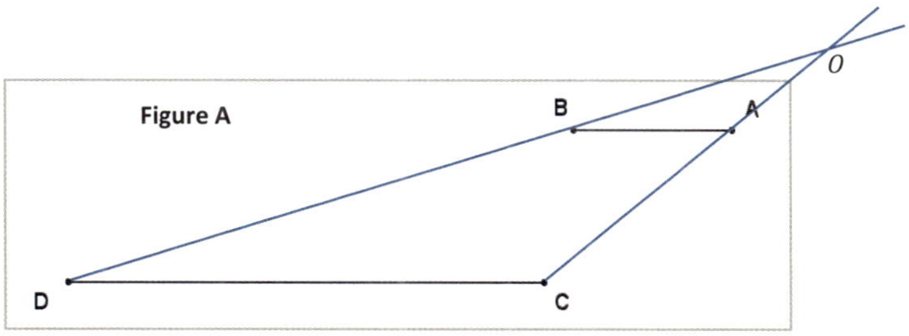

Figure A

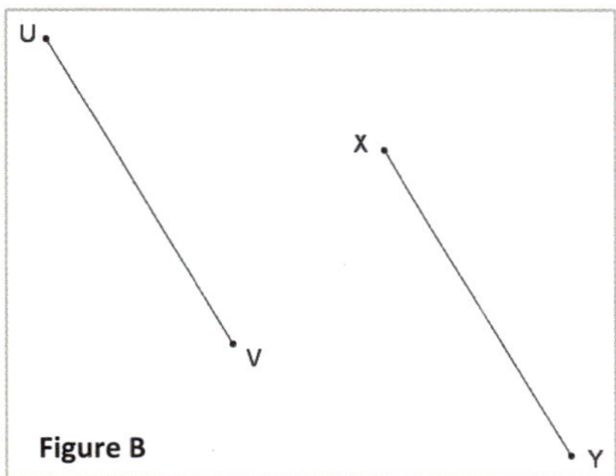

Figure B

There is no dilation that maps $\overline{UV}$ to $\overline{XY}$. If the segments are both parallel and congruent, then they form two sides of a parallelogram, which means that $\overleftrightarrow{UX} \parallel \overleftrightarrow{VY}$. If there was a dilation mapping $\overline{UV}$ to $\overline{XY}$, then $\overleftrightarrow{UX}$ and $\overleftrightarrow{VY}$ would have to intersect at the center of dilation, but they cannot intersect because they are parallel. We also showed that a directed line segment maps to a directed line segment pointing in the same direction, so it is not possible for U to map to Y and also V map to X under a dilation.

Problem Set Sample Solutions

1. **Draw the dilation of parallelogram $ABCD$ from center O using the scale factor $r = 2$, and then answer the questions that follow.**

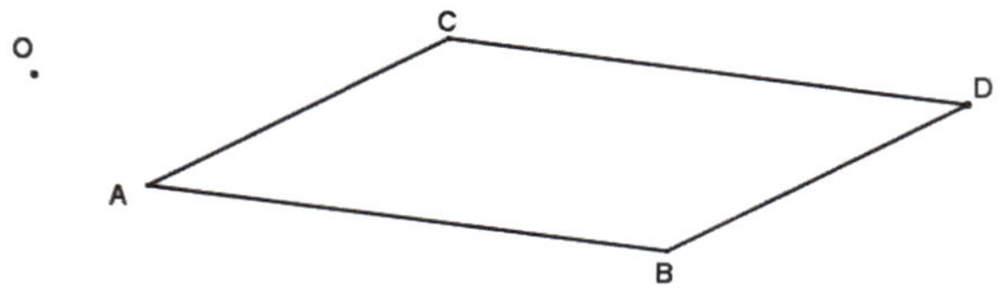

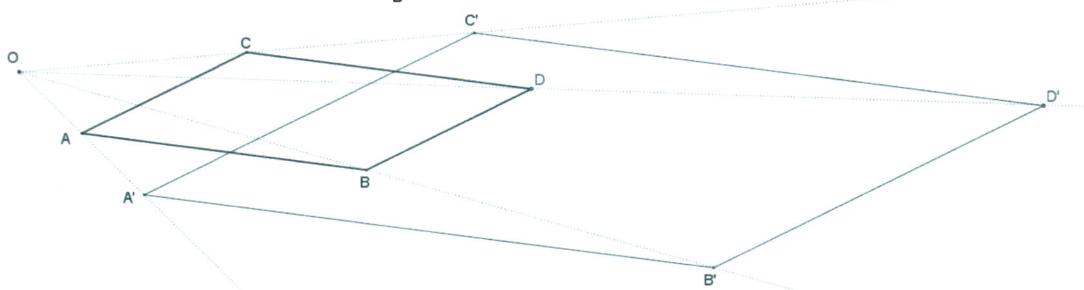

   a. **Is the image $A'B'C'D'$ also a parallelogram? Explain.**

      Yes By the dilation theorem, $\overline{AC} \parallel \overline{A'C'}$ and $\overline{BD} \parallel \overline{B'D'}$, and because $\overline{AC} \parallel \overline{BD}$, it follows that $\overline{A'C'} \parallel \overline{B'D'}$. A similar argument follows for the other pair of opposite sides, so with $\overline{A'C'} \parallel \overline{B'D'}$ and $\overline{A'B'} \parallel \overline{C'D'}$, $A'B'C'D'$ is a parallelogram.

   b. **What do parallel lines seem to map to under a dilation?**

      Parallel lines map to parallel lines under dilations.

2. **Given parallelogram $ABCD$ with $A(-8, 1)$, $B(2, -4)$, $C(-3, -6)$, and $D(-13, -1)$, perform a dilation of the plane centered at the origin using the following scale factors.**

   a. **Scale factor $\frac{1}{2}$**

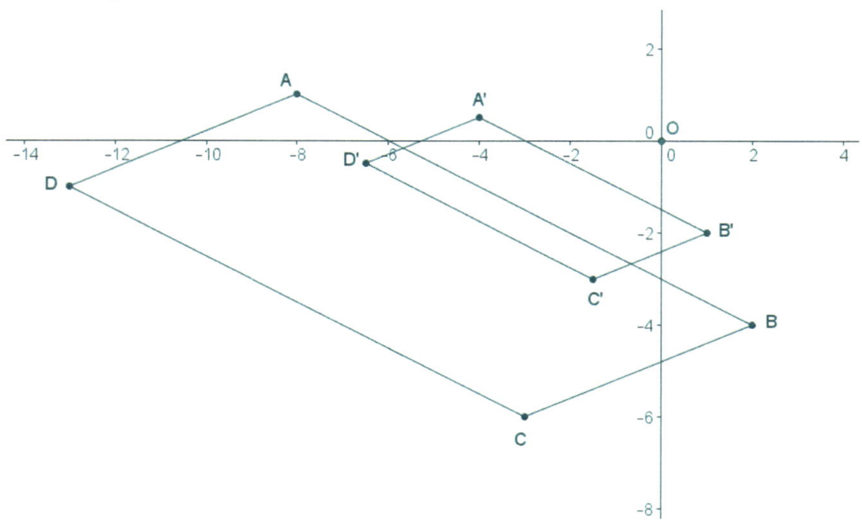

b. Scale factor 2

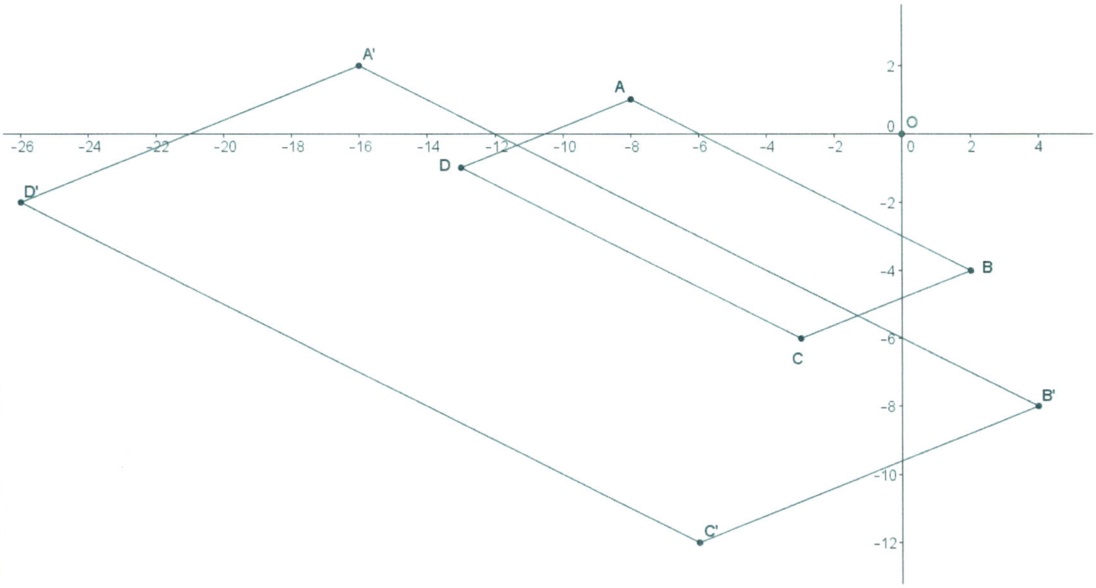

c. **Are the images of parallel line segments under a dilation also parallel? Use your graphs to support your answer.**

The slopes of sides $\overline{AB}$ and $\overline{CD}$ in the original graph are $-\frac{1}{2}$, and the slopes of each of the images of those sides under the dilations are also $-\frac{1}{2}$. The slopes of sides $\overline{AD}$ and $\overline{BC}$ in the original graph are $\frac{2}{5}$, and the slopes of each of the images of those sides under the dilations are also $\frac{2}{5}$. This informally verifies that the images of parallel line segments are also parallel line segments.

3. **In Lesson 7, Example 3, we proved that a line segment PQ, where O, P, and Q are the vertices of a triangle, maps to a line segment $P'Q'$ under a dilation with a scale factor $r < 1$. Using a similar proof, prove that for O not on $\overleftrightarrow{PQ}$, a dilation with center O and scale factor $r > 1$ maps a point R on $\overline{PQ}$ to a point R' on line PQ.**

The proof follows almost exactly the same as the proof of Example 3, but using the following diagram instead:

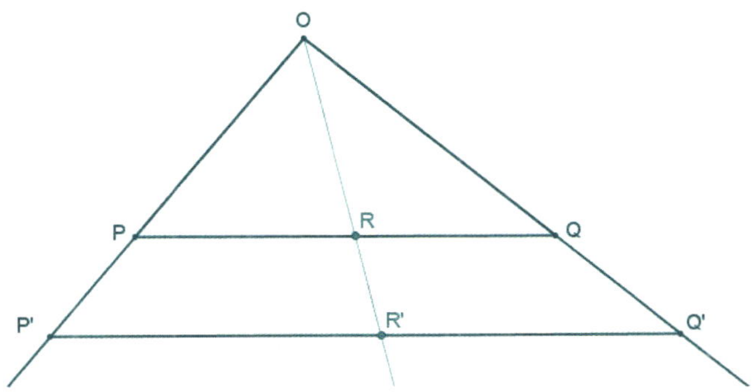

4. On the plane, $\overline{AB} \parallel \overline{A'B'}$ and $\overleftrightarrow{AB} \neq \overleftrightarrow{A'B'}$. Describe a dilation mapping $\overline{AB}$ to $\overline{A'B'}$. (Hint: There are 2 cases to consider.)

Case 1: $\overline{AB}$ and $\overline{A'B'}$ are parallel directed line segments oriented in the same direction.

By the dilation theorem for segments, there is a dilation mapping A and B to A' and B', respectively.

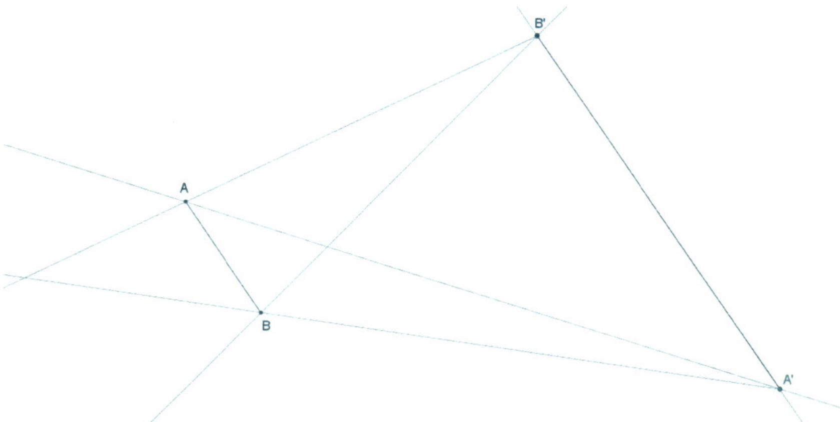

Case 2: $\overline{AB}$ and $\overline{A'B'}$ are parallel directed line segments oriented in the opposite directions.

We showed that directed line segments map to directed line segments that are oriented in the same direction, so there is a dilation mapping the parallel segments but only where the dilation maps A and B to B' and A', respectively.

Note to the teacher: Students may state that a scale factor $r < 0$ would produce the figure below where the center of a dilation is between the segments; however, this violates the definition of dilation. In such a case, discuss the fact that a scale factor must be greater than 0; otherwise, it would create negative distance, which of course does not make mathematical sense.

5. Only one of Figures A, B, or C below contains a dilation that maps A to A' and B to B'. Explain for each figure why the dilation does or does not exist. For each figure, assume that $\overleftrightarrow{AB} \neq \overleftrightarrow{A'B'}$.

a.

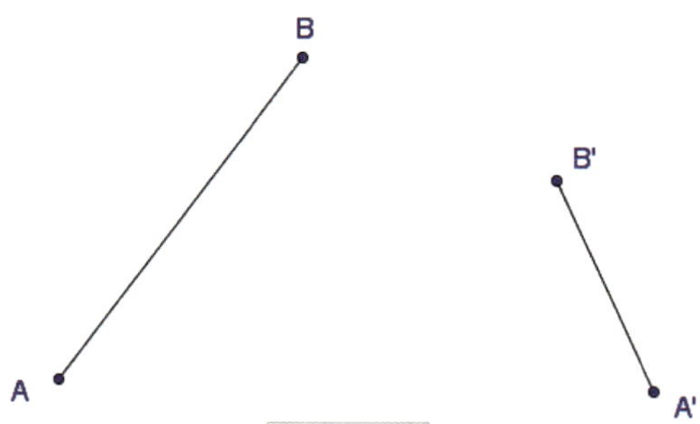

Figure A

By the dilation theorem, if $\overline{AB}$ and $\overline{A'B'}$ are line segments in the plane of different lengths, then there is a dilation that maps one to the other if and only if $\overleftrightarrow{AB} = \overleftrightarrow{A'B'}$ or $\overleftrightarrow{AB} \parallel \overleftrightarrow{A'B'}$. The segments do not lie in the same line and are also not parallel, so there is no dilation mapping A to A' and B to B'.

EUREKA MATH

b.

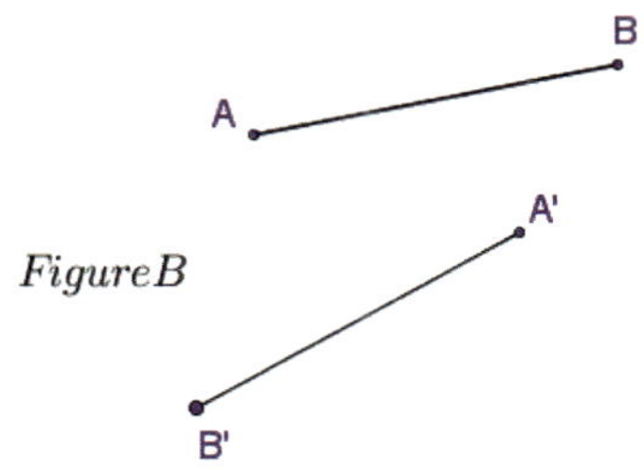

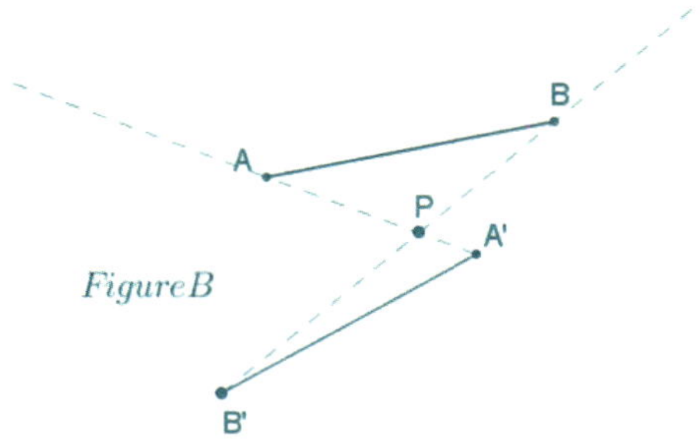

By the dilation theorem, if $\overline{AB}$ and $\overline{A'B'}$ are line segments in the plane of different lengths, then there is a dilation that maps one to the other if and only if $\overleftrightarrow{AB} = \overleftrightarrow{A'B'}$ or $\overleftrightarrow{AB} \parallel \overleftrightarrow{A'B'}$. The diagram shows that $\overline{AB}$ and $\overline{A'B'}$ do not lie in the same line, and it can also be seen that the line segments are not parallel. Furthermore, for the dilation to exist that maps A to A' and B to B', the center of dilation P would need to be between the segments, which violates the definition of dilation. Therefore, there is no dilation in Figure B mapping A and B to A' and B', respectively.

c.

Figure C

By the dilation theorem, if $\overline{AB}$ and $\overline{A'B'}$ are line segments in the plane of different lengths, then there is a dilation that maps one to the other if and only if $\overleftrightarrow{AB} = \overleftrightarrow{A'B'}$ or $\overleftrightarrow{AB} \parallel \overleftrightarrow{A'B'}$. Assuming that $\overleftrightarrow{AB} \neq \overleftrightarrow{A'B'}$, the segments are shown to lie in the same line; therefore, the dilation exists.

Lesson 8: How Do Dilations Map Lines, Rays, and Circles?

Student Outcomes

- Students prove that a dilation maps a ray to a ray, a line to a line, and a circle to a circle.

Lesson Notes

The objective in Lesson 7 was to prove that a dilation maps a segment to a segment; in Lesson 8, students prove that a dilation maps a ray to a ray, a line to a line, and a circle to a circle. An argument similar to that in Lesson 7 can be made to prove that a ray maps to a ray; allow students the opportunity to establish this argument as independently as possible.

Classwork

Opening (2 minutes)

As in Lesson 7, remind students of their work in Grade 8, when they studied the multiplicative effect that dilation has on points in the coordinate plane when the center is at the origin. Direct students to consider what happens to the dilation of a ray that is not on the coordinate plane.

- Today we are going to show how to prove that dilations map rays to rays, lines to lines, and circles to circles.
- Just as we revisited what dilating a segment on the coordinate plane is like, so we could repeat the exercise here. How would you describe the effect of a dilation on a point (x, y) on a ray about the origin by scale factor r in the coordinate plane?
  - *A point (x, y) on the coordinate plane dilated about the origin by scale factor r would then be located at (rx, ry).*
- Of course, we must now consider what happens in the plane versus the coordinate plane.

Opening Exercise (3 minutes)

MP.3

Opening Exercise

a. **Is a dilated ray still a ray? If the ray is transformed under a dilation, explain how.**

Accept any reasonable answer. The goal of this line of questioning is for students to recognize that a segment dilates to a segment that is r times the length of the original.

MP.3

b. Dilate the ray $\overrightarrow{PQ}$ by a scale factor of 2 from center O.

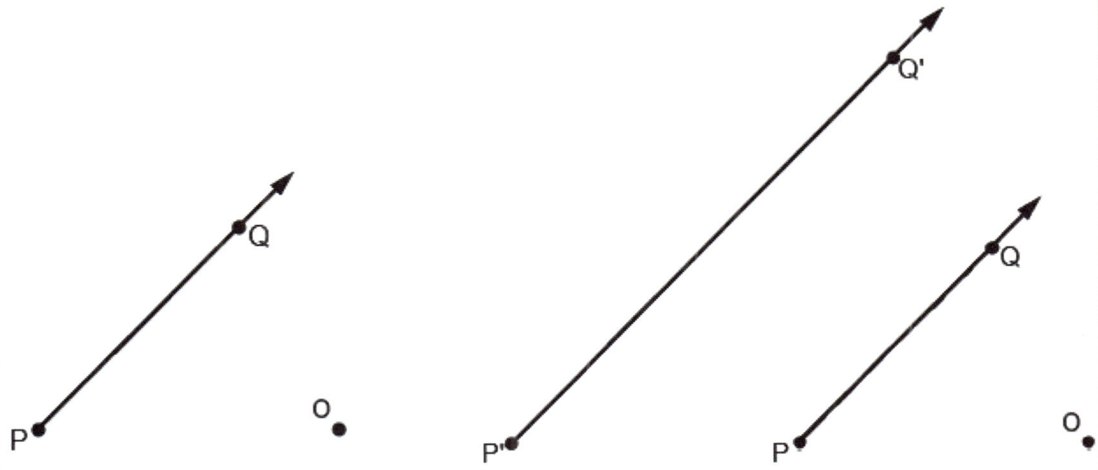

i. Is the figure $\overrightarrow{P'Q'}$ a ray?

Yes. The dilation of ray PQ produces a ray $P'Q'$.

ii. How, if at all, has the segment PQ been transformed?

The segment $P'Q'$ in ray $P'Q'$ is twice the length of the segment PQ in ray PQ. The segment has increased in length according to the scale factor of dilation.

iii. Will a ray always be mapped to a ray? Explain how you know.

Students will most likely say that a ray always maps to a ray; they may defend their answers by citing the definition of a dilation and its effect on the center and a given point.

- We are going to use our work in Lesson 7 to help guide our reasoning today.
- In proving why dilations map segments to segments, we considered three variations of the position of the center relative to the segment and value of the scale factor of dilation:
  (1) A center O and scale factor $r = 1$ and segment PQ,
  (2) A line PQ that does not contain the center O and scale factor $r \neq 1$, and
  (3) A line PQ that does contain the center O and scale factor $r \neq 1$.

Point out that the condition in case (1) does not specify the location of center O relative to the line that contains segment PQ; they can tell by this description that the condition does not impact the outcome and, therefore, is not more particularly specified.

- We use the setup of these cases to build an argument that shows dilations map rays to rays.

Examples 1–3 focus on establishing the dilation theorem of rays: A dilation maps a ray to a ray sending the endpoint to the endpoint. The arguments for Examples 1 and 2 are very similar to those of Lesson 7, Examples 1 and 3, respectively. Use discretion and student success with the Lesson 7 examples to consider allowing small-group work on the following Examples 1–2, possibly providing a handout of the solution once groups arrive at solutions of their own. Again, since the arguments are quite similar to those of Lesson 7, Examples 1 and 3, it is important to stay within the time allotments of each example. Otherwise, Examples 1–2 should be teacher led.

Example 1 (2 minutes)

Encourage students to draw upon the argument from Lesson 1, Example 1. This is intended to be a quick exercise; consider giving 30-second time warnings or using a visible timer.

> **Example 1**
>
> **Does a dilation about center O and scale factor $r = 1$ map $\overrightarrow{PQ}$ to $\overrightarrow{P'Q'}$? Explain.**
>
> ***A scale factor of $r = 1$ means that the ray and its image are equal. That is, the dilation does not enlarge or shrink the image of the figure but remains unchanged. Therefore, when the scale factor of a dilation is $r = 1$, then the dilation maps the ray to itself.***

Example 2 (6 minutes)

Encourage students to draw upon the argument for Lesson 1, Example 3.

> **Example 2**
>
> **The line that contains $\overrightarrow{PQ}$ does not contain point O. Does a dilation D about center O and scale factor $r \neq 1$ map every point of $\overrightarrow{PQ}$ onto a point of $\overrightarrow{P'Q'}$?**

> *Scaffolding:*
> - If students have difficulty following the logical progression of the proof, ask them instead to draw rays and dilate them by a series of different scale factors and then make generalizations about the results.
> - For students who are above grade level, ask them to attempt to prove the theorem independently.

MP.1

- A restatement of this problem is as follows: If R is a point on $\overrightarrow{PQ}$, then is $D(R)$ a point on $\overrightarrow{P'Q'}$? Also, if a point S' lies on the ray $\overrightarrow{P'Q'}$, then is there a point S on the ray $\overrightarrow{PQ}$ such that $D(S) = S'$?
- Consider the case where the center O is not in the line that contains $\overrightarrow{PQ}$, and the scale factor is $r \neq 1$. Then, points O, P, and Q form $\triangle OPQ$.

Draw the following figure on the board.

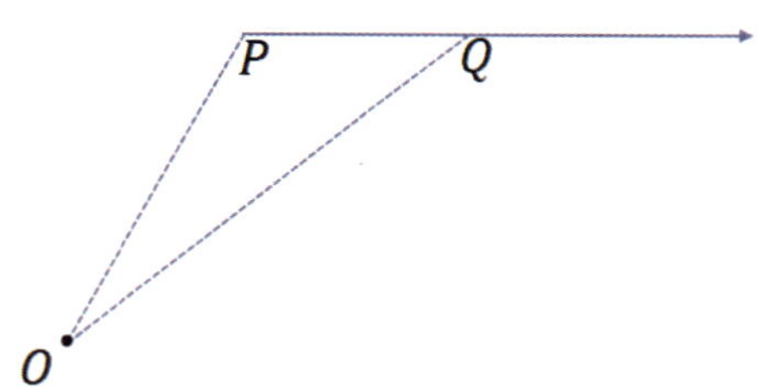

- We examine the case with a scale factor $r > 1$; the proof for $0 < r < 1$ is similar.
- Under a dilation about center O and $r > 1$, P goes to P', and Q goes to Q'; $OP' = r \cdot OP$ and $OQ' = r \cdot OQ$. What conclusion can we draw from these lengths?
- Summarize the proof and result to a partner.

Draw the following figure on the board.

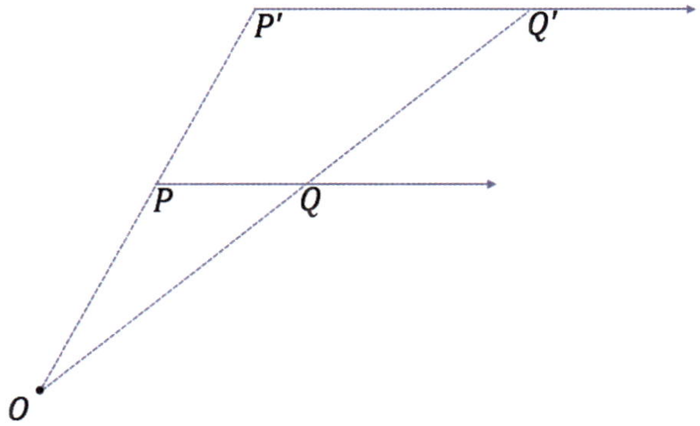

- *We can rewrite each length relationship as* $\frac{OP'}{OP} = \frac{OQ'}{OQ} = r$.

- By the triangle side splitter theorem, what else can we now conclude?
  - *The segment PQ splits $\triangle OP'Q'$ proportionally.*
  - *The lines that contain $\overrightarrow{PQ}$ and $\overrightarrow{P'Q'}$ are parallel;* $\overleftrightarrow{PQ} \parallel \overleftrightarrow{P'Q'}$; therefore, $\overrightarrow{PQ} \parallel \overrightarrow{P'Q'}$.

MP.7

- By the dilation theorem for segments (Lesson 7), the dilation from O of the segment PQ is the segment $P'Q'$, that is, $D(\overline{PQ}) = \overline{P'Q'}$ as sets of points. Therefore, we need only consider an arbitrary point R on $\overrightarrow{PQ}$ that lies outside of $\overline{PQ}$. For any such point R, what point is contained in the segment PR?
  - *The point Q*
- Let $R' = D(R)$.

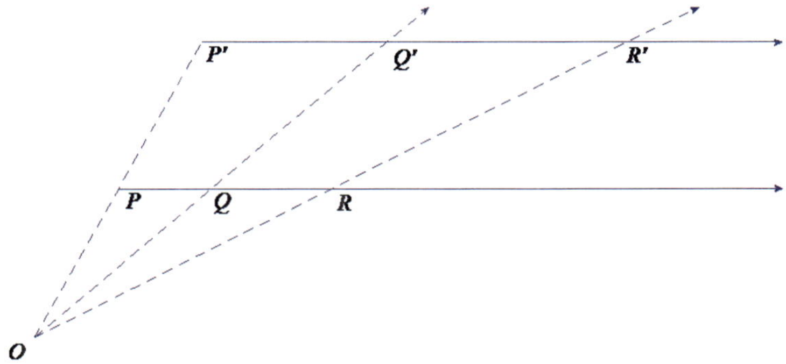

- By the dilation theorem for segments, the dilation from O of the segment PR is the segment $P'R'$. Also by the dilation theorem for segments, the point $D(Q) = Q'$ is a point on segment $P'R'$. Therefore, $\overrightarrow{P'Q'}$ and $\overrightarrow{P'R'}$ must be the same ray. In particular, R' is a point on $\overrightarrow{P'Q'}$, which was what we needed to show.

To show that for every point S' on $\overrightarrow{P'Q'}$ there is a point S on $\overrightarrow{PQ}$ such that $D(S) = S'$, consider the dilation from center O with scale factor $\frac{1}{r}$ (the inverse of the dilation D). This dilation maps S' to a point S on $\overrightarrow{PR}$ by the same reasoning as above. Then, $D(S) = S'$.

- We conclude that the points of $\overrightarrow{PQ}$ are mapped onto the points of $\overrightarrow{P'Q'}$ and, more generally, that dilations map rays to rays.

Example 3 (12 minutes)

Encourage students to draw upon the argument for Lesson 1, Example 3.

> **Example 3**
>
> **The line that contains $\overrightarrow{PQ}$ contains point O. Does a dilation about center O and scale factor r map $\overrightarrow{PQ}$ to $\overrightarrow{P'Q'}$?**

- Consider the case where the center O belongs to the line that contains $\overrightarrow{PQ}$.

> **a. Examine the case where the endpoint P of $\overrightarrow{PQ}$ coincides with the center O of the dilation.**

- If the endpoint P of $\overrightarrow{PQ}$ coincides with the center O, what can we say about $\overrightarrow{PQ}$ and $\overrightarrow{OQ}$?

Ask students to draw what this looks like, and draw the following on the board after giving them a head start.

  - *All the points on $\overrightarrow{PQ}$ also belong to $\overrightarrow{OQ}$; $\overrightarrow{PQ} = \overrightarrow{OQ}$.*
- By definition, a dilation always sends its center to itself. What are the implications for the dilation of O and P?
  - *Since a dilation always sends its center to itself, then $O = P = O' = P'$.*

MP.1

- Let X be a point on $\overrightarrow{OQ}$ so that $X \neq O$. What happens to X under a dilation about O with scale factor r?
  - *The dilation sends X to X' on $\overrightarrow{OX}$, and $OX' = r \cdot OX$.*

Ask students to draw what the position of X and X' might look like if $r > 1$ or $r < 1$. Points may move farther away ($r > 1$) or move closer to the center ($r < 1$). They do not draw this for every case; rather, it is a reminder up front.

$r > 1$

$r < 1$

- Since O, X, and Q are collinear, then O, X', and Q are also collinear; X' is on the ray OX, which coincides with $\overrightarrow{OQ}$ by definition of dilation.
- Therefore, a dilation of $\overrightarrow{OQ}$ is $\overrightarrow{PQ}$ (or when the endpoint of P of $\overrightarrow{PQ}$ coincides with the center O) about center O and scale factor r maps $\overrightarrow{O'Q'}$ to $\overrightarrow{P'Q'}$. We have answered the bigger question that a dilation maps a ray to a ray.

> **b. Examine the case where the endpoint P of $\overrightarrow{PQ}$ is between O and Q on the line containing O, P, and Q.**

Ask students to draw what this looks like, and draw the following on the board after giving them a head start.

- We already know from the previous case that the dilation of the ray $\overrightarrow{OQ}$ maps onto itself. All we need to show is that any point on the ray that is farther away from the center than P maps to a point that is farther away from the center than P'.
- Let X be a point on $\overrightarrow{PQ}$ so that $X \neq P$. What can be concluded about the relative lengths of $\overline{OP}$ and $\overline{OX}$?

Ask students to draw what this looks like, and draw the following on the board after giving them a head start. The following is one possibility.

  - *$OP < OX$*
- Describe how the lengths OP' and OX' compare once a dilation about center O and scale factor r sends P to P' and X to X'.
  - *$OP' = r \cdot OP$ and $OX' = r \cdot OX$*
  - *Multiplying both sides of $OP < OX$ by $r > 0$ gives $r \cdot OP < r \cdot OX$, so $OP' < OX'$.*

Ask students to draw what this might look like if $r > 1$, and draw the following on the board after giving them a head start.

MP.1

- Therefore, we have shown that any point on the ray $\overrightarrow{PQ}$ that is farther away from the center than P maps to a point that is farther away from the center than P'. In this case, we still see that a dilation maps a ray to a ray.

> c. **Examine the remaining case where the center O of the dilation and point Q are on the same side of P on the line containing O, P, and Q.**

- Now consider the relative position of O and Q on $\overrightarrow{PQ}$. We use an additional point R as a reference point so the O is between P and R.

Draw the following on the board; these are all the ways that O and Q are on the same side of P.

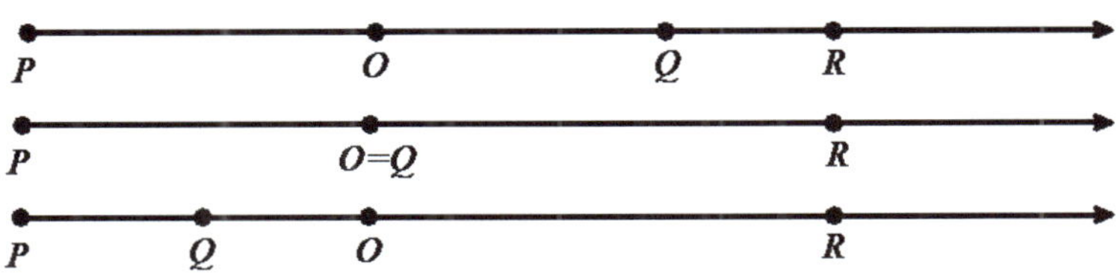

MP.1

- By case (a), we know that a dilation with center O maps $\overrightarrow{OR}$ to itself.
- Also, by our work in Lesson 7 on how dilations map segments to segments, we know that $\overline{PO}$ is taken to $\overline{P'O}$, where P' lies on $\overrightarrow{OP}$.
- The union of the $\overline{P'O}$ and $\overrightarrow{OR}$ is $\overrightarrow{P'R}$. So, the dilation maps $\overrightarrow{PQ}$ to $\overrightarrow{P'R}$.
- Since Q' is a point on $\overrightarrow{P'R}$ and $Q' \neq P'$, we see $\overrightarrow{P'R} = \overrightarrow{P'Q'}$. Therefore, in this case, we still see that $\overrightarrow{PQ}$ maps to $\overrightarrow{P'Q'}$ under a dilation.

Example 4 (6 minutes)

In Example 4, students prove the dilation theorem for lines: A dilation maps a line to a line. If the center O of the dilation lies on the line or if the scale factor r of the dilation is equal to 1, then the dilation maps the line to the same line. Otherwise, the dilation maps the line to a parallel line.

The dilation theorem for lines can be proved using arguments similar to those used in Examples 1–3 for rays and to Examples 1–3 in Lesson 7 for segments. Consider asking students to prove the theorem on their own as an exercise, especially if time is an issue, and then provide the following proof:

MP.3

- We have just seen that dilations map rays to rays. How could we use this to reason that dilations map lines to lines?
  - *$\overleftrightarrow{PQ}$ is the union of the two rays $\overrightarrow{PQ}$ and $\overrightarrow{QP}$. Dilate rays $\overrightarrow{PQ}$ and $\overrightarrow{QP}$; the dilation yields rays $\overrightarrow{P'Q'}$ and $\overrightarrow{Q'P'}$. $\overleftrightarrow{P'Q'}$ is the union of the two rays $\overrightarrow{P'Q'}$ and $\overrightarrow{Q'P'}$. Since the dilation maps the rays $\overrightarrow{PQ}$ and $\overrightarrow{QP}$ to the rays $\overrightarrow{P'Q'}$ and $\overrightarrow{Q'P'}$, respectively, then the dilation maps $\overleftrightarrow{PQ}$ to the $\overleftrightarrow{P'Q'}$.*

Example 5 (8 minutes)

In Example 5, students prove the dilation theorem for circles: A dilation maps a circle to a circle and maps the center to the center. Students need the dilation theorem for circles for proving that all circles are similar in Module 5 (**G-C.A.1**).

> **Example 5**
>
> **Does a dilation about a center O and scale factor r map a circle of radius R onto another circle?**
>
> a. **Examine the case where the center of the dilation coincides with the center of the circle.**

- We first do the case where the center of the dilation is also the center of the circle. Let C be a circle with center O and radius R.

Draw the following figure on the board.

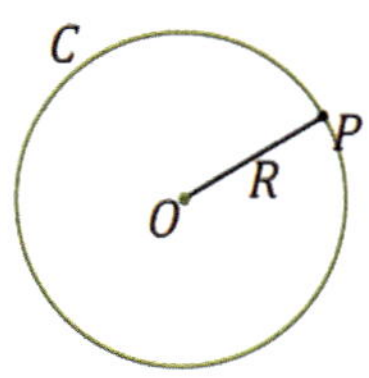

- If the center of the dilation is also O, then every point P on the circle is sent to a point P' on $\overrightarrow{OP}$ so that $OP' = r \cdot OP = rR$; that is, the point goes to the point P' on the circle C' with center O and radius rR.

- We also need to show that every point on C' is the image of a point from C: For every point P' on circle C', put a coordinate system on $\overleftrightarrow{OP'}$ such that $\overrightarrow{OP'}$ corresponds to the nonnegative real numbers with zero corresponding to point O (by the ruler axiom). Then, there exists a point P on $\overrightarrow{OP'}$ such that $OP = R$; that is, P is a point on the circle C that is mapped to P' by the dilation.

Draw the following figure on the board.

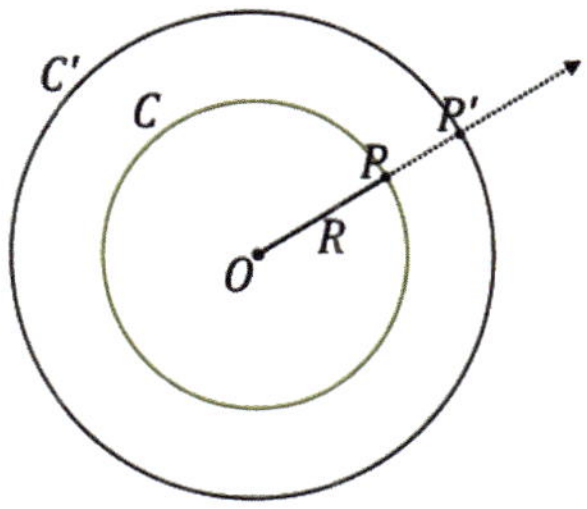

- Effectively, a dilation moves every point on a circle toward or away from the center the same amount, so the dilated image is still a circle. Thus, the dilation maps the circle C to the circle C'.
- Circles that share the same center are called *concentric circles.*

b. Examine the case where the center of the dilation is not the center of the circle; we call this the *general case*.

- The proof of the general case works no matter where the center of the dilation is. We can actually use this proof for case (a), when the center of the circle coincides with the center of dilation.
- Let C be a circle with center O and radius R. Consider a dilation with center D and scale factor r that maps O to O'. We will show that the dilation maps the circle C to the circle C' with center O' and radius rR.

Draw the following figure on the board.

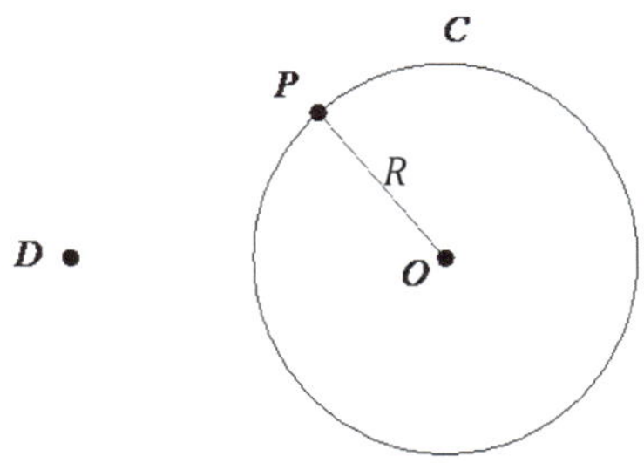

- If P is a point on circle C and the dilation maps P to P', the dilation theorem implies that $O'P' = rOP = rR$. So, P' is on circle C'.
- We also need to show that every point of C' is the image of a point from C. There are a number of ways to prove this, but we will follow the same idea that we used in part (a). For a point P' on circle C' that is not on $\overleftrightarrow{DO'}$, consider $\overrightarrow{O'P'}$ (the case when P' is on line DO' is straightforward). Construct line ℓ through O such that $\ell \parallel \overrightarrow{O'P'}$, and let A be a point on ℓ that is in the same half-plane of $\overleftrightarrow{DO'}$ as P'. Put a coordinate system on ℓ such that the ray $\overrightarrow{OA}$ corresponds to the nonnegative numbers with zero corresponding to point O (by the ruler axiom). Then, there exists a point P on $\overrightarrow{OA}$ such that $OP = R$, which implies that P is on the circle C. By the dilation theorem, P is mapped to the point P' on the circle C'.

- The diagram below shows how the dilation maps points P, Q, R, S, and T of circle C. Ask students to find point W on circle C that is mapped to point W' on circle C'.

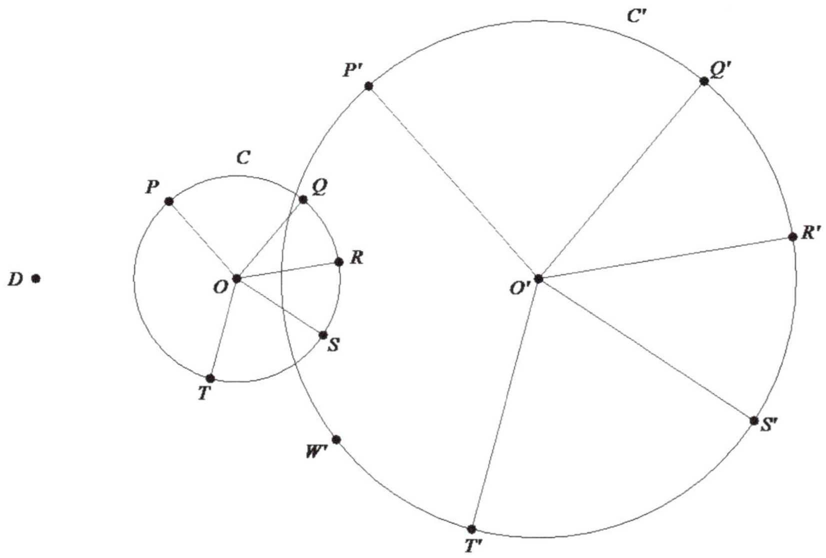

Closing (1 minute)

Ask students to respond to the following question and summarize the key points of the lesson.

- How are the proofs for the dilation theorems on segments, rays, and lines similar to each other?

Theorems addressed in this lesson:

- **DILATION THEOREM FOR RAYS**: A dilation maps a ray to a ray sending the endpoint to the endpoint.
- **DILATION THEOREM FOR LINES**: A dilation maps a line to a line. If the center O of the dilation lies on the line or if the scale factor r of the dilation is equal to 1, then the dilation maps the line to the same line. Otherwise, the dilation maps the line to a parallel line.
- **DILATION THEOREM FOR CIRCLES**: A dilation maps a circle to a circle and maps the center to the center.

> Lesson Summary
>
> - **DILATION THEOREM FOR RAYS: A dilation maps a ray to a ray sending the endpoint to the endpoint.**
> - **DILATION THEOREM FOR LINES: A dilation maps a line to a line. If the center O of the dilation lies on the line or if the scale factor r of the dilation is equal to 1, then the dilation maps the line to the same line. Otherwise, the dilation maps the line to a parallel line.**
> - **DILATION THEOREM FOR CIRCLES: A dilation maps a circle to a circle and maps the center to the center.**

Exit Ticket (5 minutes)

Name ______________________________ Date ______________

Lesson 8: How Do Dilations Map Lines, Rays, and Circles?

Exit Ticket

Given points O, S, and T below, complete parts (a)–(e):

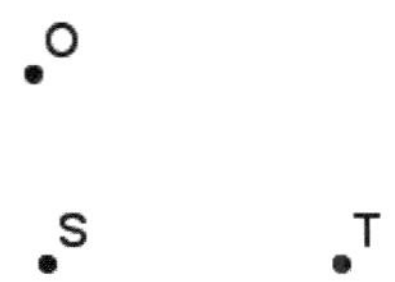

a. Draw rays $\overrightarrow{ST}$ and $\overrightarrow{TS}$. What is the union of these rays?

b. Dilate $\overrightarrow{ST}$ from O using scale factor $r = 2$. Describe the image of $\overrightarrow{ST}$.

c. Dilate $\overrightarrow{TS}$ from O using scale factor $r = 2$. Describe the image of $\overrightarrow{TS}$.

d. What does the dilation of the rays in parts (b) and (c) yield?

e. Dilate circle C with radius TS from O using scale factor $r = 2$.

Exit Ticket Sample Solutions

Given points O, S, and T below, complete parts (a)–(e):

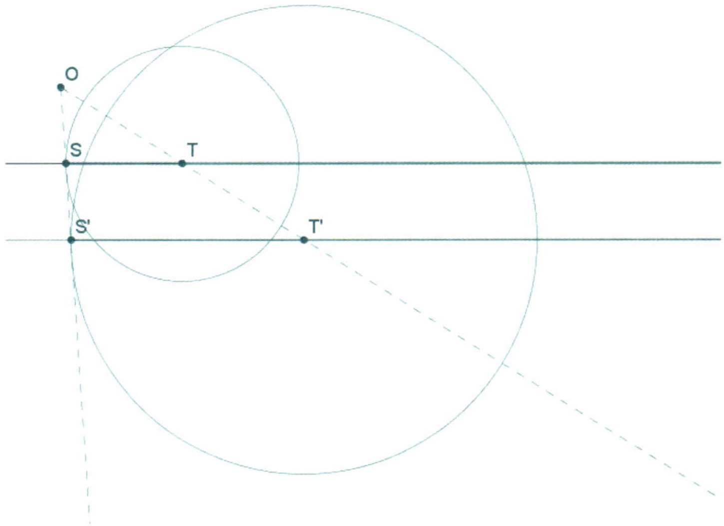

a. **Draw rays $\overrightarrow{ST}$ and $\overrightarrow{TS}$. What is the union of these rays?**

The union of $\overrightarrow{ST}$ and $\overrightarrow{TS}$ is line ST.

b. **Dilate $\overrightarrow{ST}$ from O using scale factor $r = 2$. Describe the image of $\overrightarrow{ST}$.**

The image of $\overrightarrow{ST}$ is $\overrightarrow{S'T'}$.

c. **Dilate $\overrightarrow{TS}$ from O using scale factor $r = 2$. Describe the image of $\overrightarrow{TS}$.**

The image of $\overrightarrow{TS}$ is $\overrightarrow{T'S'}$.

d. **What does the dilation of the rays in parts (b) and (c) yield?**

The dilation of rays $\overrightarrow{ST}$ and $\overrightarrow{TS}$ yields $\overleftrightarrow{S'T'}$.

e. **Dilate circle T with radius TS from O using scale factor $r = 2$.**

See diagram above.

Problem Set Sample Solutions

1. In Lesson 8, Example 2, you proved that a dilation with a scale factor $r > 1$ maps a ray PQ to a ray $P'Q'$. Prove the remaining case that a dilation with scale factor $0 < r < 1$ maps a ray PQ to a ray $P'Q'$.

   Given the dilation $D_{O,r}$, with $0 < r < 1$ maps P to P' and Q to Q', prove that $D_{O,r}$ maps $\overrightarrow{PQ}$ to $\overrightarrow{P'Q'}$.

   By the definition of dilation, $OP' = r(OP)$, and likewise, $\frac{OP'}{OP} = r$.

   By the dilation theorem, $\overline{P'Q'} \parallel \overline{PQ}$.

   Through two different points lies only one line, so $\overleftrightarrow{PQ} \parallel \overleftrightarrow{P'Q'}$.

   Draw point R on $\overrightarrow{PQ}$, and then draw $\overrightarrow{OR}$. Mark point R' at the intersection of $\overrightarrow{OR}$ and $\overrightarrow{P'Q'}$.

   By the triangle side splitter theorem, $\overline{P'R'}$ splits $\overline{OP}$ and $\overline{OR}$ proportionally, so $\frac{OR'}{OR} = \frac{OP'}{OP} = r$.
   Therefore, because R was chosen as an arbitrary point, $OR' = r(OR)$ for any point R on $\overrightarrow{PQ}$.

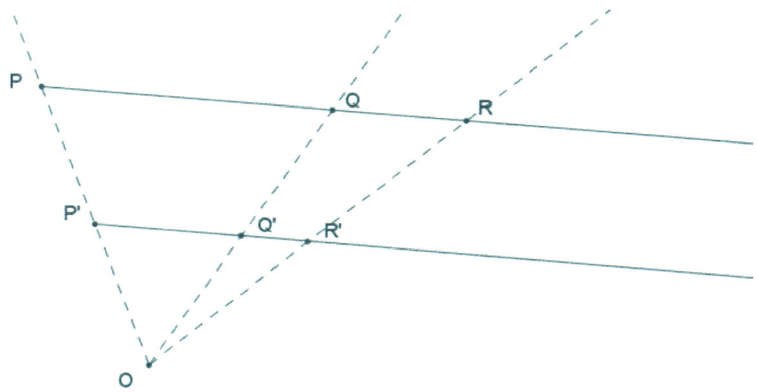

2. In the diagram below, $\overrightarrow{A'B'}$ is the image of $\overrightarrow{AB}$ under a dilation from point O with an unknown scale factor; A maps to A', and B maps to B'. Use direct measurement to determine the scale factor r, and then find the center of dilation O.

By the definition of dilation, $A'B' = r\,(AB)$, $OA' = r(OA)$, and $OB' = r(OB)$.

By direct measurement, $\frac{A'B'}{AB} = \frac{7}{4} = r$.

The images of A and B are pushed to the right on $\overleftrightarrow{AB}$ under the dilation, and $A'B' > AB$, so the center of dilation must lie on $\overleftrightarrow{AB}$ to the left of points A and B.

By the definition of dilation:

$$OA' = \frac{7}{4}(OA)$$
$$(OA + AA') = \frac{7}{4}(OA)$$
$$\frac{OA + AA'}{OA} = \frac{7}{4}$$
$$\frac{OA}{OA} + \frac{AA'}{OA} = \frac{7}{4}$$
$$1 + \frac{AA'}{OA} = \frac{7}{4}$$
$$\frac{AA'}{OA} = \frac{3}{4}$$
$$AA' = \frac{3}{4}(OA)$$
$$\frac{4}{3}(AA') = OA$$

3. Draw a line $\overleftrightarrow{AB}$, and dilate points A and B from center O where O is not on $\overleftrightarrow{AB}$. Use your diagram to explain why a line maps to a line under a dilation with scale factor r.

Two rays $\overrightarrow{AB}$ and $\overrightarrow{BA}$ on the points A and B point in opposite directions as shown in the diagram below. The union of the two rays is $\overleftrightarrow{AB}$. We showed that a ray maps to a ray under a dilation, so $\overrightarrow{AB}$ maps to $\overrightarrow{A'B'}$; likewise, $\overrightarrow{BA}$ maps to $\overrightarrow{B'A'}$. The dilation yields two rays $\overrightarrow{A'B'}$ and $\overrightarrow{B'A'}$ on the points A' and B' pointing in opposite directions. The union of the two rays is $\overleftrightarrow{A'B'}$; therefore, it is true that a dilation maps a line to a line.

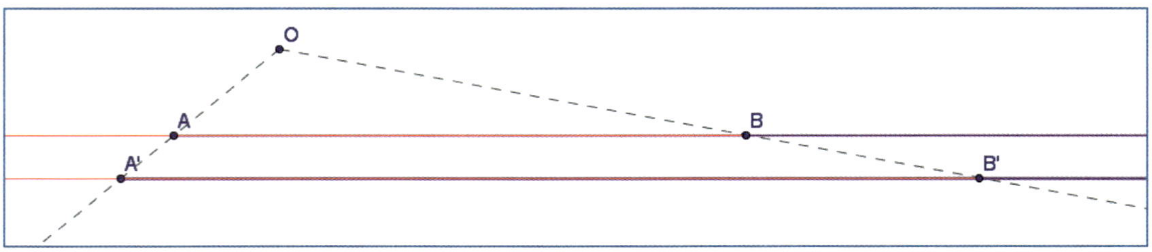

EUREKA MATH™

4. Let $\overline{AB}$ be a line segment, and let m be a line that is the perpendicular bisector of $\overline{AB}$. If a dilation with scale factor r maps $\overline{AB}$ to $\overline{A'B'}$ (sending A to A' and B to B') and also maps line m to line m', show that line m' is the perpendicular bisector of $\overline{A'B'}$.

Let P be a point on line m, and let the dilation send P to the point P' on line m'. Since P is on the perpendicular bisector of $\overline{AB}$, $PA = PB$. By the dilation theorem, $P'A' = rPA$ and $P'B' = rPB$. So, $P'A' = P'B'$, and P' is on the perpendicular bisector of $\overline{A'B'}$.

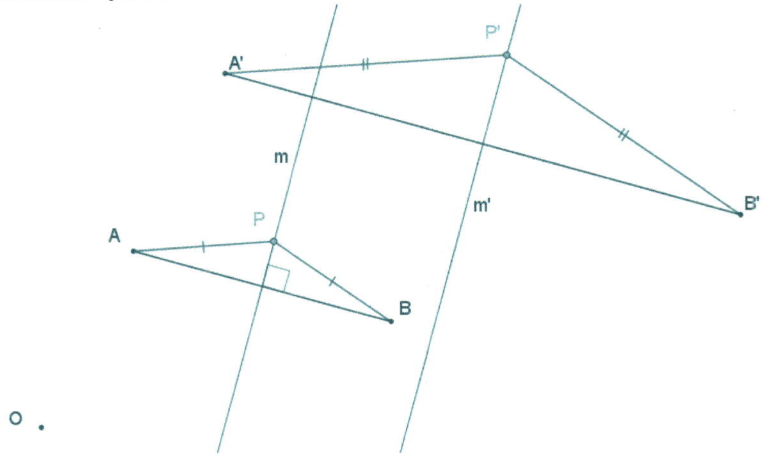

5. Dilate circle C with radius CA from center O with a scale factor $r = \frac{1}{2}$.

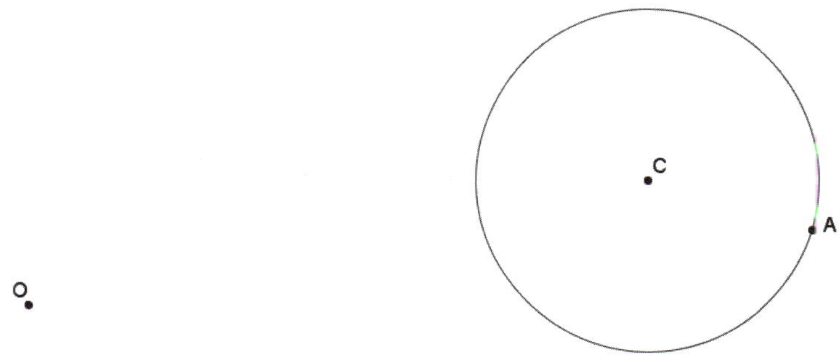

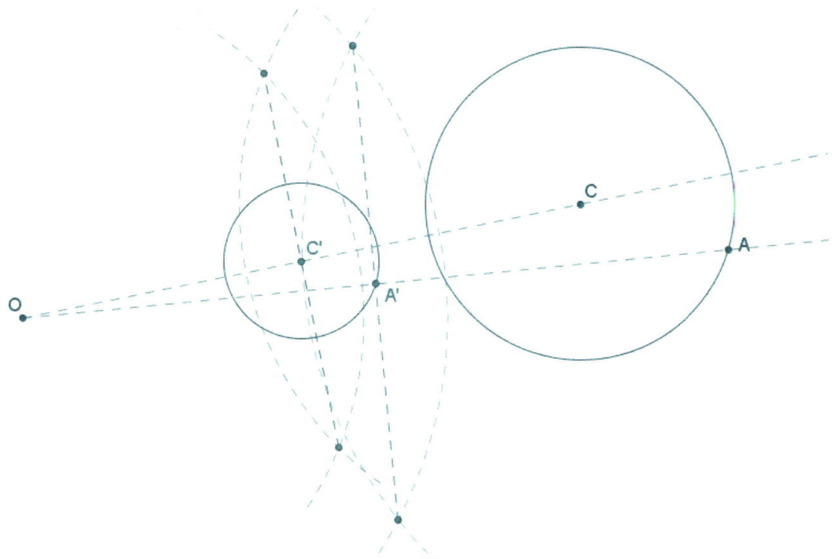

6. **In the picture below, the larger circle is a dilation of the smaller circle. Find the center of dilation O.**

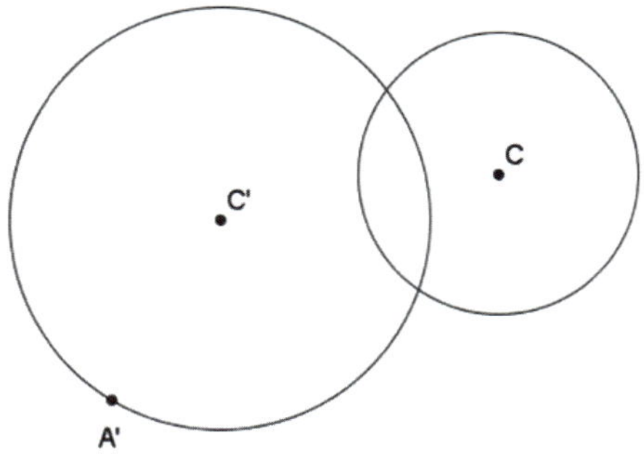

Draw $\overrightarrow{C'C}$. Center O lies on $\overrightarrow{C'C}$. Since A' is not on a line with both C and C', I can use the parallel method to find point A on circle C such that $\overleftrightarrow{CA} \parallel \overleftrightarrow{C'A'}$.

Under a dilation, a point and its image(s) lie on a ray with endpoint O, the center of dilation. Draw $\overrightarrow{A'A}$, and label the center of dilation $O = \overrightarrow{C'C} \cap \overrightarrow{A'A}$.

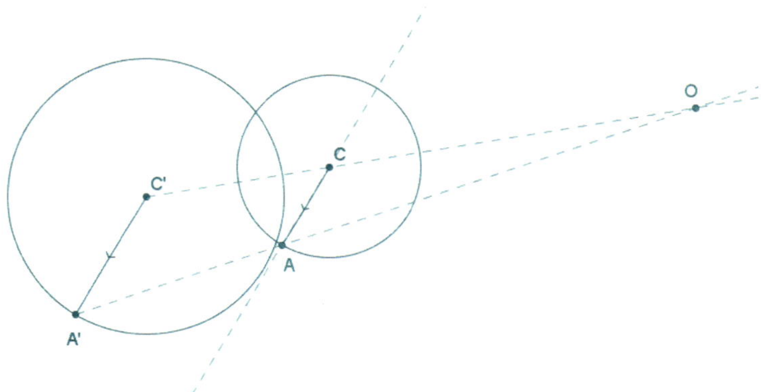

Lesson 9: How Do Dilations Map Angles?

Student Outcomes

- Students prove that dilations map an angle to an angle with equal measure.
- Students understand how dilations map triangles, squares, rectangles, trapezoids, and regular polygons.

Lesson Notes

In this lesson, students show that dilations map angles to angles of equal measure. The Exploratory Challenge requires students to make conjectures about how dilations map polygonal figures, specifically, the effect on side lengths and angle measures. The goal is for students to informally verify that dilations map angles to angles of equal measure. Further, students describe the effect dilations have on side lengths; for example, if side length $AC = 3.6$ and is dilated from a center with scale factor $r = 2$, then the dilated side $A'C' = 7.2$. The Discussion that follows the Exploratory Challenge focuses on the effect dilations have on angles. Students should already be familiar with the effect of dilation on lengths of segments, so the work with polygonal figures extends students' understanding of the effect of dilations on figures other than triangles. Consider extending the lesson over two days where on the first day students complete all parts of the Exploratory Challenge, and on the second day students work to prove their conjectures about how dilations map angles. The last Discussion of the lesson (dilating a square inscribed in a triangle) is optional and can be completed if class time permits.

This lesson highlights Mathematical Practice 3: Construct viable arguments and critique the reasoning of others. Throughout the lesson, students are asked to make a series of conjectures and justify them by experimenting with diagrams and direct measurements.

Classwork

Exploratory Challenge/Exercises 1–4 (13 minutes)

The Exploratory Challenge allows students to informally verify how dilations map polygonal figures in preparation for the Discussion that follows. Make clear to students that they must first make a conjecture about how the dilation affects the figure and then verify their conjectures by actually performing a dilation and directly measuring angles and side lengths. Consider having students share their conjectures and drawings with the class. It may be necessary to divide the class into four groups and assign each group one exercise to complete. When all groups are finished, they can share their results with the whole class.

Scaffolding:

Some groups of students may benefit from a teacher-led model of the first exercise. Additionally, use visuals or explicit examples, such as dilating triangle ABC, $A(3,1)$, $B(7,1)$, $C(3,5)$, by a variety of scale factors to help students make a conjecture.

MP.3

Exploratory Challenge/Exercises 1–4

1. **How do dilations map triangles?**
   a. **Make a conjecture.**

      A dilation maps a triangle to a triangle with the same angles, and all of the sides of the image triangle are proportional to the sides of the original triangle.

b. **Verify your conjecture by experimenting with diagrams and directly measuring angles and lengths of segments.**

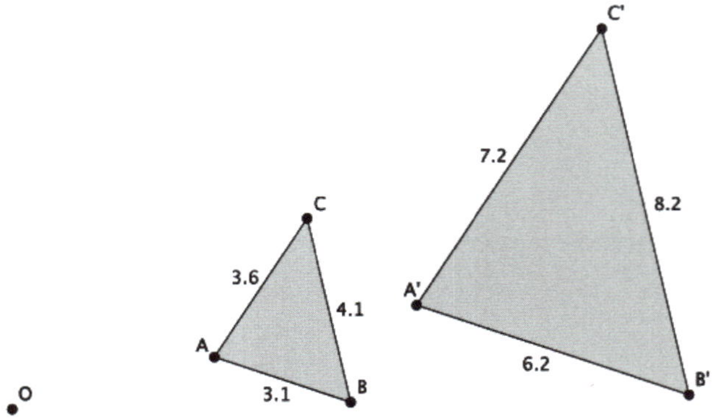

$$\frac{7.2}{3.6} = \frac{8.2}{4.1} = \frac{6.2}{3.1} = 2$$

The value of the ratios of the lengths of the dilated triangle to the original triangle is equal to the scale factor. The angles map to angles of equal measure; all of the angles in the original triangle are dilated to angles equal in measure to the corresponding angles in the dilated triangle.

2. **How do dilations map rectangles?**

a. **Make a conjecture.**

A dilation maps a rectangle to a rectangle so that the ratio of length to width is the same.

MP.3

b. **Verify your conjecture by experimenting with diagrams and directly measuring angles and lengths of segments.**

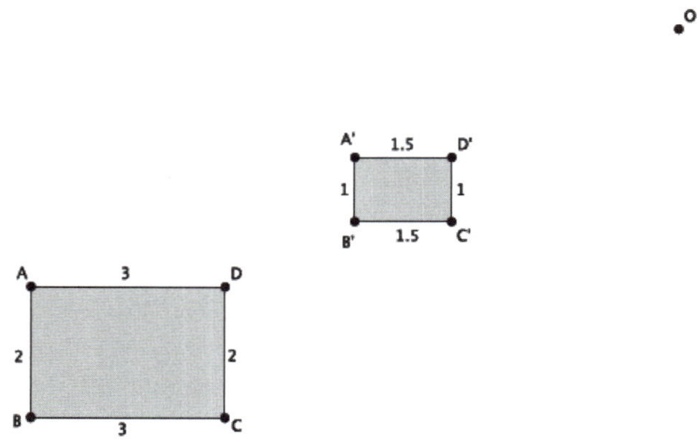

The original rectangle has a length to width ratio of $3:2$. The dilated rectangle has a length to width ratio of $1.5:1$. The values of the ratios are equal. Since angles map to angles of equal measure, all of the right angles in the original rectangle are dilated to right angles.

3. **How do dilations map squares?**

a. **Make a conjecture.**

A dilation maps a square with side length L to a square with side length rL where r is the scale factor of dilation. Since each segment of length L is mapped to a segment of length rL and each right angle is mapped to a right angle, then the dilation maps a square with side length L to a square with side length rL.

b. Verify your conjecture by experimenting with diagrams and directly measuring angles and lengths of segments.

Sample student drawing:

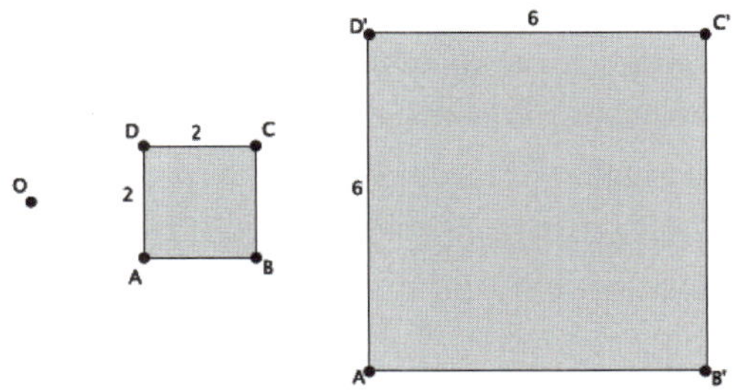

The side length of the original square is 2 units. The side length of the dilated square is 6 units. The side length of the dilated square is equal to the length of the original square multiplied by the scale factor. Since angles map to angles of equal measure, all of the right angles in the original square are dilated to right angles.

MP.3

4. How do dilations map regular polygons?

a. Make a conjecture.

A dilation maps a regular polygon with side length L to a regular polygon with side length rL where r is the scale factor of dilation. Since each segment of length L is mapped to a segment of length rL and each angle is mapped to an angle that is equal in measure, then the dilation maps a regular polygon with side length L to a regular polygon with side length rL.

b. Verify your conjecture by experimenting with diagrams and directly measuring angles and lengths of segments.

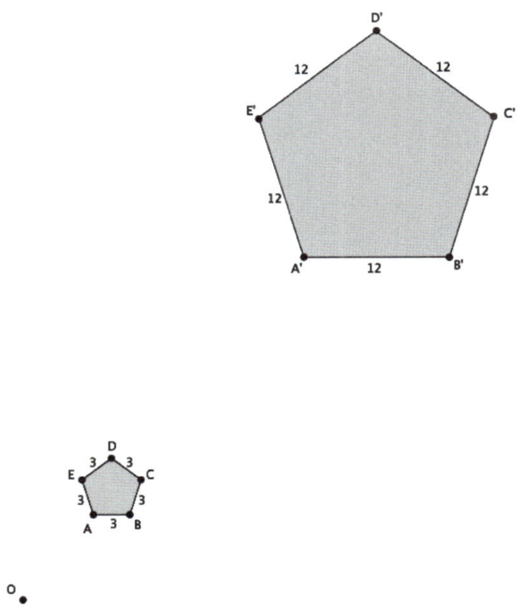

The side length of the original regular polygon is 3 units. The side length of the dilated regular polygon is 12 units. The side length of the dilated regular polygon is equal to the length of the original regular polygon multiplied by the scale factor. Since angles map to angles of equal measure, all of the angles in the original regular polygon are dilated to angles of each measure.

Discussion (7 minutes)

Begin the Discussion by debriefing the Exploratory Challenge. Elicit the information about the effect of dilations on angle measures and side lengths as described in the sample student responses above. Then, continue with the Discussion below.

- In Grade 8, we showed that under a dilation with a center at the origin and scale factor r, an angle formed by the union of two rays and the image of the angle would be equal in measure.
- The multiplicative effect that dilation has on points (when the origin is the center of dilation) was used to show that rays $\overrightarrow{QP}$ and $\overrightarrow{QR}$ map to rays $\overrightarrow{Q'P'}$ and $\overrightarrow{Q'R'}$, respectively. Then, facts about parallel lines cut by a transversal were used to prove that $m\angle PQR = m\angle P'Q'R'$.

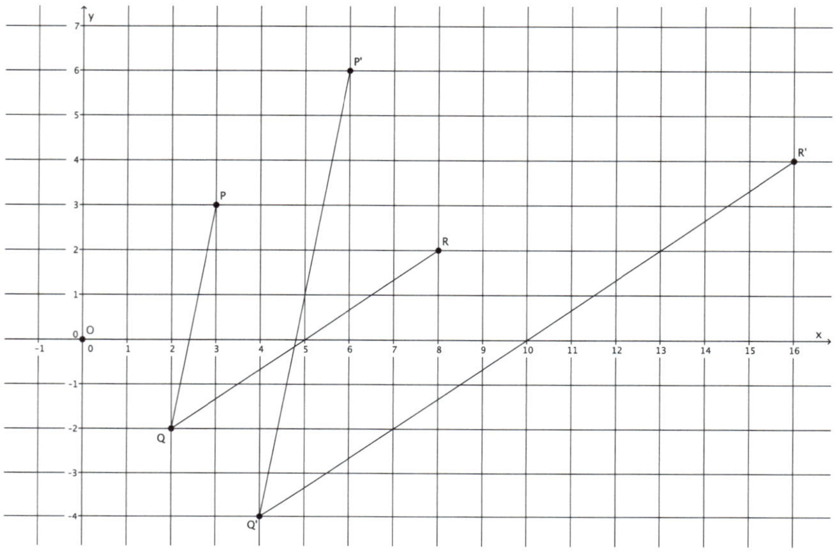

Scaffolding:

For some groups of students, a simpler example where the vertex of the angle is on the x-axis may aid understanding.

- Now that we know from the last two lessons that a dilation maps a segment to a segment, a ray to a ray, and a line to a line, we can prove that dilations map angles to angles of equal measure without the need for a coordinate system.

Exercises 5–6 (9 minutes)

MP.3

Provide students time to develop the proof that under a dilation, the measure of an angle and its dilated image are equal. Consider having students share their proofs with the class. If necessary, share the proof shown in Exercise 6 with the class as described in the Scaffolding box to the right.

Scaffolding:
- If students struggle, reference Exercise 5, particularly the corresponding angles, as a hint that guides students' thinking.
- Consider offering the proof to students as the work of a classmate and having students paraphrase the statements in the proof.

Exercises 5–6

5. **Recall what you learned about parallel lines cut by a transversal, specifically about the angles that are formed.**

   When parallel lines are cut by a transversal, then corresponding angles are equal in measure, alternate interior angles are equal in measure, and alternate exterior angles are equal in measure.

EUREKA MATH

MP.7

6. A dilation from center O by scale factor r maps $\angle BAC$ to $\angle B'A'C'$. Show that $m\angle BAC = m\angle B'A'C'$.

By properties of dilations, we know that a dilation maps a line to itself or a parallel line. We consider a case where the dilated rays of the angle are mapped to parallel rays as shown below, and two rays meet in a single point. Then, the line containing $\overrightarrow{AB}$ is parallel to the line containing $\overrightarrow{A'B'}$. $\overrightarrow{AC}$ is a transversal that cuts the parallel lines. Let D be the point of intersection of $\overrightarrow{AC}$ and $\overrightarrow{A'B'}$, and let E be a point to the right of D on $\overrightarrow{AC}$. Since corresponding angles of parallel lines are congruent and, therefore, equal in measure, then $m\angle BAC = m\angle B'DE$. The dilation maps $\overrightarrow{AC}$ to $\overrightarrow{A'C'}$, and the lines containing those rays are parallel. Then, $\overrightarrow{A'B'}$ is a transversal that cuts the parallel lines. Therefore, $m\angle B'DE = m\angle B'A'C'$. Since $m\angle BAC = m\angle B'DE$ and $m\angle B'DE = m\angle B'A'C'$, by the transitive property $m\angle BAC = m\angle B'A'C'$. Therefore, the dilation maps an angle to an angle of equal measure.

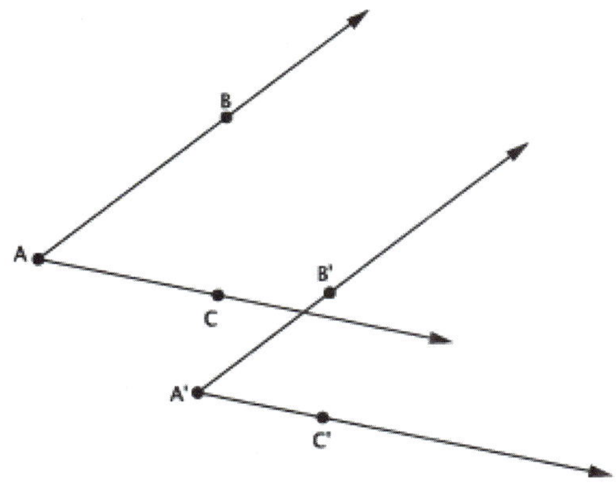

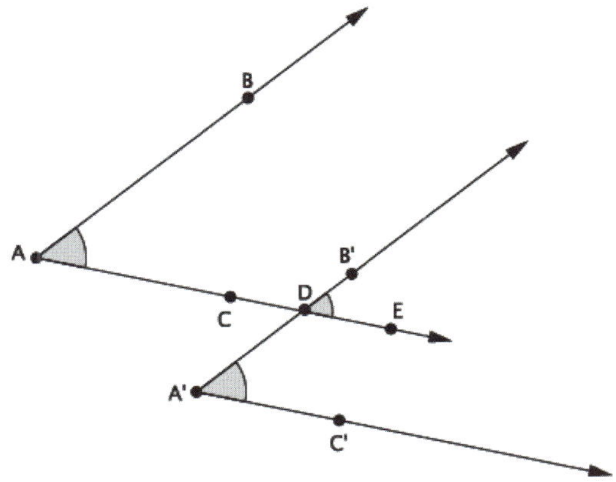

Discussion (4 minutes)

While the teacher leads the discussion, students should record the information about the dilation theorem and its proof in their student pages.

Discussion

The dilation theorem for angles is as follows:

DILATION THEOREM: A dilation from center O and scale factor r maps an angle to an angle of equal measure.

We have shown this when the angle and its image intersect at a single point, and that point of intersection is not the vertex of the angle.

- So far, we have seen the case when the angles intersect at a point. What are other possible cases that we need to consider?

Provide time for students to identify the two other cases: (1) The angles do not intersect at a point, and (2) the angles have vertices on the same ray. The teacher may need to give students this information if students cannot develop the cases on their own.

- We will cover another case where a dilation maps $\angle BAC$ to $\angle B'A'C'$. When the angle and its dilated image do not have intersecting rays (as shown below), how can we show that the angle and its dilated image are equal in measure?

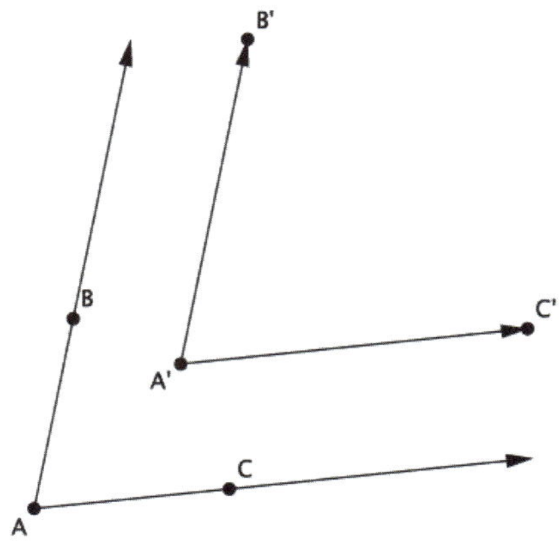

If students do not respond, the teacher may need to give the information in the bullet point below.

- *You can draw an auxiliary line and use the same reasoning to show that the angle and its image are equal in measure.*

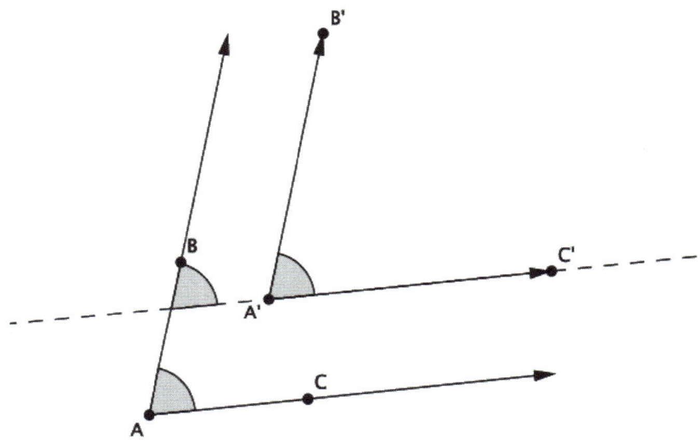

- Notice now that by drawing the auxiliary line, we have two angles that intersect at a point, much like we had in the Opening Exercise. Therefore, the same reasoning shows that $m\angle BAC = m\angle B'A'C'$.
- The only case left to consider is when the line containing a ray also contains the image of the ray. In this case, ray BC and ray $B'C'$ lie in the same line.

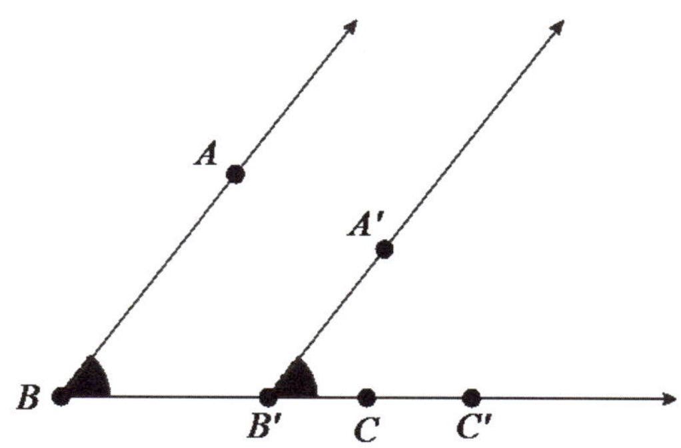

- Why does $m\angle ABC = m\angle A'B'C'$?
  - *They are corresponding angles from parallel lines cut by a transversal.*

Discussion (5 minutes)

This Discussion is optional and can be used if class time permits.

- Given $\triangle ABC$ with $\angle A$ and $\angle B$ acute, inscribe a square inside the triangle so that two vertices of the square lie on side $\overline{AB}$, and the other two vertices lie on the other two sides.

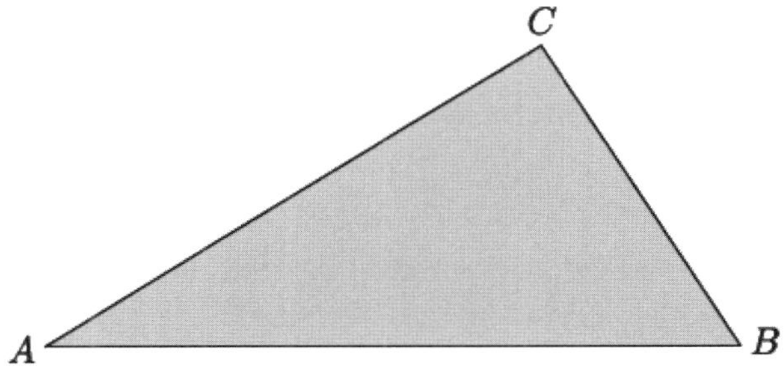

- We begin by drawing a small square near vertex A so that one side is on $\overline{AB}$ and one vertex is on $\overline{AC}$.

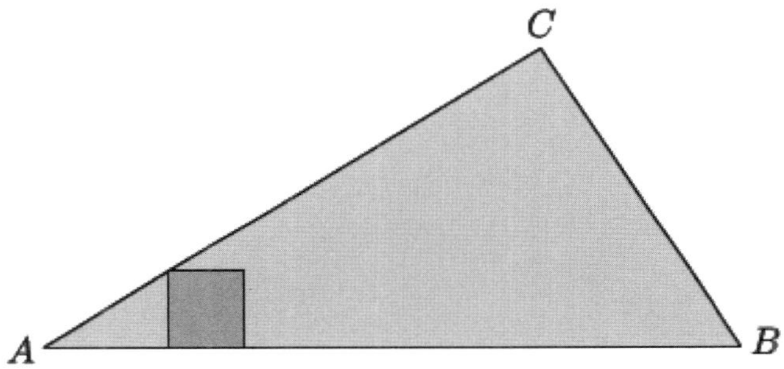

Ask students how they can dilate the square so that the other two vertices lie on the other two sides. Provide time for students to discuss this challenging question in small groups and to share their thoughts with the class. Validate any appropriate strategies, or suggest the strategy described below.

- Next, draw a ray from A through the vertex of the square that does not touch any side of the triangle. Name the point where the ray intersects the triangle T.

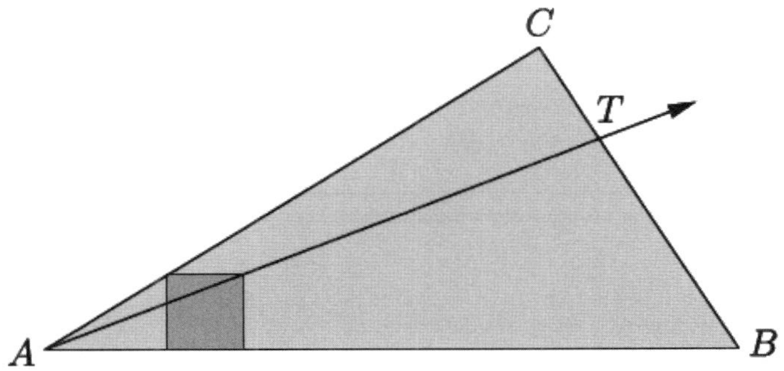

- The point T can be used as one of the vertices of the desired square. The parallel method can then be used to construct the desired square.

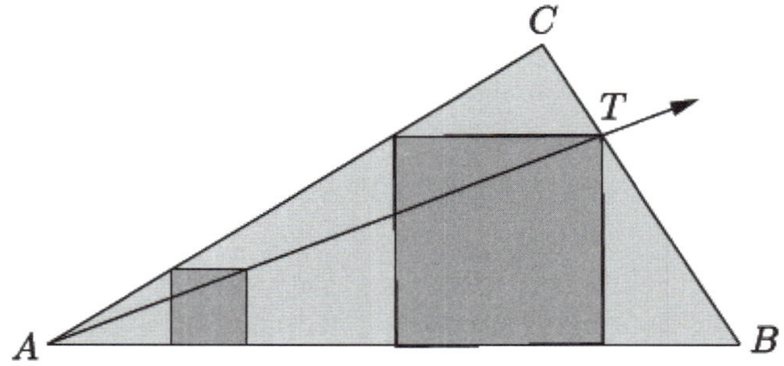

Consider trying this on another triangle before proceeding with the question below.

- Why does this work?
  - *Three of the four vertices of the square are on two sides of the triangle. The location of the fourth vertex must be on side $\overline{BC}$. Since the ray was drawn through the vertex of the small square, a scale factor r maps the dilated vertex on the ray. To inscribe it in the desired location, we note the location where the ray intersects the side of the triangle, giving us the vertex of the desired square. Since dilations map squares to squares, it is just a matter of locating the point of the vertex along the opposite side of $\angle A$ and then constructing the square.*

Closing (2 minutes)

- How do dilations map angles?
  - *Dilations map angles to angles of equal measure.*
- What foundational knowledge did we need to prove that dilations map angles to angles of equal measure?
  - *We needed to know that rays are dilated to rays that are parallel or rays that coincide with the original ray. Then, using what we know about parallel lines cut by a transversal, we could use what we knew about corresponding angles of parallel lines being equal to show that dilations map angles to angles of equal measure. We also needed to know about auxiliary lines and how to use them in order to produce a diagram where parallel lines are cut by a transversal.*
- How do dilations map polygonal figures?
  - *Dilations map polygonal figures to polygonal figures whose angles are equal in measure to the corresponding angles of the original figure and whose side lengths are equal to the corresponding side lengths multiplied by the scale factor.*

Lesson Summary

- **Dilations map angles to angles of equal measure.**
- **Dilations map polygonal figures to polygonal figures whose angles are equal in measure to the corresponding angles of the original figure and whose side lengths are equal to the corresponding side lengths multiplied by the scale factor.**

Exit Ticket (5 minutes)

Name ______________________________ Date ______________

Lesson 9: How Do Dilations Map Angles?

Exit Ticket

1. Dilate parallelogram $STUV$ from center O using a scale factor of $r = \frac{3}{4}$.

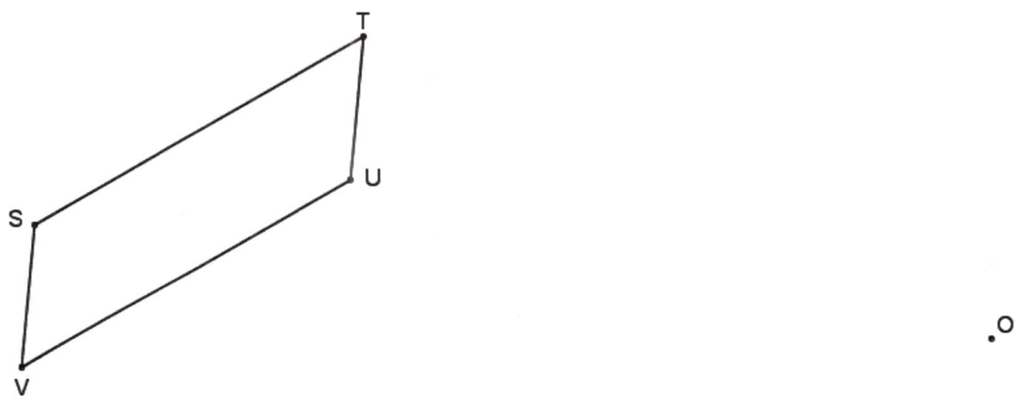

2. How does $m\angle T'$ compare to $m\angle T$?

3. Using your diagram, prove your claim from Problem 2.

Exit Ticket Sample Solutions

1. Dilate parallelogram $STUV$ from center O using a scale factor of $r = \frac{3}{4}$.

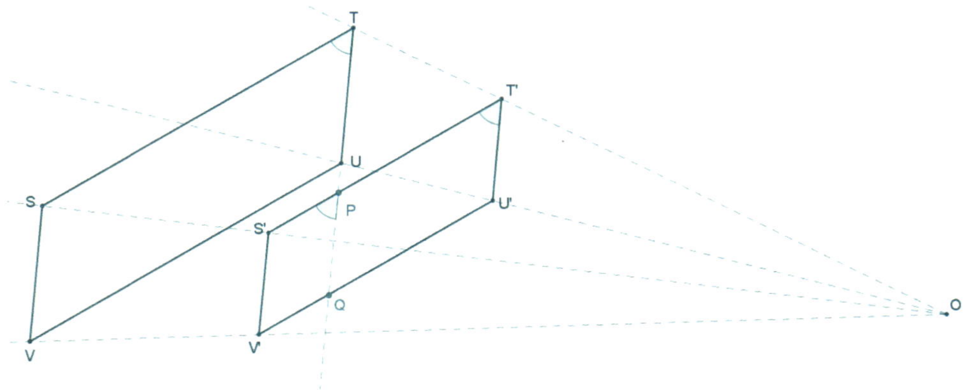

2. How does $m\angle T'$ compare to $m\angle T$?

   $m\angle T' = m\angle T$ because dilations preserve angle measure.

3. Using your diagram, prove your claim from Problem 2.

   Extend $\overrightarrow{TU}$ such that it intersects $\overline{S'T'}$ at a point P. Dilations map lines to parallel lines, so $\overline{TU} \parallel \overline{T'U'}$; therefore, by corresponding $\angle$'s, $\overline{TU} \parallel \overline{T'U'}$, $\angle S'PQ \cong \angle T'$. Under the same dilation, $\overline{S'T'} \parallel \overline{ST}$, again, by corresponding $\angle$'s, $\overline{S'T'} \parallel \overline{ST}$, $\angle S'PQ \cong \angle T$. By transitivity, $\angle T' \cong \angle T$.

Problem Set Sample Solutions

1. Shown below is $\triangle ABC$ and its image $\triangle A'B'C'$ after it has been dilated from center O by scale factor $r = \frac{5}{2}$. Prove that the dilation maps $\triangle ABC$ to $\triangle A'B'C'$ so that $m\angle A = m\angle A'$, $m\angle B = m\angle B'$, and $m\angle C = m\angle C'$.

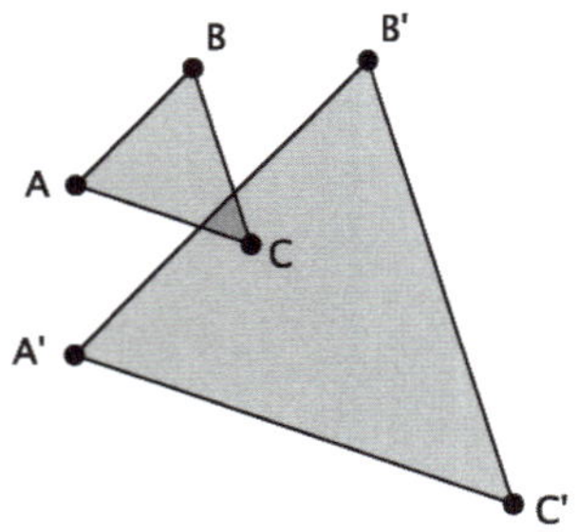

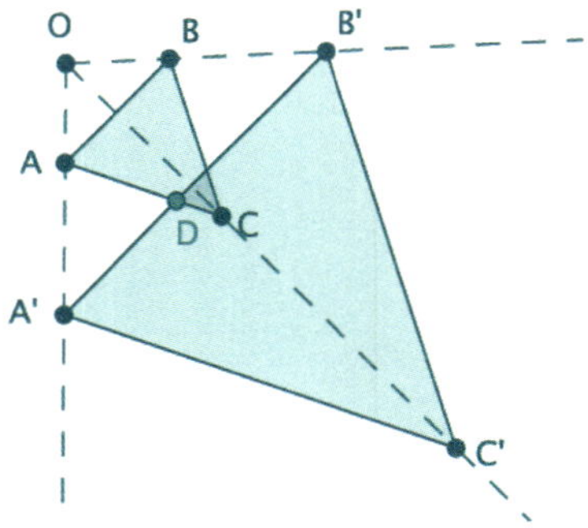

Locate the center of dilation O by drawing rays through each of the pairs of corresponding points. The intersection of the rays is the center of dilation O. Since dilations map segments to segments, and the dilated segments must either coincide with their pre-image or be parallel, then we know that $\overleftrightarrow{AB} \parallel \overleftrightarrow{A'B'}$, $\overleftrightarrow{AC} \parallel \overleftrightarrow{A'C'}$, and $\overleftrightarrow{BC} \parallel \overleftrightarrow{B'C'}$.
Let D be the point where side $\overline{AC}$ intersects with side $\overline{A'B'}$. Then, $\angle B'A'C'$ is congruent to $\angle B'DC$ because corresponding $\angle$'s, $\overleftrightarrow{AC} \parallel \overleftrightarrow{A'C'}$, cut by a transversal, $\overleftrightarrow{A'B'}$, are congruent. Then, $\angle B'DC \cong \angle BAC$ because corresponding $\angle$'s, $\overleftrightarrow{AB} \| \overleftrightarrow{A'B'}$, cut by a transversal $\overleftrightarrow{AC}$, are congruent.
By the transitive property, $\angle B'A'C' \cong \angle B'DC \cong \angle BAC$ and $\angle B'A'C' \cong \angle BAC$.
Since congruent angles are equal in measure, $m\angle B'A'C' = m\angle A'$ and $m\angle BAC = m\angle A$; then, $m\angle A = m\angle A'$.
Similar reasoning shows that $m\angle B = m\angle B'$ and $m\angle C = m\angle C'$.

EUREKA MATH

2. **Explain the effect of a dilation with scale factor r on the length of the base and height of a triangle. How is the area of the dilated image related to the area of the pre-image?**

Let P represent the endpoint of altitude $\overline{AP}$ of $\triangle ABC$, such that P lies on $\overleftrightarrow{BC}$. Thus, the base length of the triangle is BC, and the height of the triangle is AP. The area of the given triangle then is $\frac{1}{2}(BC)(AP)$. By the definition of dilation, $B'C' = r(BC)$ and $A'P' = r(AP)$, so the base and height of the dilated image are proportional to the base and height of the original just as the lengths of the sides of the triangle. The area of the dilated triangle then would be $\frac{1}{2}\left(r(BC) \cdot r(AP)\right) = r^2 \cdot \frac{1}{2}(BC)(AP)$.

The ratio of the area of the dilated image to the area of the pre-image is r^2.

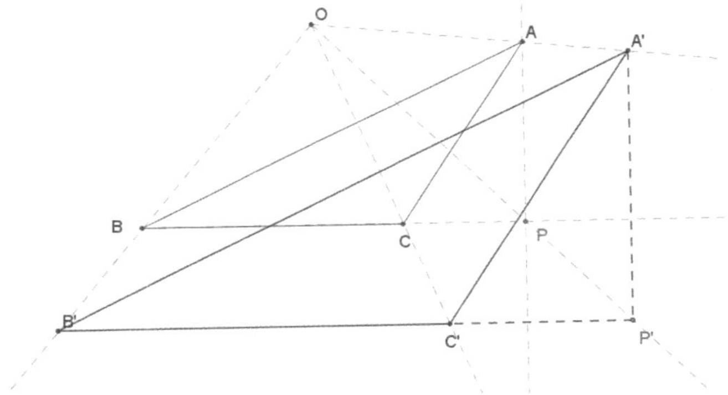

3. **Dilate trapezoid $ABED$ from center O using a scale factor of $r = \frac{1}{2}$.**

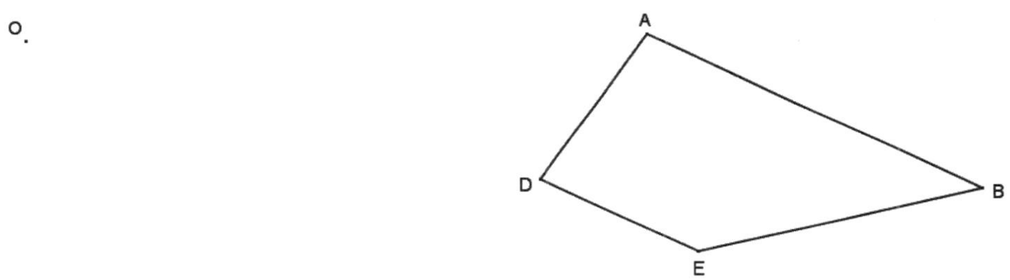

A dilation maps a trapezoid to a trapezoid so that the ratio of corresponding sides is the same and corresponding interior angles are the same measure.

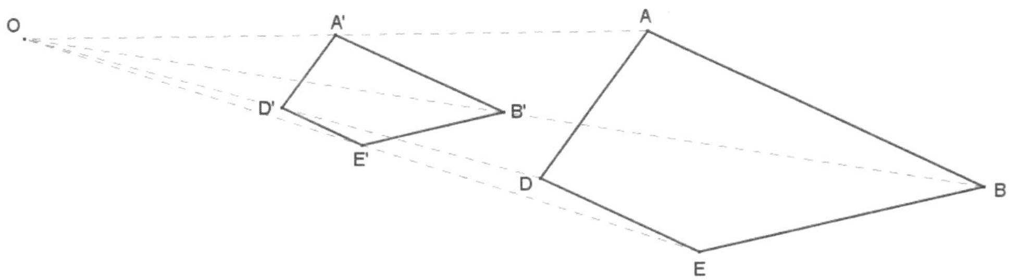

4. Dilate kite $ABCD$ from center O using a scale factor $r = 1\frac{1}{2}$.

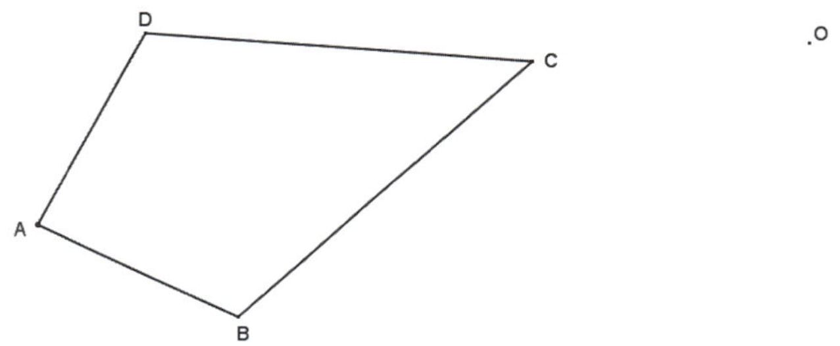

A dilation maps a kite to a kite so that the ratio of corresponding sides is the same and corresponding interior angles are the same measure.

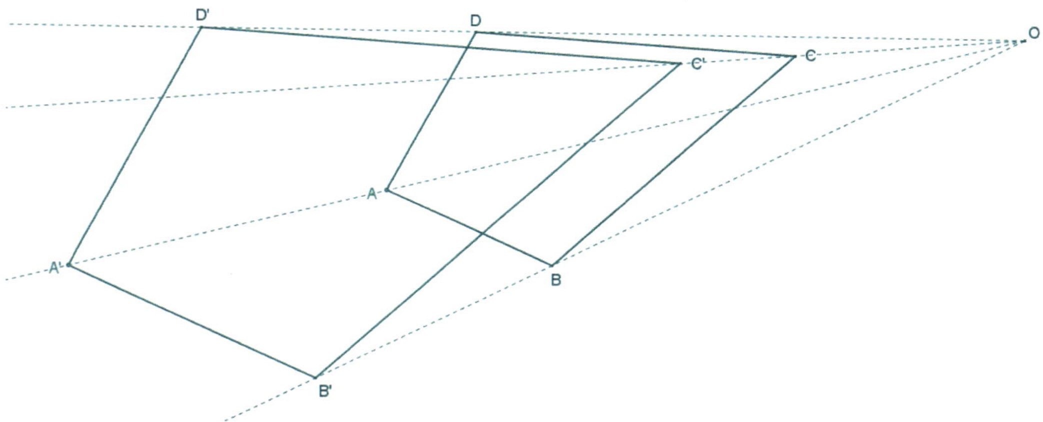

5. Dilate hexagon $DEFGHI$ from center O using a scale factor of $r = \frac{1}{4}$.

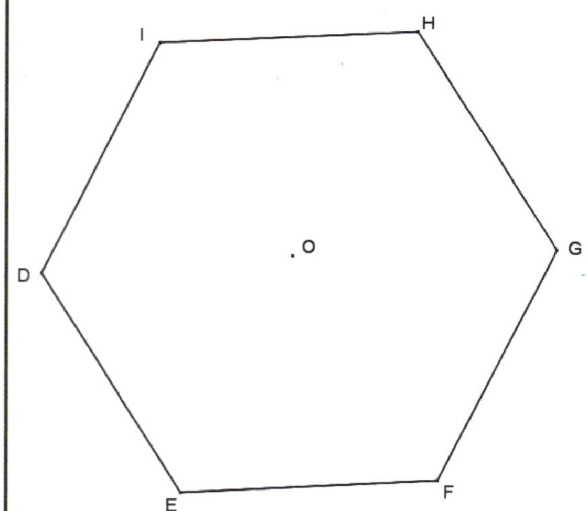

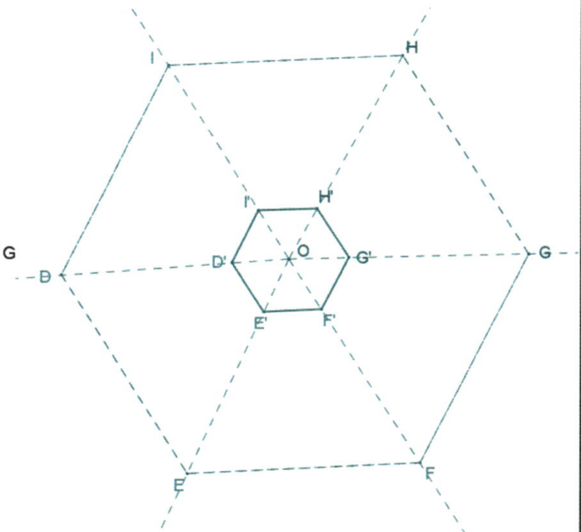

6. **Examine the dilations that you constructed in Problems 2–5, and describe how each image compares to its pre-image under the given dilation. Pay particular attention to the sizes of corresponding angles and the lengths of corresponding sides.**

   In each dilation, the angles in the image are the same size as the corresponding angles in the pre-image, as we have shown that dilation preserves angle measure. We also know that the lengths of corresponding sides are in the same ratio as the scale factor.

Lesson 10: Dividing the King's Foot into 12 Equal Pieces

Student Outcomes

- Students divide a line segment into n equal pieces by the side splitter and dilation methods.
- Students know how to locate fractions on the number line.

Materials

- Poster paper or chart paper
- Yardstick
- Compass
- Straightedge
- Set square

Lesson Notes

Students explore how their study of dilations relates to the constructions that divide a segment into a whole number of equal-length segments.

Classwork

Opening (2 minutes)

MP.3

In an age when there was no universal consensus on measurement, the human body was often used to create units of measurement. One can imagine how a king might declare the length of his foot to be what we know as the unit of a foot. How would we go about figuring out how to divide one foot into twelve equal portions, as the 12 inches that comprise a foot? Have students write or discuss their thoughts.

Opening Exercise (3 minutes)

Opening Exercise

Use a compass to mark off equally spaced points C, D, E, and F so that AB, BC, CD, DE, and EF are equal in length. Describe the steps you took.

I adjust the compass to the length of $\overline{AB}$ and then place the point of the compass on B and use the adjustment to make a mark so that it intersects with the ray. This is the location of C, and I repeat these steps until I locate point F.

- Marking off equal segments is entirely a matter of knowing how to use the compass.
- What if you knew the length of a segment but needed to divide it into equal-length intervals? For example, suppose you had a segment AB that was 10 cm in length. How could you divide it into ten 1 cm parts?
  - *We can try to create a perpendicular bisector to locate the midpoint of the segment.*
  - *Allow students time to discuss. They may try to use what they know about creating a perpendicular bisector to locate the midpoint of the segment; however, they quickly see that it does not easily lead to determining a 1 cm unit.*
- We can tackle this problem with a construction that relates to our work on dilations.

Exploratory Challenge 1 (12 minutes)

In the Exploratory Challenge, students learn a construction that divides a segment into n equal parts. They understand that the constructed parallel segments are evenly spaced proportional side splitters of $\triangle ABA_3$.

- We are going to use a compass and straightedge to divide a segment of known length by a whole number n.
- We are going to divide the following segment AB into three segments of equal length.

Exploratory Challenge 1

Divide segment AB into three segments of equal lengths.

Draw each step of the Exploratory Challenge so that students can refer to the correct steps whether they are ahead or working alongside the teacher.

- Pick a point A_1 not on $\overline{AB}$. Draw $\overrightarrow{AA_1}$.

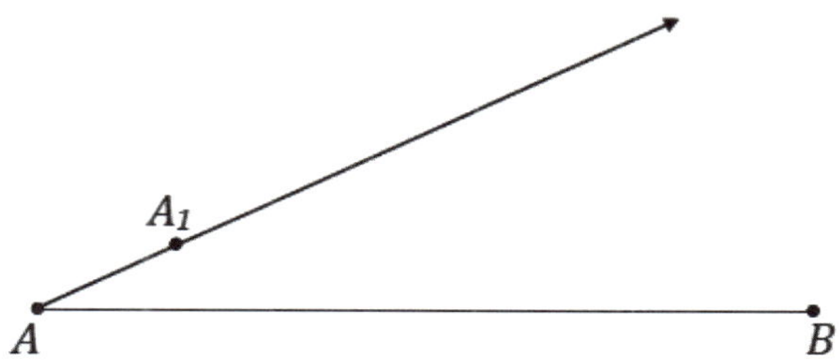

- Mark points A_2 and A_3 on the ray so that $AA_1 = A_1A_2 = A_2A_3$.

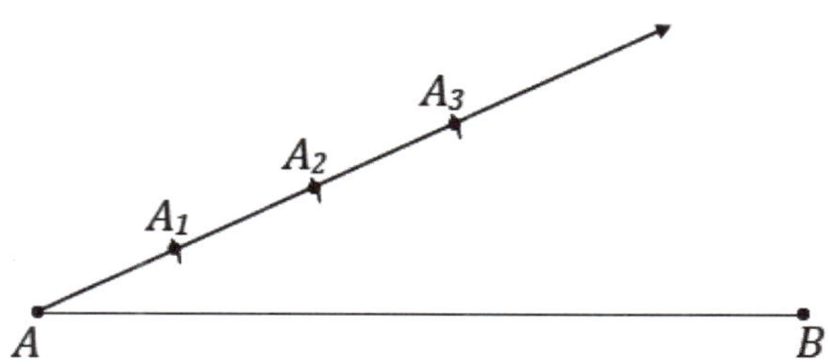

- Use a straightedge to draw $\overline{A_3B}$. Use a set square to draw segments parallel to $\overline{A_3B}$ through A_1 and A_2.

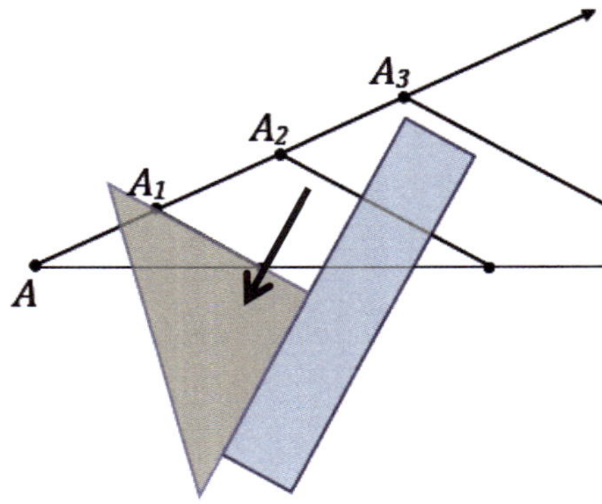

- Label the points where the constructed segments intersect $\overline{AB}$ as B_1 and B_2.

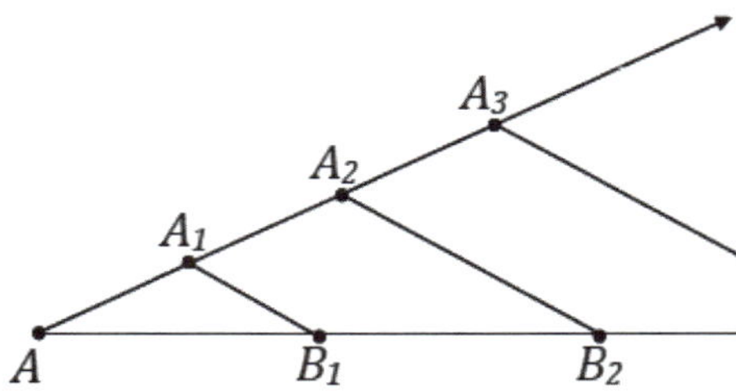

Allow students time to discuss with each other why the construction divides AB into equal parts.

- Did we succeed? Measure segments AB_1, B_1B_2, and B_2B. Are they all equal in measurement?
- After the last step of the construction, answer the following: Why does this construction divide the segment AB into equal parts?

Allow students a moment to jot down their thoughts, and then take responses.

MP.2

- ▫ *By construction, segments A_1B_1 and A_2B_2 are parallel to segment A_3B. By the triangle side splitter theorem, $\overline{A_1B_1}$ and $\overline{A_2B_2}$ are proportional side splitters of triangle ABA_3. So, $\frac{AB_1}{AB} = \frac{AA_1}{AA_3} = \frac{1}{3}$ and $\frac{AB_2}{AB} = \frac{AA_2}{AA_3} = \frac{2}{3}$. Thus, $AB_1 = \frac{1}{3}AB$ and $AB_2 = \frac{2}{3}AB$, and we can conclude that B_1 and B_2 divide line segment AB into three equal pieces.*

- Would the construction work if you had chosen a different location for A_1? Try the construction again, and choose a location different from the location in the first construction.

Scaffolding:

- Consider having students use a different color for this construction.
- Additionally, consider splitting the class so that one half begins the construction from A, while the other half begins at B.

Students should retry the construction and discover that the location of A_1 is irrelevant to dividing $\overline{AB}$ into three equal-length segments. The triangle drawn in this second attempt is different from the one initially created, so though A_1 is in a different location, the triangle drawn is also different. Therefore, the proportional side splitters are also different but achieve the same result, dividing AB into three equal-length segments. Furthermore, the location of A_1 was never specified to begin with, nor was the relative angle of $\angle BAA_1$. So, technically, this question has already been answered.

- We call this method of dividing the segment into n equal-length segments the *side splitter method* in reference to the triangle side splitter theorem.
- **SIDE SPLITTER METHOD**: If $\overline{AB}$ is a line segment, construct a ray AA_1, and mark off n equally spaced points using a compass of fixed radius to get points $A = A_0, A_1, A_2, \cdots, A_n$. Construct $\overline{A_nB}$ that is a side of $\triangle ABA_n$. Through each point $A_1, A_2, \cdots, A_{n-1}$, construct line segments $\overline{A_1B_1}, \overline{A_2B_2}, \ldots, \overline{A_{n-1}B_{n-1}}$ parallel to $\overline{A_nB}$ that connect two sides of $\triangle AA_nB$.

Exercise 1 (3 minutes)

Teachers may elect to move directly to the next Exploratory Challenge depending on time. Alternatively, if there is some time, the teacher may elect to reduce the number of divisions to 3.

Exercise 1

Divide segment AB into five segments of equal lengths.

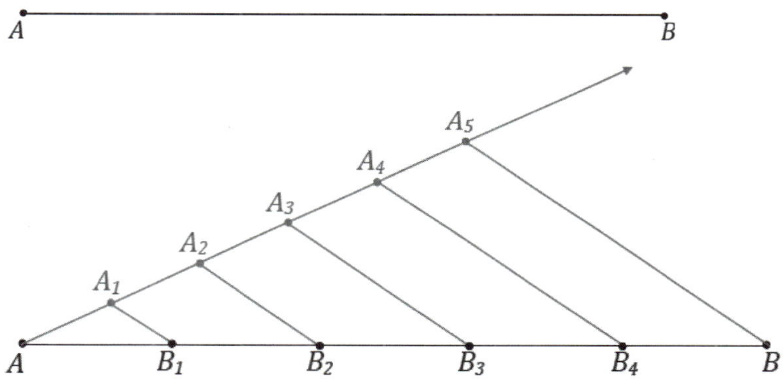

Exploratory Challenge 2 (10 minutes)

Now students try an alternate method of dividing a segment into n equal-length segments.

- Let's continue with our exploration and try a different method to divide a segment of known length by a whole number n. Again, we rely on the use of a compass, straightedge, and this time, a set square.

Scaffolding:

- Consider focusing on the side splitter method alone for struggling students.
- Using this strategy, select an early question from the Problem Set to work on in class.

Exploratory Challenge 2

Divide segment AB into four segments of equal length.

- Use the set square to create a ray XY parallel to $\overline{AB}$. Select the location of the endpoint X so that it falls to the left of A; the location of Y should be oriented in relation to X in the same manner as B is in relation to A. We construct the parallel ray below $\overline{AB}$, but it can be constructed above the segment as well.

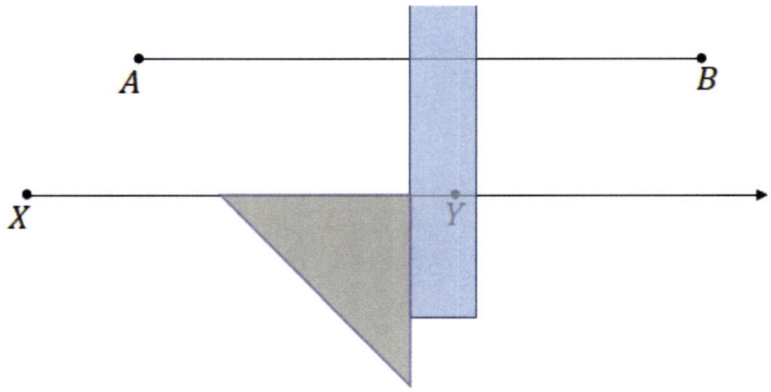

- From X, use the compass to mark off four equal segments along $\overrightarrow{XY}$. Label each intersection as X_1, X_2, X_3, and X_4. It is important that $XX_4 \neq AB$. In practice, XX_4 should be clearly more or clearly less than AB.

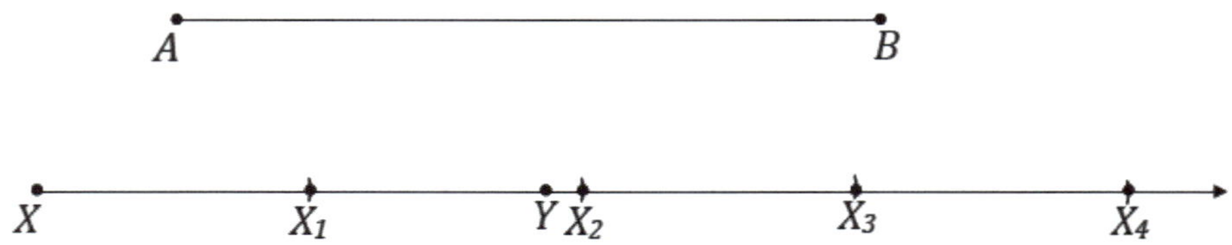

- Draw line XA and line X_4B. Mark the intersection of the two lines as O.

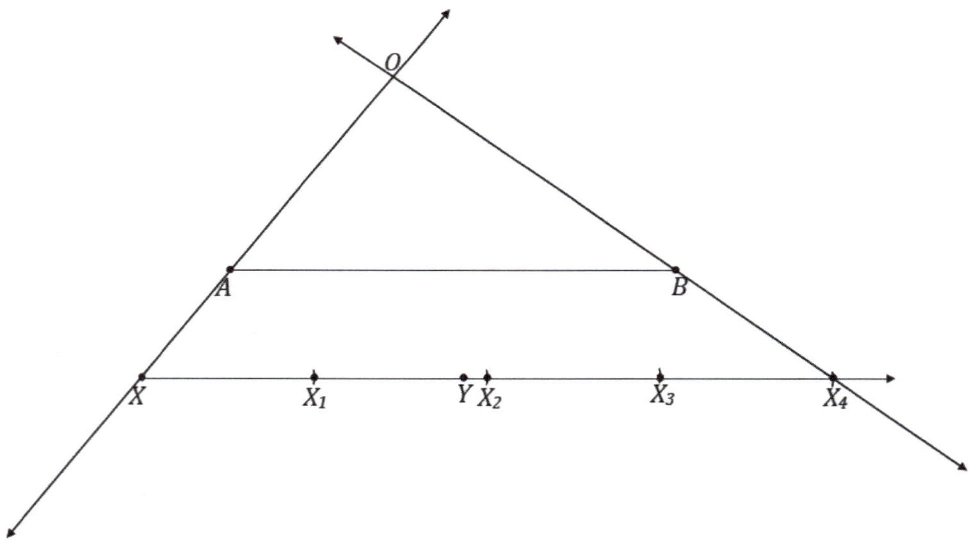

- Construct rays from O through X_1, X_2, and X_3. Label each intersection with $\overline{AB}$ as A_1, A_2, and A_3.

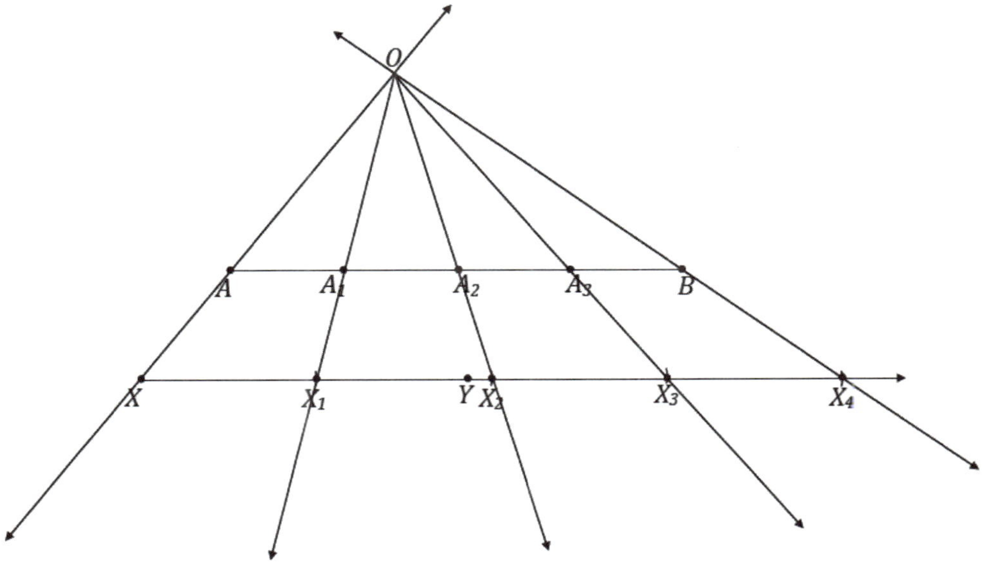

- $\overline{AB}$ should now be divided into four segments of equal length. Measure segments AA_1, A_1A_2, A_2A_3, and A_3B. Are they all equal in measurement?
- Why does this construction divide the segment AB into equal parts?

Allow students time to discuss with each other why the construction divides $\overline{AB}$ into equal parts.

  - *We constructed $\overline{AB}$ to be parallel to $\overrightarrow{XY}$. In $\triangle XOX_4$, since the side splitter $\overline{AB}$ is parallel to $\overline{XX_4}$, by the triangle side splitter theorem, it must also be a proportional side splitter. By the dilation theorem, this means that $\frac{AA_1}{XX_1} = \frac{A_1A_2}{X_1X_2} = \frac{A_2A_3}{X_2X_3} = \frac{A_3B}{X_3X_4}$. Since we know that the values of all four denominators are the same, the value of all four numerators must also be the same to make the equation true. Therefore, $\overline{AB}$ has been divided into four segments of equal length.*

- We call this method of dividing the segment into equal lengths the *dilation method*, as the points that divide $\overline{AB}$ are by definition dilated points from center O with scale factor $r = \frac{OA}{OX}$ of the evenly spaced points on $\overrightarrow{XY}$.
- **DILATION METHOD**: Construct a ray XY parallel to $\overline{AB}$. On the parallel ray, use a compass to mark off n equally spaced points $X_1, X_2, \cdots, X_n$ so that $XX_n \neq AB$. Lines $\overleftrightarrow{AX}$ and $\overleftrightarrow{BX_n}$ intersect at a point O. Construct the rays $\overrightarrow{OX_1}$, $\overrightarrow{OX_2}$, ..., $\overrightarrow{OX_n}$ that meet $\overline{AB}$ in points A_1, A_2, ..., A_n, respectively.
- What happens if line segments XX_n and AB are close to the same length?
  - *The point O is very far away.*
- That is why it is best to make XX_n clearly more or less than AB. It is also best to keep line segments XX_n and AB centered, or the point O will also be far away.

Exercise 2 (8 minutes)

Students should complete Exercise 2 in pairs. If possible, the teacher should pre-mark each piece of poster paper with a 1-foot mark to allow students to get right to the activity.

- Let's return to our opening remarks on the king's foot.

Exercise 2

On a piece of poster paper, draw a segment AB with a measurement of 1 foot. Use the dilation method to divide $\overline{AB}$ into twelve equal-length segments, or into 12 inches.

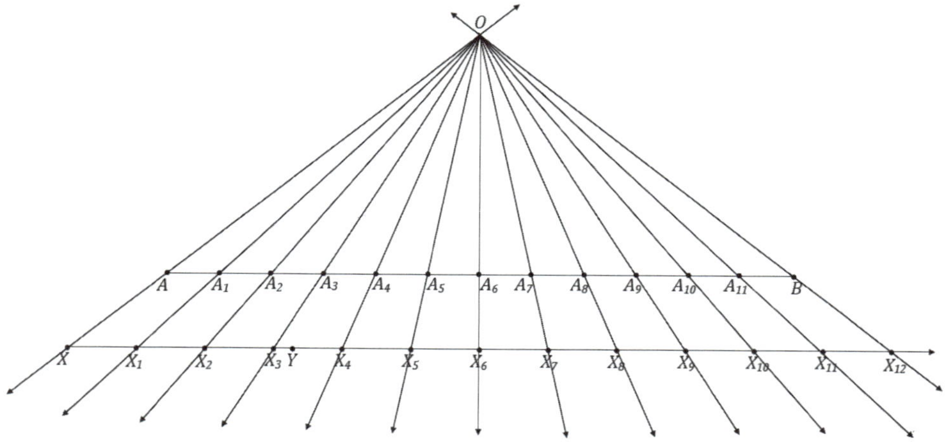

Closing (2 minutes)

- Compare the side splitter method to the dilation method: What advantage does the first method have?
  - *With the side splitter method, we do not have to worry about checking lengths (i.e., the initial point can be chosen completely arbitrarily), whereas with the dilation method, we have to choose our unit so that n units does not match the original length (or come close to matching).*
- How does either of the methods help identify fractions on the number line?
  - *We can break up whole units on a number line into any division we choose; for example, we can mimic a ruler in inches by dividing up a number line into eighths.*

Lesson Summary

SIDE SPLITTER METHOD: If $\overline{AB}$ is a line segment, construct a ray AA_1, and mark off n equally spaced points using a compass of fixed radius to get points $A = A_0, A_1, A_2, \cdots, A_n$. Construct $\overline{A_nB}$ that is a side of $\triangle ABA_n$. Through each point $A_1, A_2, \cdots, A_{n-1}$, construct line segments $\overline{A_iB_i}$ parallel to $\overline{A_nB}$ that connect two sides of $\triangle AA_nB$.

DILATION METHOD: Construct a ray XY parallel to $\overline{AB}$. On the parallel ray, use a compass to mark off n equally spaced points $X_1, X_2, \cdots, X_n$ so that $XX_n \neq AB$. Lines $\overleftrightarrow{AX}$ and $\overleftrightarrow{BX_n}$ intersect at a point O. Construct the rays $\overrightarrow{OX_i}$ that meet $\overline{AB}$ in points A_i.

Exit Ticket (5 minutes)

Name ______________________________ Date ____________________

Lesson 10: Dividing the King's Foot into 12 Equal Pieces

Exit Ticket

1. Use the side splitter method to divide $\overline{MN}$ into 7 equal-sized pieces.

2. Use the dilation method to divide $\overline{PQ}$ into 11 equal-sized pieces.

3. If the segment below represents the interval from zero to one on the number line, locate and label $\frac{4}{7}$.

Exit Ticket Sample Solutions

1. **Use the side splitter method to divide $\overline{MN}$ into 7 equal-sized pieces.**

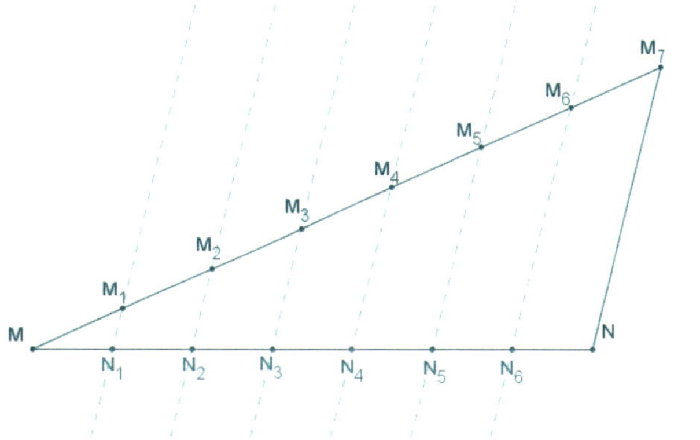

2. **Use the dilation method to divide $\overline{PQ}$ into 11 equal-sized pieces.**

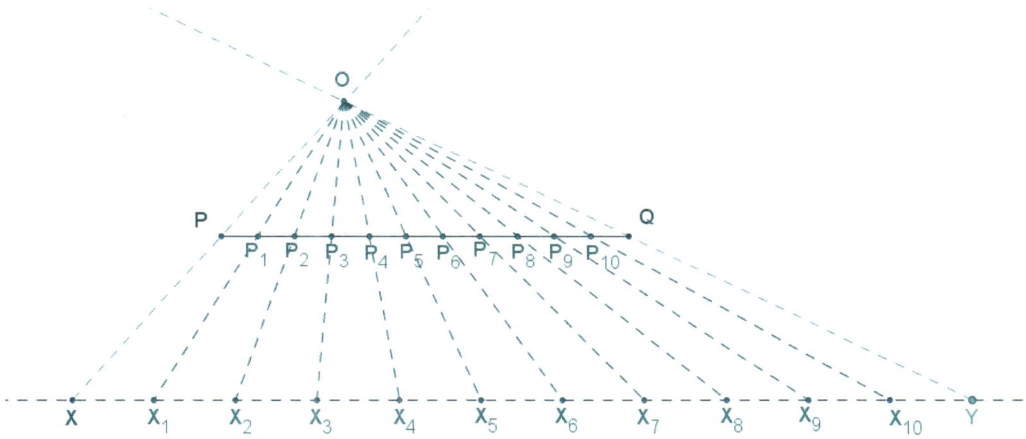

3. **If the segment below represents the interval from zero to one on the number line, locate and label $\frac{4}{7}$.**

   Students may use either the side splitter method or the dilation method and need only find the location of the fourth equal-sized piece of the segment, as shown in the diagram below.

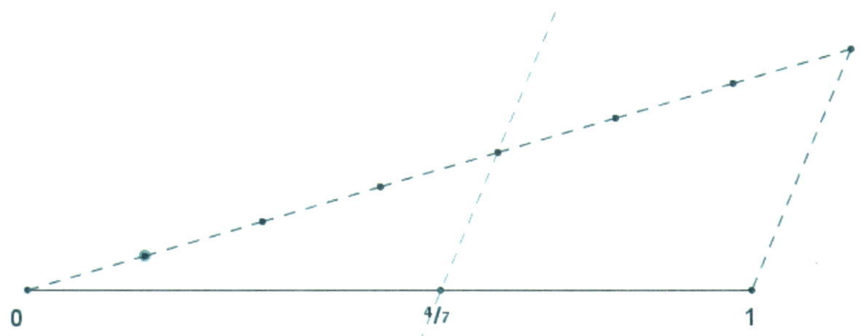

Problem Set Sample Solutions

1. Pretend you are the king or queen and that the length of your foot is the official measurement for one foot. Draw a line segment on a piece of paper that is the length of your foot. (You may have to remove your shoe.) Use the method above to find the length of 1 inch in your kingdom.

   I drew $\overline{AB}$ representing the length of my foot. I then divided $\overline{AB}$ into twelve equal pieces using the dilation method as follows:

   I constructed $\overleftrightarrow{DE}$ parallel to $\overline{AB}$ and, using a compass, marked off twelve consecutive segments on $\overleftrightarrow{DE}$, each having length DC. (Note that because $\overline{AB}$ is a large segment, students will likely choose a length DC so that the total length of all of the segments constructed on $\overleftrightarrow{DE}$ is noticeably shorter than AB.)

   I then constructed a ray from A and B through the endpoints of the composed segment, as shown in the diagram, to form a triangle ABO. Next, I constructed $\overrightarrow{OC}$, and then I marked its intersection with $\overline{AB}$ point I. This distance AI is the length of 1 inch in my kingdom.

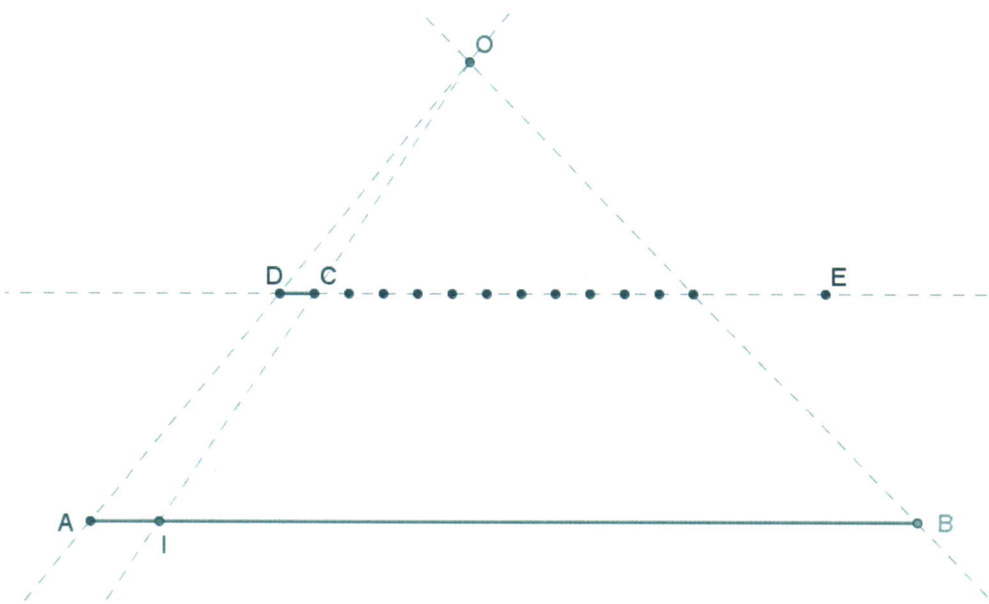

2. Using a ruler, draw a segment that is 10 cm. This length is referred to as a *decimeter.* Use the side splitter method to divide your segment into ten equal-sized pieces. What should be the length of each of those pieces based on your construction? Check the length of the pieces using a ruler. Are the lengths of the pieces accurate?

   Verify that students' diagrams show the use of the side splitter method. The length of each piece should be 1 cm.

3. Repeat Problem 2 using the dilation method. What should be the length of each of those pieces based on your construction? Check the lengths of the pieces using a ruler. Are the lengths of the pieces accurate?

   Verify that students' diagrams show the use of the dilation method. The length of each piece should be 1 cm.

4. A portion of a ruler that measured whole centimeters is shown below. Determine the location of $5\frac{2}{3}$ cm on the portion of the ruler shown.

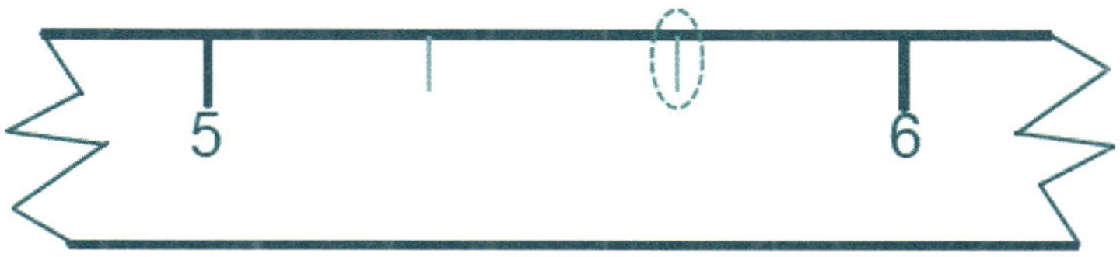

Responses should show the segment between 5 *and* 6 *divided into* 3 *equal pieces with the division point closest to* 6 *chosen as the location of* $5\frac{2}{3}$.

5. Merrick has a ruler that measures in inches only. He is measuring the length of a line segment that is between 8 in. and 9 in. Divide the one-inch section of Merrick's ruler below into eighths to help him measure the length of the segment.

Using the side splitter method, I divided the one-inch interval into eighths, labeled as B_1, B_2, *etc., on the diagram. The line segment that Merrick is measuring closely corresponds with* B_3, *which represents* $\frac{5}{8}$ *of the distance from* 8 in. *to* 9 in. *Therefore, the length of the line segment that Merrick is measuring is approximately* $8\frac{5}{8}$ in.

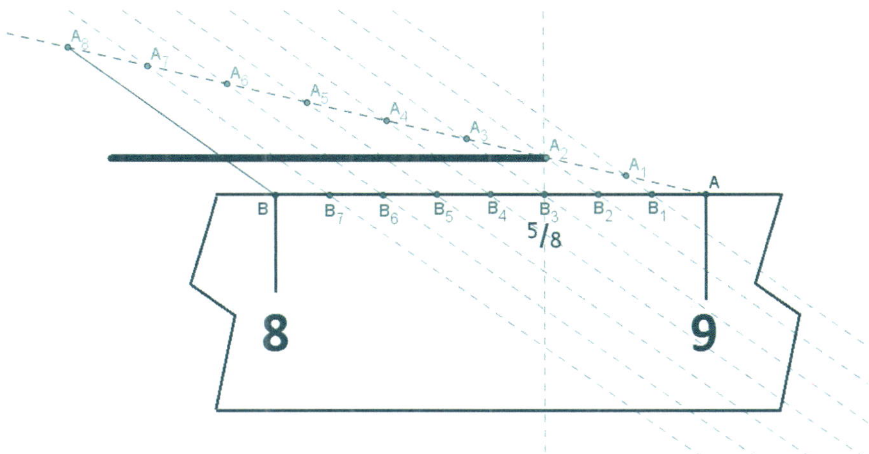

6. **Use the dilation method to create an equally spaced 3×3 grid in the following square.**

There are several ways to complete this construction. The sample below used the dilation method along two sides of the square with centers O and O_2 to divide the side into three equal-size segments. The sides of the square were extended such that $DD' = CD$, $CC' = CD$, $BB' = CB$, and $CC'_1 = CB$.

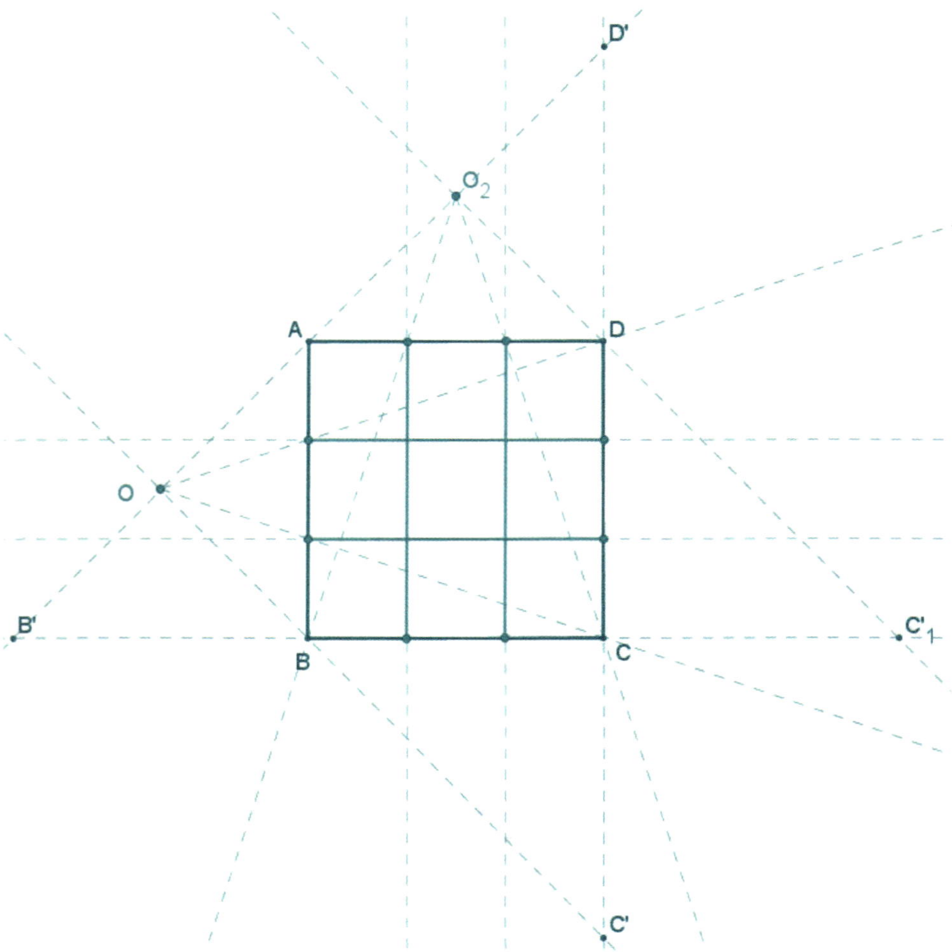

7. **Use the side splitter method to create an equally spaced 3×3 grid in the following square.**

There are several ways to complete this construction. The sample below used the side splitter method along two sides of the square to divide the side into three equal-size segments.

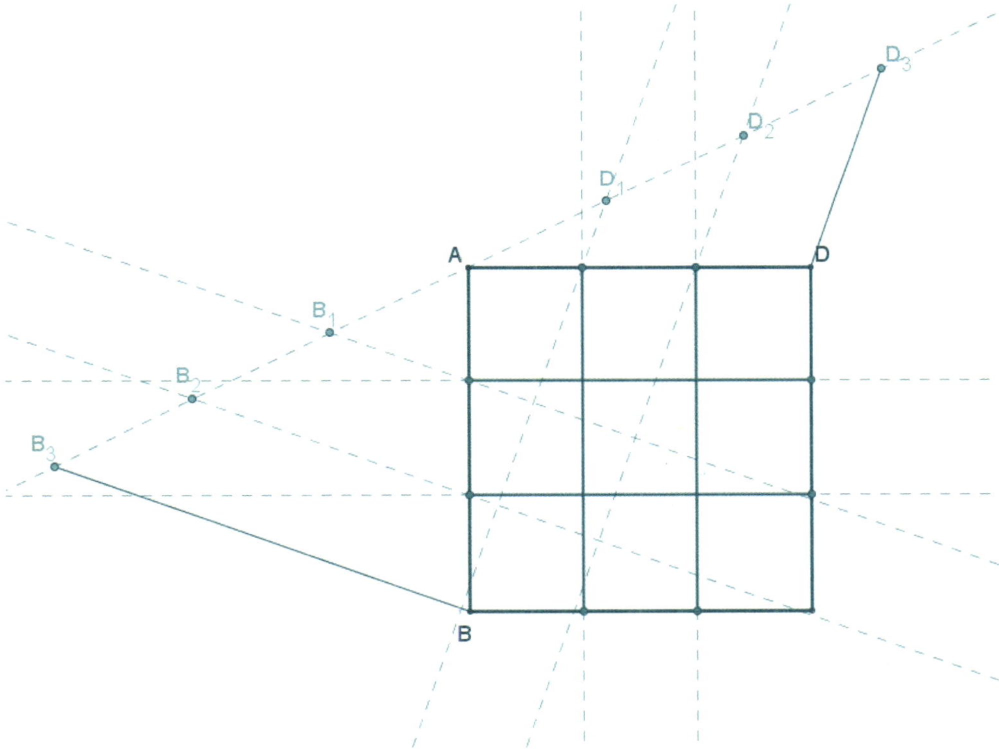

Lesson 11: Dilations from Different Centers

Student Outcomes

- Students verify experimentally that the dilation of a figure from two different centers by the same scale factor gives congruent figures. These figures are congruent by a translation along a vector that is parallel to the line through the centers.
- Students verify experimentally that the composition of dilations D_{O_1,r_1} and D_{O_2,r_2} is a dilation with scale factor r_1r_2 and center on $\overleftrightarrow{O_1O_2}$ unless $r_1r_2 = 1$.

Lesson Notes

In Lesson 11, students examine the effects of dilating figures from two different centers. By experimental verification, they examine the impact on the two dilations of having two different scale factors, the same two scale factors, and scale factors whose product equals 1. Each of the parameters of these cases provides information on the centers of the dilations, their scale factors, and the relationship between individual dilations versus the relationship between an initial figure and a composition of dilations.

Classwork

Exploratory Challenge 1 (15 minutes)

In Exploratory Challenge 1, students verify experimentally that the dilation of a figure from two different centers by the same scale factor gives congruent figures that are congruent by a translation along a vector that is parallel to the line through the centers.

- In this example, we examine scale drawings of an image from different center points.

Exploratory Challenge 1

Drawing 2 and Drawing 3 are both scale drawings of Drawing 1.

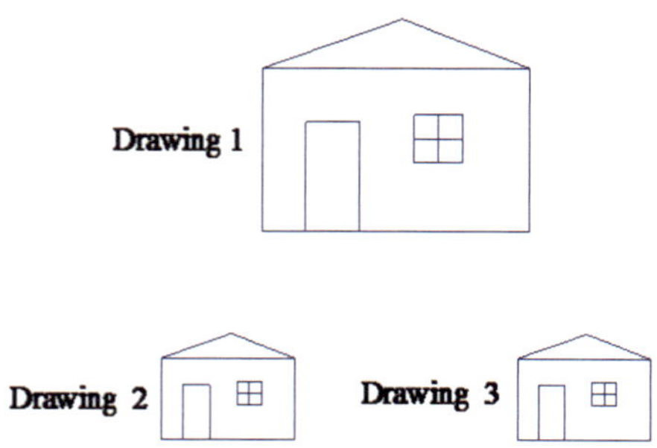

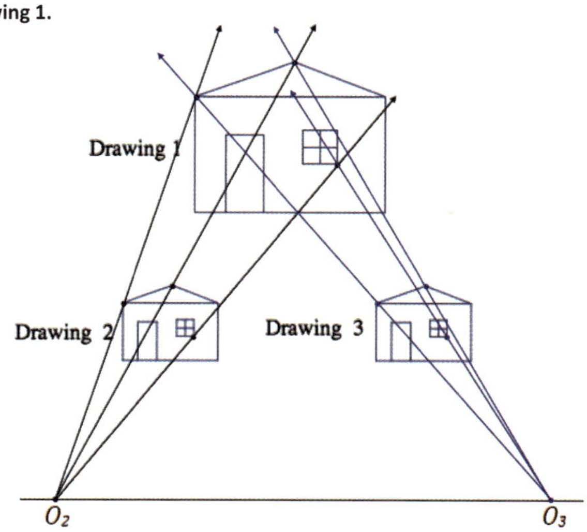

a. **Determine the scale factor and center for each scale drawing. Take measurements as needed.**

The scale factor for each drawing is the same; the scale factor for both is $r = \frac{1}{2}$. Each scale drawing has a different center.

b. **Is there a way to map Drawing 2 onto Drawing 3 or map Drawing 3 onto Drawing 2?**

Since the two drawings are identical, a translation maps either Drawing 2 onto Drawing 3 or Drawing 3 onto Drawing 2.

- What do you notice about a translation vector that maps either scale drawing onto the other and the line that passes through the centers of the dilations?
  - *A translation vector that maps either scale drawing onto the other is parallel to the line that passes through the centers of the dilations.*

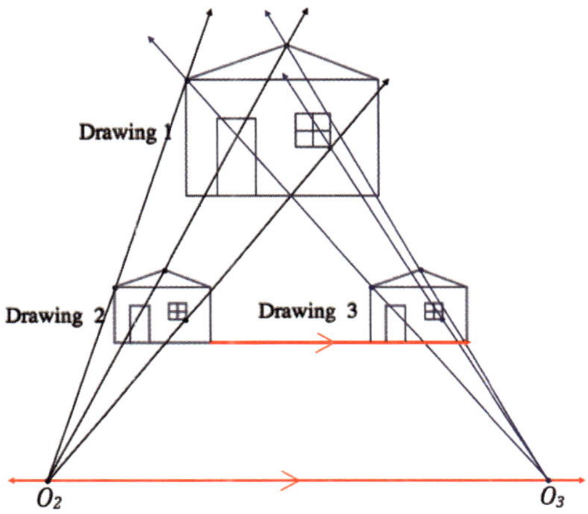

- We are not going to generally prove this, but let's experimentally verify this by dilating a simple figure (e.g., a segment) by the same scale factor from two different centers, O_1 and O_2.
- Do this twice, in two separate cases, to observe what happens.
- In the first case, dilate $\overline{AB}$ by a factor of 2, and be sure to give each dilation a different center. Label the dilation about O_1 as $\overline{A_1B_1}$ and the dilation about O_2 as $\overline{A_2B_2}$.

Scaffolding:
- Students can take responsibility for their own learning by hand-drawing the houses; however, if time is an issue, teachers can provide the drawings of the houses located at the beginning of Exploratory Challenge 1.
- Use patty paper or geometry software to help students focus on the concepts.
- Consider performing the dilation in the coordinate plane with center at the origin, for example, $\overline{AB}$ with coordinates $A(3,1)$ and $B(4,-3)$.

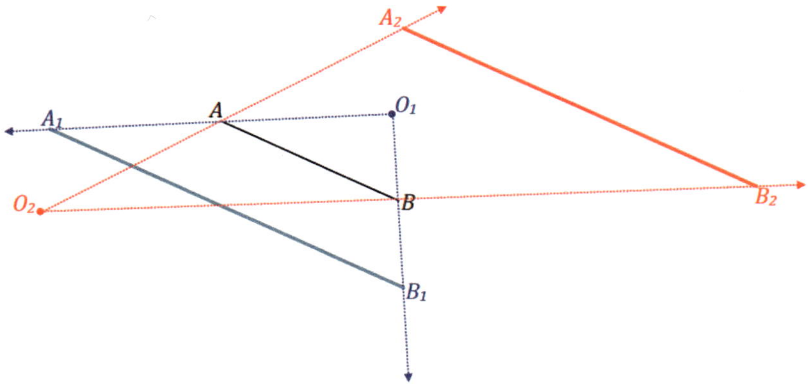

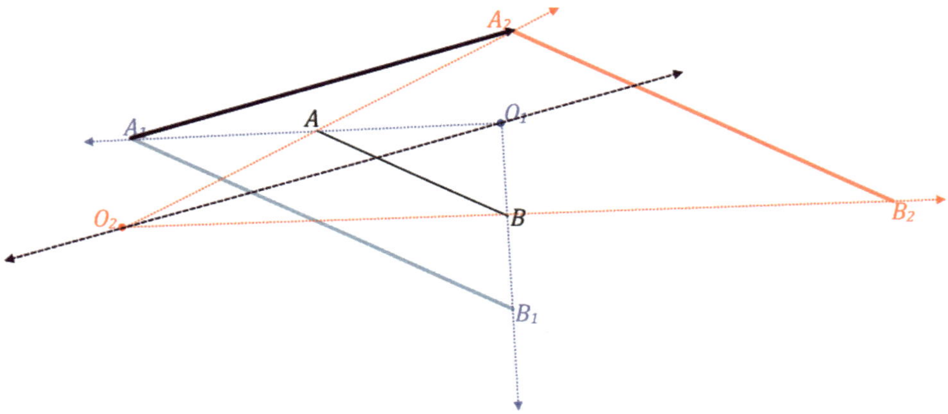

- Repeat the experiment, and create a segment, $\overline{CD}$, different from $\overline{AB}$. Dilate $\overline{CD}$ by a factor of 2, and be sure to give each dilation a different center. Label the dilation about O_1 as $\overline{C_1D_1}$ and the dilation about O_2 as $\overline{C_2D_2}$.

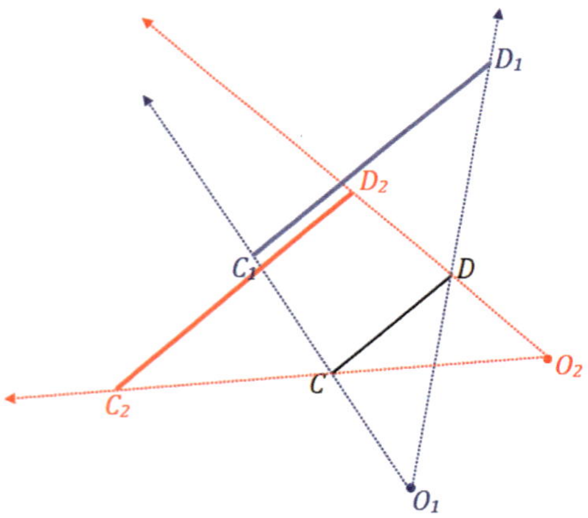

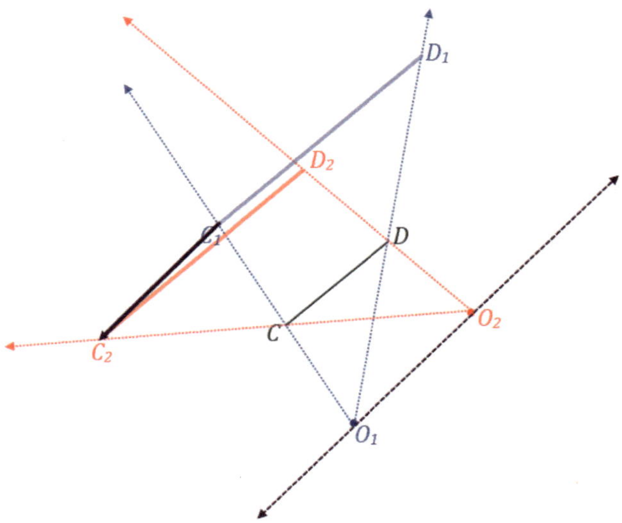

- What do you notice about the translation vector that maps the scale drawings to each other relative to the line that passes through the centers of the dilations, for example, the vector that maps $\overline{A_1B_1}$ to $\overline{A_2B_2}$?

Allow students time to complete this mini-experiment and verify that the translation vector is parallel to the line that passes through the centers of the dilations.

  - *The translation vector is always parallel to the line that passes through the centers of the dilations.*

MP 8

> c. **Generalize the parameters of this example and its results.**
>
> ***The dilation of a figure from two different centers by the same scale factor yields congruent figures that are congruent by a translation along a vector that is parallel to the line through the centers.***

Exercise 1 (4 minutes)

> **Exercise 1**
>
> **Triangle ABC has been dilated with scale factor $\frac{1}{2}$ from centers O_1 and O_2. What can you say about line segments A_1A_2, B_1B_2, and C_1C_2?**
>
> 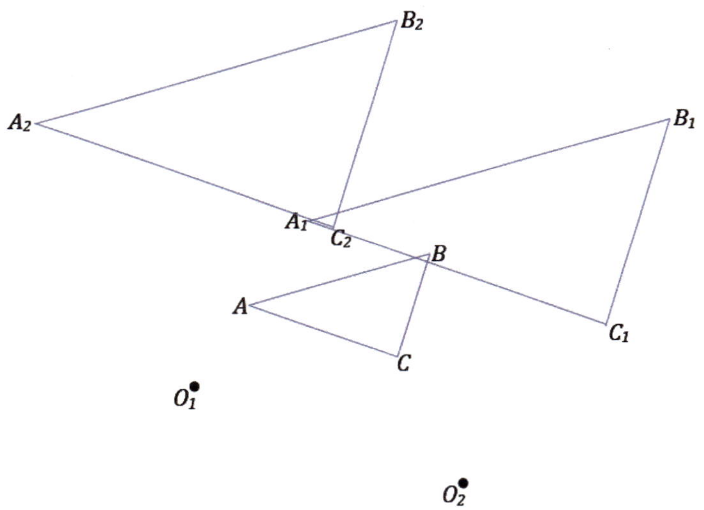
>
>
> ***They are all parallel to the line that passes through O_1O_2.***

Exploratory Challenge 2 (15 minutes)

In Exploratory Challenge 2, students verify experimentally (1) that the composition of dilations is a dilation with scale factor r_1r_2 and (2) that the center of the composition lies on the line $\overleftrightarrow{O_1O_2}$ unless $r_1r_2 = 1$. Students may need poster paper or legal-sized paper to complete part (c).

Exploratory Challenge 2

If Drawing 2 is a scale drawing of Drawing 1 with scale factor r_1 and Drawing 3 is a scale drawing of Drawing 2 with scale factor r_2, what is the relationship between Drawing 3 and Drawing 1?

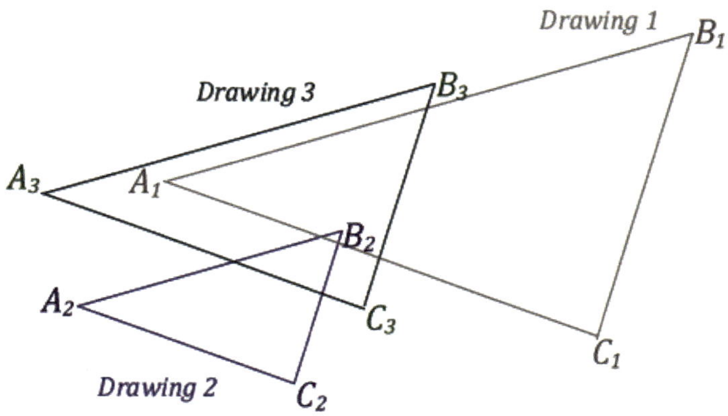

a. **Determine the scale factor and center for each scale drawing. Take measurements as needed.**

The scale factor for Drawing 2 relative to Drawing 1 is $r_1 = \frac{1}{2}$, and the scale factor for Drawing 3 relative to Drawing 2 is $r_2 = \frac{3}{2}$.

b. **What is the scale factor going from Drawing 1 to Drawing 3? Take measurements as needed.**

The scale factor to go from Drawing 1 to Drawing 3 is $r_3 = \frac{3}{4}$.

- Do you see a relationship between the value of the scale factor going from Drawing 1 to Drawing 3 and the scale factors determined going from Drawing 1 to Drawing 2 and Drawing 2 to Drawing 3?

Allow students a moment to discuss before taking responses.

  - *The scale factor to go from Drawing 1 to Drawing 3 is the same as the product of the scale factors to go from Drawing 1 to Drawing 2 and then Drawing 2 to Drawing 3. So the scale factor to go from Drawing 1 to Drawing 3 is $r_1r_2 = \left(\frac{1}{2}\right)\left(\frac{3}{2}\right) = \frac{3}{4}$.*

- To go from Drawing 1 to Drawing 3 is the same as taking a composition of the two dilations: $D_{O_2,\frac{3}{2}}\left(D_{O_1,\frac{1}{2}}\right)$.
- So, with respect to scale factor, a composition of dilations $D_{O_2,r_2}\left(D_{O_1,r_1}\right)$ results in a dilation whose scale factor is r_1r_2.

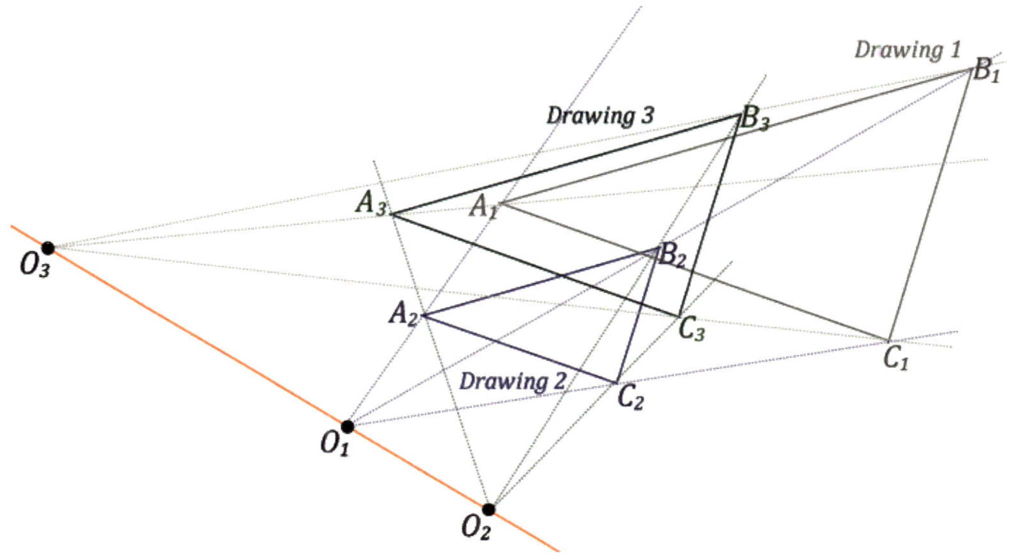

MP.1 & MP.7

c. **Compare the centers of dilations of Drawing 1 (to Drawing 2) and of Drawing 2 (to Drawing 3). What do you notice about these centers relative to the center of the composition of dilations O_3?**

The centers of each for Drawing 1 and Drawing 2 are collinear with the center of dilation of the composition of dilations.

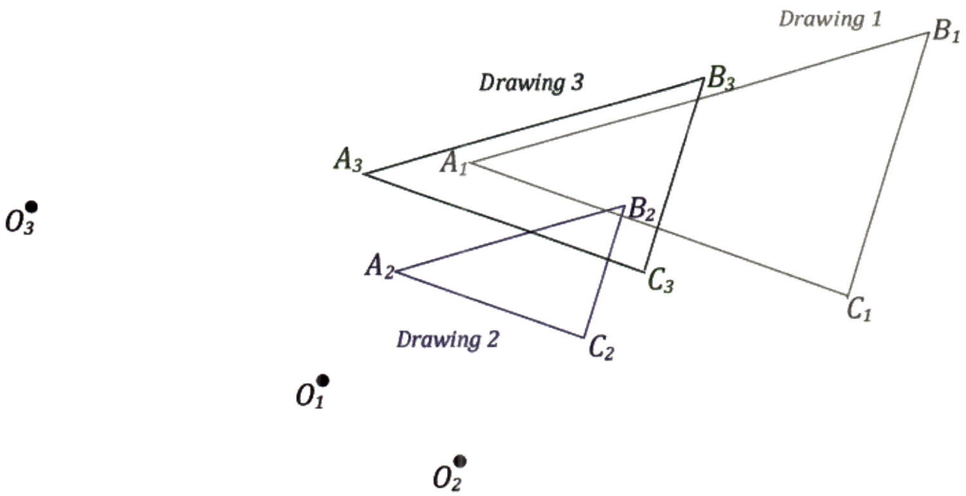

Scaffolding:

Students with difficulty in spatial reasoning can be provided this image and asked to observe what is remarkable about the centers of the dilations.

- From this example, it is tempting to generalize and say that with respect to the centers of the dilations, the center of the composition of dilations $D_{O_2,r_2}(D_{O_1,r_1})$ is collinear with the centers O_1 and O_2, but there is one situation where this is not the case.
- To observe this one case, draw a segment AB that serves as the figure of a series of dilations.
- For the first dilation D_1, select a center of dilation O_1 and scale factor $r_1 = \frac{1}{2}$. Dilate AB and label the result as $A'B'$.
- For the second dilation D_2, select a new center O_2 and scale factor $r_2 = 2$. Determine why the centers of each of these dilations cannot be collinear with the center of dilation of the composition of dilations $D_2(D_1)$.

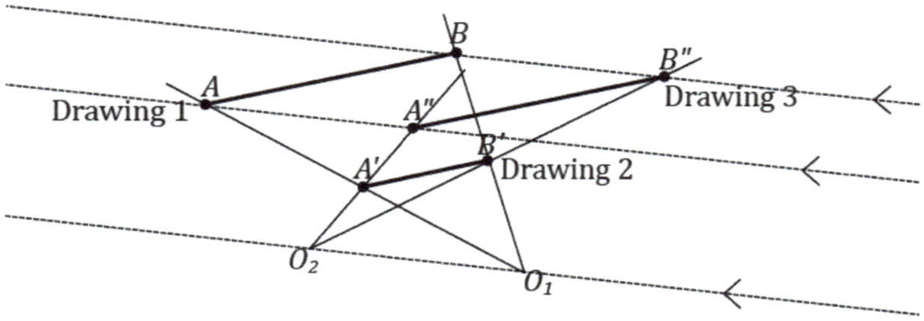

- *Since Drawing 1 and Drawing 3 are identical figures, the lines that pass through the corresponding endpoints of the segments are parallel; a translation will map Drawing 1 to Drawing 3.*

- Notice that this occurs only when $r_1 r_2 = 1$.
- Also, notice that the translation that maps $\overline{AB}$ to $\overline{A''B''}$ must be parallel to the line that passes through the centers of the two given dilations.

d. **Generalize the parameters of this example and its results.**

A composition of dilations, D_{O_1,r_1} and D_{O_2,r_2}, is a dilation with scale factor $r_1 r_2$ and center on $\overleftrightarrow{O_1O_2}$ unless $r_1 r_2 = 1$. If $r_1 r_2 = 1$, then there is no dilation that maps a figure onto the image of the composition of dilations; there is a translation parallel to the line passing through the centers of the individual dilations that maps the figure onto its image.

Exercise 2 (4 minutes)

Exercise 2

Triangle ABC has been dilated with scale factor $\frac{2}{3}$ from center O_1 to get triangle $A'B'C'$, and then triangle $A'B'C'$ is dilated from center O_2 with scale factor $\frac{1}{2}$ to get triangle $A''B''C''$. Describe the dilation that maps triangle ABC to triangle $A''B''C''$. Find the center and scale factor for that dilation.

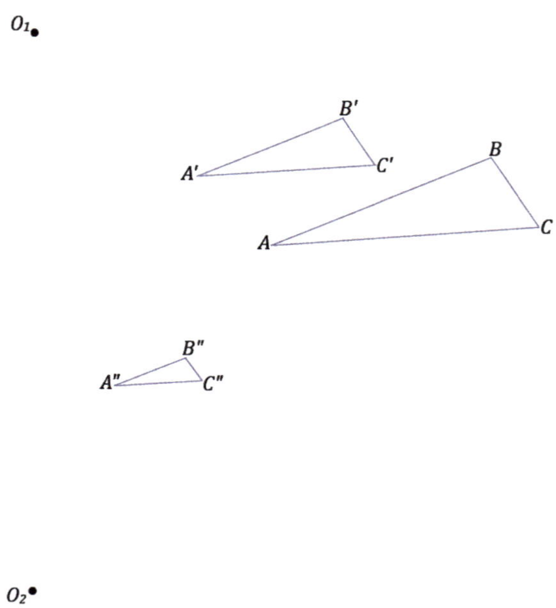

The dilation center is a point on the line segment O_1O_2, and the scale factor is $\frac{2}{3} \cdot \frac{1}{2} = \frac{1}{3}$.

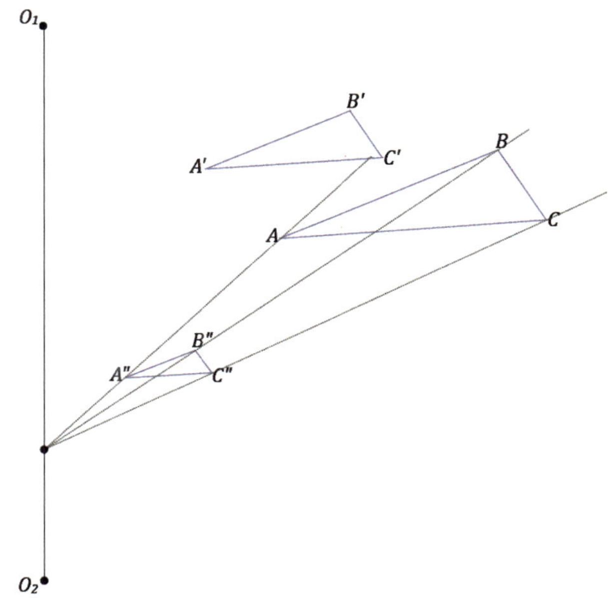

Closing (2 minutes)

- In a series of dilations, how does the scale factor that maps the original figure to the final image compare to the scale factor of each successive dilation?
  - *In a series of dilations, the scale factor that maps the original figure onto the final image is the product of all the scale factors in the series of dilations.*
- We remember here that, unlike the previous several lessons, we did not prove facts in general; we made observations through measurements.

> **Lesson Summary**
>
> **In a series of dilations, the scale factor that maps the original figure onto the final image is the product of all the scale factors in the series of dilations.**

Exit Ticket (5 minutes)

Name ______________________________ Date ______________

Lesson 11: Dilations from Different Centers

Exit Ticket

Marcos constructed the composition of dilations shown below. Drawing 2 is $\frac{3}{8}$ the size of Drawing 1, and Drawing 3 is twice the size of Drawing 2.

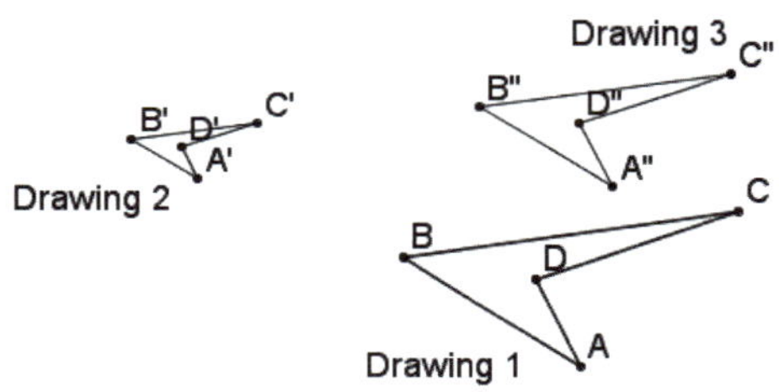

1. Determine the scale factor from Drawing 1 to Drawing 3.

2. Find the center of dilation mapping Drawing 1 to Drawing 3.

Exit Ticket Sample Solutions

Marcos constructed the composition of dilations shown below. Drawing 2 is $\frac{3}{8}$ the size of Drawing 1, and Drawing 3 is twice the size of Drawing 2.

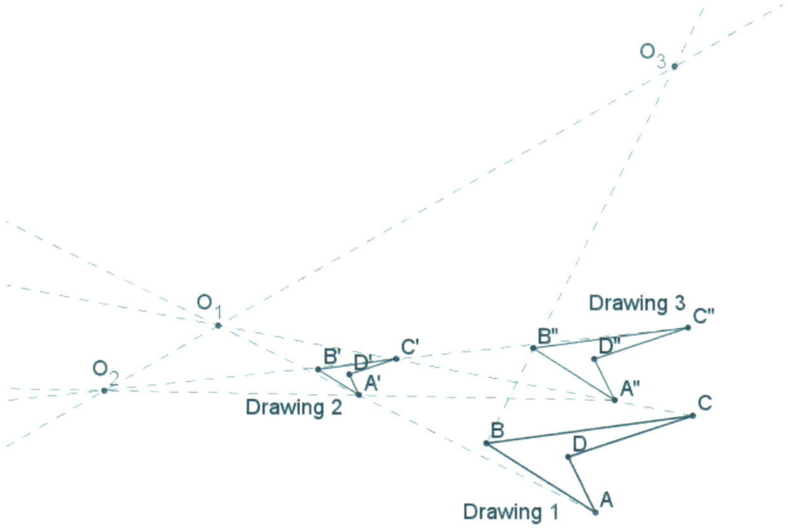

1. **Determine the scale factor from Drawing 1 to Drawing 3.**

   Drawing 2 is a $3:8$ scale drawing of Drawing 1, and Drawing 3 is a $2:1$ scale drawing of Drawing 2, so Drawing 3 is a $2:1$ scale drawing of a $3:8$ scale drawing:

   $$\textit{Drawing 3} = 2\left(\frac{3}{8}\textit{ Drawing 1}\right)$$

   $$\textit{Drawing 3} = \frac{3}{4}\ (\textit{Drawing 1})$$

   The scale factor from Drawing 1 to Drawing 3 is $\frac{3}{4}$.

2. **Find the center of dilation mapping Drawing 1 to Drawing 3.**

   See diagram: The center of dilation is O_3.

Problem Set Sample Solutions

1. **In Lesson 7, the dilation theorem for line segments said that if two different-length line segments in the plane were parallel to each other, then a dilation exists mapping one segment onto the other. Explain why the line segments must be different lengths for a dilation to exist.**

   If the line segments were of equal length, then it would have to be true that the scale factor of the supposed dilation would be $r = 1$; however, we found that any dilation with a scale factor of $r = 1$ maps any figure to itself, which implies that the line segment would have to be mapped to itself. Two different line segments that are parallel to one another implies that the line segments are not one and the same, which means that the supposed dilation does not exist.

2. Regular hexagon $A'B'C'D'E'F'$ is the image of regular hexagon $ABCDEF$ under a dilation from center O_1, and regular hexagon $A''B''C''D''E''F''$ is the image of regular hexagon $ABCDEF$ under a dilation from center O_2. Points A', B', C', D', E', and F' are also the images of points A'', B'', C'', D'', E'', and F'', respectively, under a translation along vector $\overrightarrow{D''D'}$. Find a possible regular hexagon $ABCDEF$.

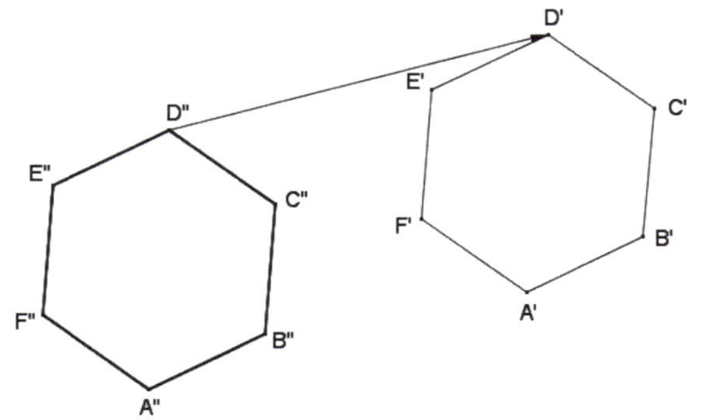

Student diagrams will vary; however, the centers of dilation O_1 and O_2 must lie on a line parallel to vector $\overrightarrow{D''D'}$.

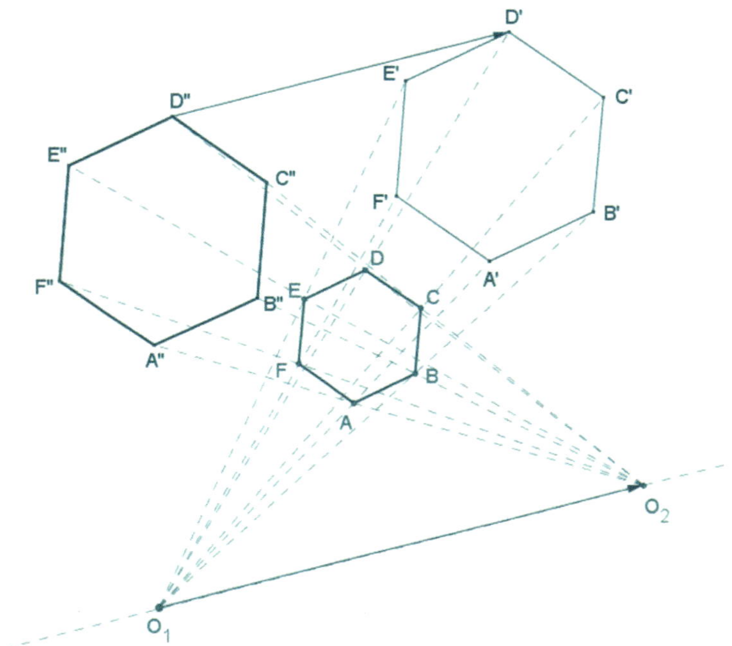

3. A dilation with center O_1 and scale factor $\frac{1}{2}$ maps figure F to figure F'. A dilation with center O_2 and scale factor $\frac{3}{2}$ maps figure F' to figure F''. Draw figures F' and F'', and then find the center O and scale factor r of the dilation that takes F to F''.

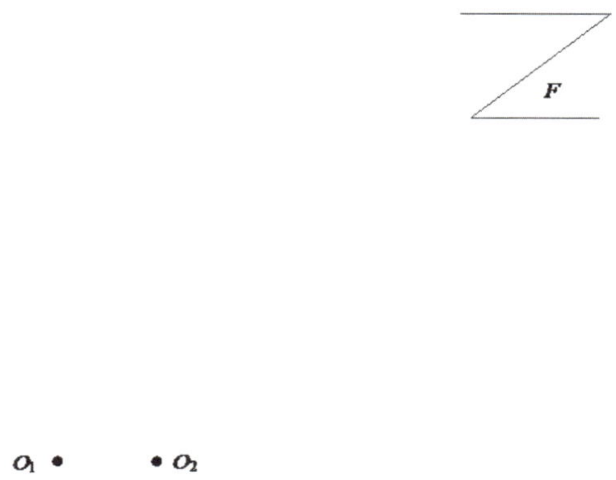

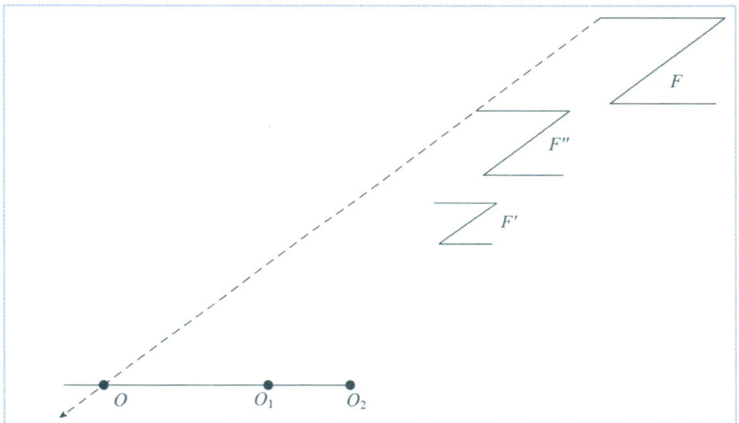

Answer: $r = \frac{3}{4}$

4. A figure T is dilated from center O_1 with a scale factor $r_1 = \frac{3}{4}$ to yield image T', and figure T' is then dilated from center O_2 with a scale factor $r_2 = \frac{4}{3}$ to yield figure T''. Explain why $T \cong T''$.

For any distance, a, between two points in figure T, the distance between corresponding points in figure T' is $\frac{3}{4}a$. For the said distance between points in T', $\frac{3}{4}a$, the distance between corresponding points in figure T'' is $\frac{4}{3}\left(\frac{3}{4}a\right) = 1a$. This implies that all distances between two points in figure T'' are equal to the distances between corresponding points in figure T. Furthermore, since dilations preserve angle measures, angles formed by any three noncollinear points in figure T'' are congruent to the angles formed by the corresponding three noncollinear points in figure T. There is then a correspondence between T and T'' in which distance is preserved and angle measures are preserved, implying that a sequence of rigid motions maps T onto T''; hence, a congruence exists between figures T and T''.

5. A dilation with center O_1 and scale factor $\frac{1}{2}$ maps figure H to figure H'. A dilation with center O_2 and scale factor 2 maps figure H' to figure H''. Draw figures H' and H''. Find a vector for a translation that maps H to H''.

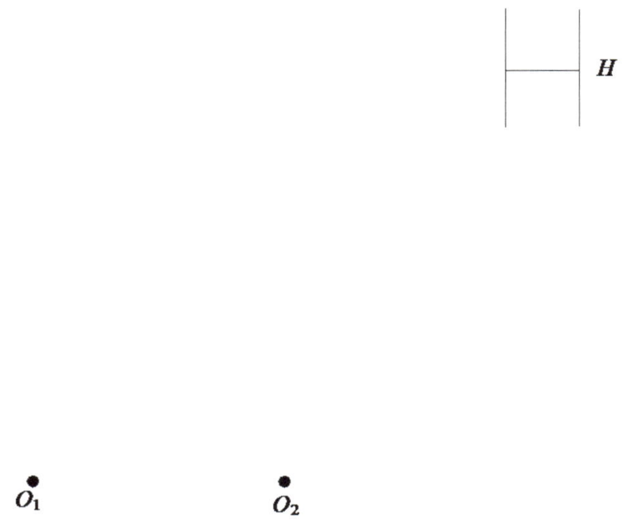

Solution:

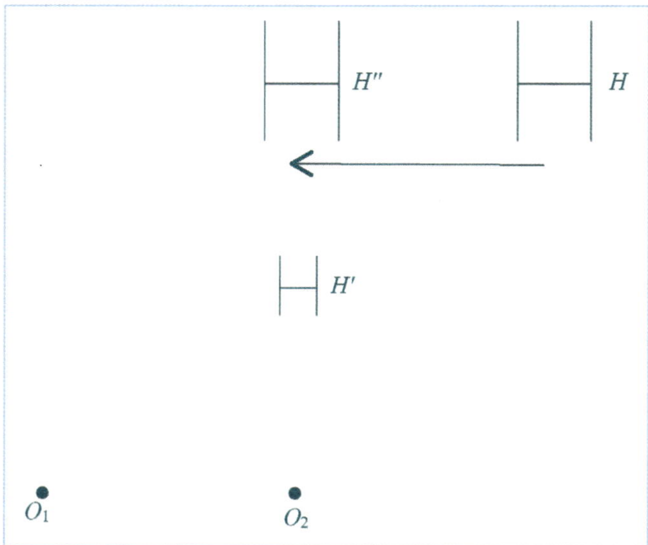

6. Figure W is dilated from O_1 with a scale factor $r_1 = 2$ to yield W'. Figure W' is then dilated from center O_2 with a scale factor $r_2 = \frac{1}{4}$ to yield W''.

a. Construct the composition of dilations of figure W described above.

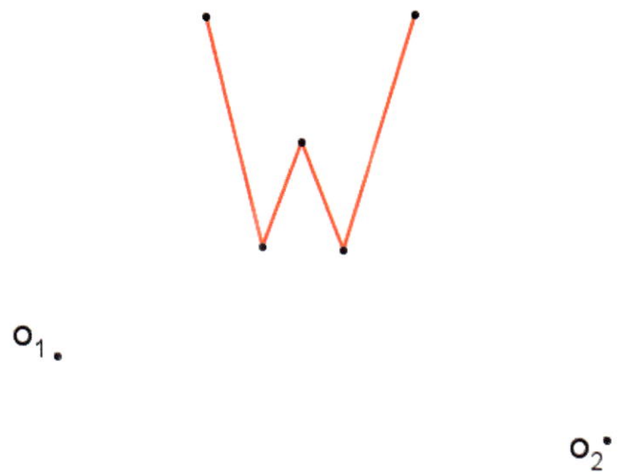

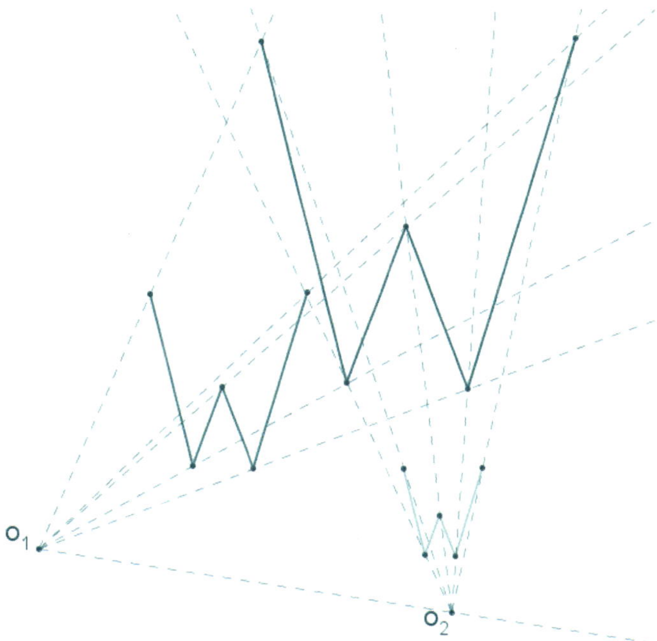

b. If you were to dilate figure W'', what scale factor would be required to yield an image that is congruent to figure W?

In a composition of dilations, for the resulting image to be congruent to the original pre-image, the product of the scale factors of the dilations must be 1.

$$r_1 \cdot r_2 \cdot r_3 = 1$$
$$2 \cdot \frac{1}{4} \cdot r_3 = 1$$
$$\frac{1}{2} \cdot r_3 = 1$$
$$r_3 = 2$$

The scale factor necessary to yield an image congruent to the original pre-image is $r_3 = 2$.

c. **Locate the center of dilation that maps W'' to W using the scale factor that you identified in part (b).**

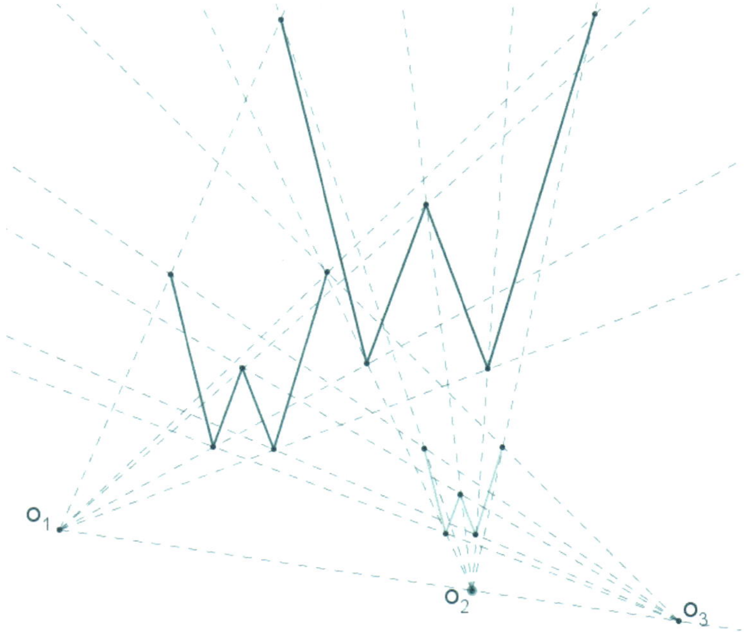

7. **Figures F_1 and F_2 in the diagram below are dilations of F from centers O_1 and O_2, respectively.**

a. **Find F.**

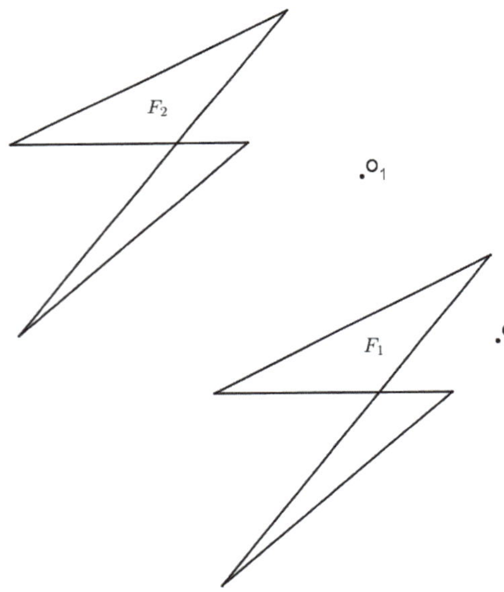

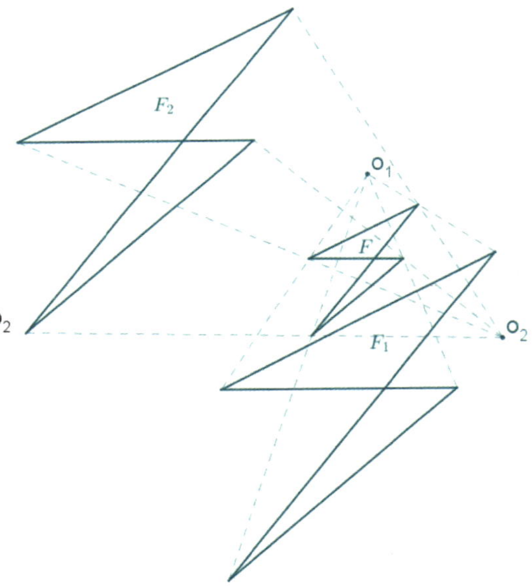

b. **If $F_1 \cong F_2$, what must be true of the scale factors r_1 and r_2 of each dilation?**

The scale factors must be equal.

c. **Use direct measurement to determine each scale factor for D_{O_1,r_1} and D_{O_2,r_2}.**

By direct measurement, the scale factor used for each dilation is $r_1 = r_2 = 2\frac{1}{2}$.

Note to the teacher: Parts of this next problem involve a great deal of mathematical reasoning and may not be suitable for all students.

8. **Use a coordinate plane to complete each part below using $U(2,3)$, $V(6,6)$, and $W(6,-1)$.**

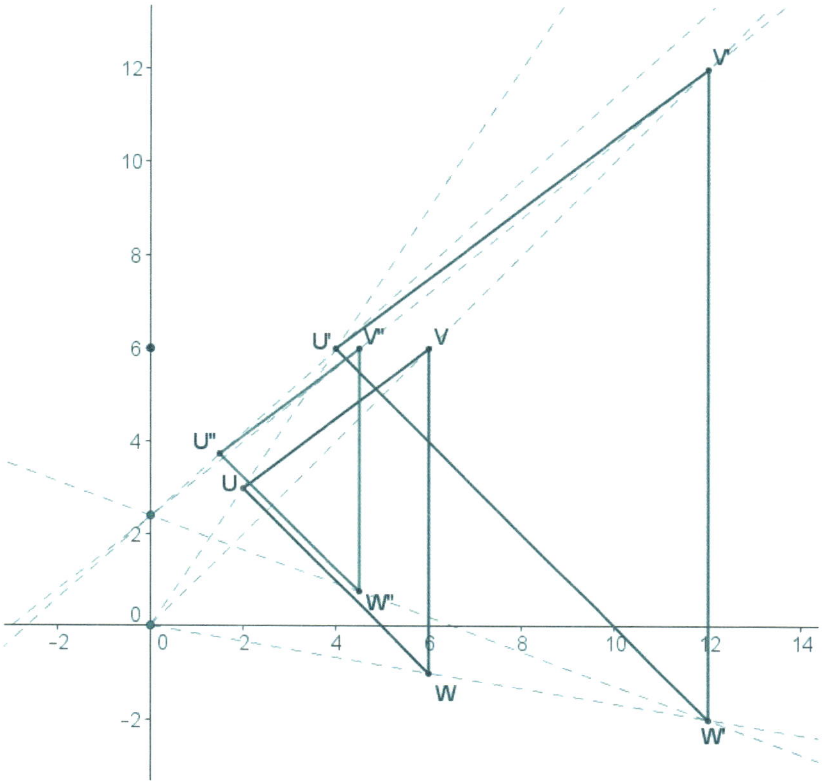

a. **Dilate $\triangle UVW$ from the origin with a scale factor $r_1 = 2$. List the coordinates of image points U', V', and W'.**

$U'(4,6)$, $V'(12,12)$, *and* $W'(12,-2)$

b. **Dilate $\triangle UVW$ from $(0,6)$ with a scale factor of $r_2 = \frac{3}{4}$. List the coordinates of image points U'', V'', and W''.**

The center of this dilation is not the origin. The x-coordinate of the center is 0, so the x-coordinates of the image points can be calculated in the same manner as in part (a). However, the y-coordinates of the pre-image must be considered as their distance from the y-coordinate of the center, 6.

Point U is 3 units below the center of dilation, point V is at the same vertical level as the center of dilation, and point W is 7 units below the center of dilation.

$y_{U''} = 6 + \left[\frac{3}{4}(-3)\right]$	$y_{V''} = 6 + \left[\frac{3}{4}(0)\right]$	$y_{W''} = 6 + \left[\frac{3}{4}(-7)\right]$
$y_{U''} = 6 + \left[-\frac{9}{4}\right]$	$y_{V''} = 6 + [0]$	$y_{W''} = 6 + \left[-\frac{21}{4}\right]$
$y_{U''} = 3\frac{3}{4}$	$y_{V''} = 6$	$y_{W''} = \frac{3}{4}$

$U''\left(1\frac{1}{2}, 3\frac{3}{4}\right)$, $V''\left(4\frac{1}{2}, 6\right)$, *and* $W''\left(4\frac{1}{2}, \frac{3}{4}\right)$

c. **Find the scale factor, r_3, from $\triangle U'V'W'$ to $\triangle U''V''W''$.**

$\triangle U'V'W'$ is the image of $\triangle UVW$ with a scale factor $r_1 = 2$, so it follows that $\triangle UVW$ can be considered the image of $\triangle U'V'W'$ with a scale factor of $r_4 = \frac{1}{2}$. Therefore, $\triangle U''V''W''$ can be considered the image of the composition of dilations $D_{(0,6),\frac{3}{4}}\left(D_{(0,0),\frac{1}{2}}\right)$ of $\triangle U'V'W'$. This means that the scale factor $r_3 = r_4 \cdot r_2$.

$$r_3 = r_4 \cdot r_2$$
$$r_3 = \frac{1}{2} \cdot \frac{3}{4}$$
$$r_3 = \frac{3}{8}$$

d. **Find the coordinates of the center of dilation that maps $\triangle U'V'W'$ to $\triangle U''V''W''$.**

The center of dilation O_3 must lie on the y-axis with centers $(0, 0)$ and $(0, 6)$. Therefore, the x-coordinate of O_3 is 0. Using the graph, it appears that the y-coordinate of O_3 is a little more than 2.

Considering the points V' and V'':

$$\frac{3}{8}(12 - y_O) = 6 - y_O$$
$$\frac{9}{2} - \frac{3}{8}y_O = 6 - y_O$$
$$\frac{9}{2} = 6 - \frac{5}{8}y_O$$
$$-\frac{3}{2} = -\frac{5}{8}y_O$$
$$\frac{24}{10} = y_O = 2.4$$

The center of dilation O_3 is $(0, 2.4)$.

Topic C

Similarity and Dilations

G-SRT.A.2, G-SRT.A.3, G-SRT.B.5, G-MG.A.1

Focus Standards:	G-SRT.A.2	Given two figures, use the definition of similarity in terms of similarity transformations to decide if they are similar; explain using similarity transformations the meaning of similarity for triangles as the equality of all corresponding pairs of angles and the proportionality of all corresponding pairs of sides.
	G-SRT.A.3	Use the properties of similarity transformations to establish the AA criterion for two triangles to be similar.
	G-SRT.B.5	Use congruence and similarity criteria for triangles to solve problems and prove relationships in geometric figures.
	G-MG.A.1	Use geometric shapes, their measures, and their properties to describe objects (e.g., modeling a tree trunk or a human torso as a cylinder).★
Instructional Days:	9	
Lesson 12:	What Are Similarity Transformations, and Why Do We Need Them? (P)[1]	
Lesson 13:	Properties of Similarity Transformations (P)	
Lesson 14:	Similarity (P)	
Lesson 15:	The Angle-Angle (AA) Criterion for Two Triangles to Be Similar (S)	
Lesson 16:	Between-Figure and Within-Figure Ratios (P)	
Lesson 17:	The Side-Angle-Side (SAS) and Side-Side-Side (SSS) Criteria for Two Triangles to Be Similar (E)	
Lesson 18:	Similarity and the Angle Bisector Theorem (P)	
Lesson 19:	Families of Parallel Lines and the Circumference of the Earth (S)	
Lesson 20:	How Far Away Is the Moon? (S)	

[1]Lesson Structure Key: **P**-Problem Set Lesson, **M**-Modeling Cycle Lesson, **E**-Exploration Lesson, **S**-Socratic Lesson

With an understanding of dilations, students are now ready to study similarity in Topic C. This is an appropriate moment to pause and reflect on the change in how the study of similarity is conducted in this curriculum versus traditional geometry curricula. It is common to open to a similarity unit in a traditional textbook and read about polygons, chiefly triangles, which are of the same shape but different size. Some may emphasize the proportional relationship between corresponding sides early in the unit. The point is that grade-school mathematics has traditionally packaged the concept of similarity into a distilled version of the bigger picture. The unpackaged view requires a more methodical journey to arrive at the concept of similarity, including the use of transformations. It is in Topic C, after a foundation of scale drawings and dilations, that we can discuss similarity.

Lesson 12 introduces the concept of a similarity transformation, which is needed to identify figures as being similar. Just as with rigid motions and congruence, the lesson intentionally presents curvilinear examples to emphasize that the use of similarity transformations allows us to compare both rectilinear and curvilinear figures. Next, in Lesson 13, students apply similarity transformations to figures by construction. This is the only lesson where students actually perform similarity transformations. The goals are to simply be able to apply a similarity as well as observe how the properties of the individual transformations that compose each similarity hold throughout construction. In Lesson 14, students observe the reflexive, symmetric, and transitive properties of similarity. The scope of figures used in Lessons 15 through 18 narrows to triangles. In these lessons, students discover and prove the AA, SSS, and SAS similarity criteria. Students use these criteria and length relationships between similar figures and within figures to solve for unknown lengths in triangles (**G-SRT.A.3**, **G-SRT.B.5**). Note that when students solve problems in Lesson 16, they are using geometric shapes, their measures and properties to describe situations (e.g., similar triangles) is work related to the modeling standard **G-MG.A.1**. Lessons 19 and 20 are modeling lessons (**G-MG.A.1**) that lead students through the reasoning the ancient Greeks used to determine the circumference of the earth (Lesson 19) and the distance from the earth to the moon (Lesson 20).

Lesson 12: What Are Similarity Transformations, and Why Do We Need Them?

Student Outcomes

- Students define a *similarity transformation* as the composition of basic rigid motions and dilations. Students define two figures to be similar if there is a similarity transformation that takes one to the other.
- Students describe a similarity transformation applied to an arbitrary figure (i.e., not just triangles) and use similarity to distinguish between figures that resemble each other versus those that are actually similar.

Lesson Notes

As noted earlier in this curriculum, congruence and similarity are presented differently here than in most previous curricula. In the past, congruence criteria have first been defined for triangles (e.g., SSS or ASA), and then, by extension, for polygonal figures, with similarity being treated in a like manner. However, the Common Core Standards approach these concepts via transformations, allowing one to accommodate not only polygonal figures, but also curvilinear figures, in one stroke.

Students begin Topic C with an understanding of what similarity transformations are and what it means for figures to be similar. They should see how similarity transformations are like the rigid motions in their use to compare figures in the plane. Unlike the work they did with similarity in Grade 8, students study similarity in the plane and not in the coordinate plane. This is, of course, because we want students to fully realize what is referred to as the "abundance" of transformations in the plane. It is not that transformations are limited in the coordinate system but rather that the coordinate system simply encourages students to see certain transformations as more natural than others (e.g., a translation parallel to the x-axis: $(x, y) \mapsto (x + a, y)$). Removing the coordinate system prevents this natural gravitation toward certain transformations over others.

Classwork

Opening Exercise (3 minutes)

Opening Exercise

Observe Figures 1 and 2 and the images of the intermediate figures between them. Figures 1 and 2 are called *similar*.

What observations can we make about Figures 1 and 2?

Answers will vary; students might say that the two figures look alike in terms of shape but not size. Accept reasonable answers at this point to start the conversation, and move on to filling out the chart below.

Figure 1 Figure 2

Definition:

A _________________ _________________ (or _________________) is a composition of a finite number of dilations or basic rigid motions. The *scale factor* of a similarity transformation is the product of the scale factors of the dilations in the composition. If there are no dilations in the composition, the scale factor is defined to be 1.

similarity transformation, similarity

Definition:

Two figures in a plane are ____________________ if there exists a similarity transformation taking one figure onto the other figure.

similar

> *Scaffolding:*
>
> Depending on student ability, consider using a fill-in-the-blank approach to the definitions listed here.

Direct students to sketch possibilities for the Examples and Non-Examples boxes, and offer them the provided example after they voice their ideas. Have them list characteristics of the Examples, and then provide them with the definitions of *similar* and *similarity transformation* (see definitions above).

<table>
<tr><td>Definition
A similarity transformation (or similarity) is a composition of a finite number of dilations or basic rigid motions. The scale factor of a similarity transformation is the product of the scale factors of the dilations in the composition. If there are no dilations in the composition, the scale factor is defined to be 1.

Two figures in a plane are similar if there exists a similarity transformation taking one figure onto the other figure.</td><td>Characteristics
Similar figures should look the same, but one is a different size, flipped, rotated, or translated relative to the other.</td></tr>
<tr><td colspan="2">similar</td></tr>
<tr><td>Examples
</td><td>Non-Examples
</td></tr>
</table>

Discussion (10 minutes)

- Consider what you know about congruence when thinking about similarity. One use of the rigid motions is to establish whether two figures are identical in the plane. How did we use rigid motions to establish this?
  - *If a series of rigid motions mapped one figure onto the other and the figures coincided, we could conclude that they were congruent.*

- We can use similarity transformations in the same way. Consider Figure 1 below.

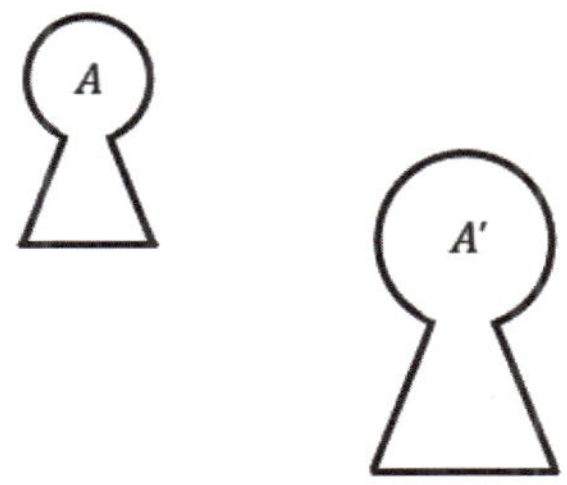

Figure 1

- From our work on dilations, we can see that there is, in fact, a dilation that would map figure A to A'. Note that a similarity transformation does not have a minimum number of dilations or rigid motions (e.g., a single reflection or a single dilation is a similarity transformation.)

Note that the similarity symbol "$\sim$" has not been mentioned in this lesson. If students remember it and have no trouble with it, feel free to discuss it. The symbol is addressed in Lesson 14.

- Now examine Figure 2. In Figure 2, the figure A' was rotated $90°$ and is now labeled as A''.

Figure 2

- Would it be correct to say that A'' is a dilation of A?
  - *No. This is not a dilation because corresponding segments are neither parallel nor collinear.*
- Yet we saw in Figure 1 that it is possible to transform A to A', which we know to be congruent to A''. So what are the necessary steps to map A to A''?

Allow a moment for students to discuss this. Confirm that either both the composition of a dilation and rotation or the composition of a dilation, rotation, and translation maps A to A''.

- The series of steps needed to map A to A'', the dilation and rotation or rotation and dilation, can be thought of as a composition of transformations, or more specifically, a similarity transformation. $A'' \cong R_{C,\theta}\left(D_{O,r}(A)\right)$
- If a similarity transformation maps one figure to another, we say the figures are similar.
- Note this important distinction. We know that it is not enough to say, "If two figures look identical, they must be congruent." We know that they are congruent only if a series of rigid motions maps one figure to the other. In the same way, it is not enough to say that two figures look like they have the same shape; we have to show that a similarity transformation maps one figure to the other to be sure that the figures really do have the same shape.

- Recall also that a scale drawing of a figure is one whose corresponding lengths are proportional and whose corresponding angles are equal in measurement. We know that a dilation produces a scale drawing. Therefore, figures that are similar must be scale drawings. Why must this be true?
  - *Any figure that maps onto another figure by similarity transformation T either has a finite number of dilations or does not have any dilations. If there are dilations involved, we have seen that dilations result in figures with proportional, corresponding lengths and corresponding angles of equal measurement. If there are no dilations, then the rigid motions that compose the similarity transformation have a scale factor of $r = 1$ by definition. Therefore, in either case, the two similar figures are scale drawings of each other.*

Note that we have not said that figures that are scale drawings must be similar. This is discussed in Lesson 14.

- We denote a similarity transformation with T. The transformations that compose a similarity transformation can be in any order; however, as a matter of convention, we usually begin a similarity transformation with the dilation (as we did in Grade 8) and follow with rigid motions.

Note that this convention is apparent in problems where students must describe the series of transformations that map one figure onto its similar image; the convention is adhered to so that in the first step the two figures become congruent, and then students are left to determine the congruence transformation that map one to the other.

- If T is a similarity transformation, then T is the composition of basic rigid motions and dilations. The scale factor r of T is the product of the scale factors of the dilations in the transformation. With respect to the above example, $T(A) = R_{C,\theta}\left(D_{O,r}(A)\right)$.
- If there is no dilation in the similarity transformation, it is a congruence. However, a congruence is simply a more specific similarity transformation, which is why the definition allows for the composition of transformations that need not include a dilation.

Example 1 (5 minutes)

Students identify the transformations that map one figure onto another. Remind students that, as a matter of convention, any dilation in a similarity transformation is identified first.

Scaffolding:

Curvilinear figures, such as that shown in Example 1, may be difficult for some students. Try using the same similarity transformations with simpler figures, such as asymmetrical letters like L and F, and then scaffold up to those involving curves such as P and Q.

Example 1

Figure Z' is similar to Figure Z. Describe a transformation that maps Figure Z onto Figure Z'.

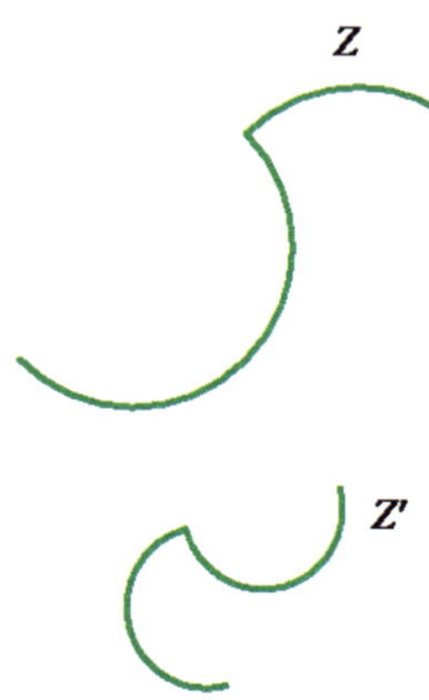

- We are not looking for specific parameters (e.g., scale factor or degree of rotation of each transformation); rather, we want to identify the series of transformations needed to map Figure Z to Figure Z'.
- Step 1: The dilation has a scale factor of $r < 1$ since Z' is smaller than Z.
- Step 2: Notice that Z' is *flipped* from Z_1. So, take a reflection of Z_1 to get Z_2 over a line l.
- Step 3: Translate the plane such that a point of Z_2 maps to a corresponding point in Z'. Call the new figure Z_3.
- Step 4: Rotate until Z_3 coincides with Z'.

Scaffolding:
Consider having more advanced students sketch the general locations of centers of rotation or dilation, lines of reflection, and translation vectors.

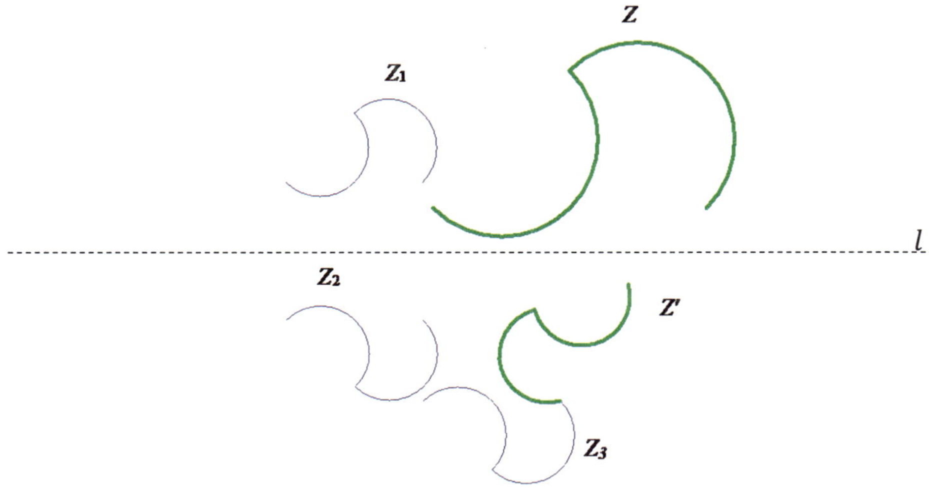

Figure Z maps to Figure Z' by first dilating by a scale factor of r until the corresponding lengths are equal in measurement and then reflecting over line l to ensure that both figures have the same orientation. Next, translate along a vector so that one point of the image corresponds to Z'. Finally, a rotation around a center C of degree θ orients Figure Z so that it coincides with Figure Z'.

Exercises 1–3 (8 minutes)

Exercises 1–3 allow students the opportunity to practice describing transformations that map an original figure onto its corresponding transformed figure. If time is an issue, have students complete one exercise that seems appropriate, and move on to Example 2.

Exercises 1–3

1. **Figure 1 is similar to Figure 2. Which transformations compose the similarity transformation that maps Figure 1 onto Figure 2?**

   First, dilate Figure 1 by a scale factor of $r > 1$ until the corresponding lengths are equal in measurement, and then reflect over a line l so that Figure 1 coincides with Figure 2.

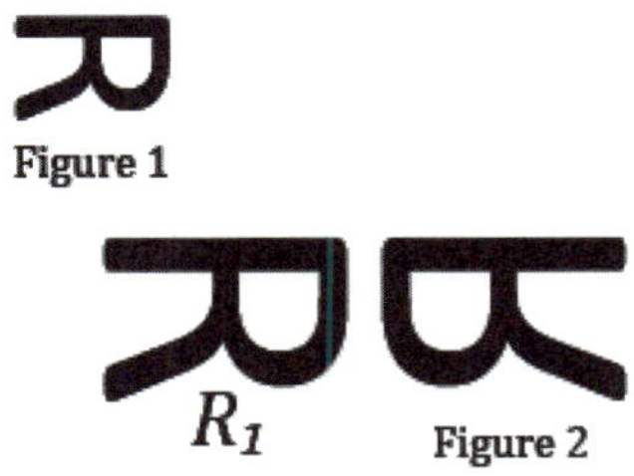

2. Figure S is similar to Figure S'. Which transformations compose the similarity transformation that maps S onto S'?

First, dilate S by a scale factor of $r > 1$ until the corresponding segment lengths are equal in measurement to those of S'. Then, S must be rotated around a center C of degree θ so that S coincides with S'.

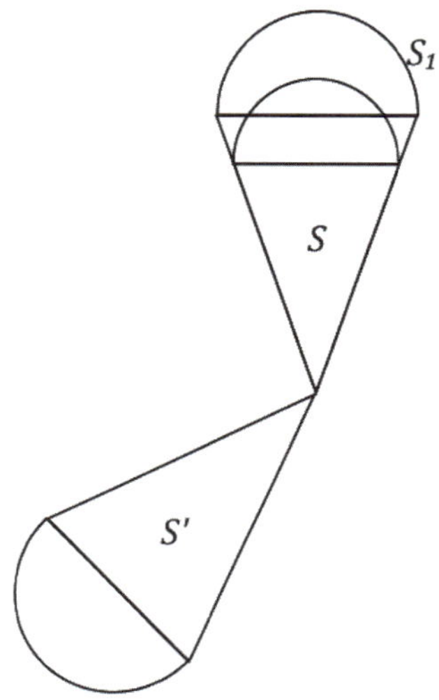

3. Figure 1 is similar to Figure 2. Which transformations compose the similarity transformation that maps Figure 1 onto Figure 2?

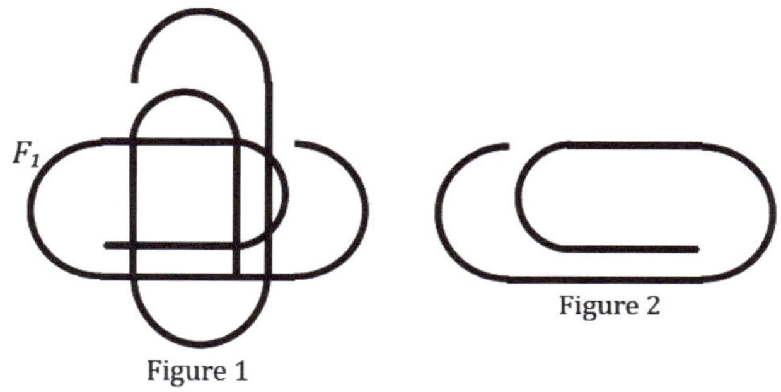

It is possible to only use two transformations: a rotation followed by a reflection. However, to do this, the correct center must be found. The solution image reflects this approach, but students may say to first rotate Figure 1 around a center C by $90°$ in the clockwise direction. Then, reflect Figure 1 over a vertical line ℓ. Finally, translate Figure 1 by a vector so that Figure 1 coincides with Figure 2.

Reemphasize to students that a similarity transformation does not *need* to have a dilation. Just as a square is a special type of rectangle, but because the relationship does not work in reverse, so is a congruence transformation. Similarity transformations generalize the notion of congruency.

Example 2 (5 minutes)

Remind students that the question asks them to take measurements in Example 2. If needed, further prompt them to consider measurements of segments, not of any curved segment. This should alert them that there are few possible measurements to make and that a relationship must exist between these measurements as well as what the question is asking.

EUREKA MATH™

Example 2

Show that no sequence of basic rigid motions and dilations takes the small figure to the large figure. Take measurements as needed.

A similarity transformation that maps the small outer circle to the large outer circle would need to have scale factor of about 2. A similarity transformation that maps the small line segment AB to the large line segment CD would need to have scale factor of about 4. So, there is no similarity transformation that maps the small figure onto the large figure.

Exercises 4–5 (7 minutes)

Exercises 4–5

4. **Is there a sequence of dilations and basic rigid motions that takes the large figure to the small figure? Take measurements as needed.**

   A similarity transformation that maps segment AB to segment WX would need to have scale factor of about $\frac{2}{3}$, but a similarity transformation that maps the small segment CD to segment YZ would need to have scale factor of about $\frac{1}{2}$. So, there is no similarity transformation that maps the large figure onto the small figure.

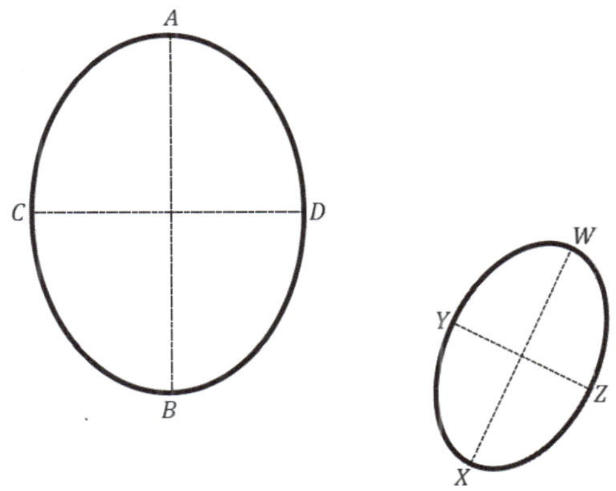

5. **What purpose do transformations serve? Compare and contrast the application of rigid motions to the application of similarity transformations.**

   We use all the transformations to compare figures in the plane. Rigid motions are distance preserving while dilations, integral to similarity transformations, are not distance preserving. We use compositions of rigid motions to determine whether two figures are congruent, and we use compositions of rigid motions and dilations, specifically similarity transformations, to determine whether figures are similar.

Closing (2 minutes)

- What does it mean for two figures to be similar?
  - *We classify two figures as similar if there exists a similarity transformation that maps one figure onto the other.*
- What are similarity transformations, and how can we use them?
  - *A similarity transformation is a composition of a finite number of dilations or rigid motions. Similarity transformations precisely determine whether two figures have the same shape (i.e., two figures are similar). If a similarity transformation does map one figure onto another, we know that one figure is a scale drawing of the other.*
- How do congruence and similarity transformations compare to each other?
  - *Both congruence and similarity transformations are a means of comparing figures in the plane. A congruence transformation is also a similarity transformation, but a similarity transformation does not need to be a congruence transformation.*

> **Lesson Summary**
>
> **Two figures are similar if there exists a similarity transformation that maps one figure onto the other.**
>
> **A similarity transformation is a composition of a finite number of dilations or rigid motions.**

Exit Ticket (5 minutes)

Name ______________________________ Date ______________

Lesson 12: What Are Similarity Transformations, and Why Do We Need Them?

Exit Ticket

1. Figure A' is similar to Figure A. Which transformations compose the similarity transformation that maps Figure A onto Figure A'?

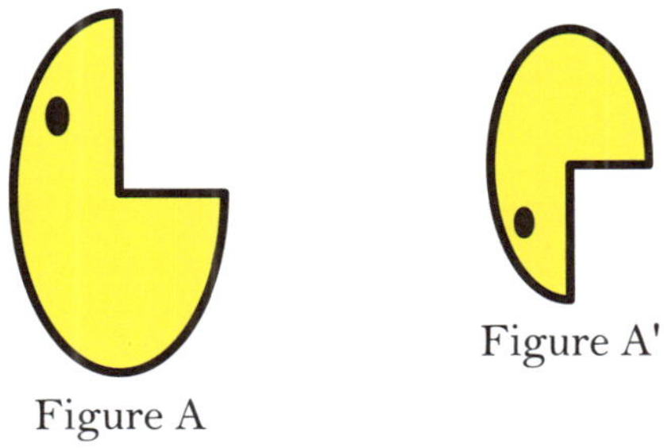

2. Is there a sequence of dilations and basic rigid motions that takes the small figure to the large figure? Take measurements as needed.

Exit Ticket Sample Solutions

1. Figure A' is similar to Figure A. Which transformations compose the similarity transformation that maps Figure A onto Figure A'?

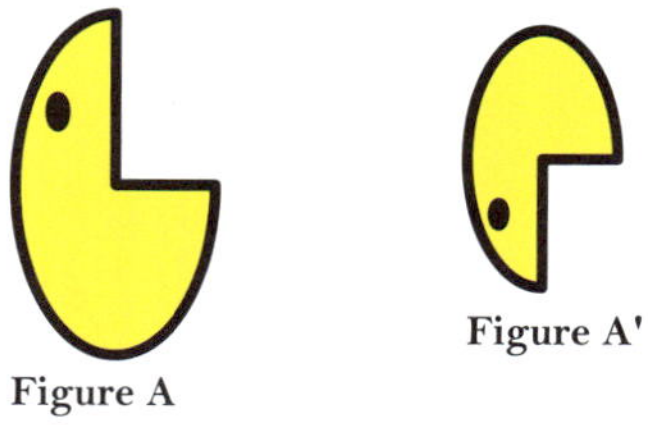

One possible solution: We first take a dilation of Figure A with a scale factor of $r < 1$ and center O, the point where the two line segments meet, until the corresponding lengths are equal to those in Figure A'. Next, take a rotation $(180°)$ about O, and then, finally, take a reflection over a (vertical) line ℓ.

2. Is there a sequence of dilations and basic rigid motions that takes the small figure to the large figure? Take measurements as needed.

No similarity transformation exists because the circled corresponding distances and the corresponding distances marked by the arrows on Figure B are not in the same ratio.

Problem Set Sample Solutions

1. What is the relationship between scale drawings, dilations, and similar figures?
   a. How are scale drawings and dilations alike?

   Scale drawings and dilated figures are alike in that all corresponding angles are congruent and all corresponding distances are in the equivalent ratio, r, called the scale factor. A dilation of a figure produces a scale drawing of that figure.

   b. How can scale drawings and dilations differ?

   Dilations are a transformation of the plane in which all corresponding points from the image and pre-image are mapped along rays that originate at the center of dilation. This is not a requirement for scale drawings.

   c. What is the relationship of similar figures to scale drawings and dilations?

   Similar figures are scale drawings because they can be mapped together by a series of dilations and rigid motions.

2. Given the diagram below, identify a similarity transformation, if one exists, that maps Figure A onto Figure B. If one does not exist, explain why.

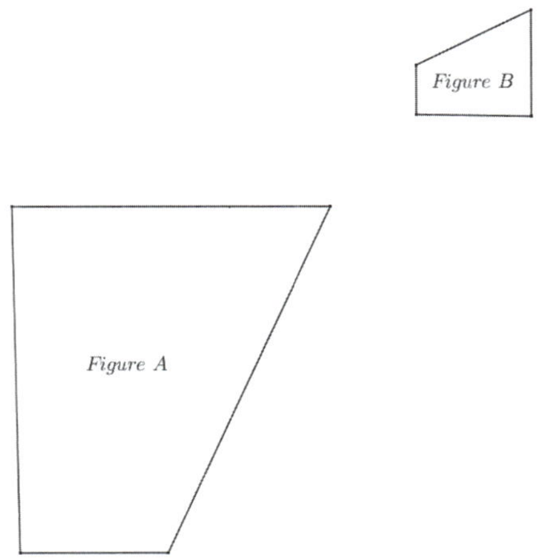

(Note to the teacher: The solution below is only one of many valid solutions to this problem.)

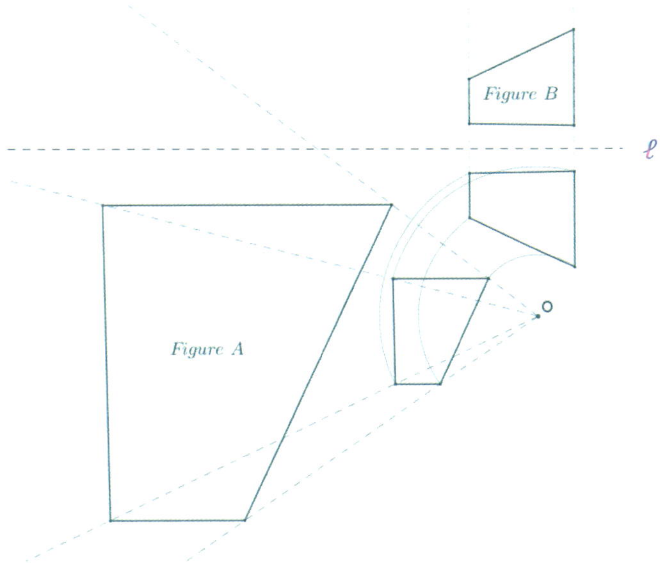

First, Figure A is dilated from center O with a scale factor of $\frac{1}{3}$. Next, the image is rotated $-90°$ about center O. Finally, the image is reflected over horizontal line ℓ onto Figure B.

3. Teddy correctly identified a similarity transformation with at least one dilation that maps Figure I onto Figure II. Megan correctly identified a congruence transformation that maps Figure I onto Figure II. What must be true about Teddy's similarity transformation?

If Megan correctly identified a congruence transformation that maps Figure I onto Figure II, then Figure I and Figure II must be congruent. Therefore, Teddy's similarity transformation must have either included a single dilation with a scale factor of 1 or must have included more than one dilation of which the product of all scale factors was 1 because it included at least one dilation.

4. **Given the coordinate plane shown, identify a similarity transformation, if one exists, that maps X onto Y. If one does not exist, explain why.**

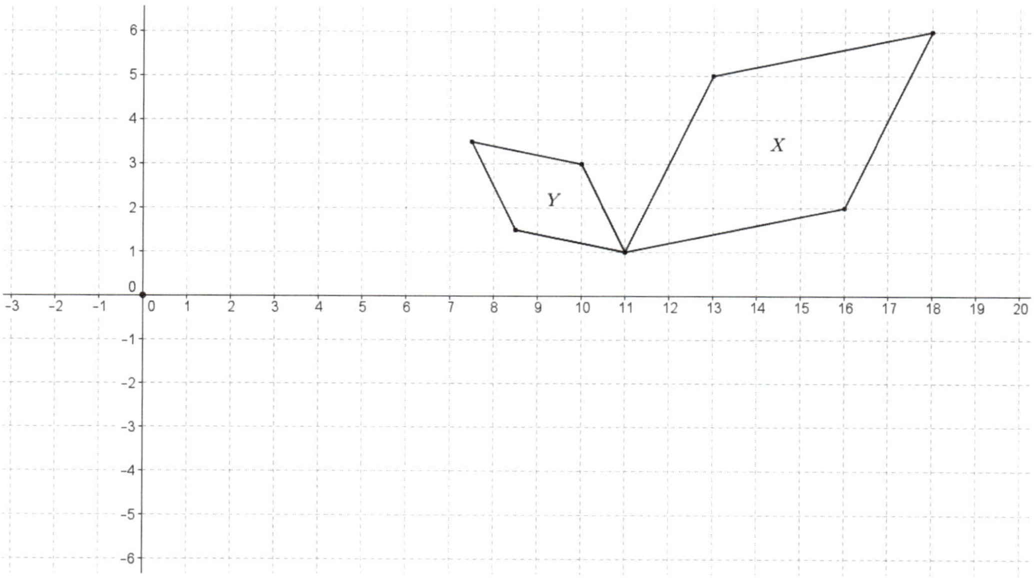

(Note to the teacher: The solution below is only one of many valid solutions to this problem.)

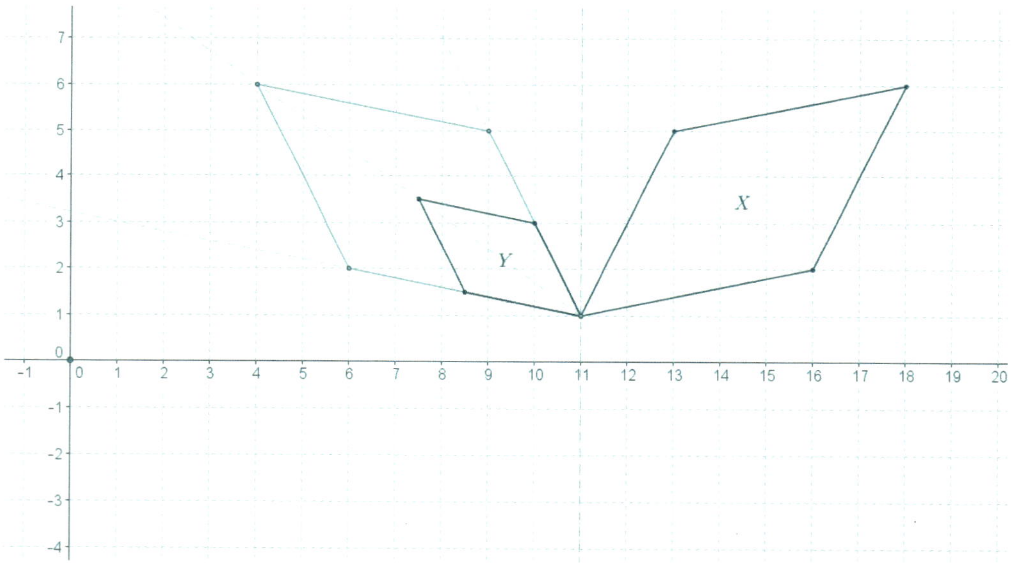

First, reflect X over line $x = 11$. Then, dilate the image from center $(11, 1)$ with a scale factor of $\frac{1}{2}$ to obtain Y.

5. **Given the diagram below, identify a similarity transformation, if one exists, that maps G onto H. If one does not exist, explain why. Provide any necessary measurements to justify your answer.**

A similarity transformation does not exist that maps G onto H because the side lengths of the figures are not all proportional. Figure G is a rectangle (not a square), whereas Figure H is a square.

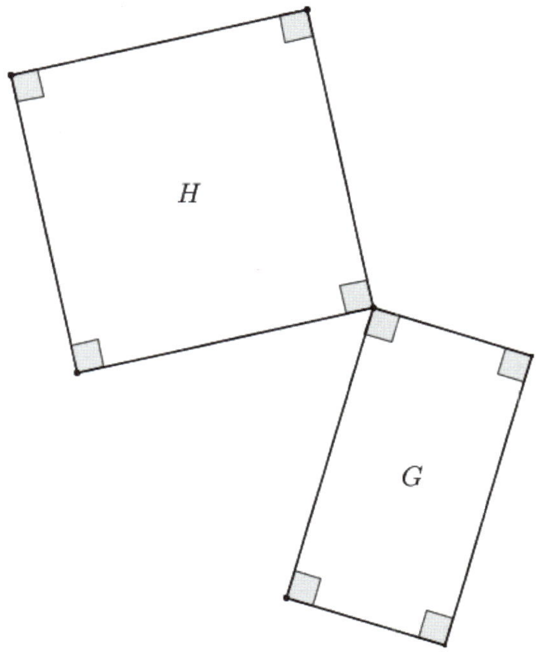

6. **Given the coordinate plane shown, identify a similarity transformation, if one exists, that maps $ABCD$ onto $A'''B'''C'''D'''$. If one does not exist, explain why.**

(Note to the teacher: Students need to use a protractor to obtain the correct degree measure of rotation. The solution below is only one of many valid solutions to this problem.)

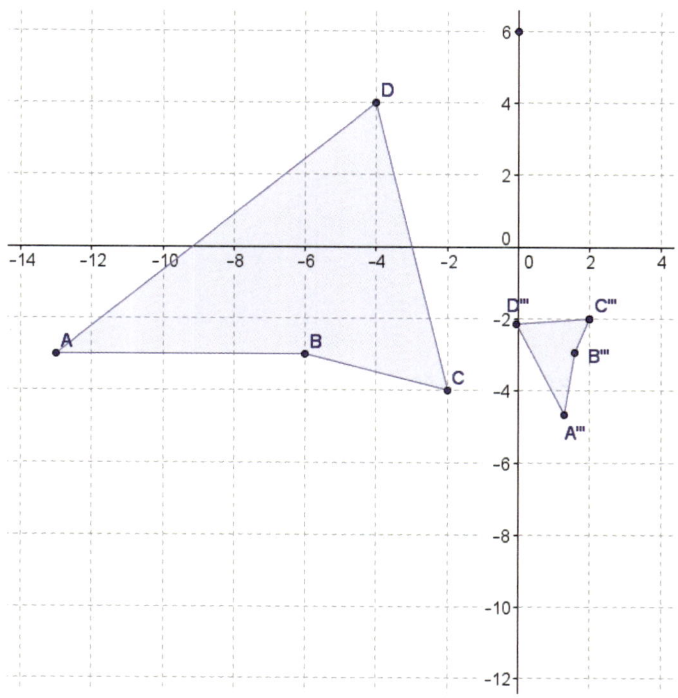

$ABCD$ can be mapped onto $A'''B'''C'''D'''$ by first translating along the vector $\overrightarrow{CC'''}$, then rotating about point C''' by $80°$, and finally dilating from point C''' using a scale factor of $\frac{1}{4}$.

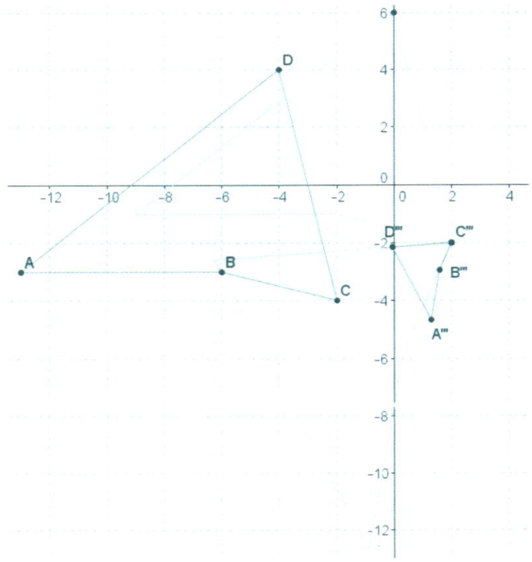

7. **The diagram below shows a dilation of the plane ... or does it? Explain your answer.**

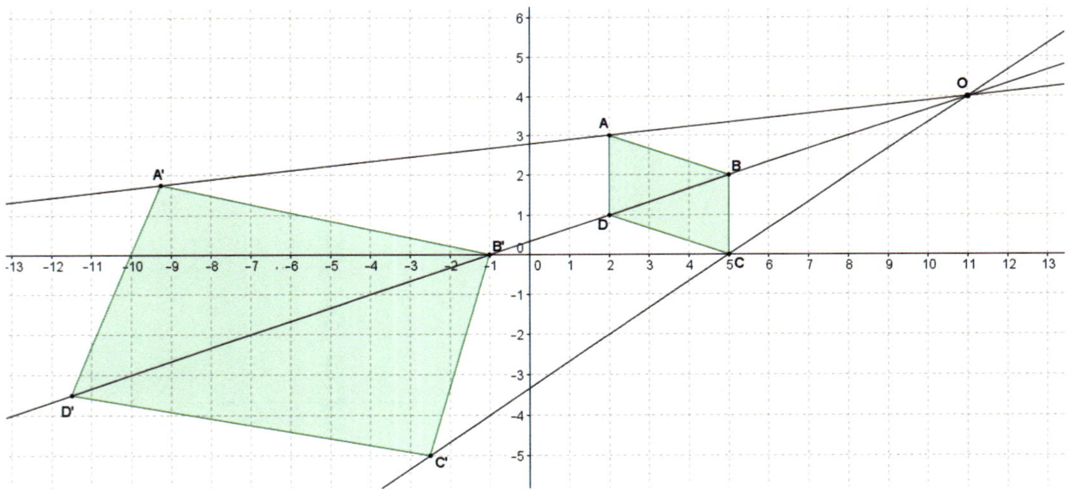

The diagram does not show a dilation of the plane from point O, even though the corresponding points are collinear with the center O. To be a dilation of the plane, a constant scale factor must be used for all points from the center of dilation; however, the scale factor relating the distances from the center in the diagram range from 2 to 2.5.

Lesson 13: Properties of Similarity Transformations

Student Outcomes

- Students know the properties of a similarity transformation are determined by the transformations that compose the similarity transformation.
- Students are able to apply a similarity transformation to a figure by construction.

Lesson Notes

In Lesson 13, students apply similarity transformations to figures by construction. It is important to note here that teachers should emphasize unhurried, methodical drawing and careful use of tools to students. Each exercise entails many construction marks, and part of students' success depends on their perseverance. This is the only lesson where students actually construct what happens to a figure that undergoes a similarity transformation; students experience this process once to witness how the points of a figure move about the plane instead of just hearing about it or describing it.

Just as part of any lesson preparation, it is a good idea to do the examples to better anticipate where students might struggle. Teachers should tailor the number of examples for their respective classes. Examples of varying difficulty have been provided so that teachers have options to differentiate for their diverse classrooms.

Finally, space is available in the student books, but teachers may prefer to work outside of the books to maximize available space. This can be done by photocopying the initial image onto blank paper. The initial image of each problem is provided at the close of the lesson.

Classwork

Opening (10 minutes)

- We have spent a good deal of time discussing the properties of transformations. For example, we know that the property that distinguishes dilations from rigid motions is that dilations do not preserve distance, whereas translations, rotations, and reflections do preserve distance.
- Take a few moments with a partner to list the properties that all transformations have in common.
- These properties that are true for all dilations, reflections, rotations, and translations (the transformations that comprise similarity transformations) also hold true for similarity transformations in general. Title your list as "Properties of Similarity Transformations."

Allow students time to develop as complete a list as possible of the properties. Develop a comprehensive class list. Consider having a premade poster with all the properties listed and keeping each property covered. This can be done on a poster

Scaffolding:

- To help organize students' thinking, write a list of all possible similarity transformations that are composed of exactly two different transformations. Consider why those that contain a dilation do not preserve distance, and keep all other properties that are consistent across each of the similarity transformations in the list as properties of similarity transformations.
- Two examples of similarity transformations are (1) a translation and reflection and (2) a reflection and dilation.

with a strip of paper or on a projector or interactive white board. As students list the properties they recall, reveal that property from the poster. After students have offered their lists, review any remaining properties that were not mentioned. A few words are mentioned below in anticipation of the properties that students may not recall.

Properties of similarity transformations:

1. Distinct points are mapped to distinct points.
   - This means that if $P \neq Q$, then for a transformation G, $G(P) \neq G(Q)$.

2. Each point P' in the plane has a pre-image.
   - If P' is a point in the plane, then $P' = G(P)$ for some point P.

> *Scaffolding:*
> Post the complete list of properties of similarity transformations in a prominent place in the classroom.

3. There is a scale factor r for G so that for any pair of points P and Q with images $P' = G(P)$ and $Q' = G(Q)$, then $P'Q' = rPQ$.
   - The scale factor for a similarity transformation is the product of the scale factors. Remember, the scale factor associated to any congruence transformation is 1. The scale factor of a similarity transformation is really the product of the scale factors of all the transformations that compose the similarity; however, since we know the scale factor of all rigid motions is 1, the scale factor of the similarity transformation is the product of the scale factors of all the dilations in the similarity.

4. A similarity transformation sends lines to lines, rays to rays, line segments to line segments, and parallel lines to parallel lines.

5. A similarity transformation sends angles to angles of equal measure.

6. A similarity transformation maps a circle of radius R to a circle of radius rR, where r is the scale factor of the similarity transformation.
   - All of the properties are satisfied by a similarity transformation consisting of a single translation, reflection, rotation, or dilation. If the similarity transformation consists of more than one such transformation, then the properties still hold because they hold one step at a time.
   - For instance, if G is the composition of three transformations G_1, G_2, G_3, where each of G_1, G_2, G_3 is a translation, reflection, rotation, or dilation, then G_1 maps a pair of parallel lines to a second pair of parallel lines that are then taken by G_2 to another pair of parallel lines that are then taken by G_3 to yet another pair of parallel lines. The composition of G_1, G_2, G_3 takes any pair of parallel lines to another pair of parallel lines.
   - We keep these properties in mind as we work on examples where multiple transformations comprise the similarity transformation.

Example 1 (10 minutes)

Students apply a similarity transformation to an initial figure and determine the image. Review the steps to apply a reflection and rotation in Example 1.

- Use a compass, protractor, and straightedge to determine the image of the triangle.

> **Example 1**
>
> **Similarity transformation G consists of a rotation about the point P by $90°$, followed by a dilation centered at P with scale factor of $r = 2$, and then followed by a reflection across line ℓ. Find the image of the triangle.**

> *Scaffolding:*
> - Depending on student ability and time, consider limiting G to the first two transformations.
> - Consider placing examples on the coordinate plane, on grid paper, and/or using transparencies, patty paper, or geometry software.

- Before getting to the actual construction process, draw a predictive sketch of the image of the triangle after transformation. You should have three sketches.

MP.1

Drawing a predictive sketch helps illuminate the path ahead before getting into the details of each construction. The sketches also provide a time to reflect on how the properties are true for each transformation. For example, ask students to select a property from their lists and describe where in their sketches they see it in the transformation. For instance, a student might select property (2), which states that each point has a pre-image P and can point to each pre-image and image with each passing transformation.

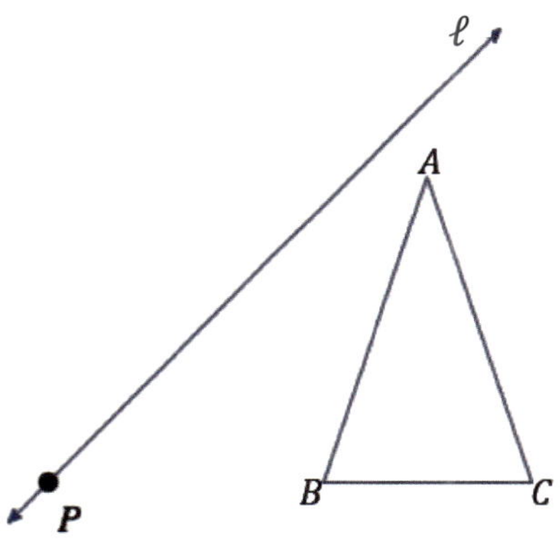

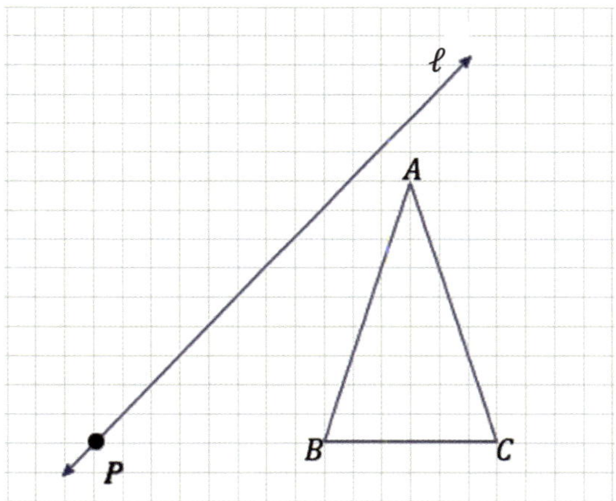

Complete this example with students. Remind them that the rotation requires all three geometry tools.

Note to the teacher: As an alternative strategy, consider using coordinate geometry by placing the image over an appropriately scaled grid. Students gained familiarity with coordinate geometry as used with transformations of the plane in Grade 8. The image on the right above has been provided for this alternative. Placement of the x- and y-axes can be determined where convenient for this example. Given this flexibility, most students likely choose point P to be the origin of the coordinate plane since two of the given transformations are centered at P.

- How will we rotate the triangle $90°$?
  - *To locate the vertices of the triangle's image, draw rays through each vertex of the triangle originating from P. Use each ray to form three $90°$ angles, and then use the compass to locate each new vertex on the corresponding ray.*

Point out to students that in their constructions, the ray through B' and C' coincides and appears to be one ray, giving an appearance of only two rays, not three.

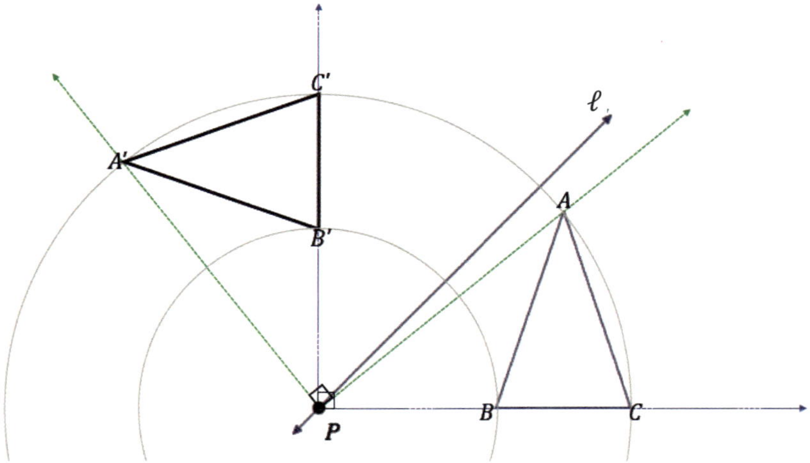

Next, allow students time to dilate $\triangle A'B'C'$. If necessary, review the steps to create a dilation.

MP.1

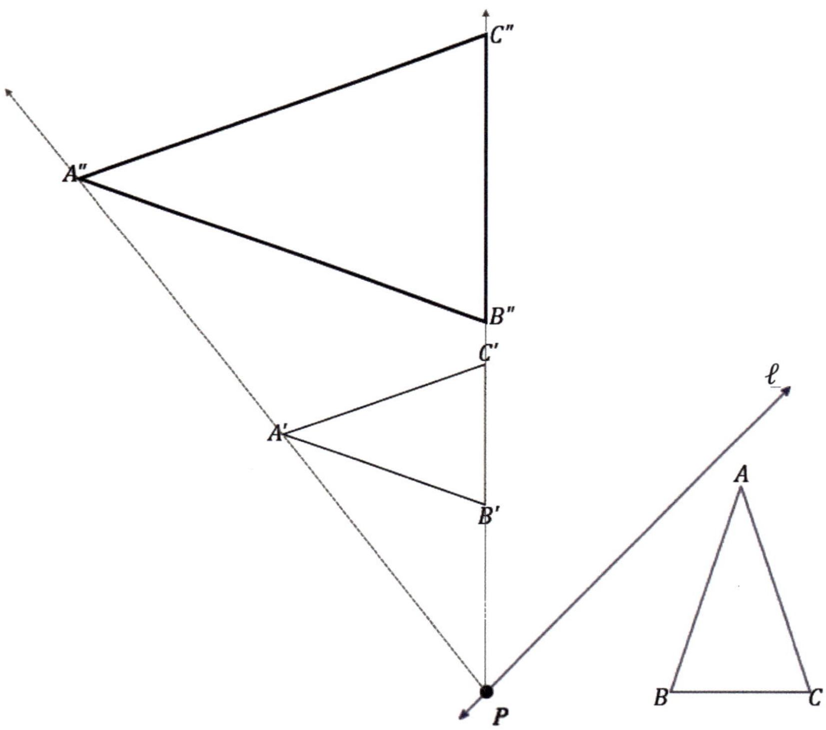

- How will we reflect the triangle over the line?
  - *Create the construction marks that determine the image of each vertex so that the line of reflection is the perpendicular bisector of the segment that joins each vertex with its image. In other words, we must create the construction marks so that the images of the vertices are located so that the line of reflection is the perpendicular bisector of* $\overline{A''A'''}$, $\overline{B''B'''}$, $\overline{C''C'''}$.

EUREKA MATH

The steps to determine B''' are shown below.

Scaffolding:

- Diagrams with more than two transformations of a figure can become cluttered very quickly. Consider allowing students to use different colored pencils (or pens) to complete each stage of the similarity transformation.
- Also, to reduce clutter, draw only construction arcs as opposed to full construction circles (as are shown in the diagram).

MP.1

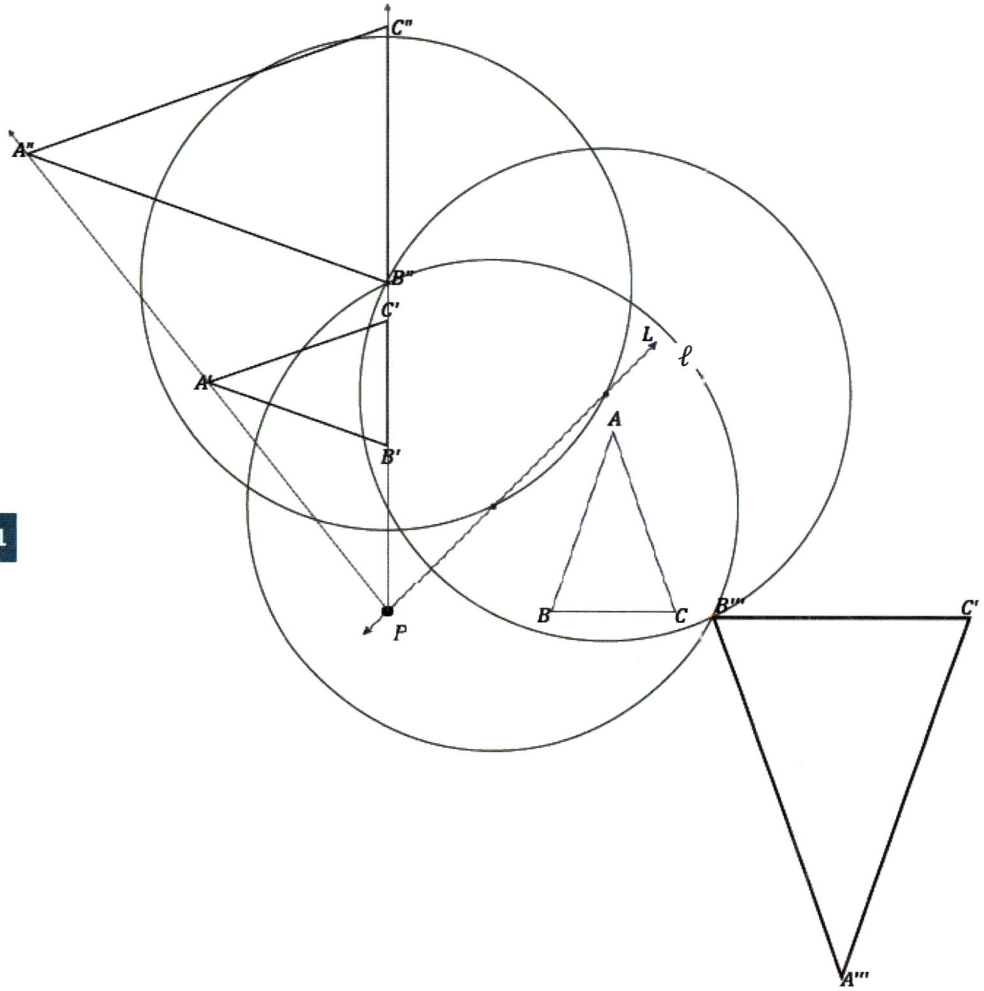

- We have applied the outlined similarity transformation to $\triangle ABC$ and found its image, that is, the similar figure $\triangle A'''B'''C'''$.
- Since G comprises three individual transformations and each of the transformations satisfies the known properties, we know that G also satisfies the properties.

Example 2 (10 minutes)

Example 2 incorporates a translation, a reflection, and a dilation in the similarity transformation. Review the steps of how to apply a translation in Example 2.

Example 2

A similarity transformation G applied to trapezoid $ABCD$ consists of a translation by vector $\overrightarrow{XY}$, followed by a reflection across line m, and then followed by a dilation centered at P with a scale factor of $r = 2$. Recall that we can describe the same sequence using the following notation: $D_{P,2}\left(r_m\left(T_{XY}(ABCD)\right)\right)$. Find the image of $ABCD$.

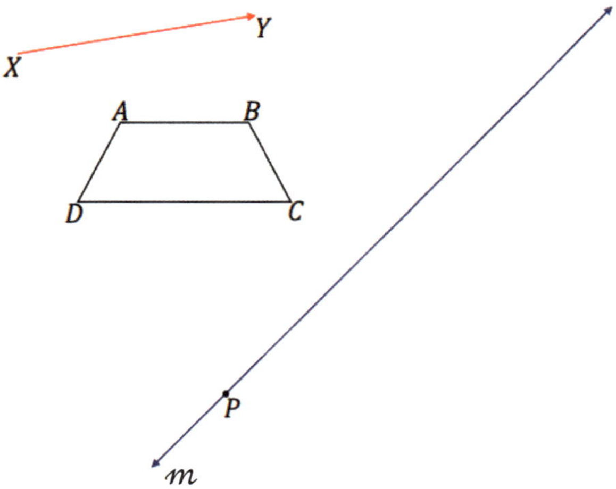

Encourage students to draw a predictive sketch for each stage of the transformation before beginning the construction.

- Describe the steps to apply the translation by vector XY to one point of the figure.
  - *To apply the translation to A, construct C_1: center A, radius XY. Then, construct C_2: center Y, radius XA.*

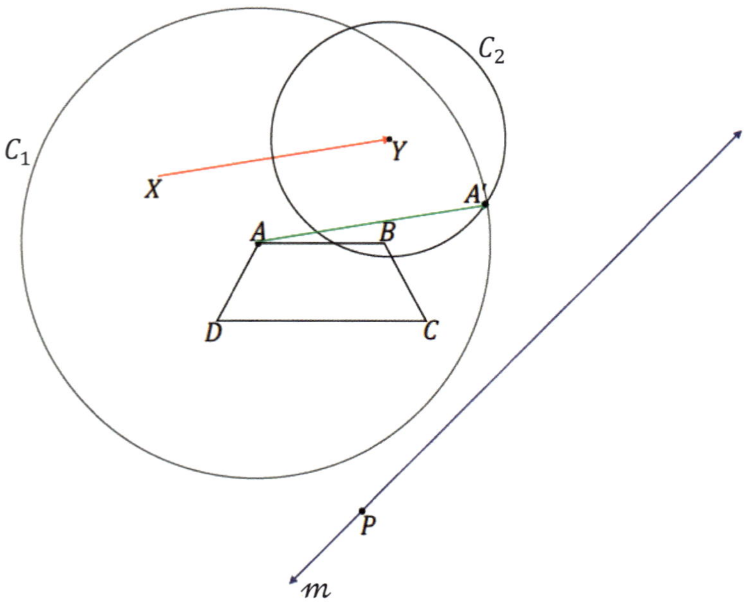

> *Scaffolding:*
> Help students get started at translating point A by finding the fourth vertex, A', of parallelogram $XAA'Y$. Finding the images of the remaining points B, C, and D under the translation can be found by following similar processes.

The process for locating the image of B under the translation is shown below:

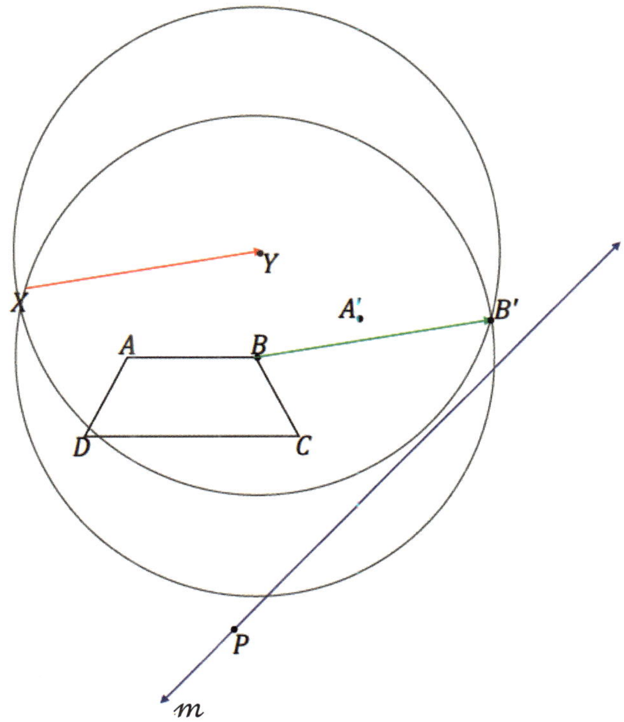

Allow students time to complete the rest of the example before reviewing it.

The following image shows the reflection of vertex C':

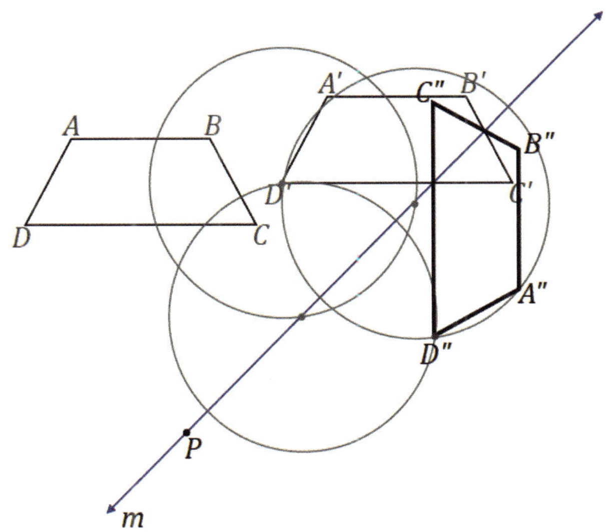

The following image shows the dilation of $A''B''C''D''$:

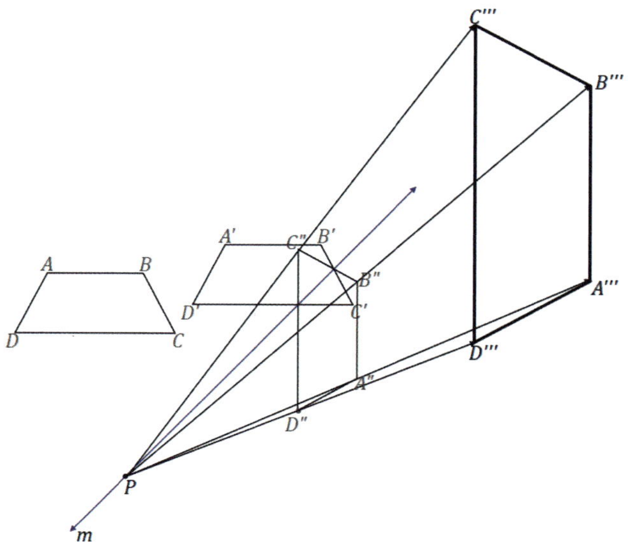

Exercise 1 (8 minutes)

Allow students to work on Exercise 1 independently. Encourage students to draw a predictive sketch for each stage of the transformation before beginning the construction.

Exercise 1

A similarity transformation for triangle DEF is described by $r_n\left(D_{A,\frac{1}{2}}\left(R_{A,90^\circ}(DEF)\right)\right)$. Locate and label the image of triangle DEF under the similarity.

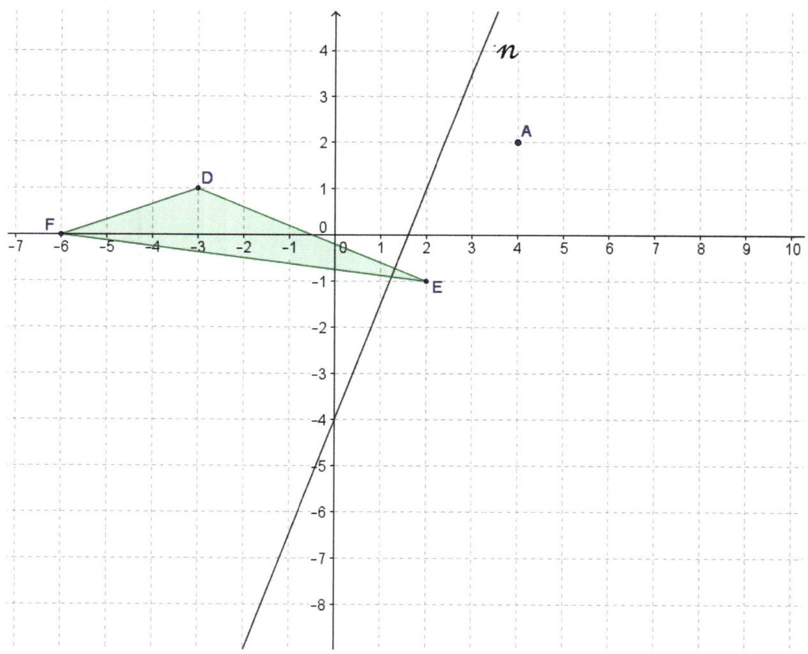

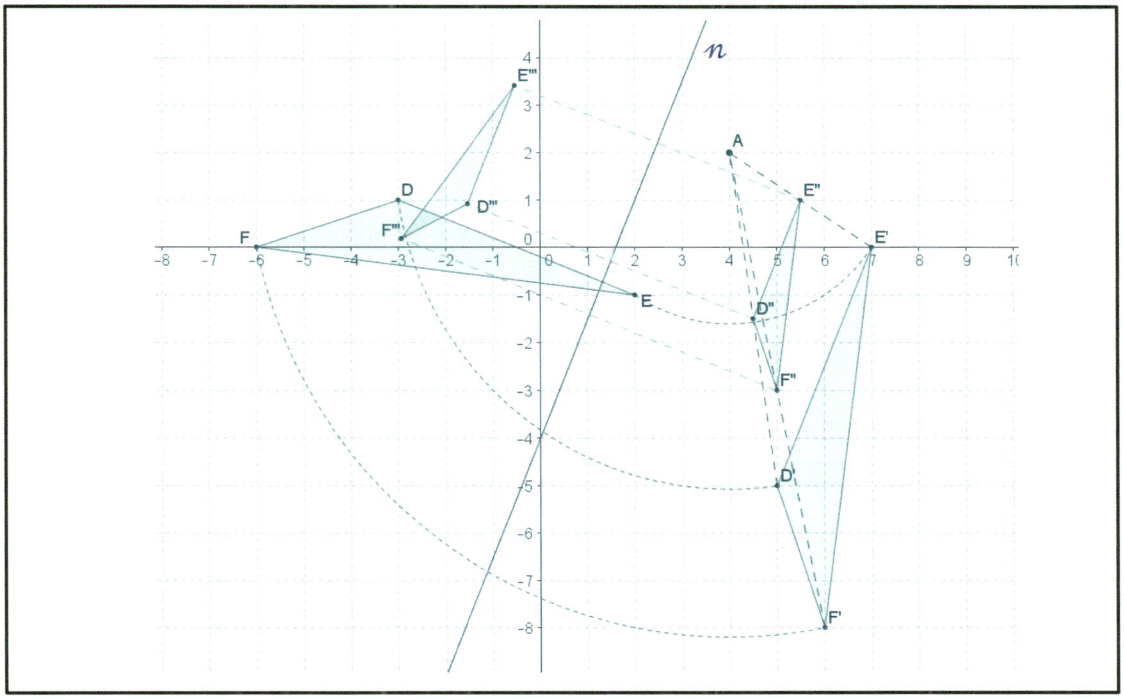

Closing (2 minutes)

- Why are the properties of a similarity transformation the same as those of both dilations and rigid motions?
  - *The properties enjoyed by individual transformations are true for a similarity transformation, as each transformation in a composition is done one transformation at a time.*

Review the properties of similarity transformations:

1. Distinct points are mapped to distinct points.
2. Each point P' in the plane has a pre-image.
3. There is a scale factor of r for G so that for any pair of points P and Q with images $P' = G(P)$ and $Q' = G(Q)$, then $P'Q' = rPQ$.
4. A similarity transformation sends lines to lines, rays to rays, line segments to line segments, and parallel lines to parallel lines.
5. A similarity transformation sends angles to angles of equal measure.
6. A similarity transformation maps a circle of radius R to a circle of radius rR, where r is the scale factor of the similarity transformation.

Lesson Summary

Properties of similarity transformations:

1. **Distinct points are mapped to distinct points.**
2. **Each point P' in the plane has a pre-image.**
3. **There is a scale factor of r for G so that for any pair of points P and Q with images $P' = G(P)$ and $Q' = G(Q)$, then $P'Q' = rPQ$.**
4. **A similarity transformation sends lines to lines, rays to rays, line segments to line segments, and parallel lines to parallel lines.**
5. **A similarity transformation sends angles to angles of equal measure.**
6. **A similarity transformation maps a circle of radius R to a circle of radius rR, where r is the scale factor of the similarity transformation.**

Exit Ticket (5 minutes)

Note to the teacher: The Exit Ticket contains a sequence of three transformations on the plane that may require more than 5 minutes to complete. The second step in the sequence is a dilation, so students should be directed to complete at least the first two transformations in the sequence to find $A''C''D''E''$.

Name ______________________________ Date ______________

Lesson 13: Properties of Similarity Transformations

Exit Ticket

A similarity transformation consists of a translation along the vector $\overrightarrow{FG}$, followed by a dilation from point P with a scale factor of $r = 2$, and finally a reflection over line m. Use construction tools to find $A'''C'''D'''E'''$.

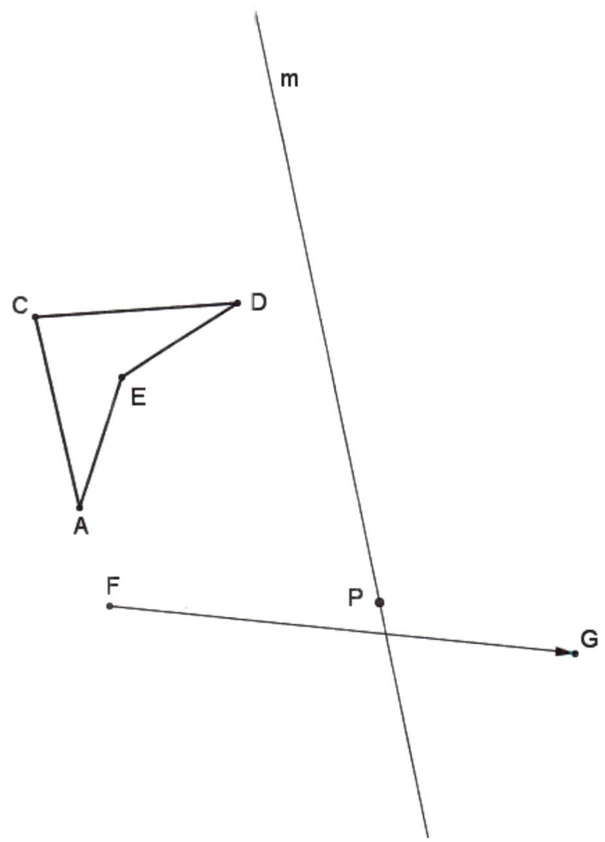

Exit Ticket Sample Solutions

A similarity transformation consists of a translation along the vector $\overrightarrow{FG}$, followed by a dilation from point P with a scale factor of $r = 2$, and finally a reflection over line m. Use construction tools to find $A'''C'''D'''E'''$.

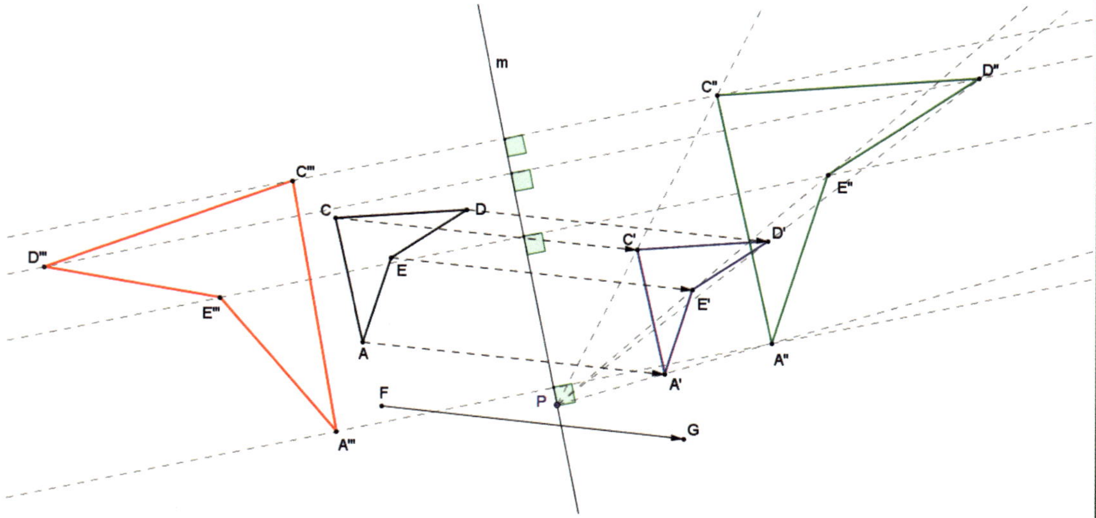

Problem Set Sample Solutions

1. A similarity transformation consists of a reflection over line ℓ, followed by a dilation from O with a scale factor of $r = \frac{3}{4}$. Use construction tools to find $\triangle G''H''I''$.

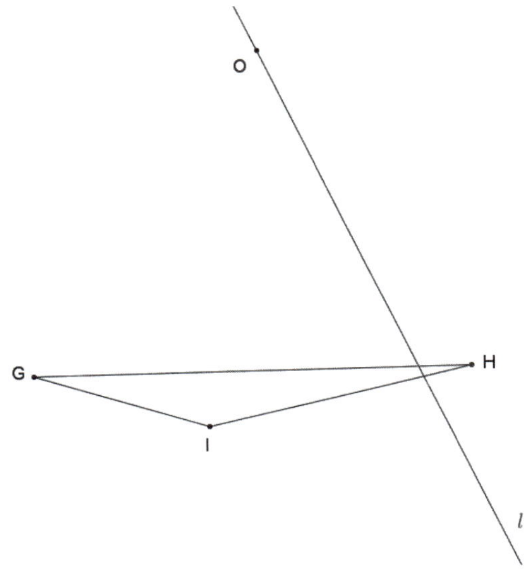

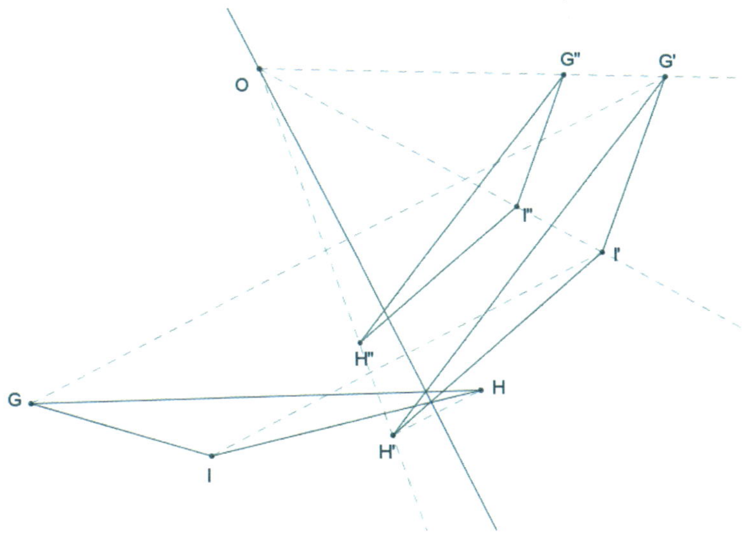

2. A similarity transformation consists of a dilation from point O with a scale factor of $r = 2\frac{1}{2}$, followed by a rotation about O of $-90°$. Use construction tools to find kite $A''B''C''D''$.

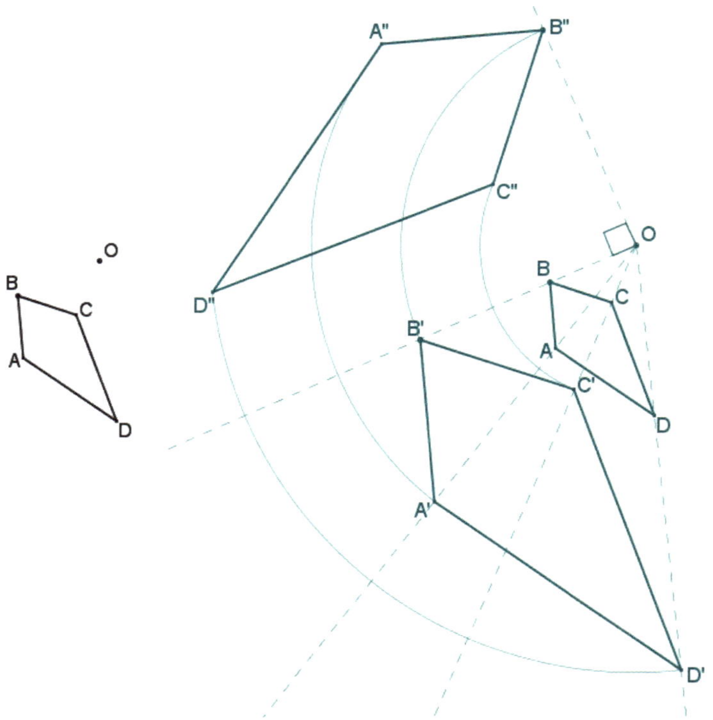

3. For the Figure Z, find the image of $r_{\ell}\left(R_{P,90^\circ}\left(D_{P,\frac{1}{2}}(Z)\right)\right)$.

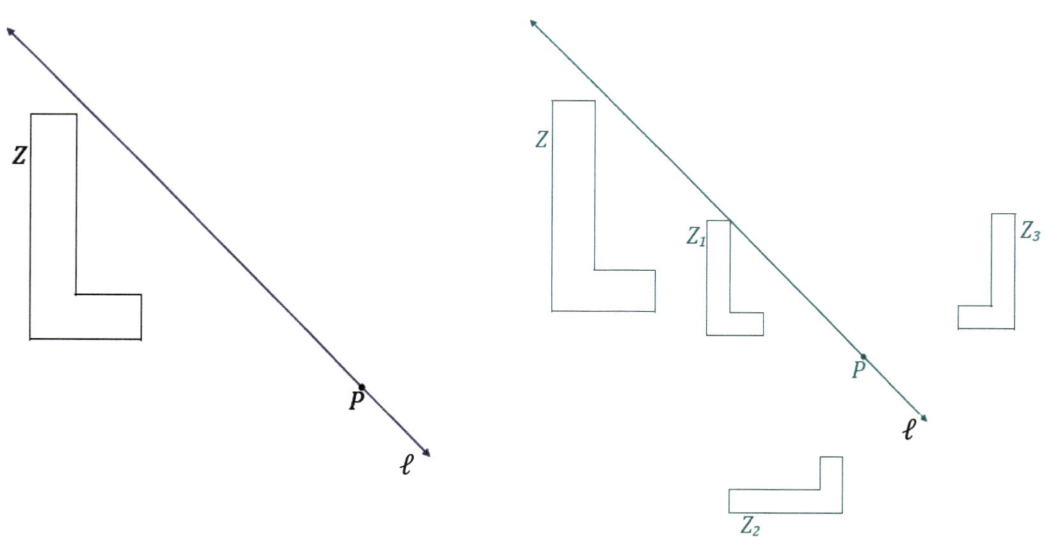

4. **A similarity transformation consists of a translation along vector $\overrightarrow{UV}$, followed by a rotation of $60°$ about P, then followed by a dilation from P with a scale factor of $r = \frac{1}{3}$. Use construction tools to find $\triangle X'''Y'''Z'''$.**

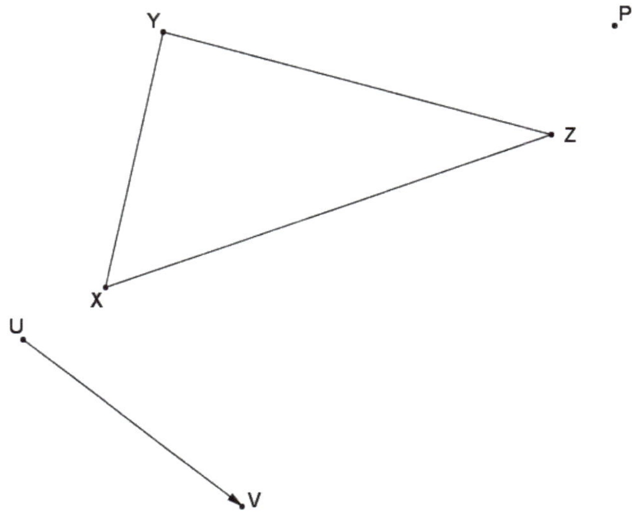

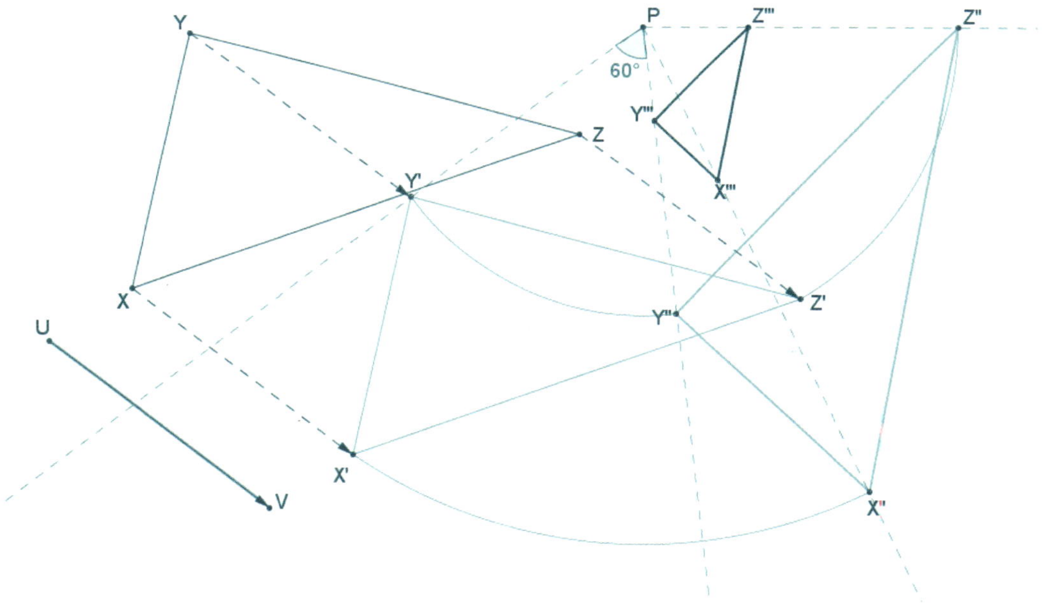

5. Given the quarter-circular figure determined by points A, B, and C, a similarity transformation consists of a $-65°$ rotation about point B, followed by a dilation from point O with a scale factor of $r = \frac{1}{2}$. Find the image of the figure determined by points A'', B'', and C''.

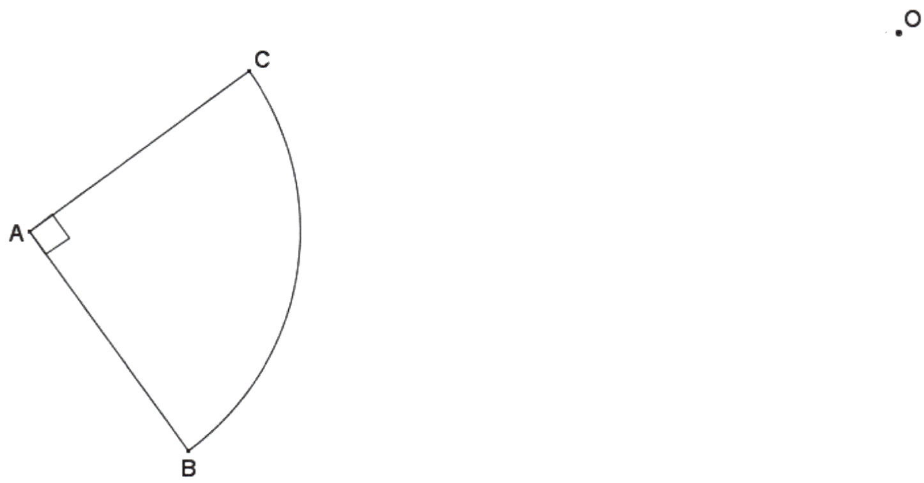

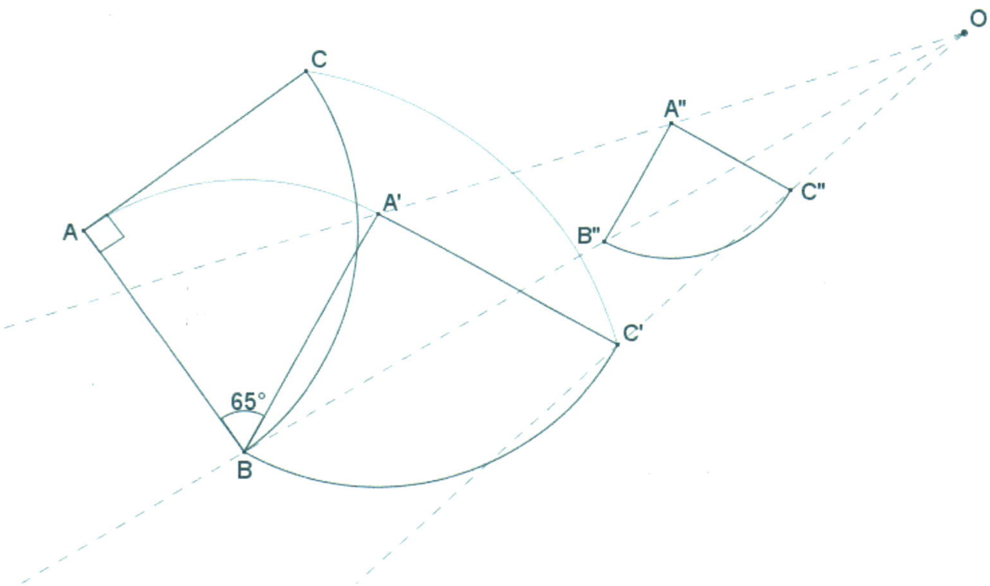

Describe a different similarity transformation that would map quarter-circle ABC to quarter-circle $A''B''C''$.

The quarter-circular region could have first been dilated from point O using a scale factor of $\frac{1}{2}$, followed by a rotation of $65°$ about point B'.

6. A similarity transformation consists of a dilation from center O with a scale factor of $\frac{1}{2}$, followed by a rotation of $60°$ about point O. Complete the similarity transformation on Figure T to complete the drawing of Figure T''.

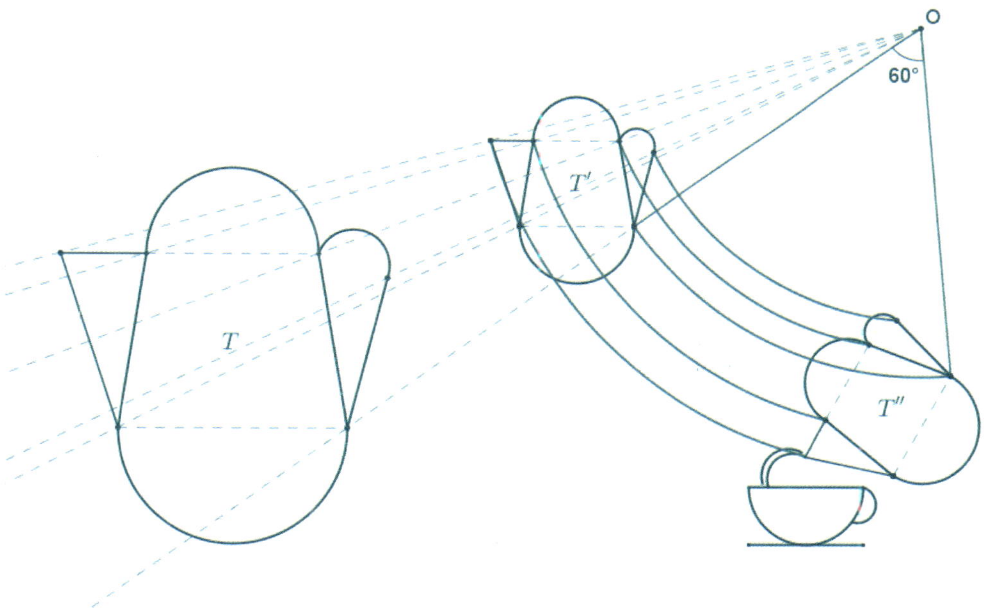

7. **Given Figure R on the coordinate plane shown below, a similarity transformation consists of a dilation from $(0, 6)$ with a scale factor of $\frac{1}{4}$, followed by a reflection over line $x = -1$, and then followed by a vertical translation of 5 units down. Find the image of Figure R.**

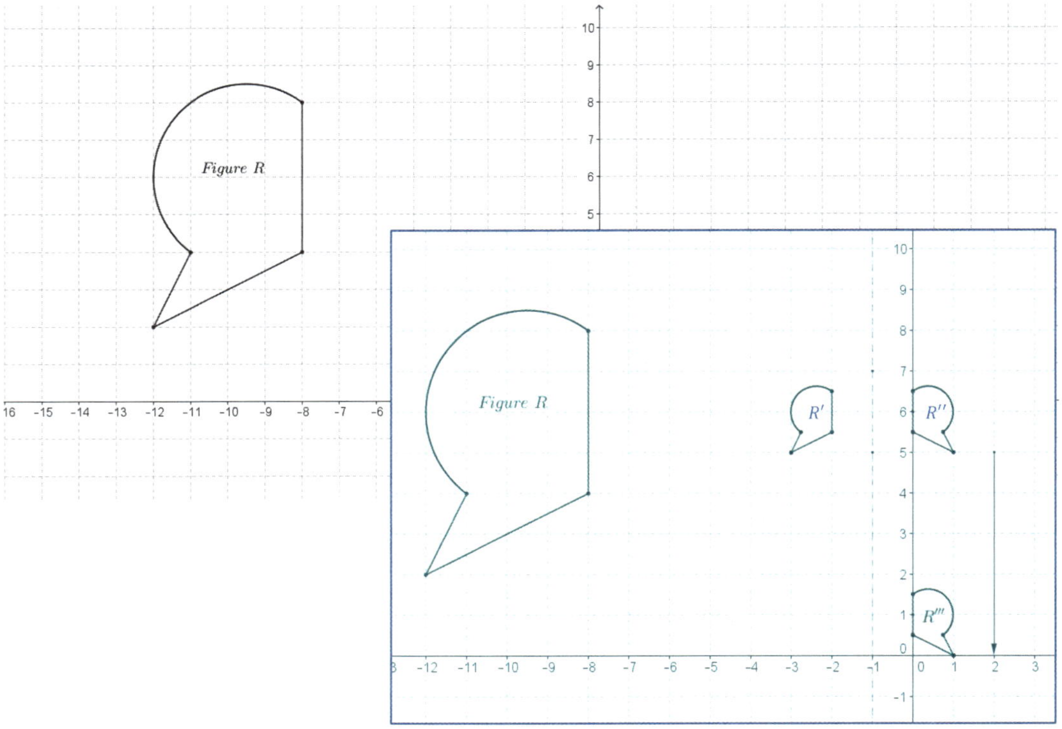

8. **Given $\triangle ABC$, with vertices $A(2, -7)$, $B(-2, -1)$, and $C(3, -4)$, locate and label the image of the triangle under the similarity transformation $D_{B', \frac{1}{2}}\left(R_{A,120°}\left(r_{x=2}(ABC)\right)\right)$.**

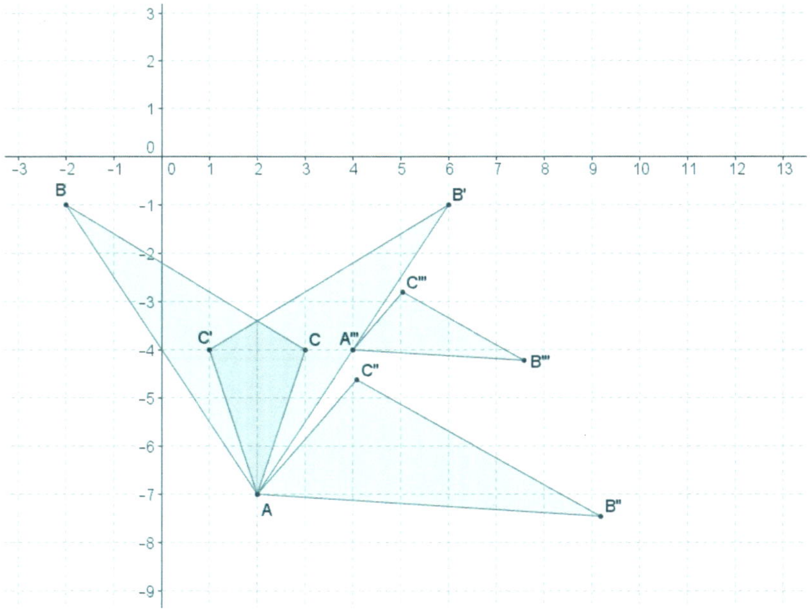

9. **In Problem 8, describe the relationship of A''' to $\overline{AB'}$, and explain your reasoning.**

A''' is the midpoint of $\overline{AB'}$. I know this because $A = A' = A''$, and since the dilation at the end of the similarity transformation was centered at B', the image of A'' must lie on the ray joining it to the center B', which means that A''' lies on $\overline{AB'}$. Furthermore, by the definition of dilation, $A'''B' = r(A''B')$, and since the given scale factor was $r = \frac{1}{2}$, it follows that $A'''B' = \frac{1}{2}(A''B')$, so A''' must, therefore, be the midpoint of $\overline{AB'}$.

10. **Given $O(-8,3)$ and quadrilateral $BCDE$, with $B(-5,1)$, $C(-6,-1)$, $D(-4,-1)$, and $E(-4,2)$, what are the coordinates of the vertices of the image of $BCDE$ under the similarity transformation $r_{x-axis}\left(D_{O,3}(BCDE)\right)$?**

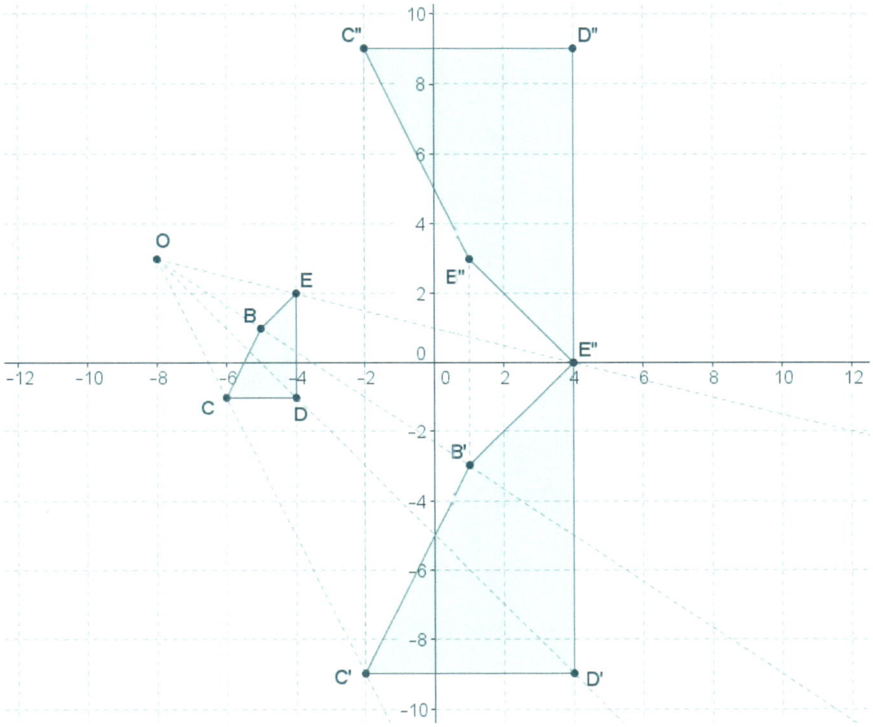

The coordinates of the image of $BCDE$ are $B''(1,3)$, $C''(-2,9)$, $D''(4,9)$, and $E''(4,0)$.

11. **Given triangle ABC as shown on the diagram of the coordinate plane:**

   a. **Perform a translation so that vertex A maps to the origin.**

   See diagram.

   b. **Next, dilate the image $A'B'C'$ from the origin using a scale factor of $\frac{1}{3}$.**

   See diagram.

c. **Finally, translate the image $A''B''C''$ so that the vertex A'' maps to the original point A.**

See diagram.

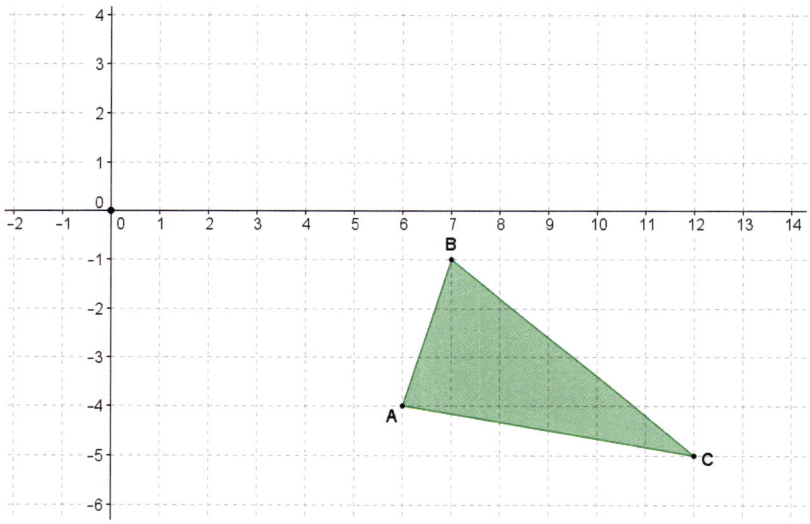

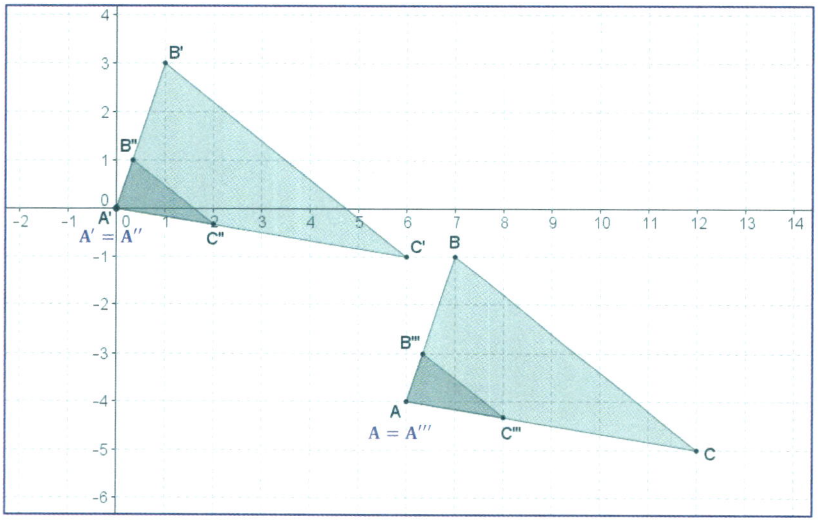

d. **Using transformations, describe how the resulting image $A'''B'''C'''$ relates to the original figure ABC.**

The image $A'''B'''C'''$ is a dilation of figure ABC from point A with a scale factor of $\frac{1}{3}$.

12.

a. In the coordinate plane, name the single transformation resulting from the composition of the two dilations shown below:

$$D_{(0,0),2} \text{ followed by } D_{(0,4),\frac{1}{2}}$$

(Hint: Try it!)

The image can be obtained by a translation two units to the right (a vector that has half the distance as the distance between the centers of dilation).

b. In the coordinate plane, name the single transformation resulting from the composition of the two dilations shown below:

$$D_{(0,0),2} \text{ followed by } D_{(4,4),\frac{1}{2}}$$

(Hint: Try it!)

The image can be obtained by a translation two units to the right (a vector that has half the distance as the distance between the centers of dilation).

c. Using the results from parts (a) and (b), describe what happens to the origin under both similarity transformations.

The origin maps to the midpoint of a segment joining the centers used for each dilation.

Example 1

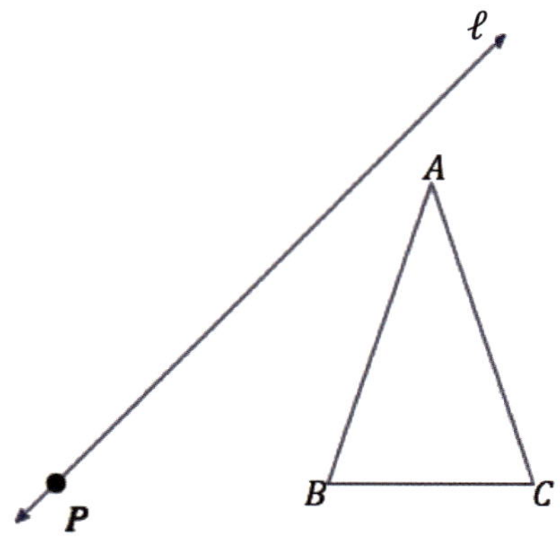

Example 2

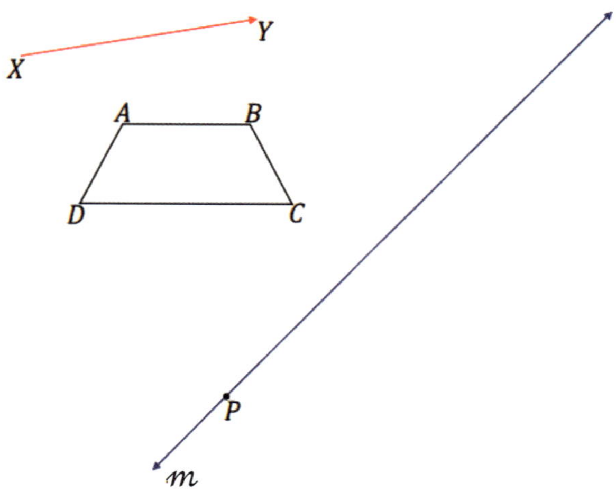

Lesson 14: Similarity

Student Outcomes

- Students understand that similarity is reflexive, symmetric, and transitive.
- Students recognize that if two triangles are similar, there is a correspondence such that corresponding pairs of angles have the same measure and corresponding sides are proportional. Conversely, they know that if there is a correspondence satisfying these conditions, then there is a similarity transformation taking one triangle to the other respecting the correspondence.

Lesson Notes

In Lesson 14, students delve more deeply into what it means for figures to be similar. Examples address the properties of similarity and also focus on circles and, finally, triangles. Lessons 12–13 were intentionally not focused on triangles to emphasize that instructors need to move away from equating similarity with triangles or strictly rectilinear figures and rather to think of and teach similarity as a broader concept. With this idea covered in Lessons 12 and 13, attention is shifted to triangles in the second half of this lesson in order to prepare students for triangle similarity criteria in Lessons 15 and 17.

Classwork

Opening (4 minutes)

- As initially mentioned in Lesson 12, two figures in the plane are similar if there is a similarity transformation that takes one to the other. If A and B are similar, we write $A \sim B$ where "$\sim$" denotes similarity.
- Which of the following properties do you believe to be true? Vote yes by raising your hand.
- For a Figure A in the plane, do you believe that $A \sim A$?

After taking the vote, show students a simple figure such as the following triangle, and ask them to reconsider if the figure is similar to itself.

MP.3 Allow students 30 seconds to justify their responses in a sentence, and then have them compare reasons with a neighbor before sharing out and moving on to the next property.

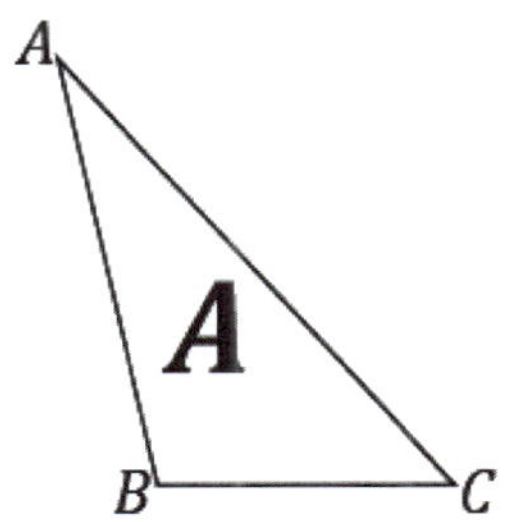

> *Scaffolding:*
> For struggling students, place the triangles in the coordinate plane.

- For two figures A and B in the plane, do you believe that if $A \sim B$, then $B \sim A$?

After taking the vote, show Figures A and B, and ask them to reconsider the notion that if A is similar to B, then is B similar to A.

MP.3 Allow students 30 seconds to justify their responses in a sentence, and then have them compare reasons with a neighbor before sharing out and moving on to the next property.

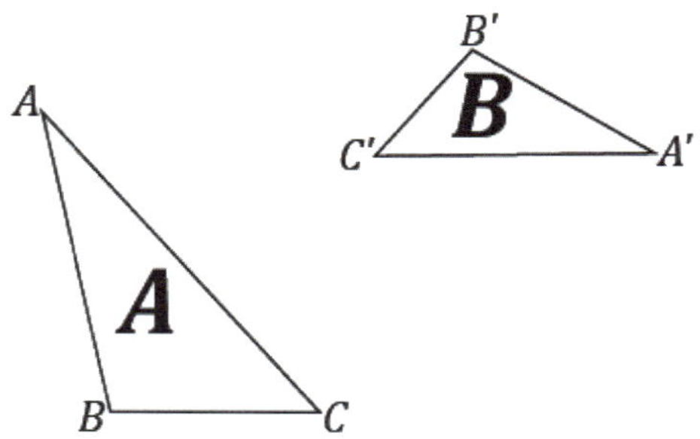

- For Figures A, B, and C in the plane, do you believe that if $A \sim B$ and $B \sim C$, then $A \sim C$?

After taking the vote, show students Figures A, B, and C, and ask them to reconsider if A is similar to B and B is similar to C, then is A similar to C.

MP.3 Allow students 30 seconds to justify their responses in a sentence, and then have them compare reasons with a neighbor before sharing out and moving on to the next property.

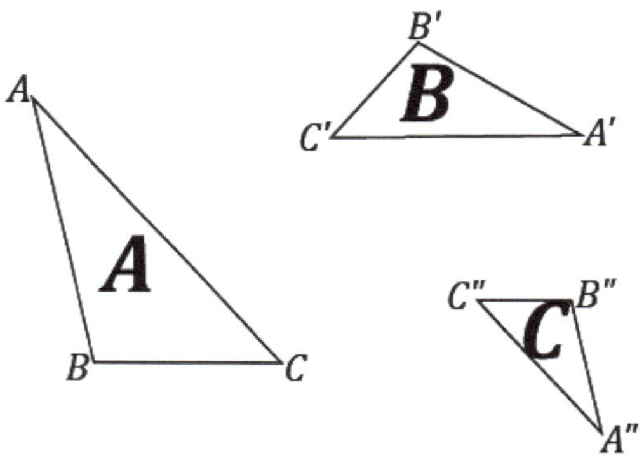

Announce that the properties are in fact true, and state them:

- For each figure A in the plane, $A \sim A$. Similarity is reflexive.
- If A and B are figures in the plane so that $A \sim B$, then $B \sim A$. Similarity is symmetric.
- If A, B, and C are figures in the plane such that $A \sim B$ and $B \sim C$, then $A \sim C$. Similarity is transitive.
- In Examples 1 and 2, we form informal arguments to prove why the conditions on similarity must be true.

Example 1 (4 minutes)

Present the question to the class, and then consider employing any discussion strategies commonly used for a brief brainstorming session, whether it is a whole-group share out, a timed talk-and-turn session with a neighbor, or a Quick Write. Allow about two minutes for whichever strategy is selected and the remaining two minutes for sharing out as a whole group, demonstrating (as needed) any students' suggestions on the board.

> **Example 1**
>
> **We said that for a figure A in the plane, it must be true that $A \sim A$. Describe why this must be true.**

- Remember, to show that for two figures to be similar, there must be a similarity transformation that maps one to the other. Are there such transformations to show that A maps to A?

MP.3

Take multiple suggestions of transformations that map A to A:

- There are several different transformations that map A onto itself such as a rotation of $0°$ or a rotation of $360°$.
- A reflection of A across a line and a reflection right back achieves the same result.
- A translation with a vector of length 0 also maps A to A.
- A dilation with scale factor 1 maps A to A, and any combination of these transformations also maps A to A.
- Therefore, A must be similar to A because there are many similarity transformations that map A to A.
- This condition is labeled as *reflexive* because every figure is similar to itself.

Example 2 (4 minutes)

Present the question to the class, and then consider employing any discussion strategies commonly used for a brief brainstorming session, whether it is a whole-group share out, a timed talk-and-turn session with a neighbor, or a Quick Write. Allow about two minutes for whichever strategy is selected and the remaining two minutes for sharing out as a whole group, demonstrating (as needed) any students' suggestions on the board.

> **Example 2**
>
> **We said that for figures A and B in the plane so that $A \sim B$, then it must be true that $B \sim A$. Describe why this must be true.**

Now that students have completed Example 1, allow them time to discuss Example 2 among themselves.

- This condition must be true because for any composition of transformations that maps A to B, there is a composition of transformations that can undo the first composition. For example, if a translation by vector $\overrightarrow{XY}$ maps A to B, then the vector $\overrightarrow{YX}$ will undo the transformation and map B to A.
- A counterclockwise rotation of $90°$ can be undone by a clockwise rotation of $90°$.
- A dilation by a scale factor of $r = 2$ can be undone with a dilation by a scale factor of $r = \frac{1}{2}$.
- A reflection across a line can be undone by a reflection back across the same line.
- Therefore, it must be true that if a figure A is similar to a figure B in the plane (i.e., if there is a similarity transformation that maps A to B), then there must also be a composition of transformations that can undo that similarity transformation and map B back to A.
- This condition is labeled as symmetric because of its likeness to the symmetric property of equality where if one number is equal to another number, then they both must have the same value (if $a = b$, then $b = a$).

We leave the third condition, that similarity is transitive, for the Problem Set.

Example 3 (10 minutes)

In Example 3, students must show that any circle is similar to any other circle. Note the term *similar* is used here, unlike in Lesson 8 where students proved the dilation theorem for circles. Encourage students to first discuss the question with a partner.

MP.3

> **Example 3**
>
> **Based on the definition of *similar*, how would you show that any two circles are similar?**

Based on their discussions, provide students with the following cases to help them along:

- Consider the different cases you must address in showing that any two circles are similar to each other:
  a. Circles with different centers but radii of equal length
     - *If two circles have different centers but have radii of equal length, then the circles are congruent, and a translation along a vector that brings one center to the other will map one circle onto the other.*
  b. Circles with the same center but radii of different lengths
     - *If two circles have the same center, but one circle has radius R and the other has radius R', then a dilation about the center with a scale factor of $r = \frac{R'}{R}$ maps one circle onto the other so that $rR = R'$.*
  c. Circles with different centers and radii of different lengths
     - *If two circles have different centers and radii of different lengths, then the composition of a dilation described in (b) and the translation described in (a) maps one onto the other.*

Scaffolding:

- For more advanced learners, have students discuss what cases must be considered in order to show similarity between circles.
- Consider assigning cases (a), (b), and (c) to small groups. Share out results after allowing for group discussion. Case (a) is likely the easiest to consider since the circles are congruent by a translation and, therefore, also similar.

Students may notice that two circles with different centers and different radii length may alternatively be mapped onto each other by a single dilation as shown in the following image.

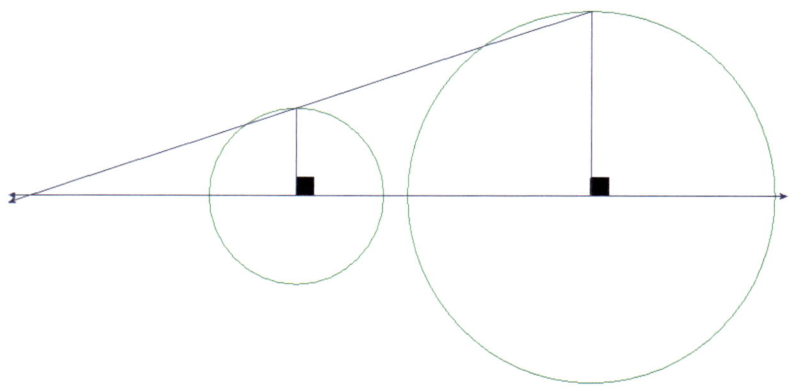

Discussion (3 minutes)

This Discussion leads to the proof of the converse of the theorem on similar triangles in Example 4.

- What do we mean when we say that two triangles are similar?
  - *Two triangles $\triangle ABC$ and $\triangle DEF$ are similar if there is a similarity transformation that maps $\triangle ABC$ to $\triangle DEF$.*
- When we write $\triangle ABC \sim \triangle DEF$, we mean, additionally, that the similarity transformation takes A to D, B to E, and C to F. So, when we see $\triangle ABC \sim \triangle DEF$, we know that a correspondence exists such that the corresponding angles are equal in measurement and the corresponding lengths of sides are proportional (i.e., $m\angle A = m\angle D$, $m\angle B = m\angle E$, $m\angle C = m\angle F$, and $\frac{DE}{AB} = \frac{EF}{BC} = \frac{DF}{AC}$).

THEOREM ON SIMILAR TRIANGLES: If $\triangle ABC \sim \triangle DEF$, then $m\angle A = m\angle D$, $m\angle B = m\angle E$, $m\angle C = m\angle F$, and $\frac{DE}{AB} = \frac{EF}{BC} = \frac{DF}{AC}$.

Example 4 (7 minutes)

Example 4 asks students whether the converse of the theorem on similar triangles is true. Students begin with the fact that a correspondence exists between two triangles or that the corresponding lengths are proportional and corresponding angles are equal in measurement. Students must argue whether this makes the triangles similar.

The argument may not be a stretch for students to make since they have worked through the original argument. Allow students a few minutes to attempt the informal proof before reviewing it with them.

Example 4

Suppose $\triangle ABC \leftrightarrow \triangle DEF$ and that, under this correspondence, corresponding angles are equal and corresponding sides are proportional. Does this guarantee that $\triangle ABC$ and $\triangle DEF$ are similar?

- We have already shown that if two figures (e.g., triangles) are similar, then corresponding angles are of equal measurement and corresponding sides are proportional in length.
- This question is asking whether the converse of the theorem is true. We know that a correspondence exists between $\triangle ABC$ and $\triangle DEF$. What does the correspondence imply?
  - *The correspondence implies that $m\angle A = m\angle D$, $m\angle B = m\angle E$, $m\angle C = m\angle F$, and $\frac{DE}{AB} = \frac{EF}{BC} = \frac{DF}{AC}$.*
- We will show there is a similarity transformation taking $\triangle ABC$ to $\triangle DEF$ that starts with a dilation. How can we use what we know about the correspondence to express the scale factor r?
  - *Since the side lengths are proportional under the correspondence, then $r = \frac{DE}{AB} = \frac{EF}{BC} = \frac{DF}{AC}$.*
- The dilation that takes $\triangle ABC$ to $\triangle DEF$ can have any point for a center. The dilation maps A to A', B to B', and C to C' so that $\triangle ABC \sim \triangle A'B'C'$.
- How would you describe the length $A'B'$?
  - $A'B' = rAB$
- We take this one step further and say that $A'B' = rAB = \frac{DE}{AB}AB = DE$. Similarly, $B'C' = EF$ and $A'C' = DF$. So, all the sides and all the angles of $\triangle A'B'C'$ and $\triangle DEF$ match up, and the triangles are congruent.

- Since they are congruent, a sequence of basic rigid motions takes $\triangle A'B'C'$ to $\triangle DEF$.
- So, a dilation takes $\triangle ABC$ to $\triangle A'B'C'$, and a congruence transformation takes $\triangle A'B'C'$ to $\triangle DEF$, and we conclude that $\triangle ABC \sim \triangle DEF$.

Example 5 (7 minutes)

The intent of Example 5 is to highlight how an efficient sequence of transformations can be found to map one figure onto its similar image. There are many sequences that get the job done, but the goal here is to show the least number of needed transformations to map one triangle onto the other.

> *Scaffolding:*
> Example 5 is directed at more advanced learners who show proficiency with similarity transformations. The question poses an interesting consideration but is not a requirement. In general, time and focus in this lesson should be placed on the content preceding Example 5.

Example 5

a. In the diagram below, $\triangle ABC \sim \triangle A'B'C'$. Describe a similarity transformation that maps $\triangle ABC$ to $\triangle A'B'C'$.

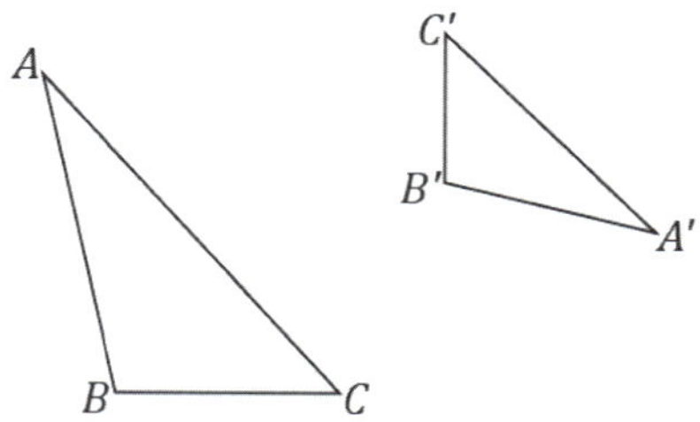

Students most likely describe some sequence that involves a dilation, a reflection, a translation, and a rotation. Direct them to part (b).

b. Joel says the sequence must require a dilation and three rigid motions, but Sharon is sure there is a similarity transformation composed of just a dilation and two rigid motions. Who is right?

Allow students to struggle with the question before revealing that it is, in fact, possible to describe a similarity composed of a dilation and just two rigid motions shown in the following sequence:

- Step 1: Dilate by a scale factor of r and center O so that (1) the resulting $\triangle ABC$ is congruent to $\triangle A'B'C'$ and (2) one pair of corresponding vertices coincide.

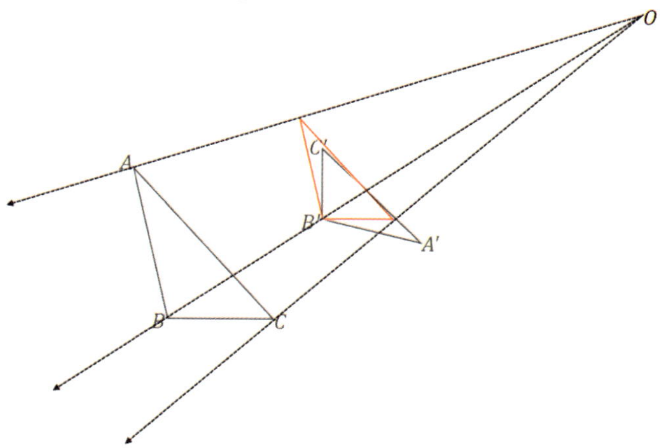

- Step 2: Reflect the dilated triangle across $\overline{B'C'}$.

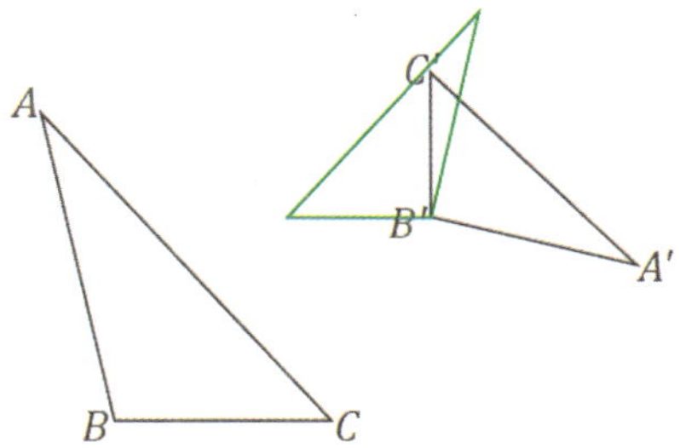

- Step 3: Rotate the reflected triangle until it coincides with $\triangle A'B'C'$.

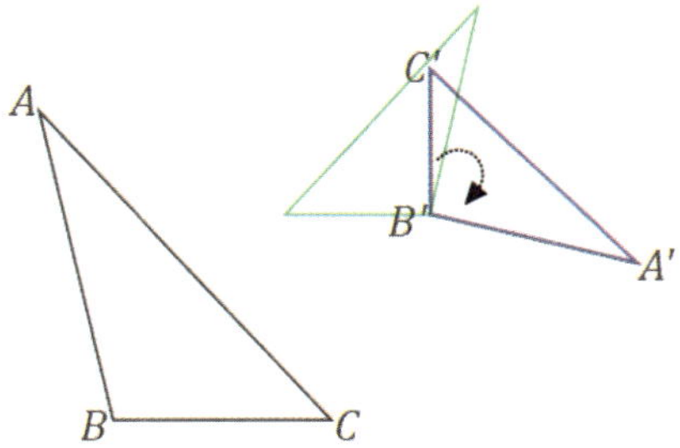

- We have found a similarity transformation that maps $\triangle ABC$ to $\triangle A'B'C'$ with just one dilation and two rigid motions instead of three.

Closing (1 minute)

- What does it mean for similarity to be reflexive? Symmetric?

Students develop an informal argument for why similarity is transitive in the Problem Set.

  - *Similarity is reflexive because a figure is similar to itself. Similarity is symmetric because once a similarity transformation is determined to take a figure to another, there are inverse transformations that can take the figure back to the original.*

- We have seen in our earlier work that if two figures (e.g., two triangles) are similar, then there exists a correspondence such that corresponding lengths are proportional and corresponding angles are equal in measurement. What is the converse of this statement? Is it true?

> **Lesson Summary**
>
> **Similarity is reflexive because a figure is similar to itself.**
>
> **Similarity is symmetric because once a similarity transformation is determined to take a figure to another, there are inverse transformations that can take the figure back to the original.**

Exit Ticket (5 minutes)

Name ______________________________ Date ______________

Lesson 14: Similarity

Exit Ticket

1. In the diagram, $\triangle ABC \sim \triangle DEF$ by the dilation with center O and a scale factor of r. Explain why $\triangle DEF \sim \triangle ABC$.

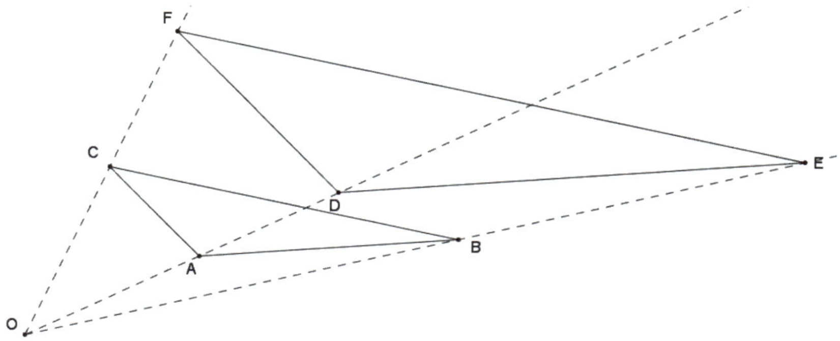

2. Radii $\overline{CA}$ and $\overline{TS}$ are parallel. Is circle $C_{C,CA}$ similar to circle $C_{T,TS}$? Explain.

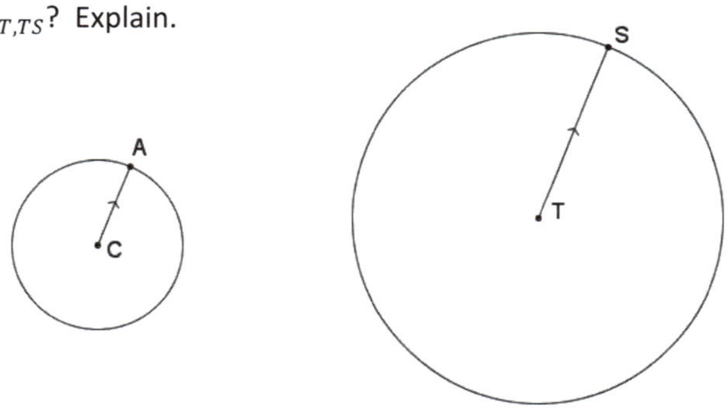

3. Two triangles, $\triangle ABC$ and $\triangle DEF$, are in the plane so that $m\angle A = m\angle D$, $m\angle B = m\angle E$, $m\angle C = m\angle F$, and $\frac{DE}{AB} = \frac{EF}{BC} = \frac{DF}{AC}$. Summarize the argument that proves that the triangles must be similar.

EUREKA MATH

Exit Ticket Sample Solutions

1. **In the diagram, $\triangle ABC \sim \triangle DEF$ by the dilation with center O and a scale factor of r. Explain why $\triangle DEF \sim \triangle ABC$.**

   We know that $\triangle ABC \sim \triangle DEF$ by a dilation with center O and a scale factor of r. A dilation with the same center O but a scale factor of $\frac{1}{r}$ maps $\triangle DEF$ onto $\triangle ABC$; this means $\triangle DEF \sim \triangle ABC$.

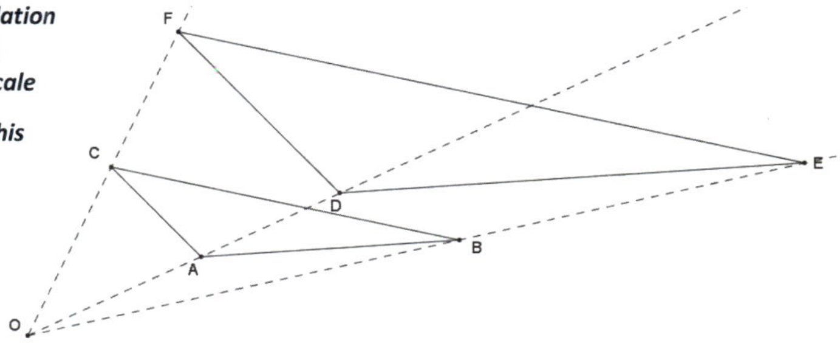

2. **Radii $\overline{CA}$ and $\overline{TS}$ are parallel. Is circle $C_{C,CA}$ similar to circle $C_{T,TS}$? Explain.**

   Yes, the circles are similar because a dilation with center O and a scale factor of r exists that maps $C_{C,CA}$ onto $C_{T,TS}$.

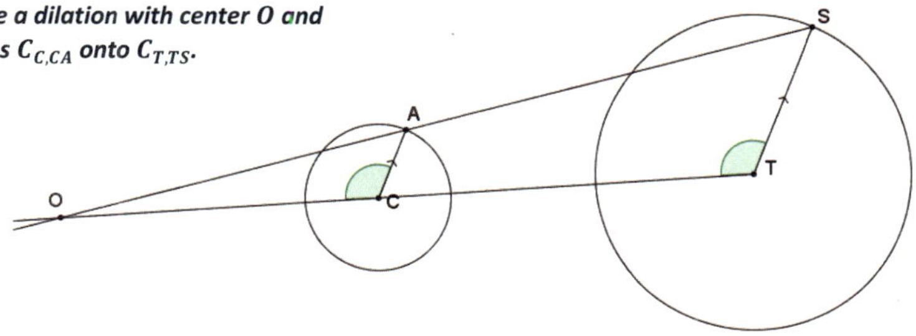

3. **Two triangles, $\triangle ABC$ and $\triangle DEF$, are in the plane so that $m\angle A = m\angle D$, $m\angle B = m\angle E$, $m\angle C = m\angle F$, and $\frac{DE}{AB} = \frac{EF}{BC} = \frac{DF}{AC}$. Summarize the argument that proves that the triangles must be similar.**

   A dilation exists such that the lengths of one triangle can be made equal to the lengths of the other triangle. Once the triangles have lengths and angles of equal measurement, or are congruent, then a sequence of rigid motions maps one triangle to the other. Therefore, a similarity transformation exists that maps one triangle onto the other, and the triangles must be similar.

Problem Set Sample Solutions

1. **If you are given any two congruent triangles, describe a sequence of basic rigid motions that takes one to the other.**

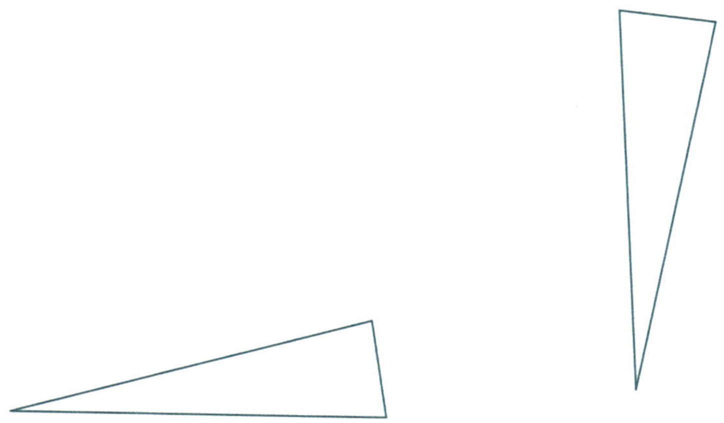

Translate one triangle to the other by a vector $\overrightarrow{XY}$ so that the triangles coincide at a vertex.

Case 1: If both triangles are of the same orientation, simply rotate about the common vertex until the triangles coincide.

Case 2: If the triangles are of opposite orientations, reflect the one triangle over one of its two sides that include the common vertex, and then rotate around the common vertex until the triangles coincide.

2. **If you are given two similar triangles that are not congruent triangles, describe a sequence of dilations and basic rigid motions that takes one to the other.**

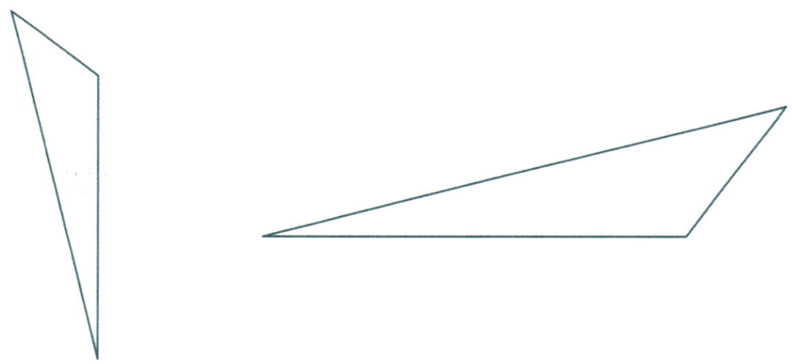

Dilate one triangle with a center O such that when the lengths of its sides are equal to the corresponding lengths of the other triangle, one pair of corresponding vertices coincide. Then, follow one of the sequences described in Case 1 or Case 2 of Problem 1.

Students may have other similarity transformations that map one triangle to the other consisting of translations, reflections, rotations, and dilations. For example, they might dilate one triangle until the corresponding lengths are equal, then translate one triangle to the other by a vector $\overrightarrow{XY}$ so that the triangles coincide at a vertex, and then follow one of the sequences described in Case 1 or Case 2 of Problem 1.

EUREKA MATH™

3. Given two line segments, $\overline{AB}$ and $\overline{CD}$, of different lengths, answer the following questions:

   a. It is always possible to find a similarity transformation that maps $\overline{AB}$ to $\overline{CD}$ sending A to C and B to D. Describe one such similarity transformation.

   Rotate $\overline{AB}$ so that it is parallel to $\overline{CD}$ with corresponding points A and C (and likewise B and D) on the same side of each line segment. Then, translate the image of $\overline{AB}$ by a vector $\overrightarrow{XY}$ so that the midpoint of $\overline{AB}$ coincides with the midpoint of $\overline{CD}$. Finally, dilate $\overline{AB}$ until the two segments are equal in length.

   b. If you are given that $\overline{AB}$ and $\overline{CD}$ are not parallel, are not congruent, do not share any points, and do not lie in the same line, what is the fewest number of transformations needed in a sequence to map $\overline{AB}$ to $\overline{CD}$? Which transformations make this work?

   Rotate $\overline{AB}$ to $\overline{A'B'}$ so that $\overline{A'B'}$ and $\overline{CD}$ are parallel and oriented in the same direction; then, use the fact that there is a dilation that takes $\overline{A'B'}$ to $\overline{CD}$.

   c. If you performed a similarity transformation that instead takes A to D and B to C, either describe what mistake was made in the similarity transformation, or describe what additional transformation is needed to fix the error so that A maps to C and B maps to D.

   The rotation in the similarity transformation was not sufficient to orient the directed line segments in the same direction, resulting in mismatched corresponding points. This error could be fixed by changing the rotation such that the desired corresponding endpoints lie on the same end of each segment.

   If the desired endpoints do not coincide after a similarity transformation, rotate the line segment about its midpoint by $180°$.

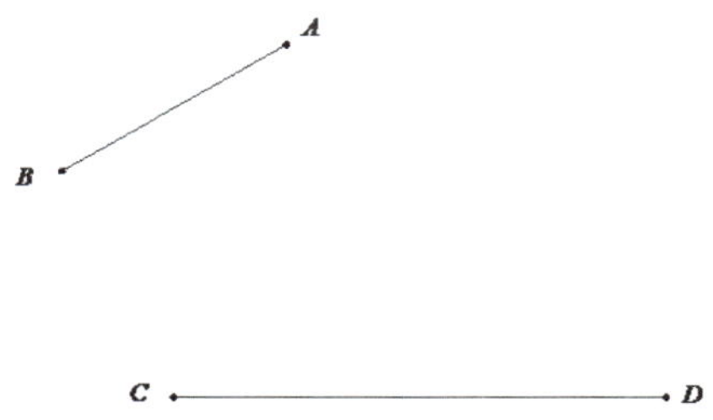

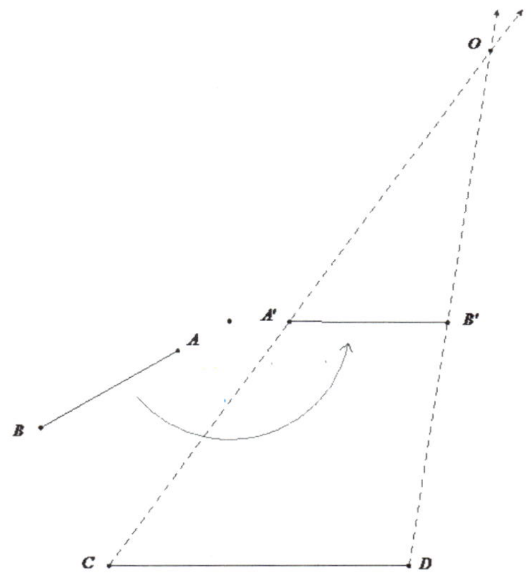

4. We claim that similarity is transitive (i.e., if A, B, and C are figures in the plane such that $A \sim B$ and $B \sim C$, then $A \sim C$). Describe why this must be true.

   If similarity transformation T_1 maps A to B and similarity transformation T_2 maps B to C, then the composition of basic rigid motions and dilations that takes A to B together with the composition of basic rigid motions and dilations that takes B to C shows that $A \sim C$.

5. Given two line segments, $\overline{AB}$ and $\overline{CD}$, of different lengths, we have seen that it is always possible to find a similarity transformation that maps $\overline{AB}$ to $\overline{CD}$, sending A to C and B to D with one rotation and one dilation. Can you do this with one reflection and one dilation?

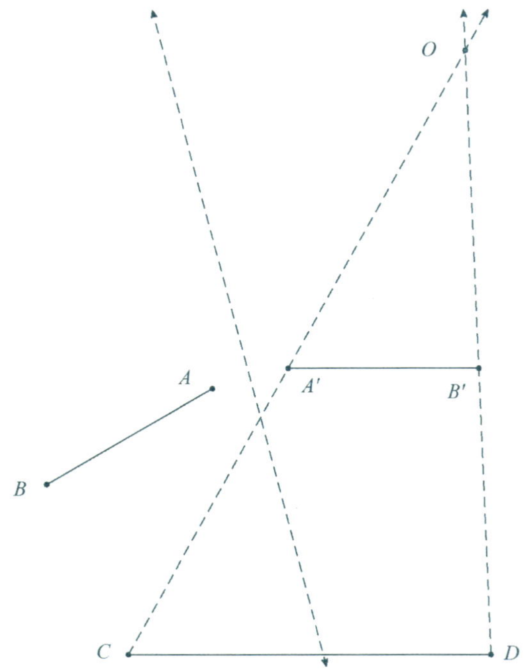

Yes. Reflect $\overline{AB}$ to $\overline{A'B'}$ so that $\overline{A'B'}$ and $\overline{CD}$ are parallel, and then use the fact that there is a dilation that takes $\overline{A'B'}$ to $\overline{CD}$.

6. Given two triangles, $\triangle ABC \sim \triangle DEF$, is it always possible to rotate $\triangle ABC$ so that the sides of $\triangle ABC$ are parallel to the corresponding sides in $\triangle DEF$ (e.g., $\overline{AB} \parallel \overline{DE}$)?

No, it is not always possible. Sometimes a reflection is necessary. If you consider the diagram to the right, $\triangle DEF$ can be rotated in various ways such that $\overline{DE}$ either coincides with or is parallel to $\overline{AB}$, where corresponding points A and D are located at the same end of the segment. However, in each case, F lies on one side of $\overline{DE}$ but is opposite the side on which C lies with regard to $\overline{AB}$. This means that a reflection is necessary to reorient the third vertex of the triangle.

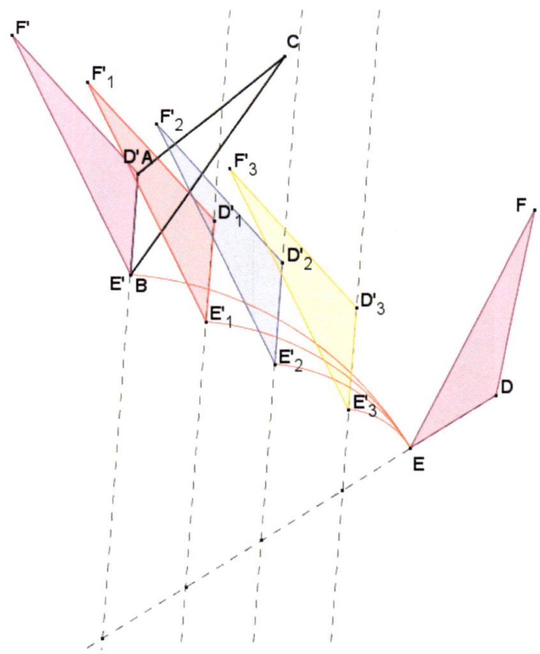

Lesson 15: The Angle-Angle (AA) Criterion for Two Triangles to Be Similar

Student Outcomes

- Students prove the angle-angle criterion for two triangles to be similar and use it to solve triangle problems.

Lesson Notes

In Lesson 4, students learned that the triangle side splitter theorem is an important theorem because it is the central ingredient in proving the AA criterion for similar triangles. In this lesson, this statement is substantiated by using the theorem (in the form of the dilation theorem) to prove the AA criterion. The AA criterion is arguably one of the most useful theorems for recognizing and proving that two triangles are similar. However, students may not need to accept that this statement is true without justification: They get plenty of opportunities in the remaining lessons (and modules) to see how useful the AA criterion is.

Classwork

Exercises 1–5 (10 minutes)

Exercises

MP.5

1. **Draw two triangles of different sizes with two pairs of equal angles. Then, measure the lengths of the corresponding sides to verify that the ratio of their lengths is proportional. Use a ruler, compass, or protractor, as necessary.**

   Students' work will vary. Verify that students have drawn a pair of triangles with two pairs of equal angles and have shown via direct measurement that the ratios of corresponding side lengths are proportional.

MP.3

2. **Are the triangles you drew in Exercise 1 similar? Explain.**

   Yes, the triangles are similar. The converse of the theorem on similar triangles states that when we have two triangles $\triangle ABC$ and $\triangle A'B'C'$ with corresponding angles that are equal and corresponding side lengths that are proportional, then the triangles are similar.

3. **Why is it that you only need to construct triangles where two pairs of angles are equal but not three?**

   If we are given the measure of two angles of a triangle, then we also know the third measure because of the triangle sum theorem. All three angles must add to $180°$, so showing two pairs of angles are equal in measure is just like showing all three pairs of angles are equal in measure.

4. **Why were the ratios of the corresponding sides proportional?**

Since the triangles are similar, we know that there exists a similarity transformation that maps one triangle onto another. Then, corresponding sides of similar triangles must be in proportion because of what we know about similarity, dilation, and scale factor. Specifically, the length of a dilated segment is equal to the length of the original segment multiplied by the scale factor. For example,

$$\frac{A'B'}{AB} = r \text{ and } A'B' = r \cdot AB.$$

5. **Do you think that what you observed will be true when you construct a pair of triangles with two pairs of equal angles? Explain.**

Accept any reasonable explanation. Use students' responses as a springboard for the Opening discussion and the presentation of the AA criterion for similarity.

Opening (4 minutes)

Debrief the work that students completed in Exercises 1–5 by having them share their responses to Exercise 5. Then, continue the discussion with the points below and the presentation of the theorem.

- Based on our understanding of similarity transformations, we know that we can show two figures in the plane are similar by describing a sequence of dilations and rigid motions that would map one figure onto another.
- Since a similarity implies the properties observed in Exercises 1–5 about corresponding side lengths and angle measures, will it be necessary to show that all 6 conditions (3 sides and 3 angles) are met before concluding that triangles are similar?

Provide time for students to discuss this question in small groups and to make conjectures about the answers. Consider having students share their conjectures with the class.

- Instead of having to check all 6 conditions, it would be nice to simplify our work by checking just two or three of the conditions. Our work in Exercises 1–5 leads us to our next theorem:

THEOREM: Two triangles with two pairs of equal corresponding angles are similar. (This is known as the *AA criterion for similarity*.)

Exercise 6 (4 minutes)

This exercise is optional and can be used if students require more time to explore whether two pairs of equal corresponding angles can produce similar triangles.

MP.5

6. **Draw another two triangles of different sizes with two pairs of equal angles. Then, measure the lengths of the corresponding sides to verify that the ratio of their lengths is proportional. Use a ruler, compass, or protractor, as necessary.**

Students' work will vary. Verify that they have drawn a pair of triangles with two pairs of equal angles and have shown via direct measurement that the ratios of corresponding side lengths are proportional.

Discussion (9 minutes)

- To prove the AA criterion, we need to show that two triangles with two pairs of equal corresponding angles are in fact similar. To do so, we apply our knowledge of both congruence and dilation.
- Recall the ASA criterion for congruent triangles. If two triangles have two pairs of equal angles and an included side equal in length, then the triangles are congruent. The proof of the AA criterion for similarity is related to the ASA criterion for congruence. Can you think of how they are related?

> *Scaffolding:*
> Prior to discussing the relationship between ASA criterion for congruence and AA criterion for similarity, it may be necessary for students to review the congruence criterion learned in Module 1.

Provide students time to discuss the relationship between ASA and AA in small groups.

- ▫ *The ASA criterion for congruence requires the included side to be equal in length between the two figures. Since AA is for similarity, we would not expect the lengths to be equal in measure but more likely proportional to the scale factor. Both ASA and AA criteria require two pairs of equal angles.*

- Given two triangles, $\triangle ABC$ and $\triangle DEF$, where $m\angle A = m\angle D$ and $m\angle B = m\angle E$, show that $\triangle ABC \sim \triangle DEF$.

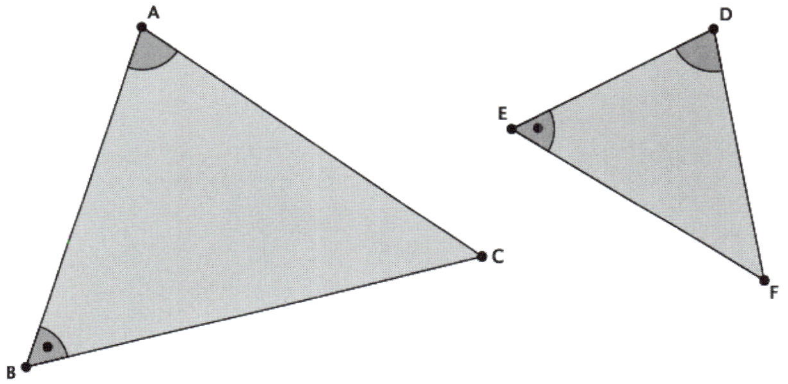

> *Scaffolding:*
> - Consider using cardboard cutouts (reproducible available at the end of the lesson) of the triangles as manipulatives to make the discussion of the proof less abstract. Cutouts can be given to small groups of students or used only by the teacher.
> - Consider asking students to make their own cutouts. Have students create triangles with the same two angles (e.g., 50° and 70°) and then compare their triangles with their neighbors' triangles.

MP.2

- If we can show that $\triangle DEF$ is congruent to a triangle that is a dilated version of $\triangle ABC$, then we can describe the similarity transformation to prove that $\triangle ABC \sim \triangle DEF$. To do so, what scale factor should we choose?
  - ▫ *We should let* $r = \frac{DE}{AB}$.
- Since $\frac{DE}{AB} = r$, we dilate triangle $\triangle ABC$ from center A by a scale factor of r to produce $\triangle AB'C'$ as shown below.

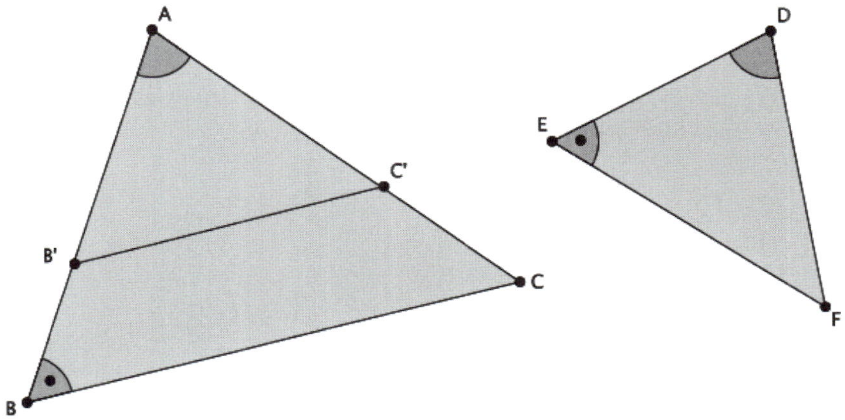

MP.5

- Have we constructed a triangle so that $\triangle AB'C' \cong \triangle DEF$? Explain. (Hint: Use ASA for congruence.)

  Provide students time to discuss this question in small groups.
  - *Proof using ASA for congruence:*
  - *Angle:* $m\angle A = m\angle D$. *Given*
  - *Side:* $AB' = rAB = DE$. *The first equality is true because* $\triangle AB'C'$ *is a dilation of* $\triangle ABC$ *by scale factor* r. *The second equality is true because* r *is defined by* $r = \frac{DE}{AB}$.

    $$AB' = rAB = \frac{DE}{AB}AB = DE$$
  - *Angle:* $m\angle AB'C' = m\angle ABC = m\angle E$. *By the dilation theorem,* $\overleftrightarrow{B'C'} \parallel \overleftrightarrow{BC}$. *Therefore,* $m\angle AB'C' = m\angle ABC$ *because corresponding angles of parallel lines are equal in measure. Finally,* $m\angle ABC = m\angle DEF$ *is given.*
  - *Therefore,* $\triangle AB'C' \cong \triangle DEF$ *by ASA.*

> *Scaffolding:*
>
> Ask a direct question: Why is $\triangle AB'C' \cong \triangle DEF$? Then, instruct students to prove the triangles are congruent using ASA. If necessary, ask students to explain why $m\angle A = m\angle D$, why $AB' = DE$, and why $m\angle AB'C' = m\angle ABC = m\angle E$.

- Now we have the following diagram:

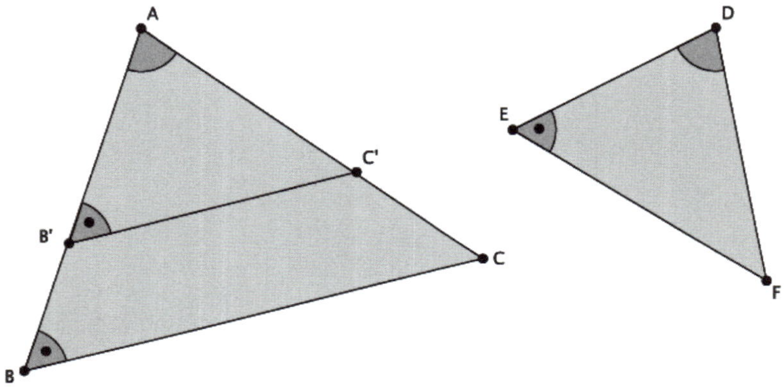

- Therefore, there is a composition of basic rigid motions that takes $\triangle AB'C'$ to $\triangle DEF$. Thus, a dilation with scale factor r composed with basic rigid motions takes $\triangle ABC$ to $\triangle DEF$. Since a similarity transformation exists that maps $\triangle ABC$ to $\triangle DEF$, then $\triangle ABC \sim \triangle DEF$.

Exercises 7–10 (9 minutes)

In Exercises 7–10, students practice using the AA criterion to determine if two triangles are similar and then determine unknown side lengths and/or angle measures of the triangles.

7. Are the triangles shown below similar? Explain. If the triangles are similar, identify any missing angle and side-length measures.

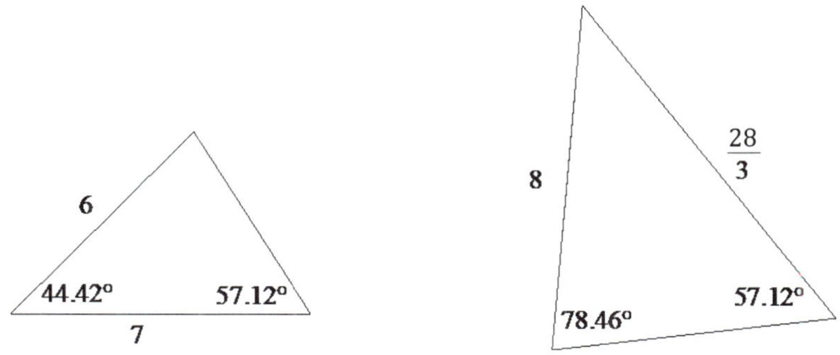

Yes, the triangles are similar because they have two pairs of equal corresponding angles. By the AA criterion, they must be similar. The angle measures and side lengths are shown below.

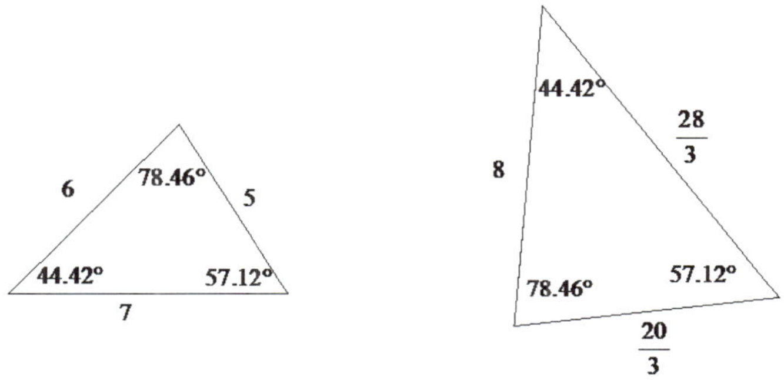

8. Are the triangles shown below similar? Explain. If the triangles are similar, identify any missing angle and side-length measures.

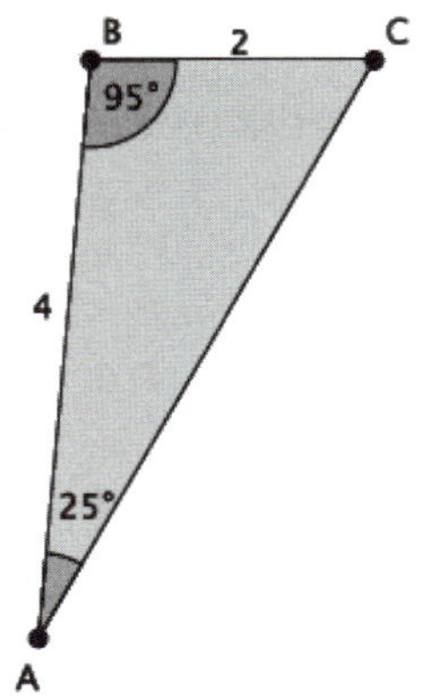

The triangles are not similar because they have just one pair of corresponding equal angles. By the triangle sum theorem, $m\angle C = 60°$ and $m\angle D = 61°$. Since similar triangles must have equal corresponding angles, we can conclude that the triangles shown are not similar.

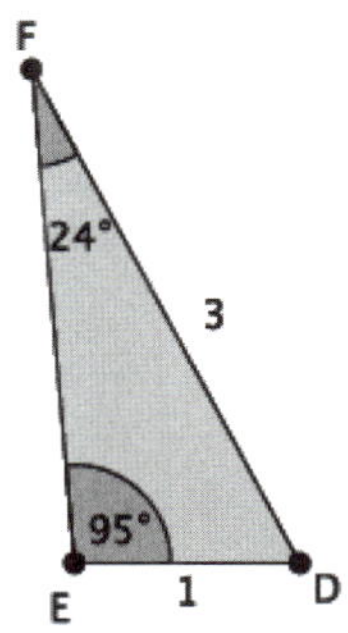

9. **The triangles shown below are similar. Use what you know about similar triangles to find the missing side lengths x and y.**

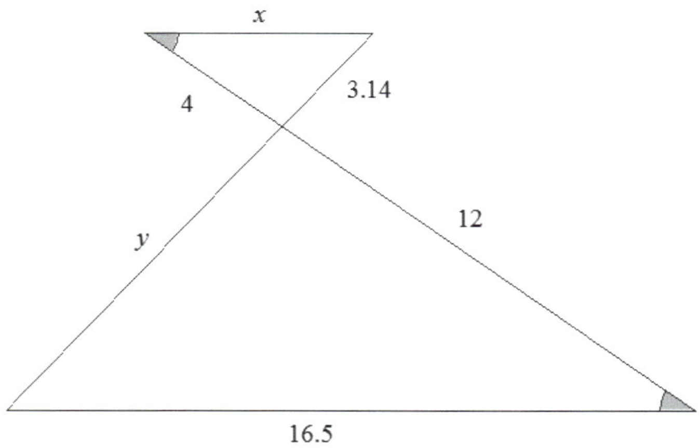

$$\frac{12}{4}=\frac{16.5}{x}$$
$$12x = 66$$
$$x = 5.5$$

$$\frac{12}{4}=\frac{y}{3.14}$$
$$37.68 = 4y$$
$$9.42 = y$$

Side length x is 5.5 units, and side length y is 9.42 units.

10. **The triangles shown below are similar. Write an explanation to a student, Claudia, of how to find the lengths of x and y.**

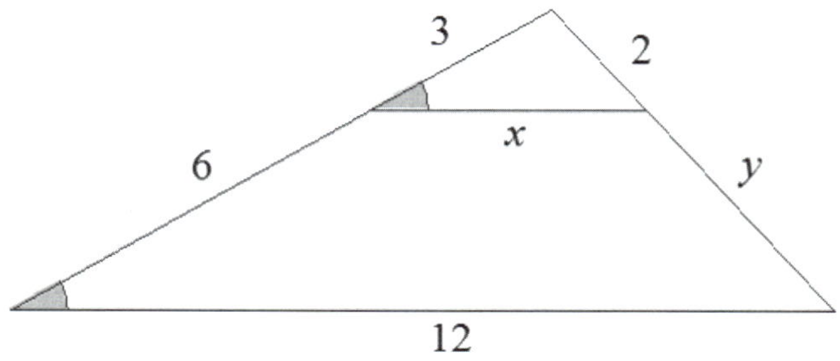

Claudia,

We are given that the triangles are similar; therefore, we know that they have equal corresponding angles and corresponding sides that are equal in ratio. For that reason, we can write $\frac{3}{9}=\frac{2}{2+y}$, which represents two pairs of corresponding sides of the two triangles. We can solve for y as follows:

$$\frac{3}{9}=\frac{2}{2+y}$$
$$6+3y = 18$$
$$3y = 12$$
$$y = 4$$

We can solve for x in a similar manner. We begin by writing two pairs of corresponding sides as equal ratios, making sure that one of the ratios contains the length x.

$$\frac{x}{12}=\frac{3}{9}$$
$$9x = 36$$
$$x = 4$$

Therefore, side length x is 4 units, and side length y is 4 units.

Closing (4 minutes)

- Explain what the AA criterion means.
  - *It means that when two pairs of corresponding angles of two triangles are equal, the triangles are similar.*
- Why is it enough to check only two conditions, two pairs of corresponding angles, as opposed to all six conditions (3 angles and 3 sides), to conclude that a pair of triangles are similar?
  - *We know that for two triangles, when two pairs of corresponding angles are equal and the included corresponding sides are equal in length, the triangles are congruent. By the triangle sum theorem, we can actually state that all three pairs of corresponding angles of the triangles are equal. Since a unique triangle is formed by two fixed angles and a fixed included side length, the other two sides are also fixed by the construction, meeting all six criteria. In the case of similarity, given two pairs of equal angles, we would expect the lengths of the corresponding included sides to be equal in ratio to the scale factor, again meeting all six conditions. For this reason, we can conclude that two triangles are similar by verifying that two pairs of corresponding angles are equal.*

Exit Ticket (5 minutes)

Name ______________________________ Date ________________

Lesson 15: The Angle-Angle (AA) Criterion for Two Triangles to Be Similar

Exit Ticket

1. Given the diagram to the right, $\overline{UX} \perp \overline{VW}$, and $\overline{WY} \perp \overline{UV}$. Show that $\triangle UXV \sim \triangle WYV$.

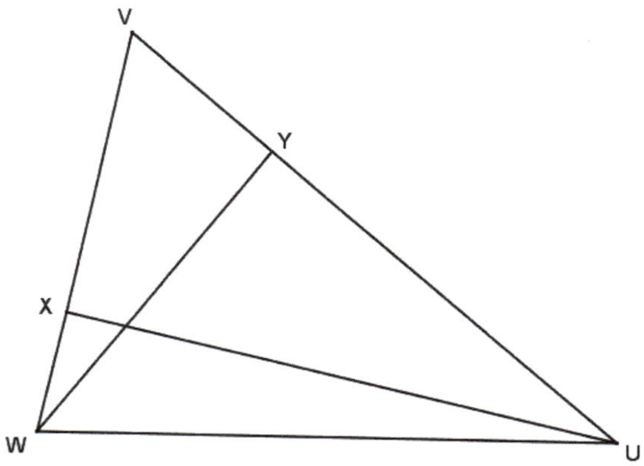

2. Given the diagram to the right and $\overline{DE} \parallel \overline{KL}$, find FE and FL.

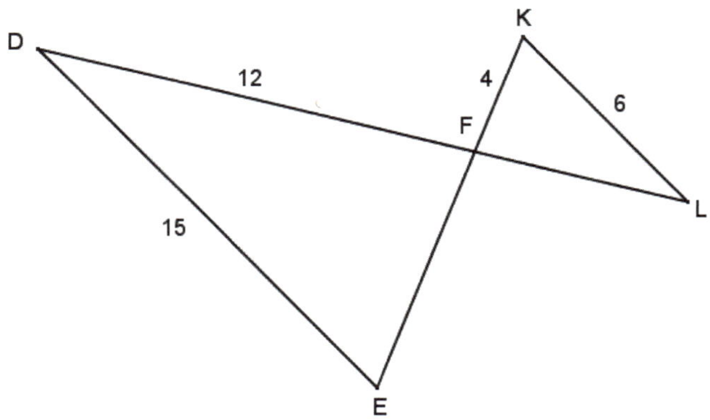

Exit Ticket Sample Solutions

1. Given the diagram to the right, $\overline{UX} \perp \overline{VW}$, and $\overline{WY} \perp \overline{UV}$. Show that $\triangle UXV \sim \triangle WYV$.

   By the definition of perpendicular lines, $\angle WYV$ and $\angle UXV$ are right angles, and all right angles are congruent, so $\angle WYV \cong \angle UXV$. Both $\triangle UXV$ and $\triangle WYV$ share $\angle V$, and by the reflexive property $\angle V \cong \angle V$. Therefore, by the AA criterion for proving similar triangles, $\triangle UXV \sim \triangle WYV$.

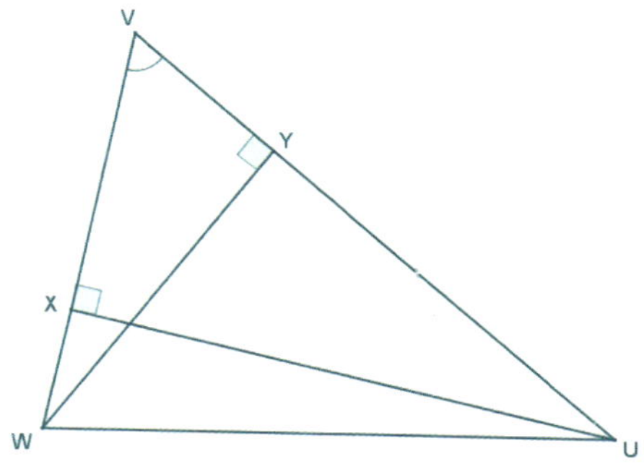

2. Given the diagram to the right and $\overline{DE} \parallel \overline{KL}$, find FE and FL.

   By alternate interior ∠'s, $\overline{DE} \parallel \overline{KL}$, it follows that $\angle K \cong \angle E$ (and by a similar argument, $\angle D \cong \angle L$). $\angle DFE$ and $\angle KFL$ are vertically opposite angles and therefore congruent. By AA criterion for proving similar triangles, $\triangle DFE \sim \triangle LFK$. Therefore,

   $$\frac{DF}{LF} = \frac{EF}{KF} = \frac{DE}{LK}.$$

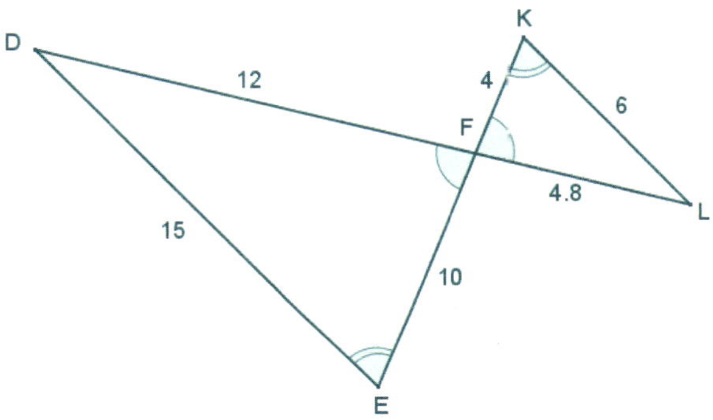

$$\frac{DE}{LK} = \frac{EF}{KF} \qquad \frac{DE}{LK} = \frac{DF}{LF}$$

$$\frac{15}{6} = \frac{EF}{4} \qquad \frac{15}{6} = \frac{12}{LF}$$

$$6EF = 60 \qquad 15LF = 72$$

$$EF = 10 \qquad LF = 4\frac{4}{5}$$

$$LF = 4.8$$

Using the relationship of similar triangles, $EF = 10$, and $LF = 4.8$.

Problem Set Sample Solutions

1. In the figure to the right, $\triangle LMN \sim \triangle MPL$.

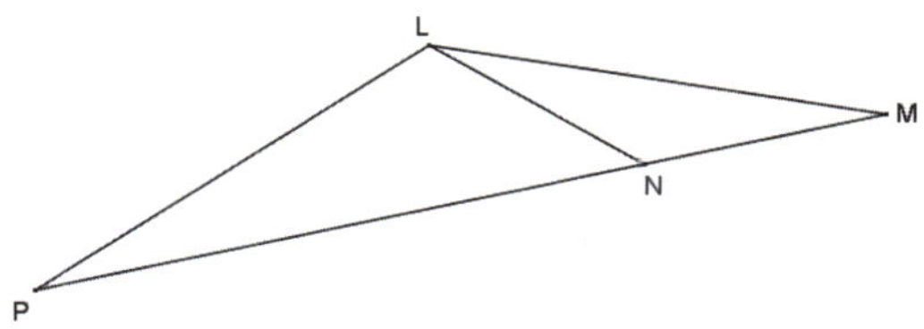

   a. Classify $\triangle LMP$ based on what you know about similar triangles, and justify your reasoning.

   By the given similarity statement, M and P are corresponding vertices; therefore, the angles at M and P must be congruent. This means that $\triangle LMP$ is an isosceles triangle by base ∠'s converse.

b. If $m\angle P = 20°$, find the remaining angles in the diagram.

$m\angle M = 20°$, $m\angle MLN = 20°$, $m\angle MNL = 140°$, $m\angle NLP = 120°$, $m\angle MLP = 140°$, and $m\angle LNP = 40°$. Triangle MNL is also isosceles.

2. In the diagram below, $\triangle ABC \sim \triangle AFD$. Determine whether the following statements must be true from the given information, and explain why.

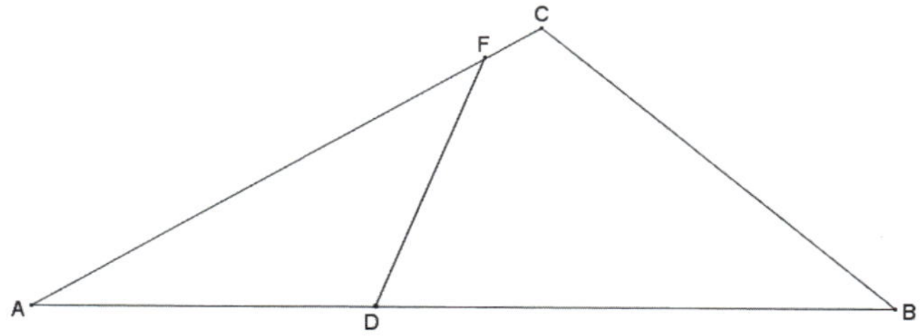

a. $\triangle CAB \sim \triangle DAF$

This statement is true because corresponding vertices are the same as in the given similarity statement but are listed in a different order.

b. $\triangle ADF \sim \triangle CAB$

There is no information given to draw this conclusion.

c. $\triangle BCA \sim \triangle ADF$

There is no information given to draw this conclusion.

d. $\triangle ADF \sim \triangle ACB$

This statement is true because corresponding vertices are the same as in the given similarity statement but are listed in a different order.

3. In the diagram below, D is the midpoint of $\overline{AB}$, F is the midpoint of $\overline{BC}$, and E is the midpoint of $\overline{AC}$. Prove that $\triangle ABC \sim \triangle FED$.

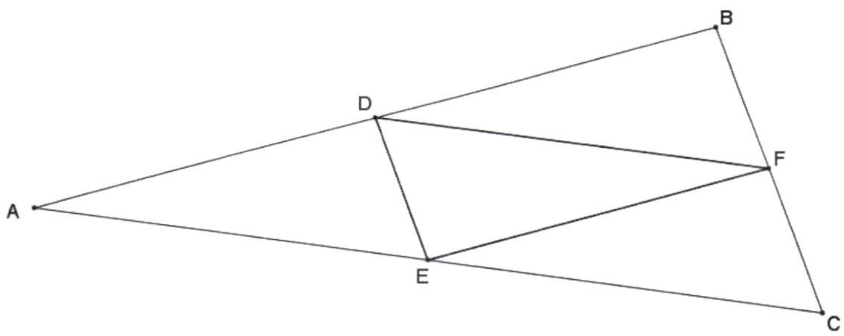

Using the triangle side splitter theorem, since D, F, and E are all midpoints of the sides of $\triangle ABC$, the sides are cut proportionally; therefore, $\overline{DF} \parallel \overline{AC}$, $\overline{DE} \parallel \overline{BC}$, and $\overline{EF} \parallel \overline{AB}$. This provides multiple pairs of parallel lines with parallel transversals.

$\angle A \cong \angle BDF$ by corresponding $\angle$'s, $\overline{DF} \parallel \overline{AC}$, and $\angle DFE \cong \angle BDF$ by alternate interior $\angle$'s, $\overline{EF} \parallel \overline{AB}$, so by transitivity, $\angle A \cong \angle DFE$.

$\angle C \cong \angle BFD$ by corresponding $\angle$'s, $\overline{DF} \parallel \overline{AC}$, and $\angle EDF \cong \angle BFD$ by alternate interior $\angle$'s, $\overline{DE} \parallel \overline{BC}$, so by transitivity, $\angle C \cong \angle EDF$.

$\triangle ABC \sim \triangle FED$ by the AA criterion for proving similar triangles.

4. Use the diagram below to answer each part.

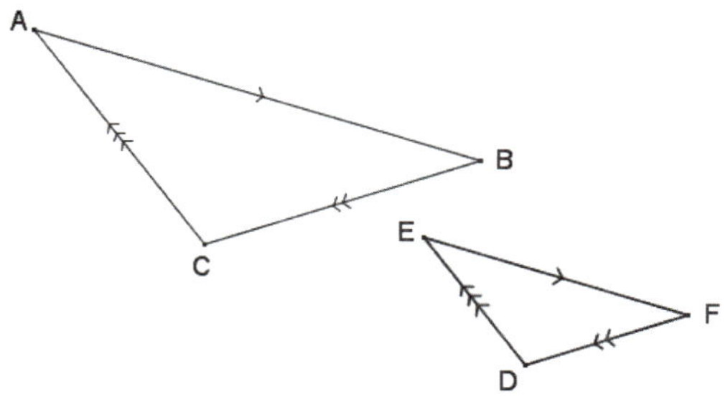

a. If $\overline{AC} \parallel \overline{ED}$, $\overline{AB} \parallel \overline{EF}$, and $\overline{CB} \parallel \overline{DF}$, prove that the triangles are similar.

By extending all sides of both triangles, there are several pairs of parallel lines cut by parallel transversals. Using corresponding angles within parallel lines and transitivity, $\triangle ABC \sim \triangle EFD$ by the AA criterion for proving similar triangles.

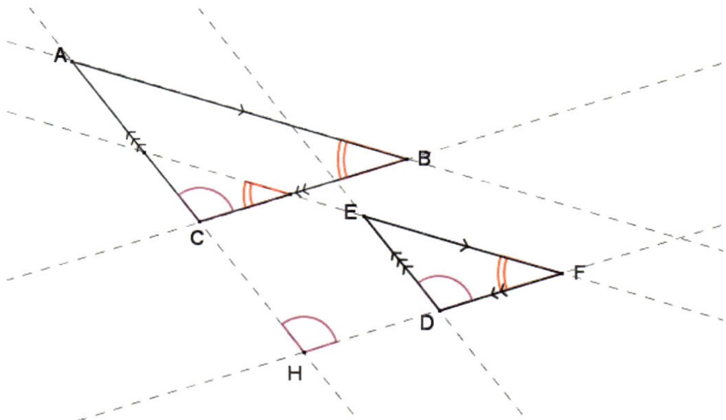

b. The triangles are not congruent. Find the dilation that takes one to the other.

Extend lines joining corresponding vertices to find their intersection O, which is the center of dilation.

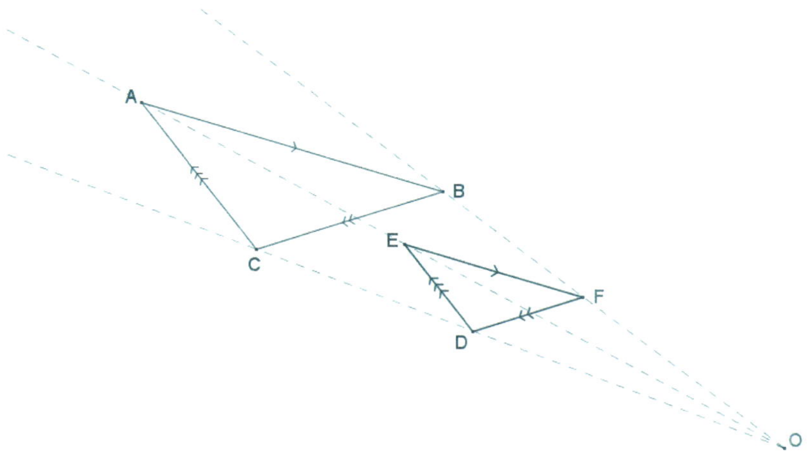

5. Given trapezoid $ABDE$, and $\overline{AB} \parallel \overline{ED}$, prove that $\triangle AFB \sim \triangle DEF$.

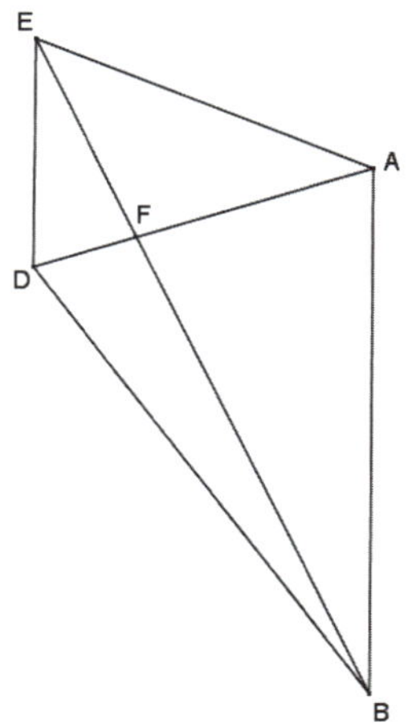

From the given information, $\angle EDA \cong \angle DAB$ by alternate interior $\angle$'s, $\overline{AB} \parallel \overline{ED}$ (by the same argument, $\angle DEB \cong \angle ABE$). Furthermore, $\angle EFD \cong \angle BFA$ because vertical angles are congruent. Therefore, $\triangle AFB \sim \triangle DEF$ by the AA criterion for proving similar triangles.

Cutouts to use for in-class discussion:

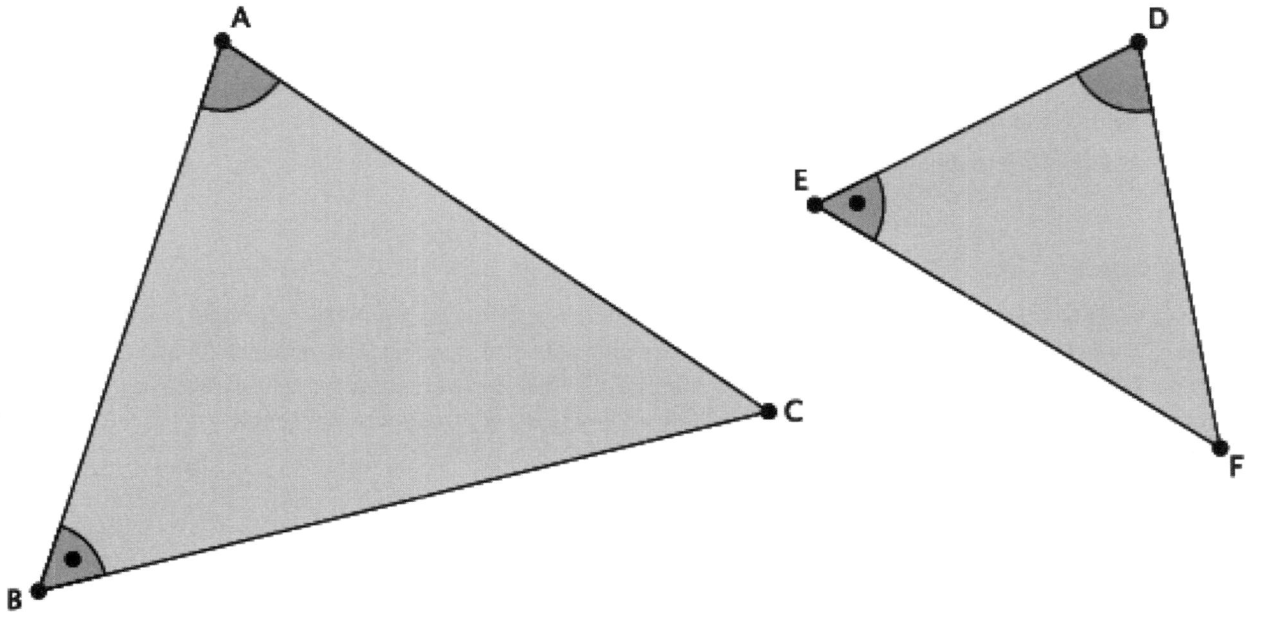

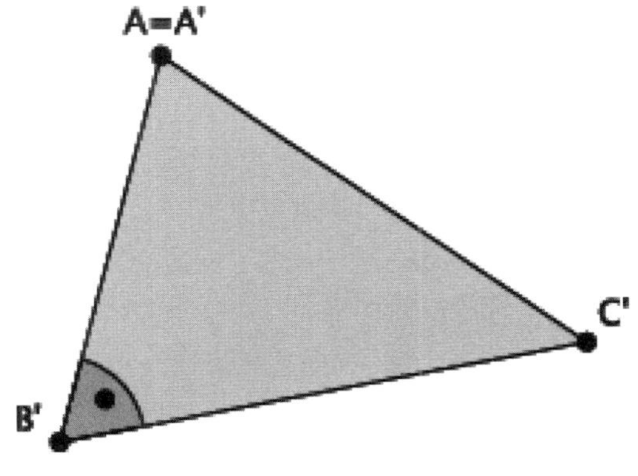

Lesson 16: Between-Figure and Within-Figure Ratios

Student Outcomes

- Students indirectly solve for measurements involving right triangles using scale factors, ratios between similar figures, and ratios within similar figures.
- Students use trigonometric ratios to solve applied problems.

Lesson Notes

By now, students are very familiar with how to use a scale factor with similar figures to determine unknown lengths of figures. This lesson demonstrates how the values of the ratios of corresponding sides between figures can be rewritten, equivalently, as ratios of corresponding lengths within figures. This work foreshadows ratios related to trigonometry. Though sine, cosine, and tangent are not formally defined in this lesson, students are essentially using the ratios as a premise for formal treatment of trigonometric ratios in Topic E.

Classwork

Opening Exercise (2 minutes)

Opening Exercise

At a certain time of day, a 12 m flagpole casts an 8 m shadow. Write an equation that would allow you to find the height, h, of the tree that uses the length, s, of the tree's shadow.

$$\frac{12}{8} = \frac{h}{s}$$
$$\frac{3}{2}s = h$$

OR

$$\frac{12}{h} = \frac{8}{s}$$
$$12s = 8h$$
$$\frac{3}{2}s = h$$

Example 1 (14 minutes)

The discussion following this exercise highlights length relationships within figures. Many students rely on ratios of lengths between figures to solve for unknown measurements of sides rather than use ratios of lengths within a figure. This example shows students that comparing ratios between and within figures is mathematically equivalent.

Example 1

Given $\triangle ABC \sim \triangle A'B'C'$, find the missing side lengths.

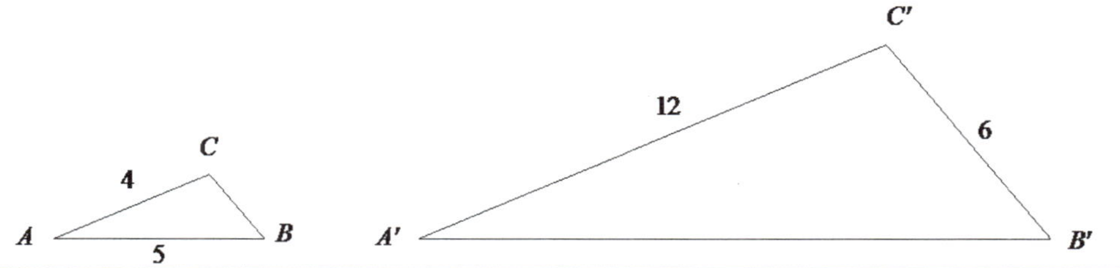

- (Take a poll.) How many people used the ratio of lengths $4:12$ or $12:4$ (i.e., the corresponding lengths of $\overline{AC}$ and $\overline{A'C'}$) to determine the measurements of $\overline{BC}$ and $\overline{A'B'}$?
- Are there any other length relationships we could use to set up an equation and solve for the missing side lengths?

MP.7

Provide students time to think about and try different equations that would lead to the same answer.

- *In addition to the ratio of corresponding lengths $AC: A'C'$, we can use a ratio of lengths within one of the triangles. For example, we could use the ratio of $AC: AB$ to find the length of $\overline{A'B'}$. Equating the values of the ratios, we get the following:*

$$\frac{4}{5} = \frac{12}{A'B'}$$
$$A'B' = 15.$$

- *To find the length of $\overline{BC}$, we can use the ratios $A'C': B'C'$ and $AC: BC$. Equating the values of the ratios, we get the following:*

$$\frac{4}{BC} = \frac{12}{6}$$
$$BC = 2.$$

- Why is it possible to use this length relationship, $AC: AB$, to solve for a missing side length?
  - *Take student responses, and then offer the following explanation:*
- There are three methods that we can use to determine the missing side lengths.

  Method 1 (Scale Factor): Find the scale factor, and use it to compute for the desired side lengths.

  Since we know $AC = 4$ and the corresponding side $A'C' = 12$, the scale factor of r satisfies $4r = 12$. So, $r = 3$, $A'B' = rAB = 3 \cdot 5 = 15$, and $BC = \frac{B'C'}{r} = \frac{6}{3} = 2$.

  Method 2 (Ratios Between-Figures): Equate the values of the ratios of the corresponding sides.

  $\frac{A'B'}{AB} = \frac{A'C'}{AC}$ implies $\frac{A'B'}{5} = \frac{12}{4}$ and $A'B' = 15$. $\frac{BC}{B'C'} = \frac{AC}{A'C'}$ implies $\frac{BC}{6} = \frac{4}{12}$ and $BC = 2$.

Method 3 (Ratios Within-Figures): Equate the values of the ratios within each triangle.

$$\frac{A'B'}{A'C'} = \frac{AB}{AC} \text{ implies } \frac{A'B'}{12} = \frac{5}{4} \text{ and } A'B' = 15. \quad \frac{BC}{AC} = \frac{B'C'}{A'C'} \text{ implies } \frac{BC}{4} = \frac{6}{12} \text{ and } BC = 2.$$

- Why does Method 3 work? That is, why can we use values of ratios *within* each triangle to find the missing side lengths?

Provide students time to explain why Method 3 works, and then offer the following explanation:

- If $\triangle ABC \sim \triangle A'B'C'$, then it is true that the ratio between any two side lengths of the first triangle is equal to the ratio between the corresponding side lengths of the second triangle; for example, $AB: BC = A'B': B'C'$. This is because if r is the scale factor, then $AB: BC = rAB: rBC = A'B': B'C'$.

Revisit the Opening Exercise. Ask students to explain which method they used to write their equations for the height of the tree. Answers may vary, but students likely have used Method 2 because that is what is most familiar to them.

Example 2 (7 minutes)

In this example, students use what they know about similar figures to indirectly measure the height of a building. If possible, use an image from the school or community to personalize the example.

- Suppose you want to know the height of a building. Would it make sense to climb to the roof and use a tape measure? Possibly. Suppose you want to know the distance between the earth and the moon. Would a tape measure work in that situation? Not likely. For now, we will take indirect measurements of trees and buildings, but we will soon learn how the Greeks measured the distance from the earth to the moon.

Example 2

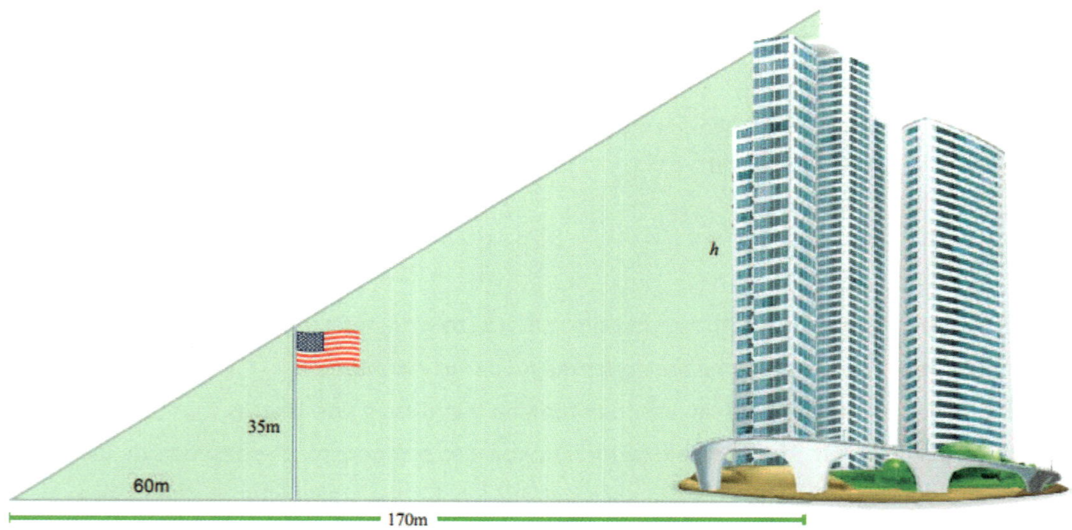

In the diagram above, a large flagpole stands outside of an office building. Marquis realizes that when he looks up from the ground 60 m away from the flagpole, the top of the flagpole and the top of the building line up. If the flagpole is 35 m tall and Marquis is 170 m from the building, how tall is the building?

a. **Are the triangles in the diagram similar? Explain.**

Yes, the triangle formed by Marquis, the ground, and the flagpole is similar to the triangle formed by Marquis, the ground, and the building. They are similar by the AA criterion. Each of the triangles has a common angle with Marquis at the vertex, and each triangle has a right angle, where the flagpole and the building meet the ground.

b. **Determine the height of the building using what you know about scale factors.**

The scale factor is $\frac{170}{60} = \frac{17}{6} = 2.8\overline{3}$. *Then, the height of the building is* $35(2.8\overline{3}) = 99.1\overline{6}\text{ m}$.

c. **Determine the height of the building using ratios between similar figures.**

$$\frac{35}{h} = \frac{60}{170}$$
$$5950 = 60h$$
$$h = 99.1\overline{6}$$

d. **Determine the height of the building using ratios within similar figures.**

$$\frac{35}{60} = \frac{h}{170}$$
$$5950 = 60h$$
$$h = 99.1\overline{6}$$

Example 3 (12 minutes)

Work through the following example with students. The goal is for them to recognize how the value of the ratio within a figure can be used to determine an unknown side length of a similar figure. Make clear to students that more than one ratio (from parts (a)–(c)) can be used to determine the requested unknown lengths in parts (d)–(f).

Example 3

The following right triangles are similar. We will determine the unknown side lengths by using ratios within the first triangle. For each of the triangles below, we define the base as the horizontal length of the triangle and the height as the vertical length.

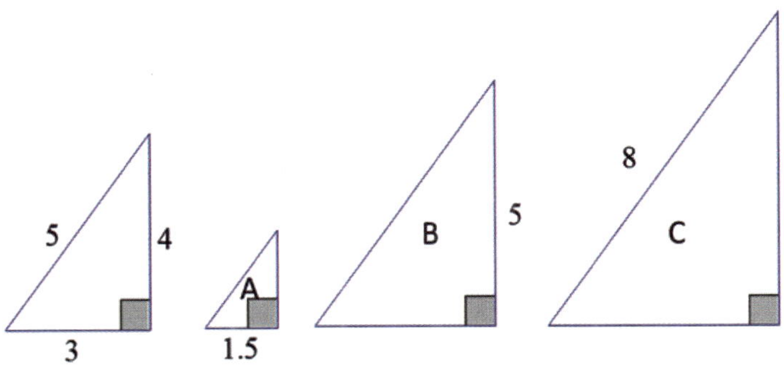

Scaffolding:

- It may benefit some groups of students to show only the 3–4–5 triangle as they complete parts (a)–(c) of the example.
- Have advanced learners consider why the triangle is called a 3–4–5 triangle and determine the significance of these numbers to any triangle that is similar to it.

a. **Write and find the value of the ratio that compares the height to the hypotenuse of the leftmost triangle.**

$\frac{4}{5} = 0.8$

b. **Write and find the value of the ratio that compares the base to the hypotenuse of the leftmost triangle.**

$\frac{3}{5} = 0.6$

c. **Write and find the value of the ratio that compares the height to the base of the leftmost triangle.**

$\frac{4}{3} = 1.\overline{3}$

d. **Use the triangle with lengths $3-4-5$ and triangle A to answer the following questions:**

i. **Which ratio can be used to determine the height of triangle A?**

The ratio that compares the height to the base can be used to determine the height of triangle A.

ii. **Which ratio can be used to determine the hypotenuse of triangle A?**

The ratio that compares the base to the hypotenuse can be used to determine the hypotenuse of triangle A.

iii. **Find the unknown lengths of triangle A.**

Let h represent the height of the triangle; then

$$\frac{h}{1.5} = 1.\overline{3}$$
$$h = 2.$$

Let c represent the length of the hypotenuse; then

$$\frac{1.5}{c} = 0.6$$
$$c = 2.5.$$

e. **Use the triangle with lengths $3-4-5$ and triangle B to answer the following questions:**

i. **Which ratio can be used to determine the base of triangle B?**

The ratio that compares the height to the base can be used to determine the base of triangle B.

ii. **Which ratio can be used to determine the hypotenuse of triangle B?**

The ratio that compares the height to the hypotenuse can be used to determine the hypotenuse of triangle B.

iii. Find the unknown lengths of triangle B.

Let b represent the length of the base; then

$\frac{5}{b} = 1\frac{1}{3}$

$b = 3.75.$

Let c represent the length of the hypotenuse; then

$\frac{5}{c} = 0.8$

$c = 6.25.$

f. Use the triangle with lengths $3-4-5$ and triangle C to answer the following questions:

i. Which ratio can be used to determine the height of triangle C?

The ratio that compares the height to the hypotenuse can be used to determine the height of triangle C.

ii. Which ratio can be used to determine the base of triangle C?

The ratio that compares the base to the hypotenuse can be used to determine the base of triangle C.

iii. Find the unknown lengths of triangle C.

Let h represent the height of the triangle; then

$\frac{h}{8} = 0.8$

$h = 6.4.$

Let b represent the length of the base; then

$\frac{b}{8} = 0.6$

$b = 4.8.$

g. Explain the relationship of the ratio of the corresponding sides within a figure to the ratio of corresponding sides within a similar figure.

Corresponding lengths of similar figures have proportional lengths, but the ratio of two lengths within a figure is equal to the corresponding ratio of two lengths in a similar figure.

h. How does the relationship you noted in part (g) allow you to determine the length of an unknown side of a triangle?

This relationship allows us to find the length of an unknown side of a triangle. We can write the ratio of corresponding sides within the figure that contains the unknown side length. This ratio will be equal to the value of the ratio of corresponding sides where the lengths are known. Since the ratios are equal, the unknown length can be found by solving a simple equation.

Closing (5 minutes)

To close the lesson, the following three questions may be asked separately. Or let students complete a Quick Write for all three prompts, and then discuss the three questions all at one time.

- What does it mean to use between-figure ratios of corresponding sides of similar triangles?
  - *Between-figure ratios are those where each ratio compares corresponding side lengths between two similar figures. That is, one number of the ratio comes from one triangle while the other number in the ratio comes from a different, similar triangle.*
- What does it mean to use within-figure ratios of corresponding sides of similar triangles?
  - *Within-figure ratios are those where each ratio is composed of side lengths from one figure of the similar figures. That is, one ratio contains numbers that represent the side lengths from one triangle, and a second ratio contains numbers that represent the side lengths from a second, similar triangle.*
- How can within-figure ratios be used to find unknown side lengths of similar triangles?
  - *If a triangle is similar to another, then the within-figure ratios can be used to find the unknown lengths of the other triangle. For example, if the ratio that compares the height to the hypotenuse is known for one triangle and that ratio is equal to* 0.6*, then a ratio that compares the height to the hypotenuse of a similar triangle will also be equal to* 0.6*.*

Exit Ticket (5 minutes)

Name ______________________________________ Date ____________________

Lesson 16: Between-Figure and Within-Figure Ratios

Exit Ticket

Dennis needs to fix a leaky roof on his house but does not own a ladder. He thinks that a 25 ft. ladder will be long enough to reach the roof, but he needs to be sure before he spends the money to buy one. He chooses a point P on the ground where he can visually align the roof of his car with the edge of the house roof. Help Dennis determine if a 25 ft. ladder will be long enough for him to safely reach his roof.

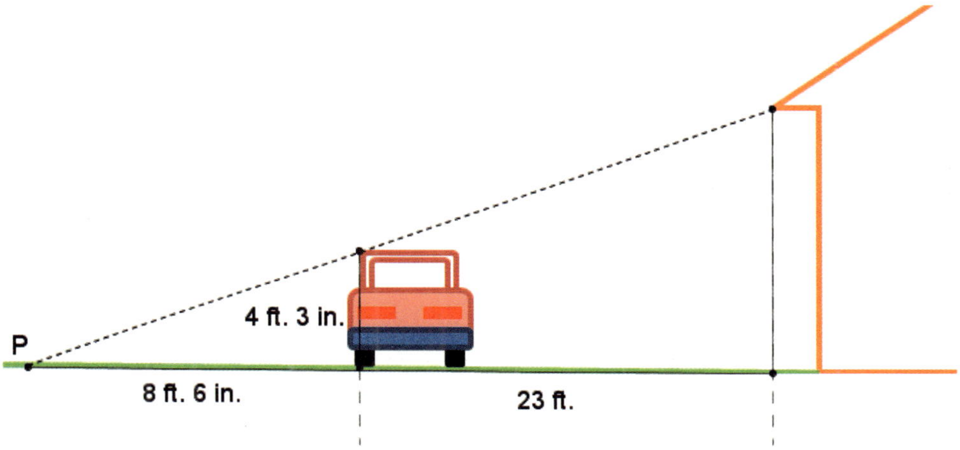

Exit Ticket Sample Solutions

Dennis needs to fix a leaky roof on his house but does not own a ladder. He thinks that a 25 ft. ladder will be long enough to reach the roof, but he needs to be sure before he spends the money to buy one. He chooses a point P on the ground where he can visually align the roof of his car with the edge of the house roof. Help Dennis determine if a 25 ft. ladder will be long enough for him to safely reach his roof.

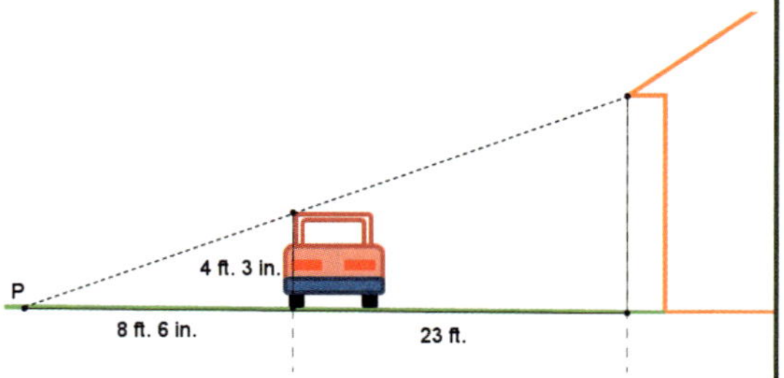

The height of the edge of the roof from the ground is unknown, so let x represent the distance from the ground to the edge of the roof. The nested triangles are similar triangles by the AA criterion for similar triangles because they share $\angle P$, and the height of the car and the distance from the ground to the edge of the house roof are both measured perpendicular to the ground. Therefore, the following is true:

$$\frac{8.5}{4.25} = \frac{(8.5 + 23)}{x}$$

$$8.5x = 133.85$$

$$x = 15.74705\ldots$$

$$x \approx 15.75.$$

The distance from the ground to the edge of the roof is 15.75 ft. (or 15 ft. 9 in.) The 25 ft. ladder is clearly long enough for Dennis to safely reach his roof.

Problem Set Sample Solutions

1. $\triangle DEF \sim \triangle ABC$ All side length measurements are in centimeters. Use between-figure ratios and/or within-figure ratios to determine the unknown side lengths.

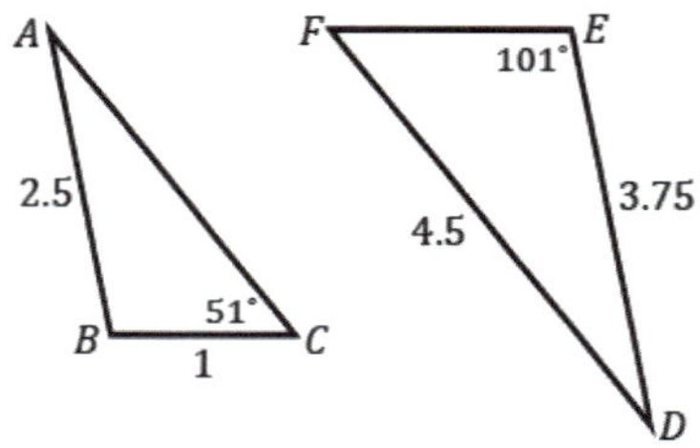

Using the given similarity statement, $\angle D$ corresponds with $\angle A$, and $\angle C$ corresponds with $\angle F$, so it follows that $\overline{AB}$ corresponds with $\overline{DE}$, $\overline{AC}$ with $\overline{DF}$, and $\overline{BC}$ with $\overline{EF}$.

$$\frac{AB}{BC} = \frac{DE}{EF} \qquad \frac{AB}{DE} = \frac{AC}{DF}$$

$$\frac{2.5}{1} = \frac{3.75}{EF} \qquad \frac{2.5}{3.75} = \frac{AC}{4.5}$$

$$2.5(EF) = 3.75 \qquad 3.75(AC) = 11.25$$

$$EF = 1.5 \qquad AC = 3$$

2. Given $\triangle ABC \sim \triangle XYZ$, answer the following questions:

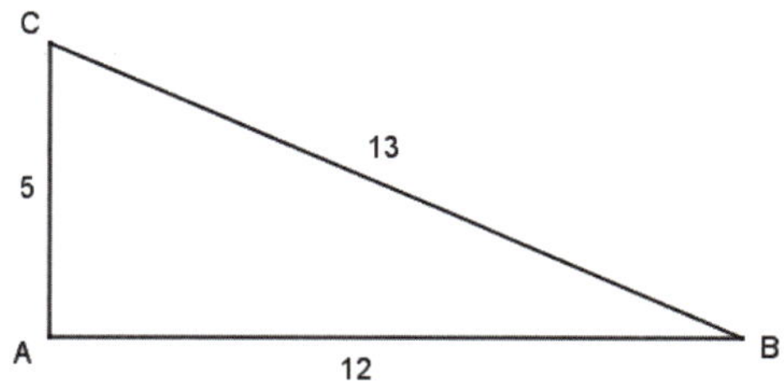

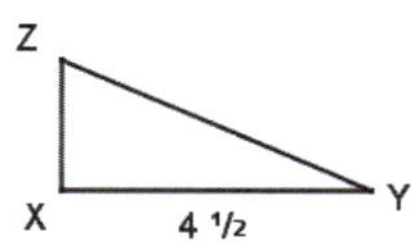

   a. Write and find the value of the ratio that compares the height $\overline{AC}$ to the hypotenuse of $\triangle ABC$.

   $$\frac{5}{13}$$

   b. Write and find the value of the ratio that compares the base $\overline{AB}$ to the hypotenuse of $\triangle ABC$.

   $$\frac{12}{13}$$

EUREKA MATH™

c. **Write and find the value of the ratio that compares the height AC to the base AB of $\triangle ABC$.**

$\frac{5}{12}$

d. **Use within-figure ratios to find the corresponding height of $\triangle XYZ$.**

$\frac{5}{12} = \frac{XZ}{4.5}$

$XZ = 1\frac{7}{8}$

e. **Use within-figure ratios to find the hypotenuse of $\triangle XYZ$.**

$\frac{12}{13} = \frac{4.5}{YZ}$

$YZ = 4\frac{7}{8}$

3. **Right triangles A, B, C, and D are similar. Determine the unknown side lengths of each triangle by using ratios of side lengths within triangle A.**

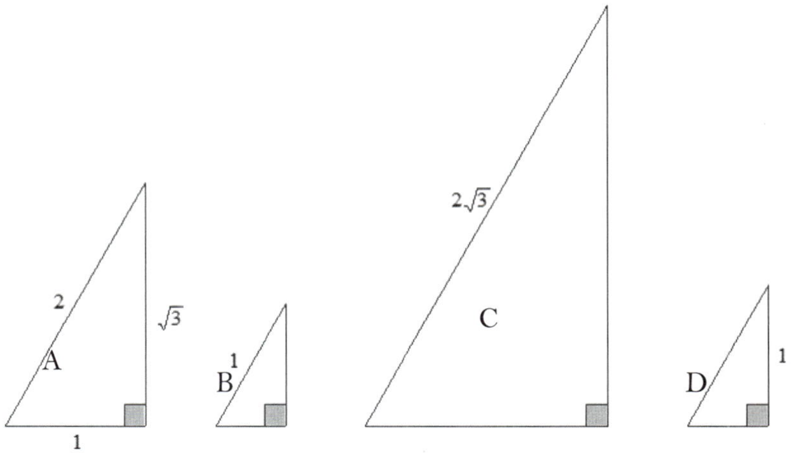

a. **Write and find the value of the ratio that compares the height to the hypotenuse of triangle A.**

$\frac{\sqrt{3}}{2} \approx 0.866$

b. **Write and find the value of the ratio that compares the base to the hypotenuse of triangle A.**

$\frac{1}{2} = 0.5$

c. **Write and find the value of the ratio that compares the height to the base of triangle A.**

$\frac{\sqrt{3}}{1} = \sqrt{3} \approx 1.73$

d. **Which ratio can be used to determine the height of triangle B? Find the height of triangle B.**

The ratio that compares the height to the hypotenuse of triangle A can be used to determine the height of triangle B. Let the unknown height of triangle B be m.

$$\frac{\sqrt{3}}{2} = \frac{m}{1}$$

$$m = \frac{\sqrt{3}}{2}$$

e. **Which ratio can be used to determine the base of triangle B? Find the base of triangle B.**

The ratio that compares the base to the hypotenuse of triangle A can be used to determine the base of triangle B. Let the unknown base of triangle B be n.

$$\frac{1}{2} = \frac{n}{1}$$

$$n = \frac{1}{2}$$

f. **Find the unknown lengths of triangle C.**

The base of triangle C is $\sqrt{3}$.

The height of triangle C is 3.

g. **Find the unknown lengths of triangle D.**

The base of triangle D is $\frac{\sqrt{3}}{3}$.

The hypotenuse of triangle D is $\frac{2\sqrt{3}}{3}$.

h. **Triangle E is also similar to triangles A, B, C, and D. Find the lengths of the missing sides in terms of x.**

The base of triangle E is x.

The height of triangle E is $x\sqrt{3}$.

The hypotenuse of triangle E is $2x$.

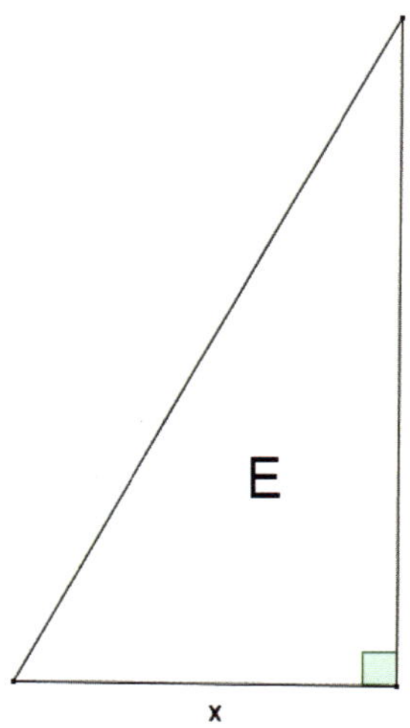

4. **Brian is photographing the Washington Monument and wonders how tall the monument is. Brian places his 5 ft. camera tripod approximately 100 yd. from the base of the monument. Lying on the ground, he visually aligns the top of his tripod with the top of the monument and marks his location on the ground approximately 2 ft. 9 in. from the center of his tripod. Use Brian's measurements to approximate the height of the Washington Monument.**

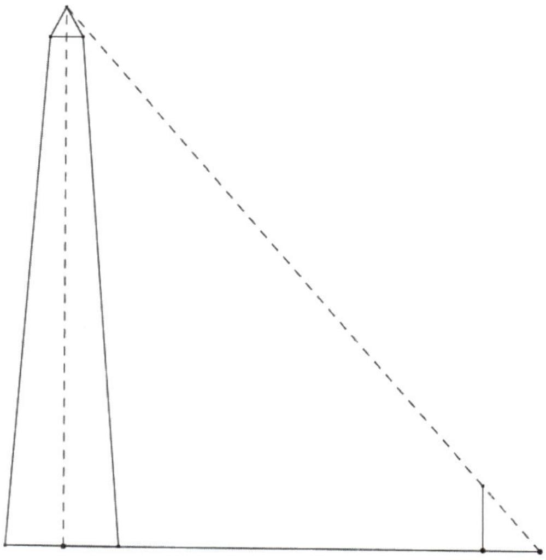

Brian's location on the ground is approximately 302.75 ft. from the base of the monument. His visual line forms two similar right triangles with the height of the monument and the height of his camera tripod.

$$\frac{5}{2.75} = \frac{h}{302.75}$$
$$1513.75 = 2.75h$$
$$550.5 \approx h$$

According to Brian's measurements, the height of the Washington Monument is approximately 550.5 ft.

Students may want to check the accuracy of this problem. The actual height of the Washington Monument is 555 ft.

5. **Catarina's boat has come untied and floated away on the lake. She is standing atop a cliff that is 35 ft. above the water in a lake. If she stands 10 ft. from the edge of the cliff, she can visually align the top of the cliff with the water at the back of her boat. Her eye level is $5\frac{1}{2}$ ft. above the ground. Approximately how far out from the cliff is Catarina's boat?**

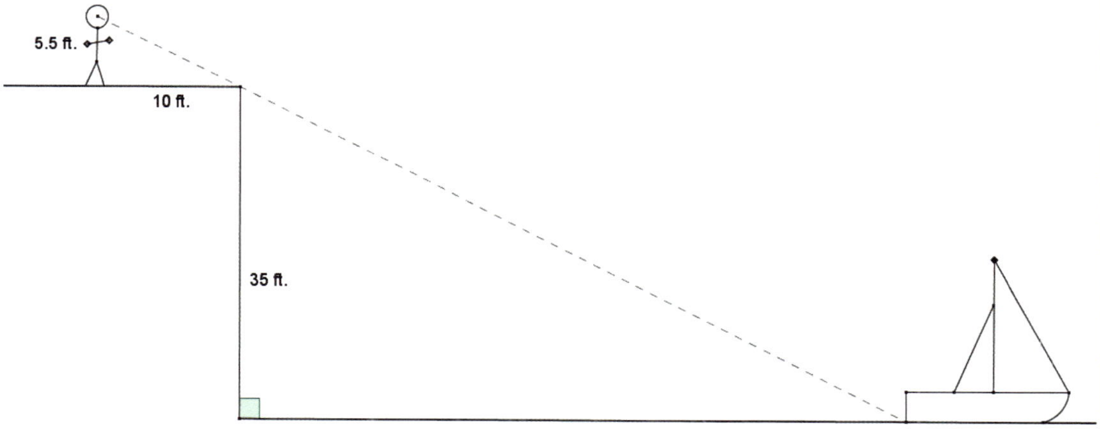

Catarina's height and the height of the cliff both are measured perpendicularly to the ground (and water), so both triangles formed by her visual path are right triangles. Assuming that the ground level on the cliff and the water are parallel, Catarina's visual path forms the same angle with the cliff as it does with the surface of the water (corresponding $\angle$'s, parallel lines). So, the right triangles are similar triangles by the AA criterion for showing triangle similarity, which means that the ratios within triangles will be equal. Let d represent the distance from the boat to the cliff:

$$\frac{5.5}{10} = \frac{35}{d}$$
$$5.5d = 350$$
$$d \approx 63.6$$

Catarina's boat is approximately 63.6 ft. away from the cliff.

6. **Given the diagram below and $\triangle ABC \sim \triangle XYZ$, find the unknown lengths x, $2x$, and $3x$.**

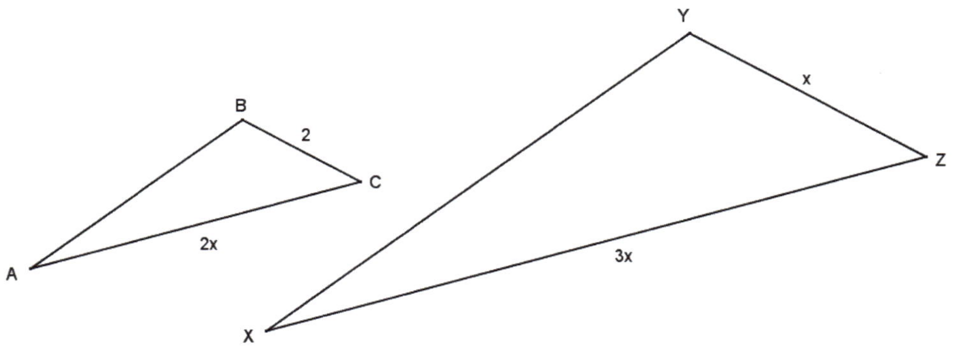

The triangles are given as similar, so the values of the ratios of the sides within each triangle must be equal.

$$\frac{2}{2x} = \frac{x}{3x}$$

The sides of the larger triangle are unknown; however, the lengths include the same factor x, which is clearly nonzero, so the ratio of the sides must then be $\frac{1}{3}$. Similarly, the sides of the smaller triangle have the same factor 2, so the value of the ratio can be rewritten as $\frac{1}{x}$.

$$\frac{1}{x} = \frac{1}{3}$$
$$x = 3$$

Since the value of x is 3, it follows that $YZ = 3$, $AC = 6$, and $XZ = 9$.

Lesson 17: The Side-Angle-Side (SAS) and Side-Side-Side (SSS) Criteria for Two Triangles to Be Similar

Student Outcomes

- Students prove the side-angle-side criterion for two triangles to be similar and use it to solve triangle problems.
- Students prove the side-side-side criterion for two triangles to be similar and use it to solve triangle problems.

Lesson Notes

At this point, students know that two triangles can be considered similar if they have two pairs of corresponding equal angles. In this lesson, students learn the other conditions that can be used to deem two triangles similar, specifically the SAS and SSS criteria.

Classwork

Opening Exercise (3 minutes)

Opening Exercise

a. **Choose three lengths that represent the sides of a triangle. Draw the triangle with your chosen lengths using construction tools.**

Answers will vary. Sample response: 6 cm, 7 cm, *and* 8 cm.

b. **Multiply each length in your original triangle by 2 to get three corresponding lengths of sides for a second triangle. Draw your second triangle using construction tools.**

Answers will be twice the lengths given in part (a). Sample response: 12 cm, 14 cm, *and* 16 cm.

MP.3

c. **Do your constructed triangles appear to be similar? Explain your answer.**

The triangles appear to be similar. Their corresponding sides are given as having lengths in the ratio $2:1$, *and the corresponding angles appear to be equal in measure. (This can be verified using either a protractor or patty paper.)*

d. **Do you think that the triangles can be shown to be similar without knowing the angle measures?**

Answers will vary.

- In Lesson 16, we discovered that if two triangles have two pairs of angles with equal measures, it is not necessary to check all 6 conditions (3 sides and 3 angles) to show that the two triangles are similar. We called it the *AA criterion*. In this lesson, we will use other combinations of conditions to conclude that two triangles are similar.

Exploratory Challenge 1/Exercises 1–2 (8 minutes)

This challenge gives students the opportunity to discover the SAS criterion for similar triangles. The Discussion that follows solidifies this fact. Consider splitting the class into two groups so that one group works on Exercise 1 and the other group completes Exercise 2. Then, have students share their work with the whole class.

Exploratory Challenge 1/Exercises 1–2

1. **Examine the figure, and answer the questions to determine whether or not the triangles shown are similar.**

Figure 1

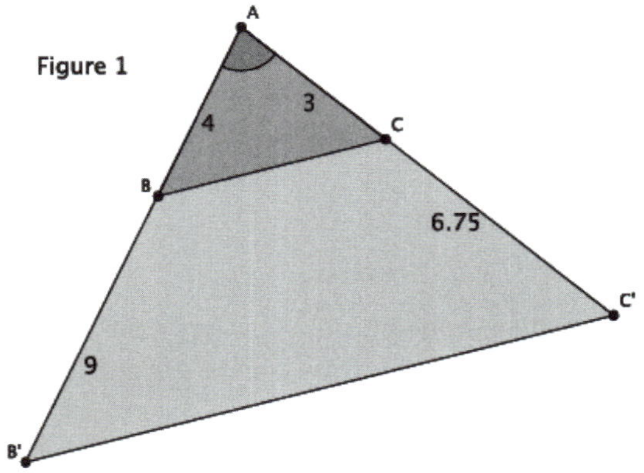

Scaffolding:

- It may be helpful to ask students how many triangles they see in Figure 1, and then have them identify each of the triangles.
- Have students create triangles that meet specific criteria. Given a starting triangle with lengths $3, 6, 7$, generate similar triangles by multiplying the sides of the "starting" triangle by a single factor, such as $1, 1.5, 2, 3$, and so on.

a. **What information is given about the triangles in Figure 1?**

We are given that $\angle A$ is common to both $\triangle ABC$ and $\triangle AB'C'$. We are also given information about some of the side lengths.

b. **How can the information provided be used to determine whether $\triangle ABC$ is similar to $\triangle AB'C'$?**

We know that similar triangles will have ratios of corresponding sides that are proportional; therefore, we can use the side lengths to check for proportionality.

c. **Compare the corresponding side lengths of $\triangle ABC$ and $\triangle AB'C'$. What do you notice?**

$$\frac{4}{13} = \frac{3}{9.75}$$

$$39 = 39$$

The cross-products are equal; therefore, the side lengths are proportional.

d. **Based on your work in parts (a)–(c), draw a conclusion about the relationship between $\triangle ABC$ and $\triangle AB'C'$. Explain your reasoning.**

The triangles are similar. By the triangle side splitter theorem, I know that when the sides of a triangle are cut proportionally, then $\overline{BC} \parallel \overline{B'C'}$. Then, I can conclude that the triangles are similar because they have two pairs of corresponding angles that are equal.

2. **Examine the figure, and answer the questions to determine whether or not the triangles shown are similar.**

Figure 2

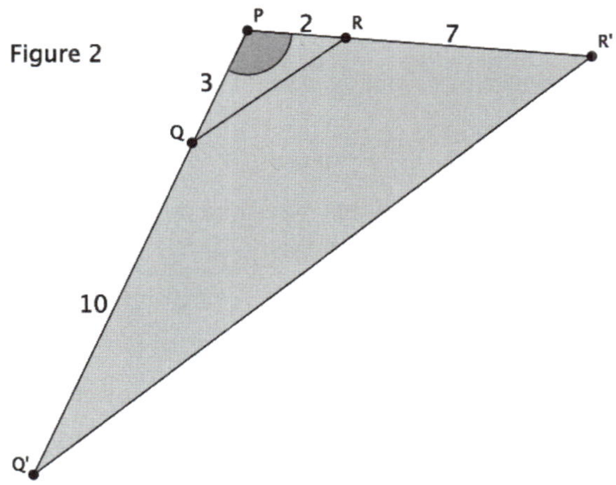

> **Scaffolding:**
> It may be helpful to ask students how many triangles they see in Figure 2; then, have them identify each of the triangles.

a. **What information is given about the triangles in Figure 2?**

We are given that $\angle P$ is common to both $\triangle PQR$ and $\triangle PQ'R'$. We are also given information about some of the side lengths.

b. **How can the information provided be used to determine whether $\triangle PQR$ is similar to $\triangle PQ'R'$?**

We know that similar triangles will have ratios of corresponding sides that are proportional; therefore, we can use the side lengths to check for proportionality.

c. **Compare the corresponding side lengths of $\triangle PQR$ and $\triangle PQ'R'$. What do you notice?**

$$\frac{3}{13} \neq \frac{2}{9}$$

$$27 \neq 26$$

The side lengths are not proportional.

d. **Based on your work in parts (a)–(c), draw a conclusion about the relationship between $\triangle PQR$ and $\triangle PQ'R'$. Explain your reasoning.**

The triangles are not similar. The side lengths are not proportional, which is what I would expect in similar triangles. I know that the triangles have one common angle, but I cannot determine from the information given whether there is another pair of equal angles. Therefore, I conclude that the triangles are not similar.

Discussion (5 minutes)

Have students reference their work in part (d) of Exercises 1 and 2 while leading the discussion below.

- Which figure contained triangles that are similar? How do you know?
  - *Figure 1 has the similar triangles. I know because the triangles have side lengths that are proportional. Since the side lengths are split proportionally by segment BC, then $BC \parallel B'C'$, and the triangles are similar by the AA criterion. The same cannot be said about Figure 2.*
- The triangles in Figure 1 are similar. Take note of the information that is given about the triangles in the diagrams, not what you are able to deduce about the angles.
  - *The diagram shows information related to the side lengths and the angle between the sides.*

Once the information contained in the above two bullets is clear to students, explain the side-angle-side criterion for triangle similarity below.

THEOREM: The *side-angle-side criterion* for two triangles to be similar is as follows:

> Given two triangles ABC and $A'B'C'$ so that $\frac{A'B'}{AB} = \frac{A'C'}{AC}$ and $m\angle A = m\angle A'$, the triangles are similar; $\triangle ABC \sim \triangle A'B'C'$. That is, two triangles are similar if they have one pair of corresponding angles that are congruent and the sides adjacent to that angle are proportional.

- The proof of this theorem is simply to take any dilation with a scale factor of $r = \frac{A'B'}{AB} = \frac{A'C'}{AC}$. This dilation maps $\triangle ABC$ to a triangle that is congruent to $\triangle A'B'C'$ by the side-angle-side congruence criterion.
- We refer to the angle between the two sides as the included angle. Or we can say the sides are adjacent to the given angle. When the side lengths adjacent to the angle are in proportion, then we can conclude that the triangles are similar by the side-angle-side criterion.

The question below requires students to apply their new knowledge of the SAS criterion to the triangles in Figure 2. Students should have concluded that they were not similar in part (d) of Exercise 2. The question below pushes students to apply the SAS and AA theorems related to triangle similarity to show definitively that the triangles in Figure 2 are not similar.

- Does Figure 2 have side lengths that are proportional? What can you conclude about the triangles in Figure 2?
  - *No, the side lengths are not proportional, so we cannot use the SAS criterion for similarity to say that the triangles are similar. If the side lengths were proportional, we could conclude that the lines containing $\overline{PQ}$ and $\overline{P'Q'}$ are parallel, but the side lengths are not proportional; therefore, the lines containing $\overline{PQ}$ and $\overline{P'Q'}$ are not parallel. This then means that the corresponding angles are not equal, and the AA criterion cannot be used to say that the triangles are similar.*
- We can use the SAS criterion to determine if a pair of triangles are similar. Since the side lengths in Figure 2 were not proportional, we can conclude that the triangles are not similar.

Exploratory Challenge 2/Exercises 3–4 (8 minutes)

This challenge gives students the opportunity to discover the SSS criterion for similar triangles. The Discussion that follows solidifies this fact. Consider splitting the class into two groups so that one group works on Exercise 3 and the other group completes Exercise 4. Then, have students share their work with the whole class.

Exploratory Challenge 2/Exercises 3–4

3. Examine the figure, and answer the questions to determine whether or not the triangles shown are similar.

Figure 3

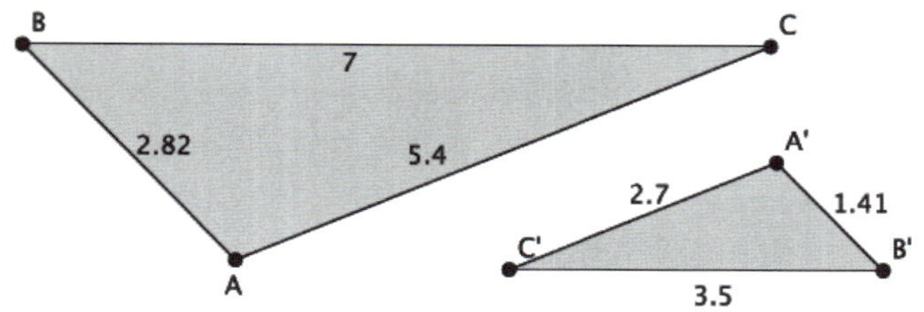

a. **What information is given about the triangles in Figure 3?**

We are only given information related to the side lengths of the triangles.

b. **How can the information provided be used to determine whether $\triangle ABC$ is similar to $\triangle A'B'C'$?**

We know that similar triangles will have ratios of corresponding sides that are proportional; therefore, we can use the side lengths to check for proportionality.

c. **Compare the corresponding side lengths of $\triangle ABC$ and $\triangle A'B'C'$. What do you notice?**

$$\frac{1.41}{2.82} = \frac{3.5}{7} = \frac{2.7}{5.4} = \frac{1}{2}$$

The side lengths are proportional.

d. **Based on your work in parts (a)–(c), make a conjecture about the relationship between $\triangle ABC$ and $\triangle A'B'C'$. Explain your reasoning.**

I think that the triangles are similar. The side lengths are proportional, which is what I would expect in similar triangles.

4. **Examine the figure, and answer the questions to determine whether or not the triangles shown are similar.**

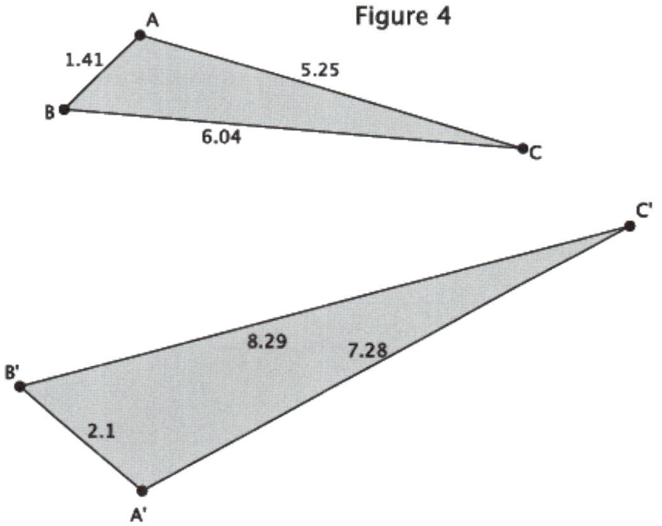

a. **What information is given about the triangles in Figure 4?**

We are only given information related to the side lengths of the triangles.

b. **How can the information provided be used to determine whether $\triangle ABC$ is similar to $\triangle A'B'C'$?**

We know that similar triangles will have ratios of corresponding sides that are proportional; therefore, we can use the side lengths to check for proportionality.

c. **Compare the corresponding side lengths of $\triangle ABC$ and $\triangle A'B'C'$. What do you notice?**

$$\frac{2.1}{1.41} \neq \frac{8.29}{6.04} \neq \frac{7.28}{5.25}$$

The side lengths are not proportional.

d. **Based on your work in parts (a)–(c), make a conjecture about the relationship between $\triangle ABC$ and $\triangle A'B'C'$. Explain your reasoning.**

I think that the triangles are not similar. I would expect the side lengths to be proportional if the triangles were similar.

Discussion (5 minutes)

MP.8

- Which figure contains triangles that are similar? What made you think they are similar?
  - *Figure 3 has the similar triangles. I think Figure 3 has the similar triangles because all three pairs of corresponding sides are in proportion. That is not the case for Figure 4.*
- The triangles in Figure 3 are similar. Take note of the information given about the triangles in the diagrams.
  - *The diagram shows information related only to the side lengths of each of the triangles.*
- **THEOREM:** The *side-side-side criterion* for two triangles to be similar is as follows.

  Given two triangles ABC and $A'B'C'$ so that $\frac{A'B'}{AB} = \frac{B'C'}{BC} = \frac{A'C'}{AC}$, the triangles are similar; $\triangle ABC \sim \triangle A'B'C'$. In other words, two triangles are similar if their corresponding sides are proportional.
- What would the scale factor, r, need to be to show that these triangles are similar? Explain.
  - *A scale factor of $r = \frac{A'B'}{AB}$ or $r = \frac{B'C'}{BC}$ or $r = \frac{A'C'}{AC}$ would show that these triangles are similar. Dilating by one of those scale factors guarantees that the corresponding sides of the triangles will be proportional.*
- Then, a dilation by scale factor $r = \frac{A'B'}{AB} = \frac{B'C'}{BC} = \frac{A'C'}{AC}$ maps $\triangle ABC$ to a triangle that is congruent to $\triangle A'B'C'$ by the side-side-side congruence criterion.
- When all three pairs of corresponding sides are in proportion, we can conclude that the triangles are similar by the side-side-side criterion.
- Does Figure 4 have side lengths that are proportional? What can you conclude about the triangles in Figure 4?
  - *No, the side lengths are not proportional; therefore, the triangles are not similar.*
- We can use the SSS criterion to determine if two triangles are similar. Since the side lengths in Figure 4 are not proportional, we can conclude that the triangles are not similar.

Exercises 5–10 (8 minutes)

Students identify whether or not the triangles are similar. For pairs of triangles that are similar, students identify the criterion used: AA, SAS, or SSS.

Exercises 5–10

5. **Are the triangles shown below similar? Explain. If the triangles are similar, write the similarity statement.**

$\frac{1}{0.68} \neq \frac{6.4}{2.42} \neq \frac{5.83}{2.13}$

There is no information about the angle measures, so we cannot use AA or SAS to conclude the triangles are similar. Since the side lengths are not proportional, we cannot use SSS to conclude the triangles are similar. Therefore, the triangles shown are not similar.

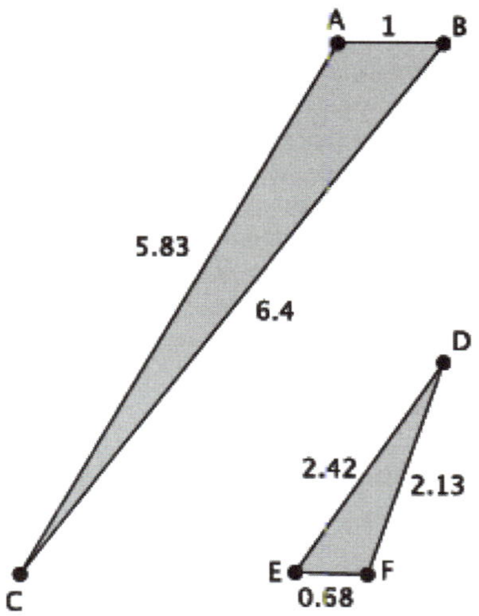

6. **Are the triangles shown below similar? Explain. If the triangles are similar, write the similarity statement.**

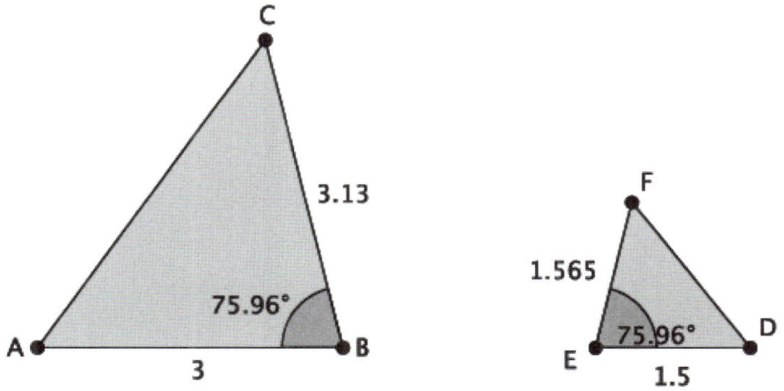

$\frac{3.13}{1.565} = \frac{3}{1.5}$

Yes, the triangles shown are similar. $\triangle ABC \sim \triangle DEF$ ***by SAS because*** $m\angle B = m\angle E$***, and the adjacent sides are proportional.***

7. **Are the triangles shown below similar? Explain. If the triangles are similar, write the similarity statement.**

Yes, the triangles shown are similar. $\triangle ABC \sim \triangle ADE$ ***by AA because*** $m\angle ADE = m\angle ABC$***, and both triangles share*** $\angle A$***.***

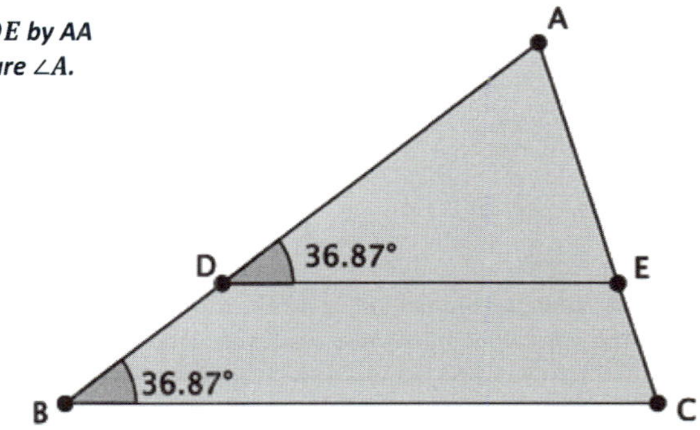

8. **Are the triangles shown below similar? Explain. If the triangles are similar, write the similarity statement.**

$\frac{5}{3} \neq \frac{3}{1}$

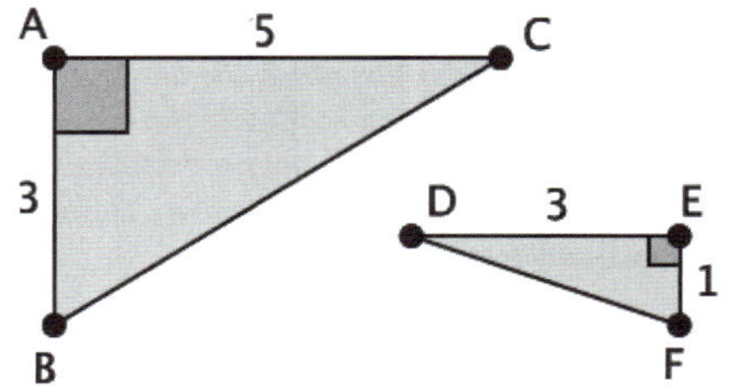

There is no information about the angle measures other than the right angle, so we cannot use AA to conclude the triangles are similar. We only have information about two of the three side lengths for each triangle, so we cannot use SSS to conclude they are similar. If the triangles are similar, we would have to use the SAS criterion, and since the side lengths are not proportional, the triangles shown are not similar. (Note that students could also utilize the Pythagorean theorem to determine the lengths of the hypotenuses and then use SSS similarity criterion to answer the question.)

9. **Are the triangles shown below similar? Explain. If the triangles are similar, write the similarity statement.**

$\frac{3.6}{2.7} = \frac{7.28}{5.46} = \frac{10}{7.5}$

Yes, the triangles are similar. $\triangle PQR \sim \triangle XYZ$ ***by SSS.***

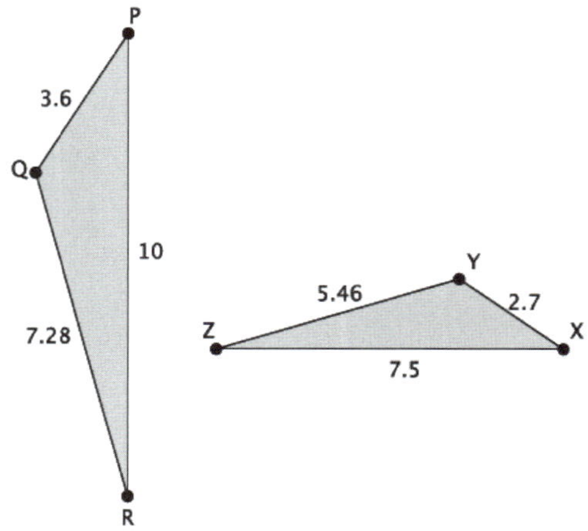

10. **Are the triangles shown below similar? Explain. If the triangles are similar, write the similarity statement.**

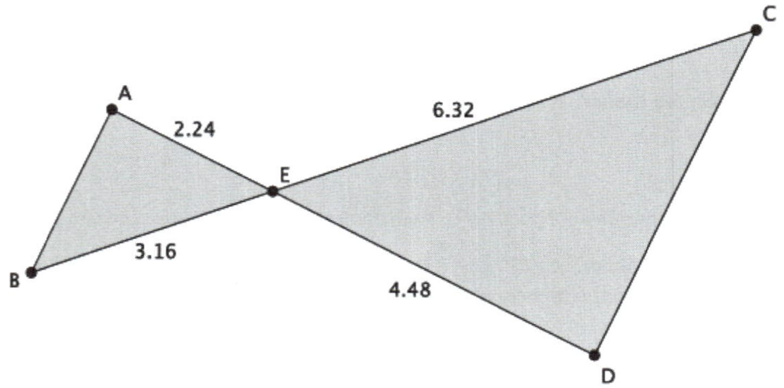

$\frac{2.24}{4.48} = \frac{3.16}{6.32}$

Yes, the triangles are similar. $\triangle ABE \sim \triangle DCE$ ***by SAS because*** $m\angle AEB = m\angle DEC$ ***(vertical angles are congruent), and the sides adjacent to those angles are proportional.***

Closing (3 minutes)

Ask the following three questions to informally assess students' understanding of the similarity criterion for triangles:

- Given only information about the angles of a pair of triangles, how can you determine if the given triangles are similar?
  - *The AA criterion can be used to determine if two triangles are similar. The triangles must have two pairs of corresponding angles that are equal in measure.*
- Given only information about one pair of angles for two triangles, how can you determine if the given triangles are similar?
  - *The SAS criterion can be used to determine if two triangles are similar. The triangles must have one pair of corresponding angles that are equal in measure, and the ratios of the corresponding adjacent sides must be in proportion.*
- Given no information about the angles of a pair of triangles, how can you determine if the given triangles are similar?
  - *The SSS criterion can be used to determine if two triangles are similar. The triangles must have three pairs of corresponding side lengths in proportion.*

Exit Ticket (5 minutes)

Name ______________________________ Date ______________

Lesson 17: The Side-Angle-Side (SAS) and Side-Side-Side (SSS) Criteria for Two Triangles to Be Similar

Exit Ticket

1. Given $\triangle ABC$ and $\triangle MLN$ in the diagram below and $\angle B \cong \angle L$, determine if the triangles are similar. If so, write a similarity statement, and state the criterion used to support your claim.

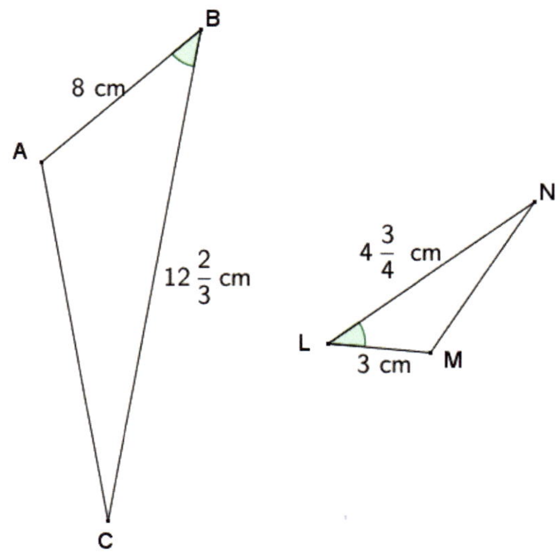

2. Given $\triangle DEF$ and $\triangle EGF$ in the diagram below, determine if the triangles are similar. If so, write a similarity statement, and state the criterion used to support your claim.

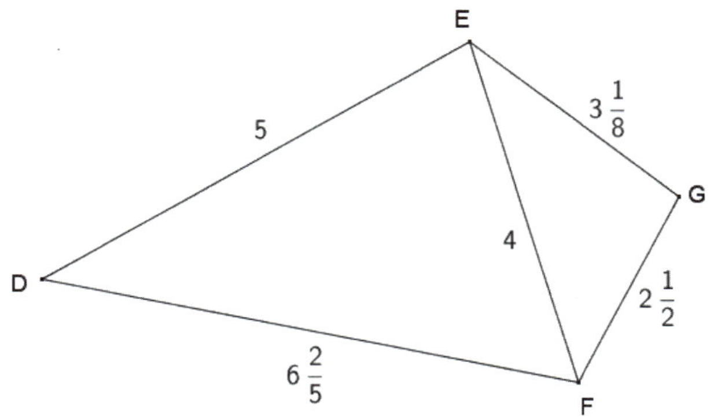

Exit Ticket Sample Solutions

1. Given $\triangle ABC$ and $\triangle MLN$ in the diagram below, determine if the triangles are similar. If so, write a similarity statement, and state the criterion used to support your claim.

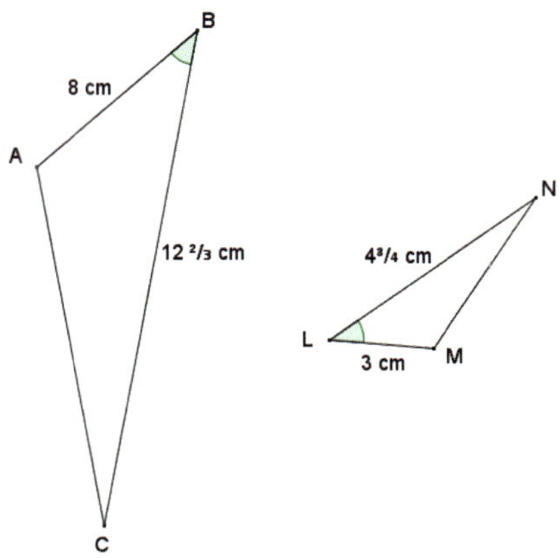

In comparing the ratios of sides between figures, I found that $\frac{AB}{ML} = \frac{CB}{NL}$ because the cross-products of the proportion $\frac{8}{3} = \frac{12\frac{2}{3}}{4\frac{3}{4}}$ are both 38. We are given that $\angle L \cong \angle B$. Therefore, $\triangle ABC \sim \triangle MLN$ by the SAS criterion for proving similar triangles.

2. Given $\triangle DEF$ and $\triangle EGF$ in the diagram below, determine if the triangles are similar. If so, write a similarity statement, and state the criterion used to support your claim.

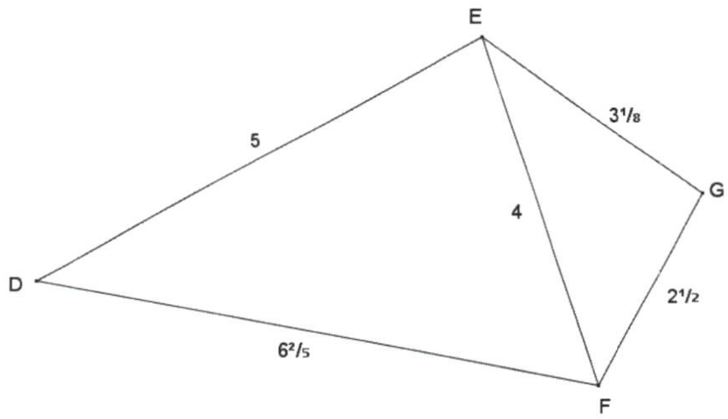

By comparison, if the triangles are in fact similar, then the longest sides of each triangle will correspond, and likewise the shortest sides will correspond. The corresponding sides from each triangle are proportional since

$\frac{5}{3\frac{1}{8}} = \frac{4}{2\frac{1}{2}} = \frac{6\frac{2}{5}}{4} = \frac{8}{5}$, *so* $\frac{DE}{EG} = \frac{EF}{GF} = \frac{DF}{EF}$.

Therefore, by the SSS criterion for showing triangle similarity, $\triangle DEF \sim \triangle EGF$.

Problem Set Sample Solutions

1. For parts (a) through (d) below, state which of the three triangles, if any, are similar and why.

   a.

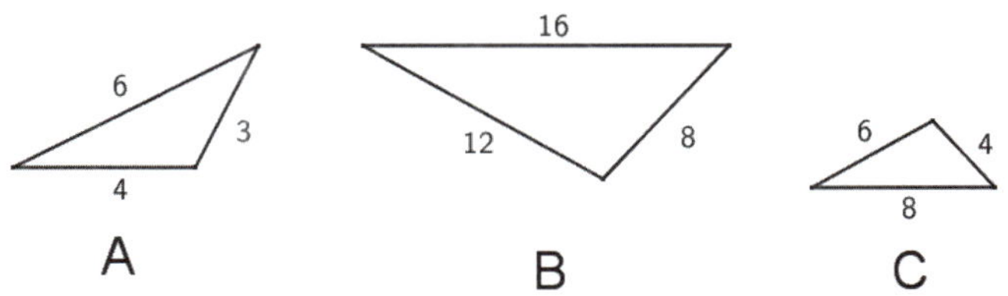

Triangles B and C are similar because they share three pairs of corresponding sides that are in the same ratio.

$$\frac{4}{8} = \frac{6}{12} = \frac{8}{16} = \frac{1}{2}$$

Triangles A and B are not similar because the ratios of their corresponding sides are not in the same ratio.

$$\frac{6}{16} \neq \frac{4}{12}$$

Further, if triangle A is not similar to triangle B, then triangle A is not similar to triangle C.

b.

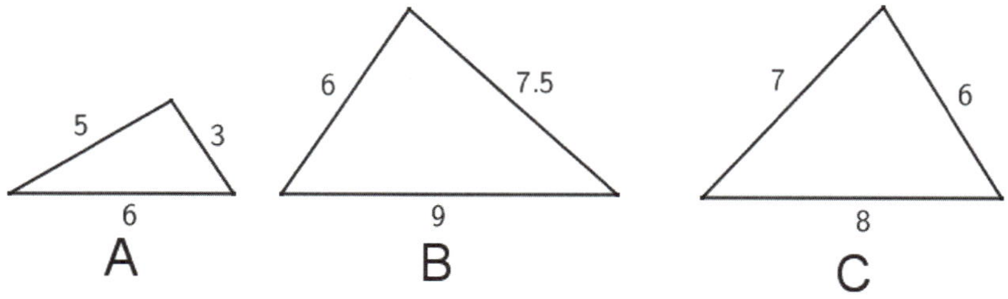

Triangles A and B are not similar because their corresponding sides are not all in the same ratio. Two pairs of corresponding sides are proportional, but the third pair of corresponding sides is not.

$$\frac{3}{6} \neq \frac{5}{7.5} = \frac{6}{9}$$

Triangles B and C are not similar because their corresponding sides are not in the same ratio.

$$\frac{6}{6} \neq \frac{7}{7.5} \neq \frac{8}{9}$$

Triangles A and C are not similar because their corresponding sides are not in the same ratio.

$$\frac{3}{6} \neq \frac{5}{7} \neq \frac{6}{8}$$

c.

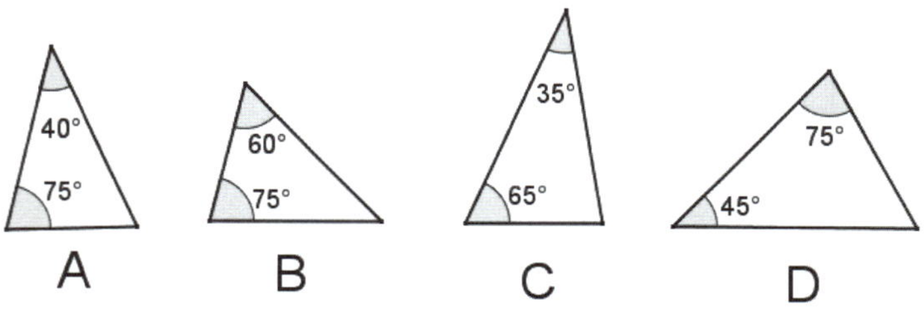

Triangles B and D are the only similar triangles because they have the same angle measures. Using the angle sum of a triangle, both triangles B and D have angles of $75°$, $60°$, and $45°$.

d.

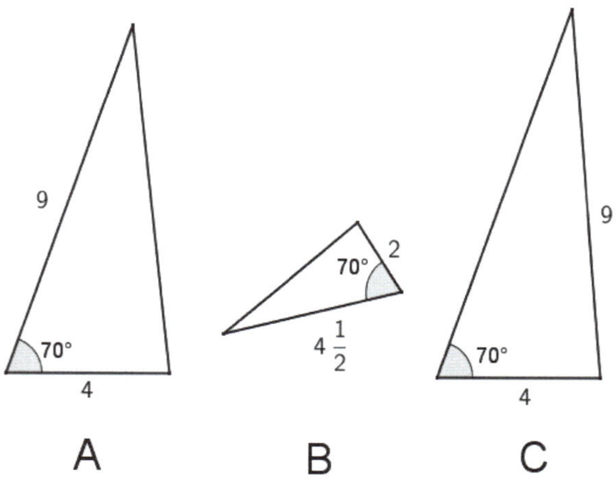

Triangles A and B are similar because they have two pairs of corresponding sides that are in the same ratio, and their included angles are equal measures. Triangle C cannot be shown to be similar because, even though it has two sides that are the same length as two sides of triangle A, the $70°$ angle in triangle C is not the included angle and, therefore, does not correspond to the $70°$ angle in triangle A.

2. **For each given pair of triangles, determine if the triangles are similar or not, and provide your reasoning. If the triangles are similar, write a similarity statement relating the triangles.**

a.

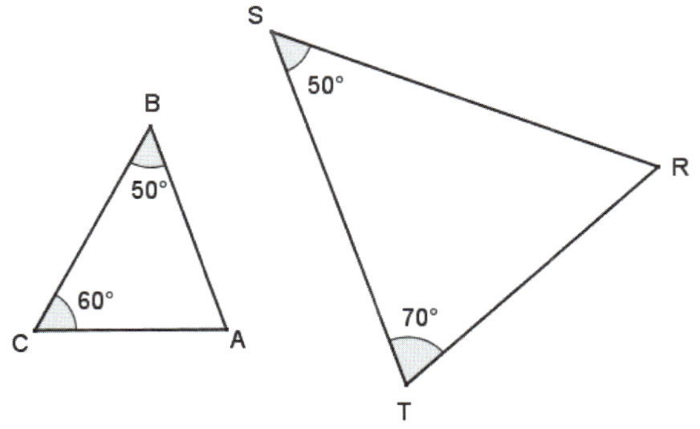

The triangles are similar because, using the angle sum of a triangle, each triangle has angle measures of $50°$, $60°$, *and* $70°$. *Therefore,* $\triangle ABC \sim \triangle TSR$.

b.

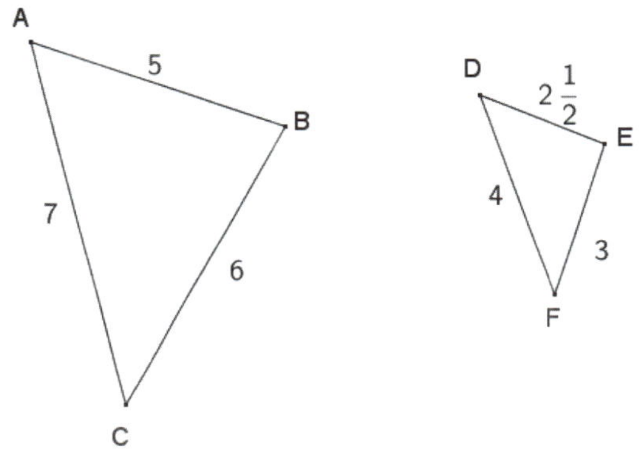

The triangles are not similar, because the ratios of corresponding sides are not all in proportion.
$\frac{AB}{DE} = \frac{BC}{EF} = 2$; *however,* $\frac{AC}{DF} = \frac{7}{4} \neq 2$.

c.

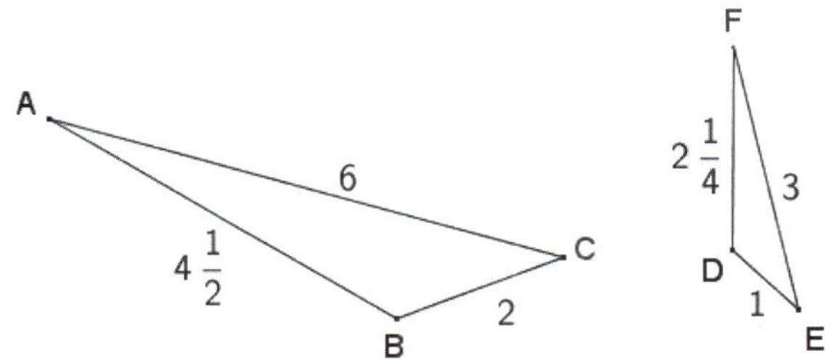

$\triangle ABC \sim \triangle FDE$ *by the SSS criterion for showing similar triangles because the ratio of all pairs of corresponding sides* $\frac{AC}{FE} = \frac{AB}{FD} = \frac{BC}{DE} = 2$.

d.

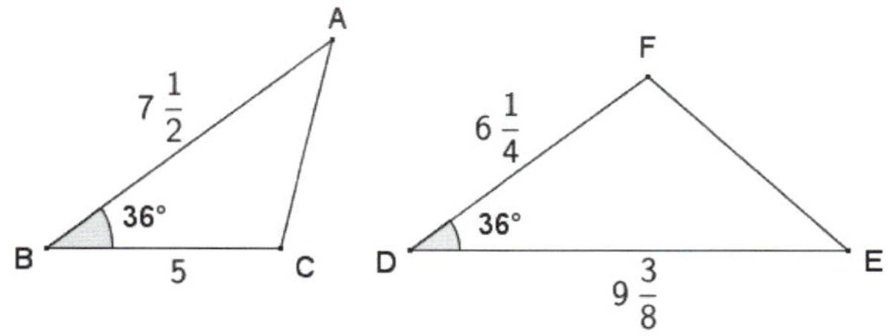

$\frac{AB}{DE} = \frac{BC}{DF} = \frac{4}{5}$, and included angles B and D are both $36°$ and, therefore, congruent, so $\triangle ABC \sim \triangle EDF$ by the SAS criterion for showing similar triangles.

3. **For each pair of similar triangles below, determine the unknown lengths of the sides labeled with letters.**

a.

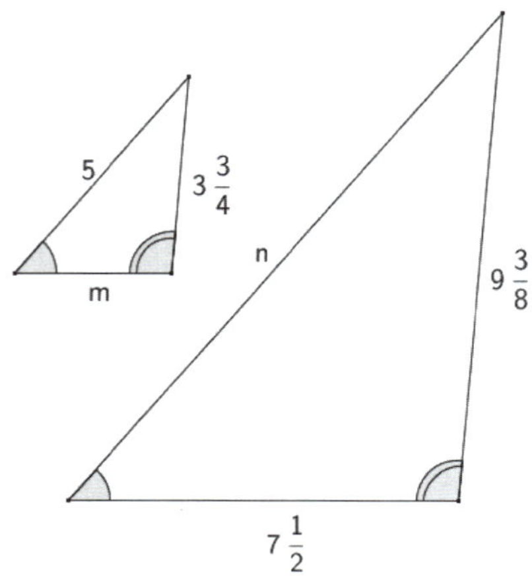

The ratios of corresponding sides must be equal, so $\frac{5}{n} = \frac{3\frac{3}{4}}{9\frac{3}{8}}$, giving $n = 12\frac{1}{2}$. Likewise, $\frac{m}{7\frac{1}{2}} = \frac{3\frac{3}{4}}{9\frac{3}{8}}$, giving $m = 3$.

b.

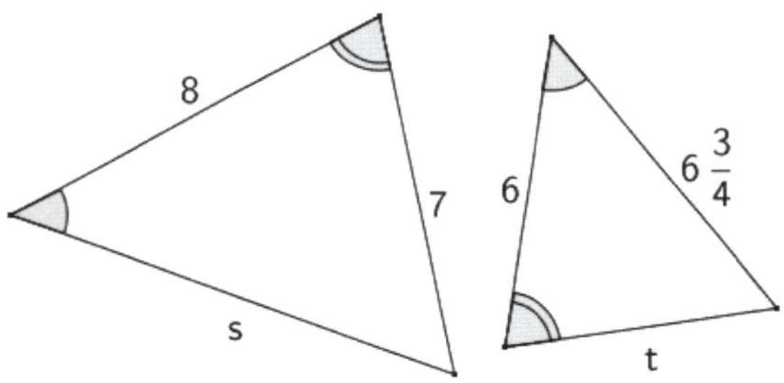

The ratios of corresponding sides must be equal, so $\frac{8}{6} = \frac{s}{6\frac{3}{4}}$, giving $s = 9$. Likewise, $\frac{8}{6} = \frac{7}{t}$, giving $t = 5\frac{1}{4}$.

4. **Given that $\overline{AD}$ and $\overline{BC}$ intersect at E and $\overline{AB} \parallel \overline{CD}$, show that $\triangle ABE \sim \triangle DCE$.**

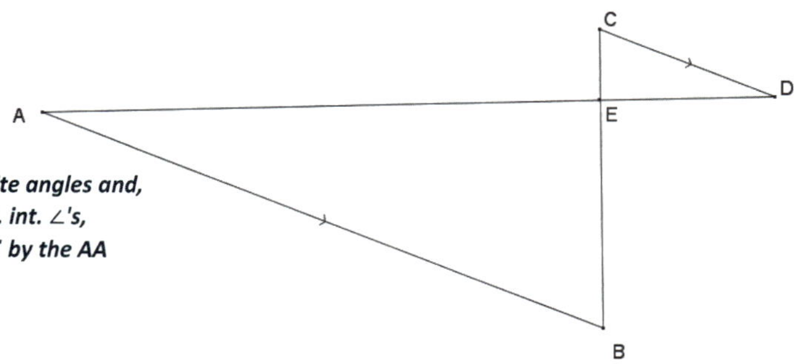

$\angle AEB$ and $\angle DEC$ are vertically opposite angles and, therefore, congruent. $\angle A \cong \angle D$ by alt. int. $\angle$'s, $\overline{AB} \parallel \overline{CD}$. Therefore, $\triangle ABE \sim \triangle DCE$ by the AA criterion for showing similar triangles.

5. Given $BE = 11$, $EA = 11$, $BD = 7$, and $DC = 7$, show that $\triangle BED \sim \triangle BAC$.

Both triangles share angle B, and by the reflexive property, $\angle B \cong \angle B$. $BA = BE + EA$, so $BA = 22$, and $BC = BD + DC$, so $BC = 14$. The ratios of corresponding sides $\frac{BE}{BA} = \frac{BD}{BC} = \frac{1}{2}$. Therefore, $\triangle BED \sim \triangle BAC$ by the SAS criterion for triangle similarity.

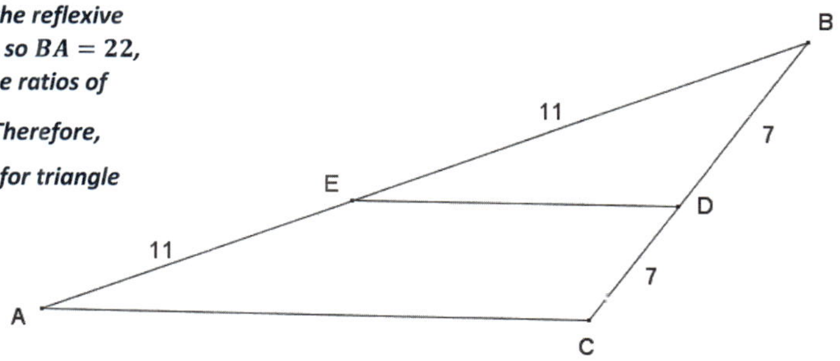

6. Given the diagram below, X is on $\overline{RS}$ and Y is on $\overline{RT}$, $XS = 2$, $XY = 6$, $ST = 9$, and $YT = 4$.

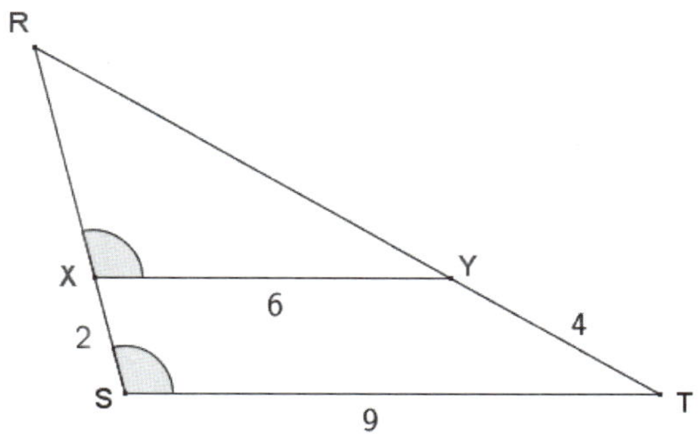

a. Show that $\triangle RXY \sim \triangle RST$.

The diagram shows $\angle RST \cong \angle RXY$. Both triangle RXY and RST share angle R, and by the reflexive property, $\angle R \cong \angle R$, so $\triangle RXY \sim \triangle RST$ by the AA criterion show triangle similarity.

b. Find RX and RY.

Since the triangles are similar, their corresponding sides must be in the same ratio.

$$\frac{RX}{RS} = \frac{RY}{RT} = \frac{XY}{ST} = \frac{6}{9}$$

$RS = RX + XS$, so $RS = RX + 2$ and $RT = RY + YT$, so $RT = RY + 4$.

$$\frac{RX}{RX+2} = \frac{6}{9} \qquad \frac{RY}{RY+4} = \frac{6}{9}$$

$$9(RX) = 6(RX) + 12 \qquad 9(RY) = 6(RY) + 24$$

$$RX = 4 \qquad RY = 8$$

7. One triangle has a $120°$ angle, and a second triangle has a $65°$ angle. Is it possible that the two triangles are similar? Explain why or why not.

No The triangles cannot be similar because in the first triangle, the sum of the remaining angles is $60°$, which means that it is not possible for the triangle to have a $65°$ angle. For the triangles to be similar, both triangles would have to have angles measuring $120°$ and $65°$, but this is impossible due to the angle sum of a triangle.

8. A right triangle has a leg that is 12 cm, and another right triangle has a leg that is 6 cm. Can you tell whether the two triangles are similar? If so, explain why. If not, what other information would be needed to show they are similar?

The two triangles may or may not be similar. There is not enough information to make this claim. If the second leg of the first triangle is twice the length of the second leg of the first triangle, then the triangles are similar by SAS criterion for showing similar triangles.

9. Given the diagram below, $JH = 7.5$, $HK = 6$, and $KL = 9$, is there a pair of similar triangles? If so, write a similarity statement, and explain why. If not, explain your reasoning.

$\triangle LKJ \sim \triangle HKL$ by the SAS criterion for showing triangle similarity. Both triangles share $\angle K$, and by the reflexive property, $\angle K \cong \angle K$. Furthermore, $\frac{LK}{JK} = \frac{HK}{LK} = \frac{2}{3}$, giving two pairs of corresponding sides in the same ratio and included angles of the same size.

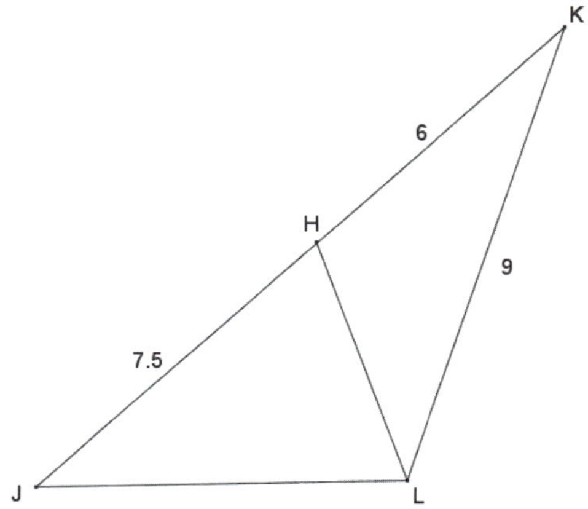

Lesson 18: Similarity and the Angle Bisector Theorem

Student Outcomes

- Students state, understand, and prove the angle bisector theorem.
- Students use the angle bisector theorem to solve problems.

Classwork

Opening Exercise (5 minutes)

The Opening Exercise should activate students' prior knowledge acquired in Module 1 that is helpful in proving the angle bisector theorem.

> **Opening Exercise**
>
> a. **What is an angle bisector?**
>
> ***The bisector of an angle is a ray in the interior of the angle such that the two adjacent angles formed by it have equal measures.***
>
> b. **Describe the angle relationships formed when parallel lines are cut by a transversal.**
>
> ***When parallel lines are cut by a transversal, the corresponding, alternate interior, and alternate exterior angles are all congruent; the same-side interior angles are supplementary.***
>
> c. **What are the properties of an isosceles triangle?**
>
> ***An isosceles triangle has at least two congruent sides, and the angles opposite to the congruent sides (i.e., the base angles) are also congruent.***

Scaffolding:

- If necessary, provide visuals to accompany these questions.
- For part (a):

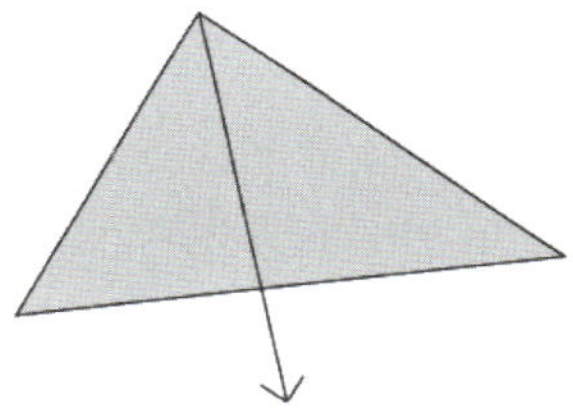

- For part (b):

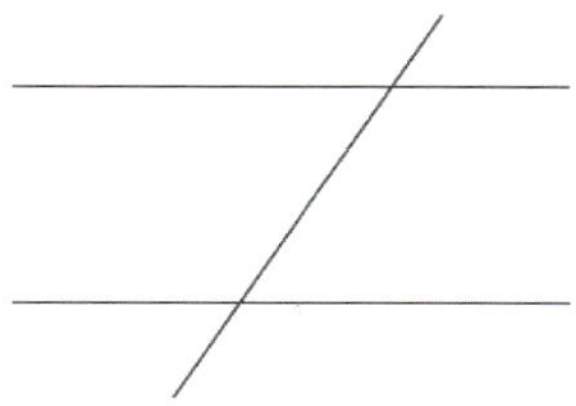

- For part (c):

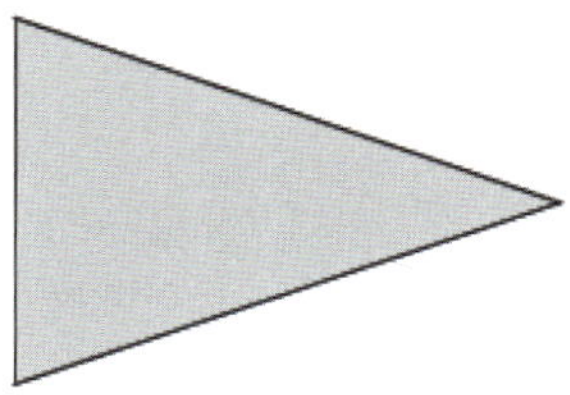

Discussion (20 minutes)

Prior to proving the angle bisector theorem, students observe the length relationships of the sides of a triangle when one of the angles of the triangle has been bisected.

- In this lesson, we investigate the length relationships of the sides of a triangle when one angle of the triangle has been bisected.

Provide students time to look for relationships between the side lengths. This requires trial and error on the part of students and may take several minutes.

MP.7

Discussion

In the diagram below, the angle bisector of $\angle A$ in $\triangle ABC$ meets side $\overline{BC}$ at point D. Does the angle bisector create any observable relationships with respect to the side lengths of the triangle?

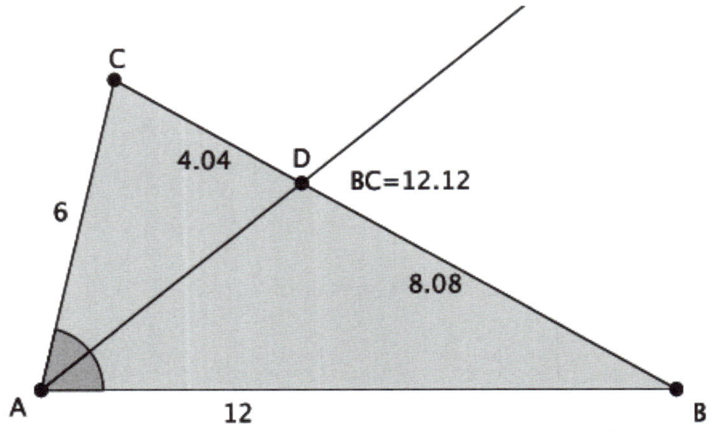

Acknowledge any relationships students may find, but highlight the relationship $BD{:}\,CD = BA{:}\,CA$. Then, continue the discussion below that proves this relationship.

- The following theorem generalizes our observation:

  THEOREM: The angle bisector theorem: In $\triangle ABC$, if the angle bisector of $\angle A$ meets side $\overline{BC}$ at point D, then $BD{:}\,CD = BA{:}\,CA$.

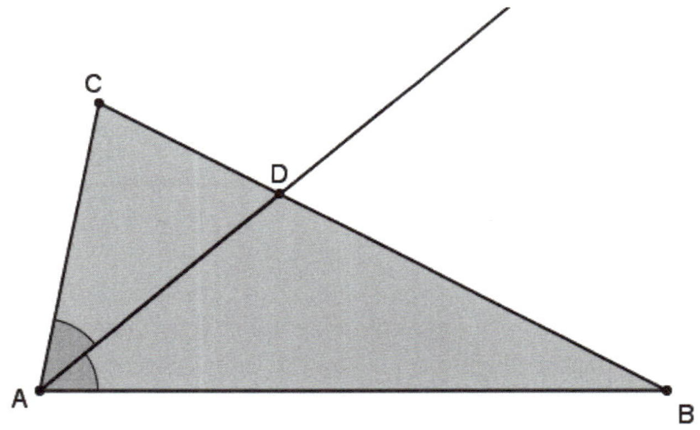

In words, the bisector of an angle of a triangle splits the opposite side into segments that have the same ratio as the adjacent sides.

- Our goal now is to prove this relationship for all triangles. We begin by constructing a line through vertex C that is parallel to side $\overline{AB}$. Let E be the point where this parallel line meets the angle bisector, as shown.

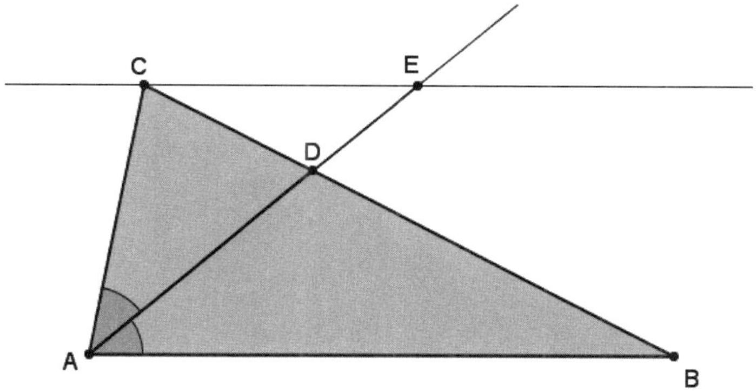

> *Scaffolding:*
> In place of the formal proof, students may construct angle bisectors for a series of triangles and take measurements to verify the relationship inductively.

If we can show that $\triangle ABD \sim \triangle ECD$, then we can use what we know about similar triangles to prove the relationship $BD: CD = BA: CA$.

Consider asking students why it is that if they can show $\triangle ABD \sim \triangle ECD$, they are closer to the goal of showing $BD: CD = BA: CA$. Students should respond that similar triangles have proportional length relationships. The triangles ABD and ECD, if shown similar, would give $BD: BA = CD: CE$, three out of the four lengths needed in the ratio $BD: CD = BA: CA$.

- How can we show that $\triangle ABD \sim \triangle ECD$?

Provide students time to discuss how to show that the triangles are similar. Elicit student responses; then, continue with the discussion below.

- It is true that $\triangle ABD \sim \triangle ECD$ by the AA criterion for similarity. Vertical angles ADB and EDC are congruent and, therefore, equal. $\angle DEC$ and $\angle BAD$ are congruent and equal because they are alternate interior angles of parallel lines cut by a transversal. (Show the diagram below.)

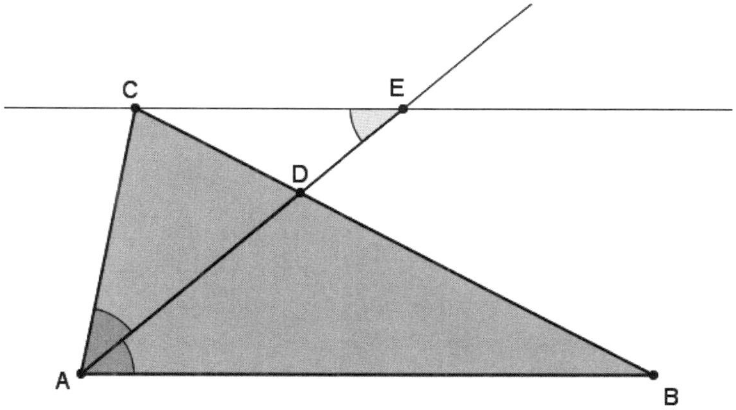

- Since the triangles are similar, we know that $BD: CD = BA: CE$. This is very close to what we are trying to show: $BD: CD = BA: CA$. What must we do now to prove the theorem?
  - *We have to show that $CE = CA$.*

Once students have identified what needs to be done, that is, show that $CE = CA$, provide them time to discuss how to show it. The prompts below can be used to guide students' thinking.

- We need to show that $CE = CA$. Notice that the segments CE and CA are two sides of $\triangle ACE$. How might that be useful?
  - *If we could show that $\triangle ACE$ is an isosceles triangle, then we would know that $CE = CA$.*
- How can we show that $\triangle ACE$ is an isosceles triangle?
  - *We were first given that angle A was bisected by $\overline{AD}$, which means that $\angle BAD \cong \angle CAD$. Then, by alternate interior $\angle$'s, $\overline{CE} \parallel \overline{AB}$, it follows that $\angle CAD = \angle CEA$. We can use the converse of the base angles of isosceles triangle theorem (i.e., base $\angle$'s converse). Since $\angle CAE = \angle CEA$, then $\triangle ACE$ must be an isosceles triangle.*
- Now that we know $\triangle ACE$ is isosceles, we can conclude that $CE = CA$ and finish the proof of the angle bisector theorem. All we must do now is substitute CA for CE in $BD: CD = BA: CE$. Therefore, $BD: CD = BA: CA$, and the theorem is proved.

Consider asking students to restate what was just proved and summarize the steps of the proof. Students should respond that the bisector of an angle of a triangle splits the opposite side into segments that have the same ratio as the adjacent sides.

Exercises 1–4 (10 minutes)

Students complete Exercises 1–4 independently.

Exercises 1–4

1. **The sides of a triangle are 8, 12, and 15. An angle bisector meets the side of length 15. Find the lengths x and y. Explain how you arrived at your answers.**

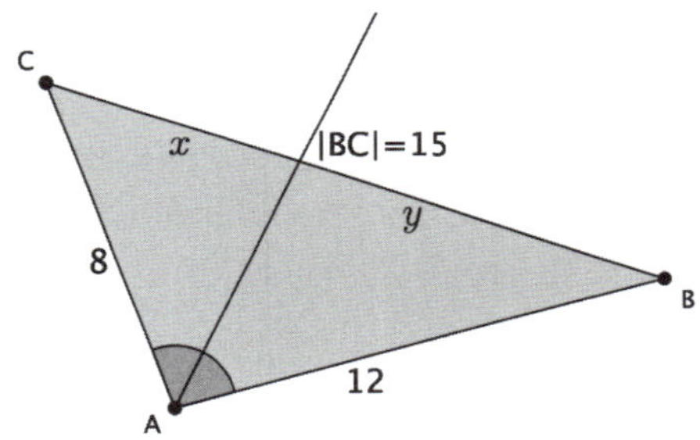

$$\frac{y}{x} = \frac{12}{8}$$
$$x = 15 - y$$

$$8y = 12x$$
$$8y = 12(15 - y)$$
$$8y = 180 - 12y$$
$$20y = 180$$
$$y = 9$$

$$x = 15 - 9$$
$$x = 6$$

The length x is 6, and the length y is 9.

Since I know that $\angle A$ is bisected, I applied what I knew about the angle bisector theorem to determine the lengths x and y. Specifically, the angle bisector cuts the side that is opposite the bisected angle so that $y: x = 12: 8$. I set up an equation using the values of the ratios, which could be solved once I rewrote one of the variables x or y. I rewrote x as $15 - y$ and then solved for y. Once I had a value for y, I could replace it in the equation $x = 15 - y$ to determine the value of x.

2. The sides of a triangle are 8, 12, and 15. An angle bisector meets the side of length 12. Find the lengths x and y.

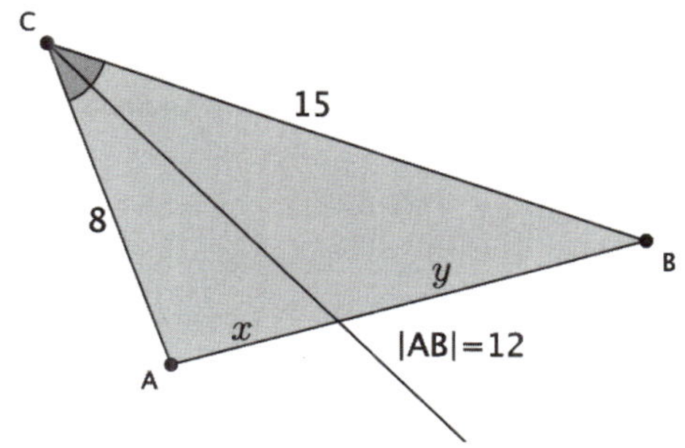

$$y = 12 - x$$

$$\frac{x}{12 - x} = \frac{8}{15}$$

$$15x = 8(12 - x)$$

$$15x = 96 - 8x$$

$$23x = 96$$

$$x = \frac{96}{23} = 4\frac{4}{23}$$

$$y = 12 - 4\frac{4}{23}$$

$$y = 7\frac{19}{23}$$

The length of x is $4\frac{4}{23}$, and the length of y is $7\frac{19}{23}$.

3. The sides of a triangle are 8, 12, and 15. An angle bisector meets the side of length 8. Find the lengths x and y.

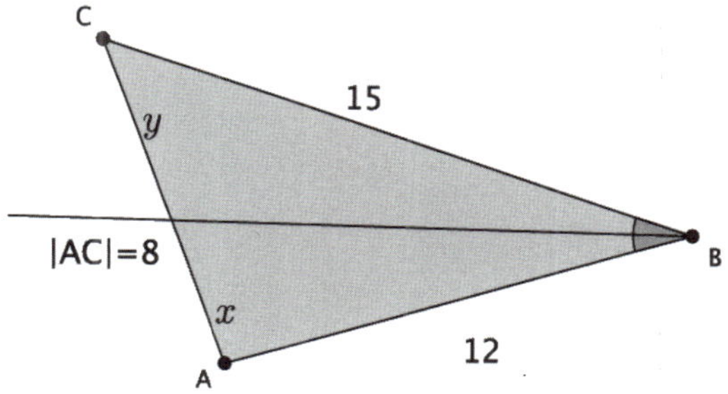

$$\frac{y}{x} = \frac{15}{12}$$

$$y = 8 - x$$

$$15x = 12y$$

$$15x = 12(8 - x)$$

$$15x = 96 - 12x$$

$$27x = 96$$

$$x = \frac{96}{27} = 3\frac{15}{27} = 3\frac{5}{9}$$

$$y = 8 - 3\frac{5}{9}$$

$$y = 4\frac{4}{9}$$

The length of x is $3\frac{5}{9}$, and the length of y is $4\frac{4}{9}$.

4. **The angle bisector of an angle splits the opposite side of a triangle into lengths 5 and 6. The perimeter of the triangle is 33. Find the lengths of the other two sides.**

Let z be the scale factor of a similarity. By the angle bisector theorem, the side of the triangle adjacent to the segment of length 5 has length of $5z$, and the side of the triangle adjacent to the segment of length 6 has length of $6z$. The sum of the sides is equal to the perimeter.

$$5 + 6 + 5z + 6z = 33$$
$$11 + 11z = 33$$
$$11z = 22$$
$$z = 2$$

$5(2) = 10$, ***and*** $6(2) = 12$ ***The lengths of the other two sides are*** 10 ***and*** 12.

Closing (5 minutes)

- Explain the angle bisector theorem in your own words.
- Explain how knowing that one of the angles of a triangle has been bisected allows you to determine unknown side lengths of a triangle.

Exit Ticket (5 minutes)

Name ______________________________ Date ____________

Lesson 18: Similarity and the Angle Bisector Theorem

Exit Ticket

1. The sides of a triangle have lengths of 12, 16, and 21. An angle bisector meets the side of length 21. Find the lengths x and y.

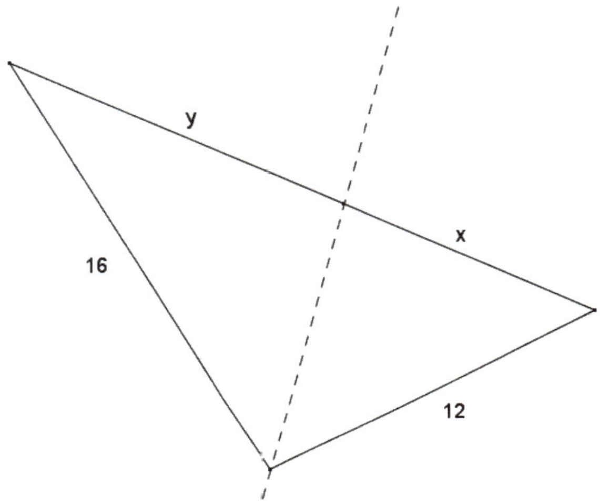

2. The perimeter of $\triangle UVW$ is $22\frac{1}{2}$. $\overrightarrow{WZ}$ bisects $\angle UWV$, $UZ = 2$, and $VZ = 2\frac{1}{2}$. Find UW and VW.

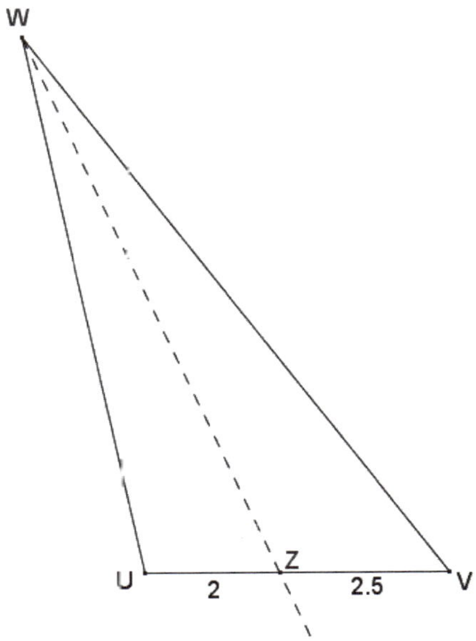

Exit Ticket Sample Solutions

1. The sides of a triangle have lengths of 12, 16, and 21. An angle bisector meets the side of length 21. Find the lengths x and y.

By the angle bisector theorem, $\frac{y}{x} = \frac{16}{12}$, *and* $y = 21 - x$, *so*

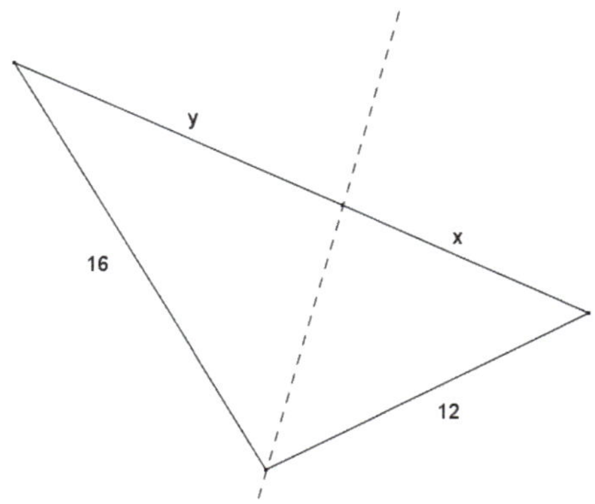

$$\frac{21-x}{x} = \frac{16}{12}$$
$$12(21-x) = 16x$$
$$252 - 12x = 16x$$
$$252 = 28x$$
$$9 = x$$

$$y = 21 - x$$
$$y = 21 - 9$$
$$y = 12$$

2. The perimeter of $\triangle UVW$ is $22\frac{1}{2}$. $\overrightarrow{WZ}$ bisects $\angle UWV$, $UZ = 2$, and $VZ = 2\frac{1}{2}$. Find UW and VW.

By the angle bisector theorem, $\frac{2}{2.5} = \frac{UW}{VW}$, *so* $UW = 2x$ *and* $VW = 2.5x$ *for some positive number* x. *The perimeter of the triangle is* $22\frac{1}{2}$, *so*

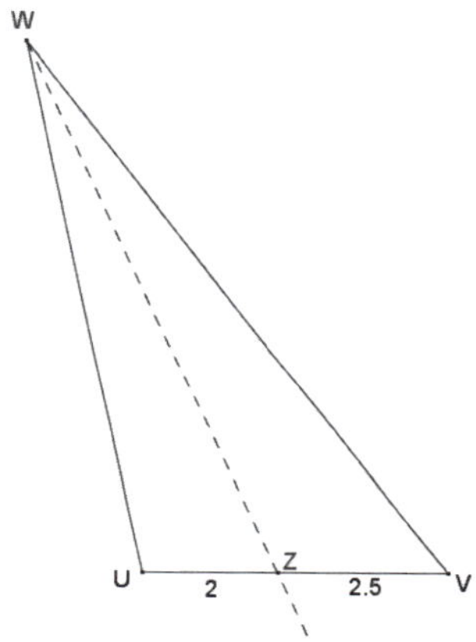

$$2 + 2.5 + 2x + 2.5x = 22.5$$
$$4.5 + 4.5x = 22.5$$
$$4.5x = 18$$
$$x = 4$$

$UW = 2x = 2(4) = 8$

$VW = 2.5x = 2.5(4) = 10$

Problem Set Sample Solutions

1. The sides of a triangle have lengths of 5, 8, and $6\frac{1}{2}$. An angle bisector meets the side of length $6\frac{1}{2}$. Find the lengths x and y.

Using the angle bisector theorem, $\frac{x}{y} = \frac{5}{8}$, *and* $y = 6\frac{1}{2} - x$, *so*

$$\frac{x}{6\frac{1}{2} - x} = \frac{5}{8}$$
$$5\left(6\frac{1}{2} - x\right) = 8x$$
$$32\frac{1}{2} - 5x = 8x$$
$$32\frac{1}{2} = 13x$$
$$x = 2\frac{1}{2}$$

$$y = 6\frac{1}{2} - x$$
$$y = 6\frac{1}{2} - 2\frac{1}{2}$$
$$y = 4$$

2. The sides of a triangle are $10\frac{1}{2}$, $16\frac{1}{2}$, and 9. An angle bisector meets the side of length 9. Find the lengths x and y.

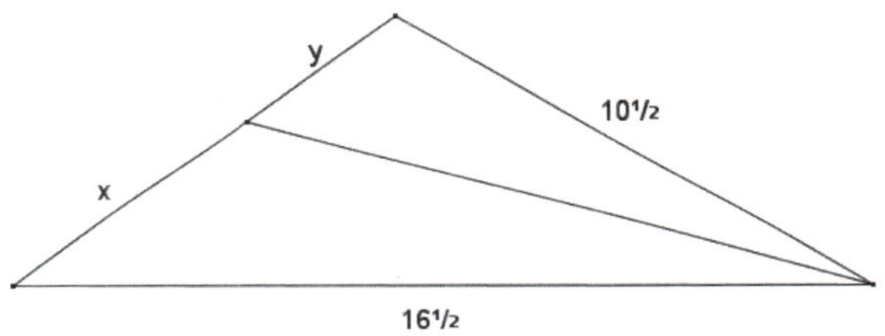

By the angle bisector theorem, $\frac{x}{y} = \frac{16\frac{1}{2}}{10\frac{1}{2}}$, *and* $y = 9 - x$, *so*

$$\frac{x}{9 - x} = \frac{16.5}{10.5}$$
$$10.5x = 16.5(9 - x)$$
$$10.5x = 148.5 - 16.5x$$
$$27x = 148.5$$
$$x = 5.5$$

$$y = 9 - x$$
$$y = 9 - 5.5$$
$$y = 3.5$$

3. In the diagram of triangle DEF below, $\overline{DG}$ is an angle bisector, $DE = 8$, $DF = 6$, and $EF = 8\frac{1}{6}$. Find FG and EG.

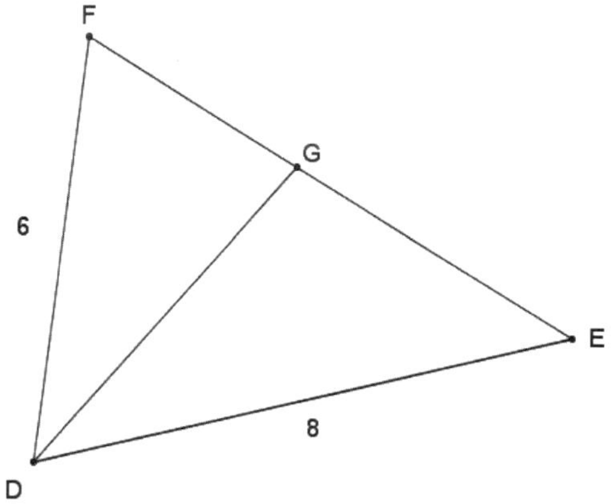

Since $\overline{DG}$ is the angle bisector of angle D, $FG: GE = FD: ED$ by the angle bisector theorem. If $EF = 8\frac{1}{6}$, then $FG = 8\frac{1}{6} - GE$.

$$\frac{\left(8\frac{1}{6} - GE\right)}{GE} = \frac{6}{8}$$

$$8\frac{1}{6} - GE = \frac{6}{8} \cdot GE$$

$$8\frac{1}{6} = \frac{14}{8} \cdot GE$$

$$\frac{49}{6} \cdot \frac{8}{14} = GE$$

$$\frac{28}{6} = GE$$

$$GE = 4\frac{2}{3}$$

$$FE = GE + FG$$

$$8\frac{1}{6} = 4\frac{2}{3} + FG$$

$$FG = 3\frac{1}{2}$$

4. Given the diagram below and $\angle BAD \cong \angle DAC$, show that $BD: BA = CD: CA$.

Using the given information, $\overline{AD}$ is the angle bisector of angle A. By the angle bisector theorem, $BD: CD = BA: CA$, so

$$\frac{BD}{CD} = \frac{BA}{CA}$$

$$BD \cdot CA = CD \cdot BA$$

$$\frac{BD}{BA} = \frac{CD}{CA}$$

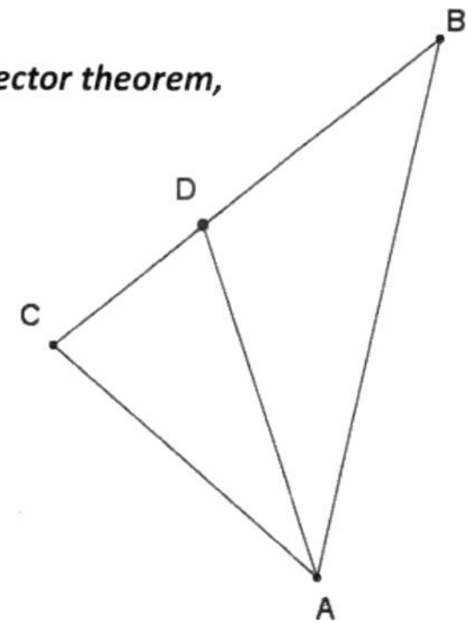

5. The perimeter of triangle LMN is 32 cm. $\overline{NX}$ is the angle bisector of angle N, $LX = 3$ cm, and $XM = 5$ cm. Find LN and MN.

Since $\overline{NX}$ is an angle bisector of angle N, by the angle bisector theorem, $XL : XM = LN : MN$; thus, $LN : MN = 3 : 5$. Therefore, $LN = 3x$ and $MN = 5x$ for some positive number x. The perimeter of the triangle is 32 cm, so

$$XL + XM + MN + LN = 32$$
$$3 + 5 + 3x + 5x = 32$$
$$8 + 8x = 32$$
$$8x = 24$$
$$x = 3$$

$3x = 3(3) = 9$ *and* $5x = 5(3) = 15$

$LN = 9$ cm *and* $MN = 15$ cm

6. Given $CD = 3$, $DB = 4$, $BF = 4$, $FE = 5$, $AB = 6$, and $\angle CAD \cong \angle DAB \cong \angle BAF \cong \angle FAE$, find the perimeter of quadrilateral $AEBC$.

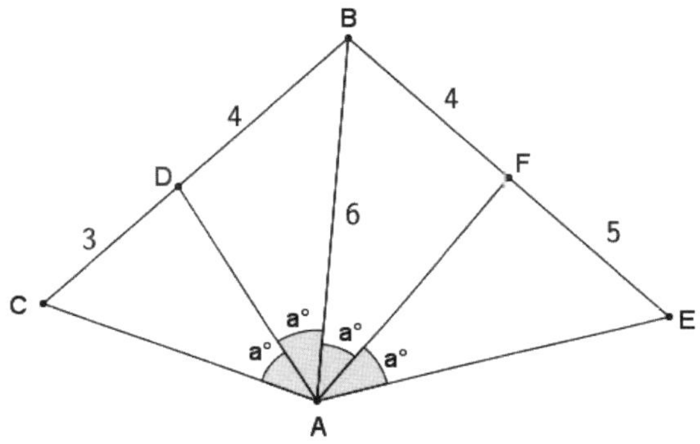

$\overline{AD}$ is the angle bisector of angle CAB, so by the angle bisector theorem, $CD : BD = CA : BA$.

$$\frac{3}{4} = \frac{CA}{6}$$
$$4 \cdot CA = 18$$
$$CA = 4.5$$

$\overline{AF}$ is the angle bisector of angle BAE, so by the angle bisector theorem, $BF : EF = BA : EA$.

$$\frac{4}{5} = \frac{6}{EA}$$
$$4 \cdot EA = 30$$
$$EA = 7.5$$

The perimeter of quadrilateral $AEBC$:

$$AEBC = CD + DB + BF + FE + EA + AC$$
$$AEBC = 3 + 4 + 4 + 5 + 7.5 + 4.5$$
$$AEBC = 28$$

7. If $\overline{AE}$ meets $\overline{BC}$ at D such that $CD: BD = CA: BA$, show that $\angle CAD \cong \angle BAD$. Explain how this proof relates to the angle bisector theorem.

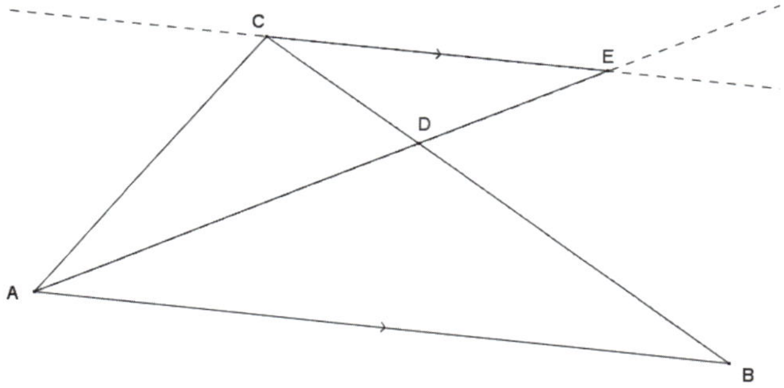

This is a proof of the converse to the angle bisector theorem.

$\angle CEA \cong \angle BAD$	*Alt. Int.* $\angle$*'s* $\overline{CE} \parallel \overline{AB}$
$\angle ADB \cong \angle EDC$	*Vert.* $\angle$*'s*
$\Delta CDE \sim \Delta BDA$	*AA* $\sim$
$\dfrac{CD}{DB} = \dfrac{CE}{BA}$	*Corr. sides of* $\sim \triangle$*'s*

Given $\dfrac{CD}{BD} = \dfrac{CA}{BA}$ *implies then that*

$CE = CA$*, so* $\triangle AEC$ *is isosceles.*

$\angle CAD \cong \angle CEA$	*Base* $\angle$*'s*
$\angle CAD \cong \angle BAD$	*Substitution*

8. In the diagram below, $\overline{ED} \cong \overline{DB}$, $\overline{BE}$ bisects $\angle ABC$, $AD = 4$, and $DC = 8$. Prove that $\triangle ADB \sim \triangle CEB$.

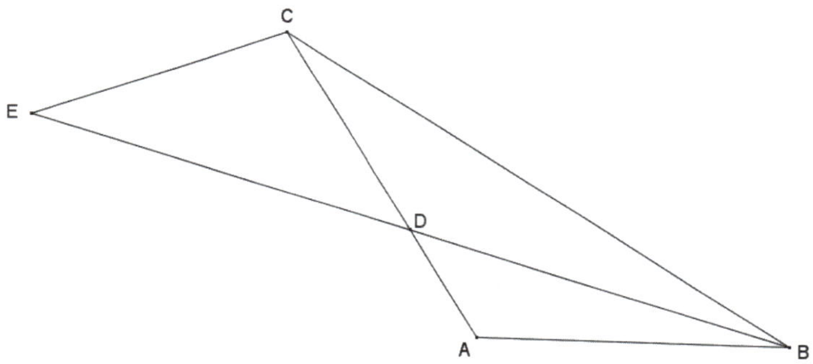

Using given information, $ED = DB$*, so it follows that* $EB = 2DB$*. Since* $\overline{BE}$ *bisects* $\angle ABC$*,* $\angle ABD \cong \angle CBD$*. Also, by the angle bisector theorem,* $\dfrac{CD}{AD} = \dfrac{BC}{BA}$*, which means* $\dfrac{BC}{BA} = \dfrac{8}{4} = \dfrac{2}{1}$.

Since $\dfrac{EB}{DB} = \dfrac{CB}{AB} = \dfrac{2}{1}$*, and* $\angle ABD \cong \angle CBD$*,* $\triangle ADB \sim \triangle CEB$ *by the SAS criterion for showing similar triangles.*

Lesson 19: Families of Parallel Lines and the Circumference of the Earth

Student Outcomes

- Students understand that parallel lines cut transversals into proportional segments. They use ratios between corresponding line segments in different transversals and ratios within line segments on the same transversal.
- Students understand Eratosthenes' method for measuring the earth and solve related problems.

Lesson Notes

Students revisit their study of side splitters and the triangle side-splitter theorem to understand how parallel lines cut transversals into proportional segments. The theorem is a natural consequence of side splitters; allow students the opportunity to make as many connections of their own while guiding them forward. The second half of the lesson is teacher-led and describes Eratosthenes' calculation of the earth's circumference. The segment lays the foundation for Lesson 20, where students study another application of geometry by the ancient Greeks.

Classwork

Opening (7 minutes)

- Consider $\triangle OAB$ below with side splitter CD.

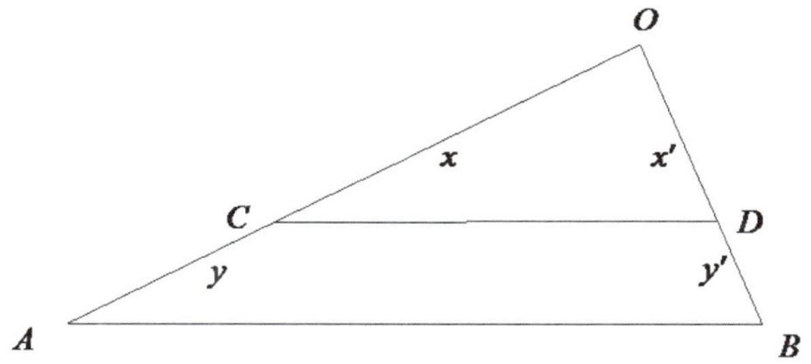

- Recall that we say line segment CD splits sides $\overline{OA}$ and $\overline{OB}$ proportionally if $\frac{OA}{OC} = \frac{OB}{OD}$ or equivalently if $\frac{OC}{OA} = \frac{OD}{OB}$.

MP.7

- Using x, y, x', y' as the lengths of the indicated segments, how can we rewrite $\frac{OA}{OC} = \frac{OB}{OD}$? Simplify as much as possible.
  - $\frac{OA}{OC} = \frac{OB}{OD}$ *is the same as* $\frac{x+y}{x} = \frac{x'+y'}{x'}$.

$$\frac{x+y}{x} = \frac{x'+y'}{x'}$$
$$1 + \frac{y}{x} = 1 + \frac{y'}{x'}$$
$$\frac{y}{x} = \frac{y'}{x'}$$

- Thus, another way to say that segment CD splits the sides proportionally is to say that the ratios $x:y$ and $x':y'$ are equal.

As scaffolding, consider reviewing a parallel example with numerical values:

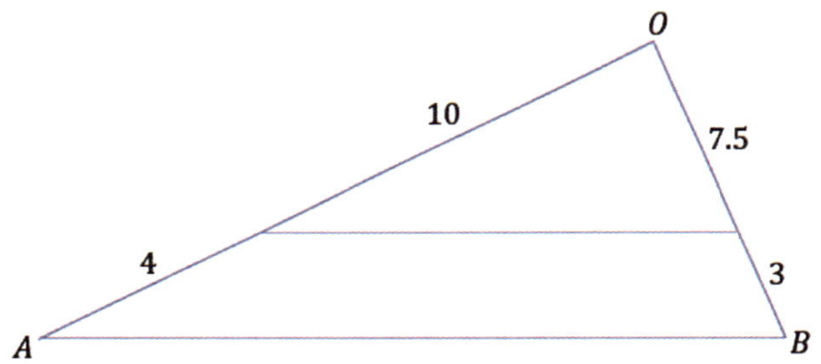

Opening Exercise (4 minutes)

Suggest students use the numerical example from the Opening to help them with the Opening Exercise.

Opening Exercise

Show $x: y = x': y'$ is equivalent to $x: x' = y: y'$.

$$\frac{x}{y} = \frac{x'}{y'}$$
$$xy' = x'y$$
$$\frac{xy'}{x'y'} = \frac{x'y}{x'y'}$$
$$\frac{x}{x'} = \frac{y}{y'}$$

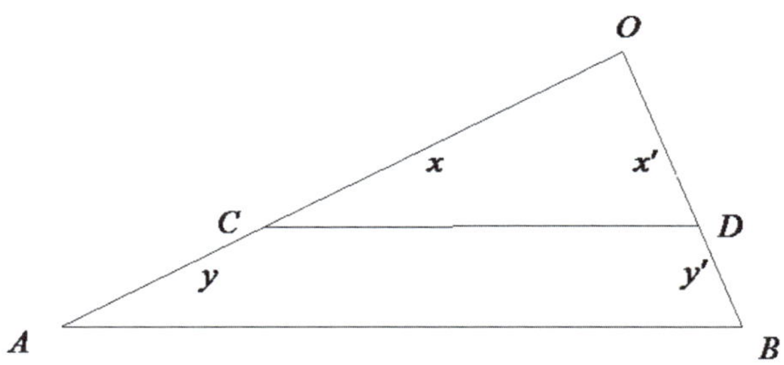

Discussion (10 minutes)

Lead students through a discussion to prove that parallel lines cut transversals into proportional segments.

- We will use our understanding of side splitters to prove the following theorem.

- **THEOREM**: Parallel lines cut transversals into proportional segments. If parallel lines are intersected by two transversals, then the ratios of the segments determined along each transversal between the parallel lines are equal.
- Draw three parallel lines that are cut by two transversals. Label the following lengths of the line segments as x, y, x', and y'.

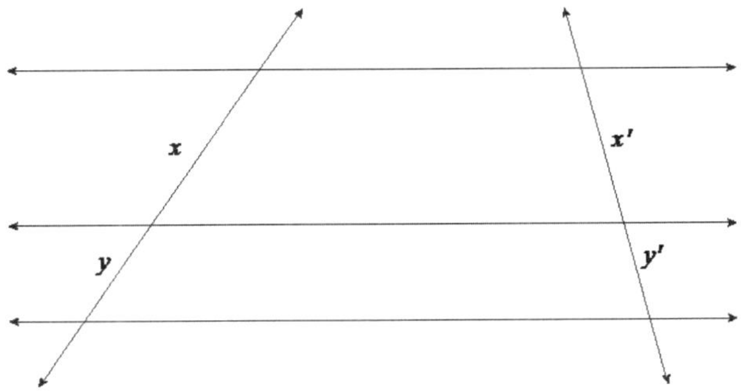

- To prove the theorem, we must show that $x: y = x': y'$. Why would this be enough to show that the ratios of the segments along each transversal between the parallel lines are equal?
  - *This would be enough because the relationship* $x: y = x': y'$ *implies that* $x: x' = y: y'$.
- Draw a segment so that two triangles are formed between the parallel lines and between the transversals.

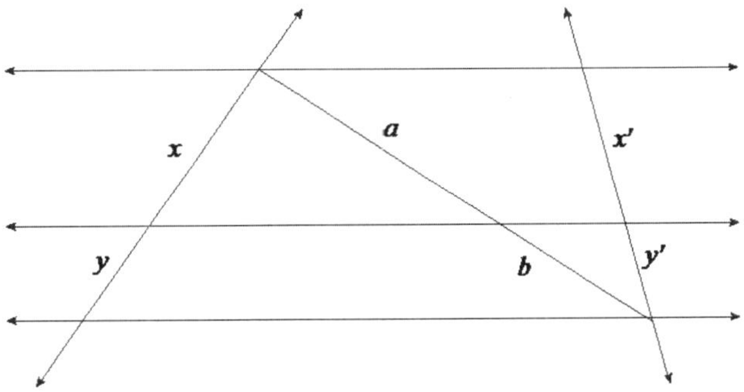

- Label each portion of the segment separated by a pair of parallel lines as a and b.
- Are there any conclusions we can draw based on the diagram?
  - *We can apply the triangle side-splitter theorem twice to see that* $x: y = a: b$ *and* $a: b = x': y'$.
  - *So,* $x: y = a: b$, *and* $x': y' = a: b$. *Thus,* $x: y = x': y'$.
- Therefore, we have proved the theorem: Parallel lines cut transversals into proportional segments.
- Notice that the two equations $x: y = x': y'$ and $x: x' = y: y'$ are equivalent as described above.

Exercises 1–2 (4 minutes)

Students apply their understanding that parallel lines cut transversals into proportional segments to determine the unknown length in each problem.

Exercises 1–2

Lines that appear to be parallel are in fact parallel.

1.

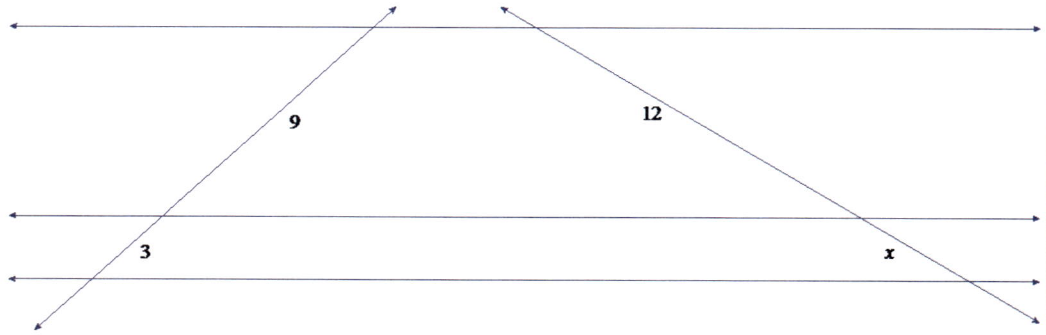

$x = 4$

2.

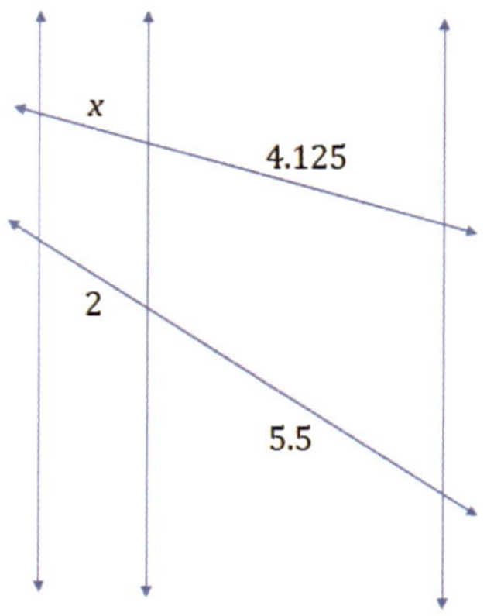

$x = 1.5$

Discussion (13 minutes)

- The word *geometry* is Greek for *geos* (earth) and *metron* (measure). A Greek named Eratosthenes, who lived over 2,200 years ago, used geometry to measure the circumference of the earth. The Greeks knew the earth was a sphere. They also knew the sun was so far away that rays from the sun (as they met the earth) were, for all practical purposes, parallel to each other. Using these two facts, here is how Eratosthenes calculated the circumference of the earth.
- Eratosthenes lived in Egypt. He calculated the circumference of the earth without ever leaving Egypt. Every summer solstice in the city of Syene, the sun was directly overhead at noon. Eratosthenes knew this because at noon on that day alone, the sun would reflect directly (perpendicularly) off the bottom of a deep well.
- On the same day at noon in Alexandria, a city north of Syene, the angle between the perpendicular to the ground and the rays of the sun was about 7.2°. We do not have a record of how Eratosthenes found this measurement, but here is one possible explanation:
- Imagine a pole perpendicular to the ground (in Alexandria), as well as its shadow. With a tool such as a protractor, the shadow can be used to determine the measurement of the angle between the ray and the pole, regardless of the height of the pole.

- You might argue that we cannot theorize such a method because we do not know the height of the pole. However, because of the angle the rays make with the ground and the $90°$ angle of the pole with the ground, the triangle formed by the ray, the pole, and the shadow are all similar triangles, regardless of the height of the pole. Why must the triangles all be similar?
    - *The situation satisfies the AA criterion; therefore, the triangles must be similar.*

Remind students that the discussion points here are similar to the work done in Lesson 16 on indirect measurement.

- Therefore, if the pole had a height of 10 meters, the shadow had a length of 1.26 meters, or if the pole had a height of 1 meter, the shadow had a length of 0.126 meters.

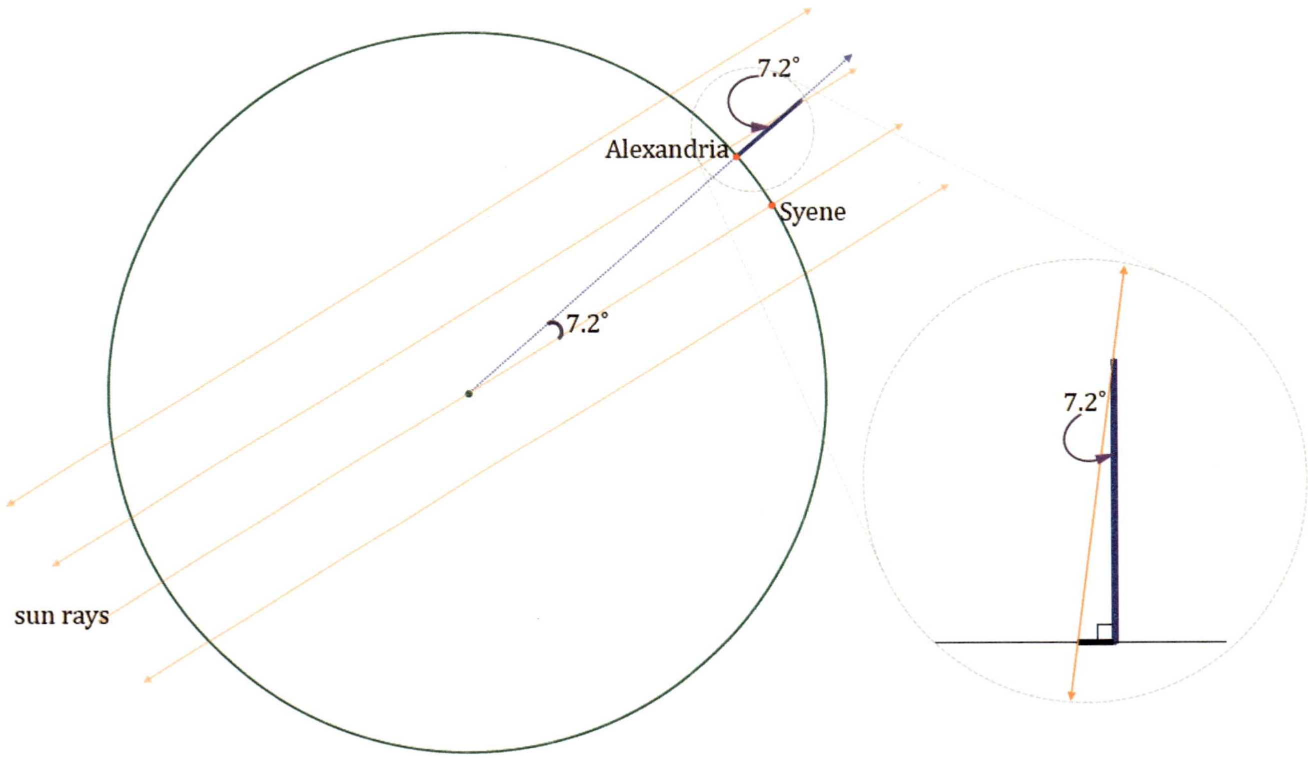

- This measurement of the angle between the sun's rays and the pole was instrumental to the calculation of the circumference. Eratosthenes used it to calculate the angle between the two cities from the center of the earth.
- You might think it necessary to go to the center of the earth to determine this measurement, but it is not. Eratosthenes extrapolated both the sun's rays and the ray perpendicular to the ground in Alexandria. Notice that the perpendicular at Alexandria acts as a transversal to the parallel rays of the sun, namely, the sun ray that passes through Syene and the center of the earth, and the sun ray that forms the triangle with the top of the pole and the shadow of the pole. Using the alternate interior angles determined by the transversal (the extrapolated pole) that intersects the parallel lines (extrapolated sun rays), Eratosthenes found the angle between the two cities to be $7.2°$.
- How can this measurement be critical to finding the entire circumference?
    - *We can divide $7.2°$ into $360°$, which gives us a fraction of the circumference, or how many times we have to multiply the distance between Alexandria to Syene to get the whole circumference.*

- Eratosthenes divided $360°$ by $7.2°$, which yielded 50. So, the distance from Syene to Alexandria is $\frac{1}{50}$ of the circumference of the earth. The only thing that is missing is that distance between Syene and Alexandria, which was known to be about $5{,}000$ stades; the stade was a Greek unit of measurement, and 1 stade ≈ 600 feet.
- So, Eratosthenes' estimate was about $50 \cdot 5{,}000 \cdot 600$ feet, or about $28{,}400$ miles. A modern-day estimate for the circumference of the earth at the equator is about $24{,}900$ miles.
- It is remarkable that around 240 B.C.E., basic geometry helped determine a very close approximation of the circumference of the earth.

Closing (2 minutes)

- In your own words, explain how parallel lines cut transversals into proportional segments.
- What are some assumptions that Eratosthenes must have made as part of his calculation?
  - *The rays of the sun are all parallel.*
  - *The earth is perfectly spherical.*

Share the following link to a video on Eratosthenes and his calculation of the earth's circumference. http://www.youtube.com/watch?v=wnElDaV4esg&feature=youtu.be (The same video was used in Module 1, Lesson 11.)

Exit Ticket (5 minutes)

Name ______________________________ Date ______________

Lesson 19: Families of Parallel Lines and the Circumference of the Earth

Exit Ticket

1. Given the diagram shown, $\overline{AG} \parallel \overline{BH} \parallel \overline{CI}$, $AB = 6.5$ cm, $GH = 7.5$ cm, and $HI = 18$ cm. Find BC.

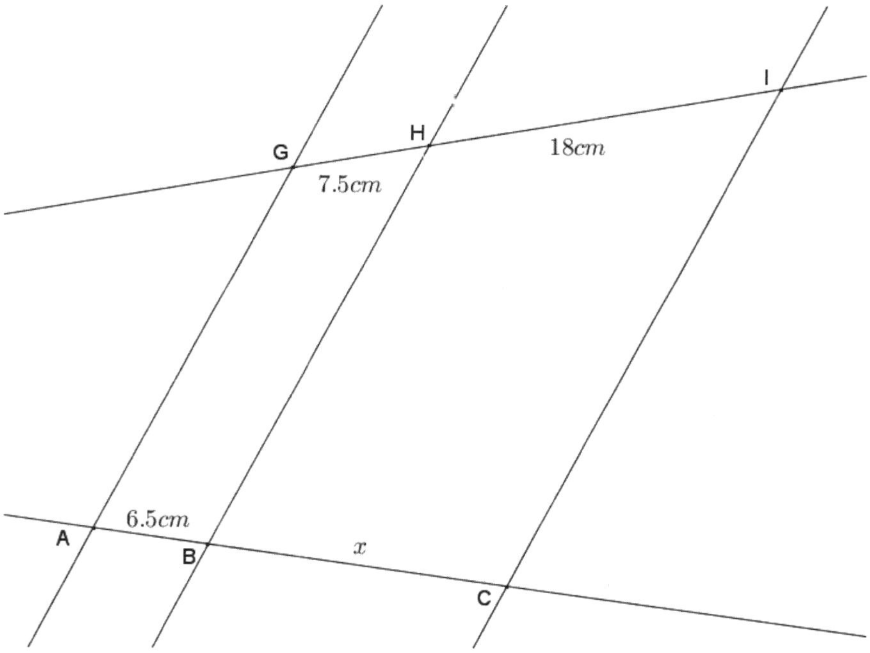

2. Martin the Martian lives on Planet Mart. Martin wants to know the circumference of Planet Mart, but it is too large to measure directly. He uses the same method as Eratosthenes by measuring the angle of the sun's rays in two locations. The sun shines on a flagpole in Martinsburg, but there is no shadow. At the same time, the sun shines on a flagpole in Martville, and a shadow forms a $10°$ angle with the pole. The distance from Martville to Martinsburg is 294 miles. What is the circumference of Planet Mart?

Exit Ticket Sample Solutions

1. **Given the diagram shown, $\overline{AG} \parallel \overline{BH} \parallel \overline{CI}$, $AB = 6.5$ cm, $GH = 7.5$ cm, and $HI = 18$ cm. Find BC.**

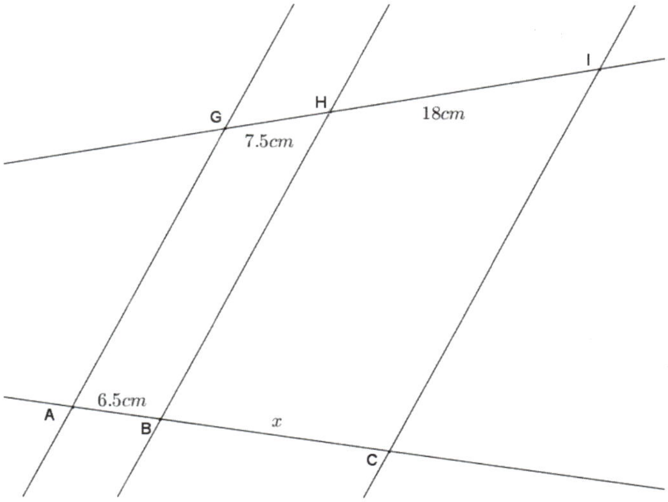

Parallel lines cut transversals proportionally; therefore, $\frac{BC}{AB} = \frac{HI}{GH}$, *and thus,*

$$BC = \frac{HI}{GH}(AB)$$
$$x = \frac{18}{7.5}(6.5)$$
$$x = 15.6.$$

The length of $\overline{BC}$ *is* 15.6 cm.

2. **Martin the Martian lives on Planet Mart. Martin wants to know the circumference of Planet Mart, but it is too large to measure directly. He uses the same method as Eratosthenes by measuring the angle of the sun's rays in two locations. The sun shines on a flagpole in Martinsburg, but there is no shadow. At the same time, the sun shines on a flagpole in Martville, and a shadow forms a $10°$ angle with the pole. The distance from Martville to Martinsburg is 294 miles. What is the circumference of Planet Mart?**

The distance from Martinsburg to Martville makes up only $10°$ of the total rotation about the planet. There are $360°$ in the complete circumference of the planet, and $36 \cdot 10° = 360°$, so $36 \cdot 294$ miles $= 10584$ miles.

The circumference of planet Mart is $10,584$ miles.

Problem Set Sample Solutions

1. Given the diagram shown, $\overline{AD} \parallel \overline{GJ} \parallel \overline{LO} \parallel \overline{QT}$, and $\overline{AQ} \parallel \overline{BR} \parallel \overline{CS} \parallel \overline{DT}$. Use the additional information given in each part below to answer the questions:

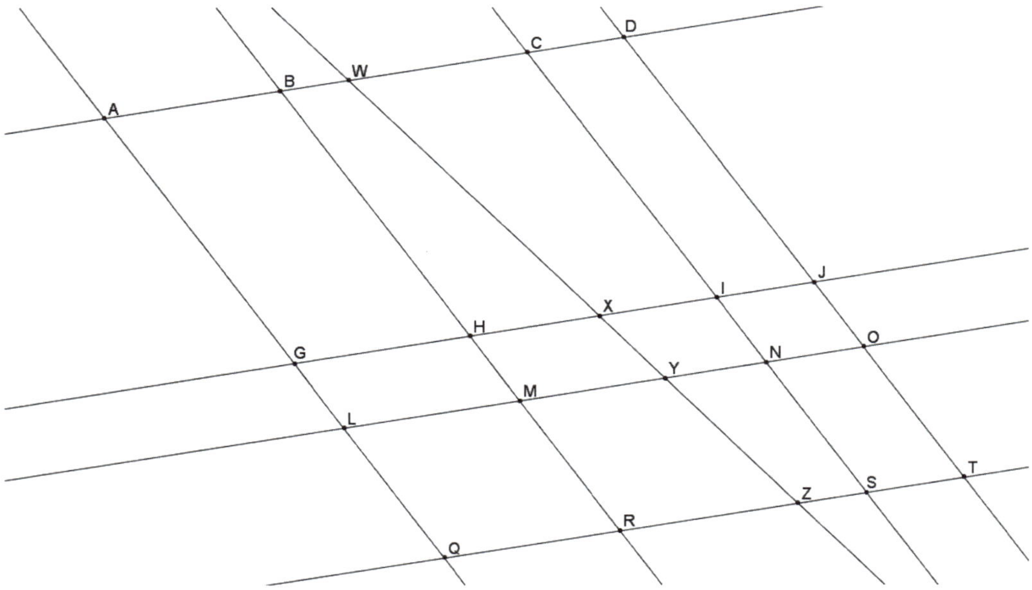

a. If $GL = 4$, what is HM?

$GHML$ forms a parallelogram since opposite sides are parallel, and opposite sides of a parallelogram are equal in length; therefore, $HM = GL = 4$.

b. If $GL = 4$, $LQ = 9$, and $XY = 5$, what is YZ?

Parallel lines cut transversals proportionally; therefore, it is true that $\frac{GL}{LQ} = \frac{XY}{YZ}$, and likewise $YZ = \frac{LQ}{GL}(XY)$.

$$YZ = \frac{9}{4}(5)$$

$$YZ = 11\frac{1}{4}$$

c. Using information from part (b), if $CI = 18$, what is WX?

By the same argument used in part (a), $IN = GL = 4$. Parallel lines cut transversals proportionally; therefore, it is true that $\frac{CI}{IN} = \frac{WX}{XY}$, and likewise $WX = \frac{CI}{IN}(XY)$.

$$WX = \frac{18}{4}(5)$$

$$WX = 22\frac{1}{2}$$

2. **Use your knowledge about families of parallel lines to find the coordinates of point P on the coordinate plane below.**

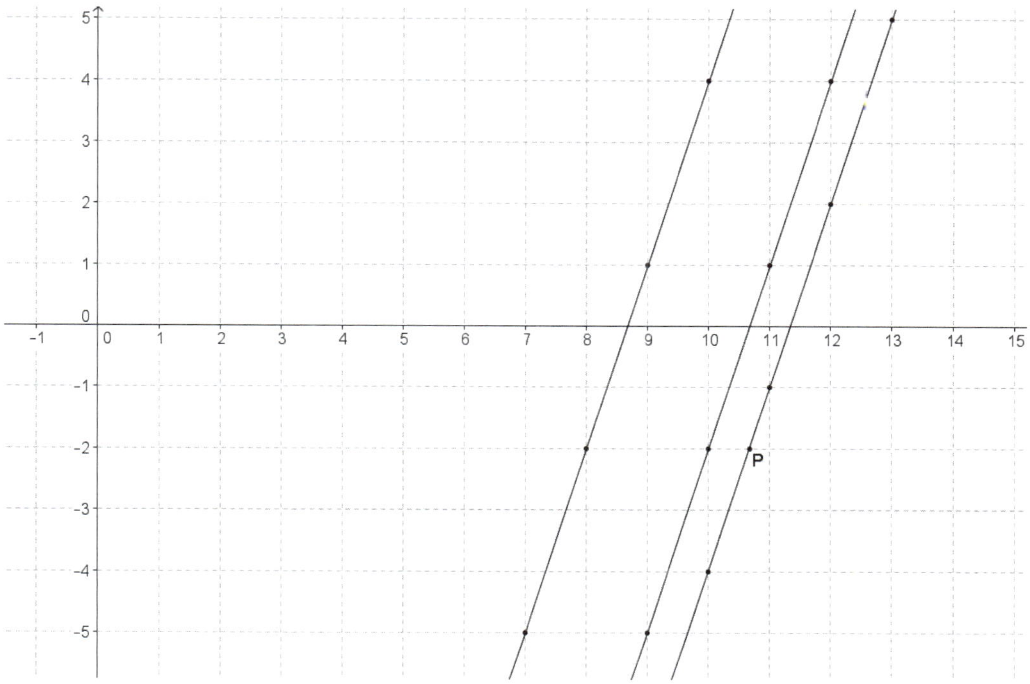

The given lines on the coordinate plane are parallel because they have the same slope $m = 3$. First, draw a horizontal transversal through points $(8, -2)$ and $(10, -2)$ and a second transversal through points $(10, 4)$ and $(10, -4)$. The transversals intersect at $(10, -2)$. Parallel lines cut transversals proportionally, so using horizontal and vertical distances, $\frac{6}{2} = \frac{2}{x}$, where x represents the distance from point $(10, -2)$ to P.

$$x = \frac{2}{6}(2)$$

$$x = \frac{4}{6} = \frac{2}{3}$$

Point P is $\frac{2}{3}$ unit more than 10, or $10\frac{2}{3}$, so the coordinates of point P are $\left(10\frac{2}{3}, -2\right)$.

3. $ACDB$ and $FCDE$ are both trapezoids with bases $\overline{AB}$, $\overline{FE}$, and $\overline{CD}$. The perimeter of trapezoid $ACDB$ is $24\frac{1}{2}$. If the ratio of $AF:FC$ is $1:3$, $AB = 7$, and $ED = 5\frac{5}{8}$, find AF, FC, and BE.

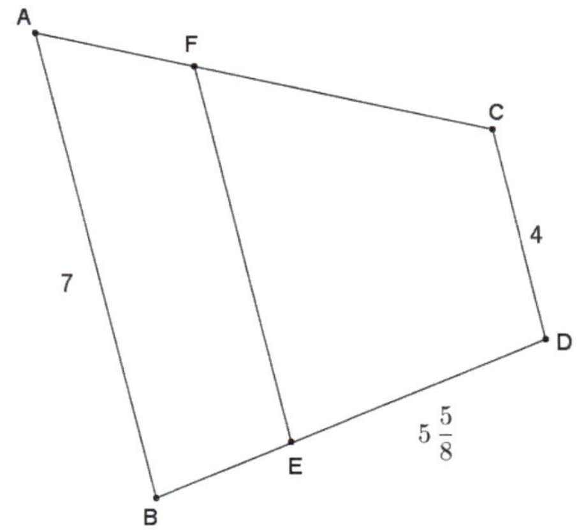

The bases of a trapezoid are parallel, and since $\overline{CD}$ serves as a base for both trapezoids, it follows that $\overline{AB}$, $\overline{FE}$, and $\overline{CD}$ are parallel line segments. Parallel lines cut transversals proportionally, so it must be true that $\frac{AF}{FC} = \frac{BE}{ED} = \frac{1}{3}$.

$$\frac{BE}{5\frac{5}{8}} = \frac{1}{3}$$

$$BE = \frac{1}{3}\left(5\frac{5}{8}\right)$$

$$BE = \frac{15}{8} = 1\frac{7}{8}$$

By the given information, $\frac{AF}{FC} = \frac{1}{3}$, it follows that $FC = 3AF$. Also, $BD = BE + ED$, so $BD = 1\frac{7}{8} + 5\frac{5}{8} = 7\frac{1}{2}$.

$$Perimeter(ACDB) = AB + BD + DC + AC$$

$AF + FC = AC$, and, by the given ratio, $FC = 3AF$, so $AC = AF + 3AF = 4AF$.

$$Perimeter(ACDB) = 7 + 7\frac{1}{2} + 4 + 4AF$$

$$24\frac{1}{2} = 18\frac{1}{2} + 4AF$$

$$6 = 4AF$$

$$\frac{3}{2} = AF$$

By substituting $\frac{3}{2}$ for AF,

$$FC = 3\left(\frac{3}{2}\right)$$

$$FC = \frac{9}{2} = 4\frac{1}{2}.$$

4. **Given the diagram and the ratio of $a: b$ is $3: 2$, answer each question below:**

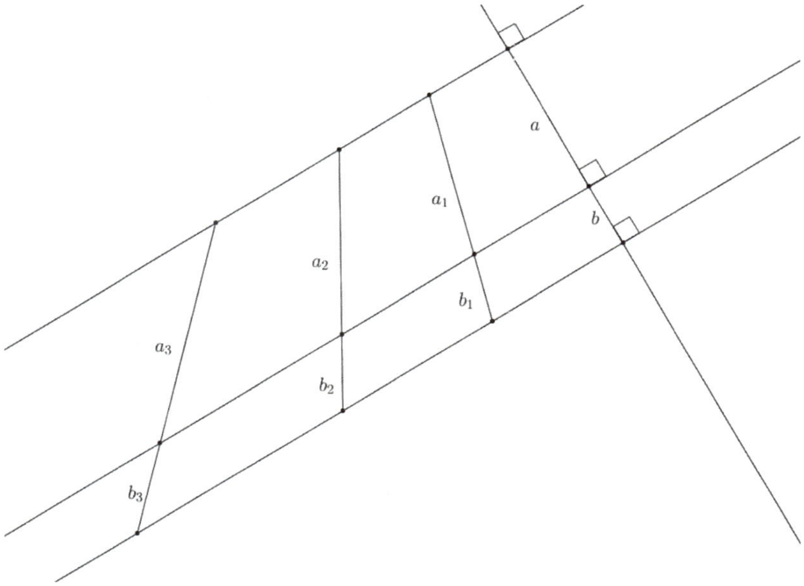

a. **Write an equation for a_n in terms of b_n.**

$a_n = \frac{3}{2}b_n$

b. **Write an equation for b_n in terms of a_n.**

$b_n = \frac{2}{3}a_n$

c. **Use one of your equations to find b_1 in terms of a if $a_1 = 1.2(a)$.**

$b_n = \frac{2}{3}a_n$

$b_1 = \frac{2}{3}\left(\frac{12}{10}a\right)$

$b_1 = \frac{4}{5}a$

d. **What is the relationship between b_1 and b?**

Using the equation from parts (a) and (c), $a = \frac{3}{2}b$, so $b_1 = \frac{4}{5}\left(\frac{3}{2}b\right)$; thus, $b_1 = \frac{6}{5}b$.

e. **What constant, c, relates b_1 and b? Is this surprising? Why or why not?**

The constant relating b_1 and b is the same constant relating a_1 to a, $c = \frac{12}{10} = 1.2$.

f. Using the formula $a_n = c \cdot a_{n-1}$, find a_3 in terms of a.

$$a_1 = \frac{12}{10}(a)$$

$$a_2 = \frac{12}{10}\left(\frac{12}{10}a\right) \qquad a_3 = \frac{12}{10}\left(\frac{36}{25}a\right)$$

$$a_2 = \frac{144}{100}a \qquad a_3 = \frac{432}{250}a$$

$$a_2 = \frac{36}{25}a \qquad a_3 = \frac{216}{125}a$$

g. Using the formula $b_n = c \cdot b_{n-1}$, find b_3 in terms of b.

$$b_3 = \frac{216}{125}b$$

h. Use your answers from parts (f) and (g) to calculate the value of the ratio of $a_3 : b_3$.

$$\frac{a_3}{b_3} = \frac{\frac{216}{125}a}{\frac{216}{125}b} = \frac{a}{b} = \frac{3}{2}$$

5. Julius wants to try to estimate the circumference of the earth based on measurements made near his home. He cannot find a location near his home where the sun is straight overhead. Will he be able to calculate the circumference of the earth? If so, explain and draw a diagram to support your claim.

***Note to the teacher:** This problem is very open-ended, requires critical thinking, and may not be suitable for all students. The problem may be scaffolded by providing a diagram with possible measurements that Julius made based on the description in the student solution below.*

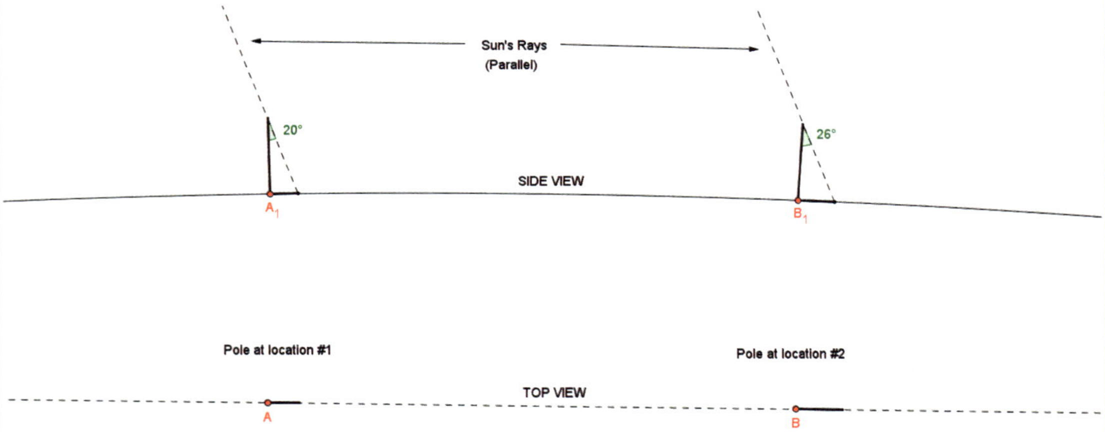

***Possible solution:** If Julius can find two locations such that those locations and their shadows lie in the same straight path, then the difference of the shadows' angles can be used as part of the $360°$ in the earth's circumference. The distance between those two locations corresponds with that difference of angles.*

Lesson 20: How Far Away Is the Moon?

Student Outcomes

- Students understand how the Greeks measured the distance from the earth to the moon and solve related problems.

Lesson Notes

In Lesson 20, students learn how to approximate the distance of the moon from the earth. Around the same time that Eratosthenes approximated the circumference of the earth, Greek astronomer Aristarchus found a way to determine the distance to the moon. Less of the history is provided in the lesson, as a more comprehensive look at eclipses had to be included alongside the central calculations. The objective of the lesson is to provide an understandable method of how the distance was calculated; assumptions are folded in to expedite the complete story, but teachers are encouraged to bring to light details that are left in the notes as they see fit for students.

Classwork

Opening Exercise (4 minutes)

Opening Exercise

What is a solar eclipse? What is a lunar eclipse?

A solar eclipse occurs when the moon passes between the earth and the sun, and a lunar eclipse occurs when the earth passes between the moon and the sun.

- In fact, we should imagine a solar eclipse occurring when the moon passes between the earth and the sun, and the earth, sun, and moon lie on a straight line, and similarly so for a lunar eclipse.

Discussion (30 minutes)

Lead students through a conversation regarding the details of solar and lunar eclipses.

- A *total* solar eclipse lasts only a few minutes because the sun and moon appear to be the same size.
- How would appearances change if the moon were closer to the earth?
  - *The moon would appear larger.*
- The moon would appear larger, and the eclipse would last longer. What if the moon were farther away from the earth? Would we experience a total solar eclipse?
  - *The moon would appear smaller, and it would not be possible for a total solar eclipse to occur, because the moon would appear as a dark dot blocking only part of the sun.*
- Sketch a diagram of a solar eclipse.

Students' knowledge on eclipses varies. This is an opportunity for students to share what they know regarding eclipses. Allow a minute of discussion, and then guide them through a basic description of the moon's shadow and how it is conical (in 3D view), but on paper in a profile view, the shadow appears as an isosceles triangle whose base coincides with the diameter of the earth. Discuss what makes the shadows of celestial bodies similar. Describe the two parts of the moon's shadow, the umbra and penumbra.

Note that the distances are not drawn to scale in the following image.

Discussion

Solar Eclipse

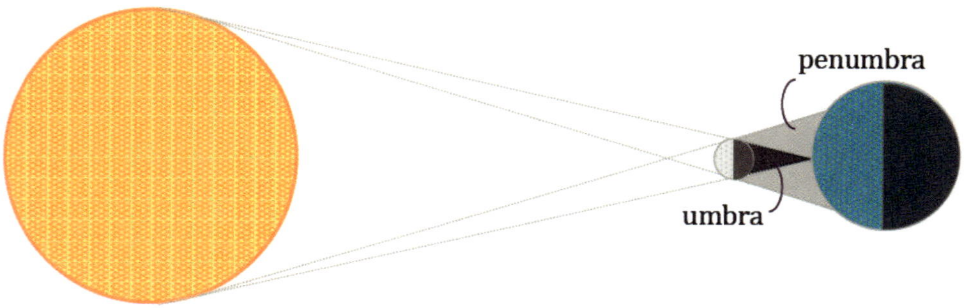

3D view:

- The umbra is the portion of the shadow where all sunlight is blocked, while the penumbra is the part of the shadow where light is only partially blocked. For the purposes of our discussion today, we will be simplifying the situation and considering only the umbra.
- What is remarkable about the full shadow caused by the eclipse? That is, what is remarkable about the umbra and the portion of the moon that is dark? Consider the relationship between the 3D and 2D image of it.
  - *If a cone represents the portion of the moon that is dark as well as the umbra, then the part that is entirely dark in the 2D image is an isosceles triangle.*
- We assume that shadows from the moon and the earth are all similar isosceles triangles.

This is, in fact, not the case for all planetary objects, as the shadow formed depends on how far away the light source is from the celestial body and the size of the planet. For the sake of simplification as well as approximation, this assumption is made. Then, shadows created by the moon and the earth will have the following relationship:

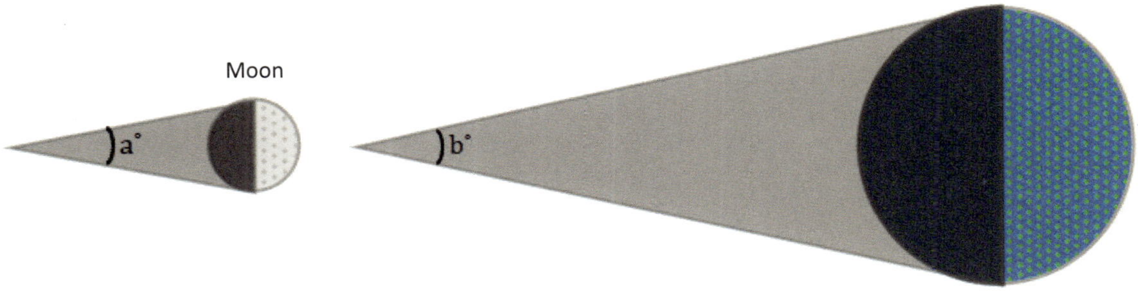

$$a = b$$

- You can imagine simulating a solar eclipse using a marble that is one inch in diameter. If you hold it one arm length away, it will easily block (more than block) the sun from one eye. Do not try this; you will damage your eye!
- To make the marble just *barely* block the sun, it must be about 9 feet (i.e., 108 inches) away from your eye. So the cone of shadow behind the marble tapers to a point, which is where your eye is, and at this point, the marble just blocks out the sun. This means that the ratio of the length of the shadow of the marble to the diameter of the marble is about $108: 1$.
- In fact, by experimenting with different-size spheres, we find that this ratio holds true regardless of the size of the sphere (or circular object) as long as the sphere is at the point where it *just* blocks the sun from our vantage point. In other words, whether you are using a marble, a tennis ball, or a basketball to model the eclipse, the distance the sphere must be held from the eye is 108 times the diameter of the sphere.
- We conclude that this ratio also holds for the moon and the earth during a solar eclipse. Since the moon and sun appear to be the same size in the sky (the moon *just* blocks the sun in a solar eclipse), we can conclude that the distance from the earth to the moon must also be roughly 108 times the diameter of the moon because the earth is at the tip of the moon's shadow.

> *Scaffolding:*
> Emphasize the altitude:base ratio of lengths by citing other everyday objects, such as a tennis ball; for example, a tennis ball just blocks out the sun if the tennis ball is viewed from 108 tennis ball diameters away.

- Let us now consider what is happening in a lunar eclipse. A lunar eclipse occurs when the moon passes behind the earth. Once again, the earth, sun, and moon lie on a straight line, but this time the earth is between the moon and the sun.
- Consequently, during a lunar eclipse, the moon is still faintly visible from the earth because of light reflected off the earth.
- Sketch a diagram of a lunar eclipse.

Lunar Eclipse

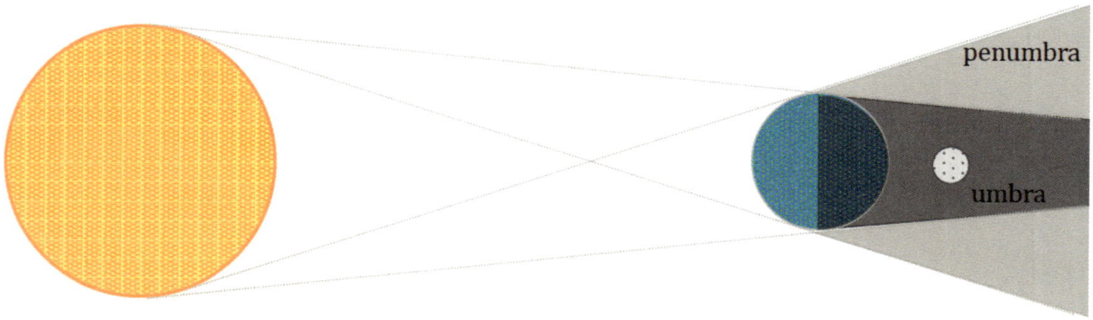

3D view:

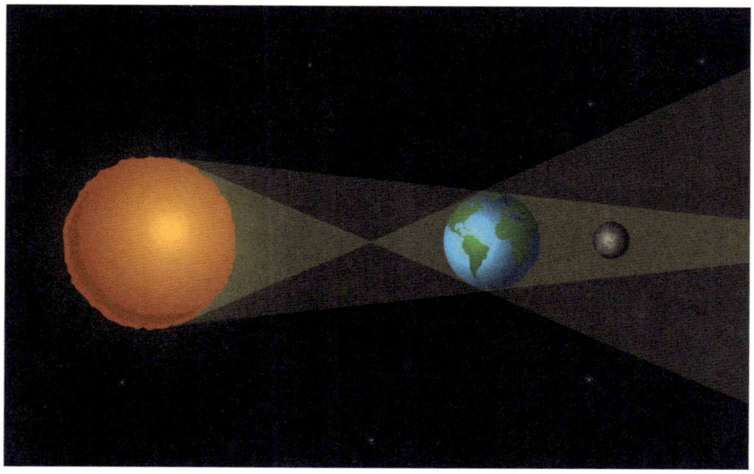

- Based on what we know about the ratio of distances for an object to just block the sun, we can conclude that the moon must be within 108 earth diameters; if it were not within that distance, it would not pass through the earth's shadow, and the earth could not block the sun out completely.
- Studying total lunar eclipses was critical to finding the distance to the moon. Other types of eclipses exist, but they involve the penumbra. The following measurement required a total eclipse, or one involving the umbra.
- By carefully examining the shadow of the earth during a total lunar eclipse, it was determined that the width, or more specifically, the diameter of the cross section of the earth's conical shadow at the distance of the moon, is about $2\frac{1}{2}$ moon diameters.

Share the following animation of a lunar eclipse: http://galileoandeinstein.physics.virginia.edu/lectures/eclipse3.htm

- With this measurement, a diagram like the following was constructed, where the shadow of the moon was reversed along the shadow of the earth:

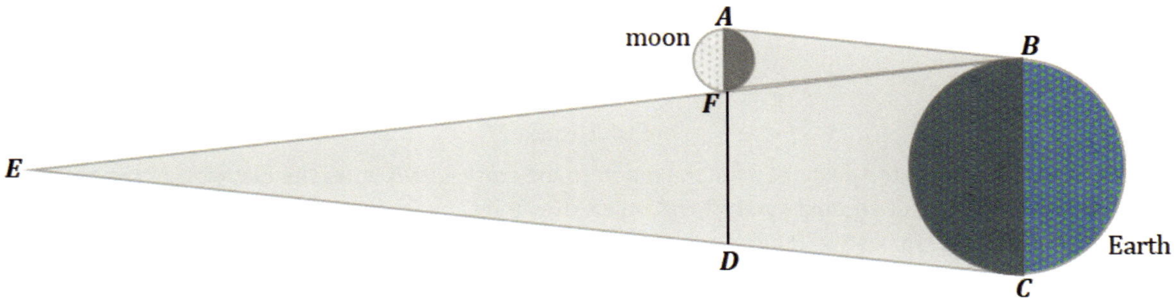

- What length and angle relationships can we label in this diagram?

Pose the following questions one at a time, and allow students to discuss and mark their diagrams.

- Based on what is known about the shadows of the moon and the earth, what do we know about the measures of $\angle ABF$ and $\angle BEC$?
  - *The angle measures are equal since we are assuming the shadows can be modeled by similar isosceles triangles.*
- What is the relationship between AF and AB?
  - *Since the distance an object must be from the eye to block out the sun is 108 times the object's diameter, the length of AB must be 108 times that of AF.*
- What is the relationship between AF and FD?
  - *We know that the diameter of the cross section of the earth's conical shadow at the distance of the moon is about $2\frac{1}{2}$ moon diameters, so the length of FD is 2.5 times the length of AF.*

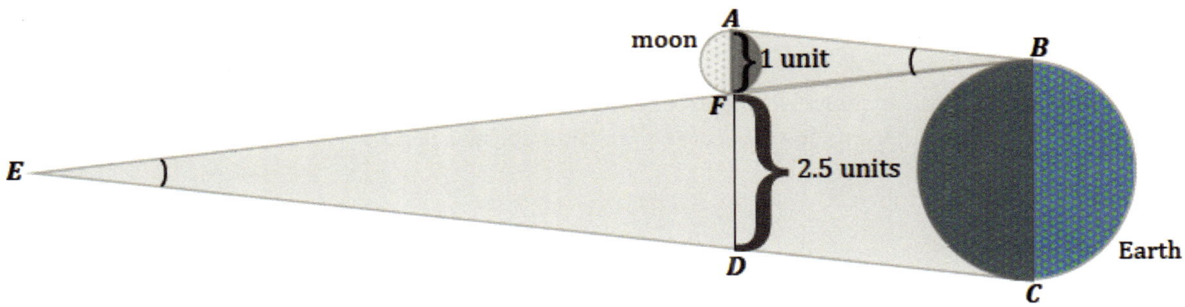

- Reversing the moon's shadow completes parallelogram $ABCD$. How can we be sure that $ABCD$ is a parallelogram?

Allow students time to discuss why this must be true, using the relationships determined in the diagram. Share out ideas before explaining further.

- $\overline{AD}$ must be parallel to $\overline{BC}$ by construction. The diameter of the earth is parallel to the segment formed by the moon's diameter and the diameter of the shadow at the distance of the moon.
- We know that $m\angle ABF = m\angle BEC$ since the shadows are similar triangles. This means that $\overline{AB}$ is parallel to $\overline{DC}$ (alternate interior angles).

- Since we know that $ABCD$ is a parallelogram, then $AD = BC$, or $BC = 3.5$ units.
- From our work in Lesson 19, we know the Greeks had the circumference of the earth at this time, which means they also had the earth's diameter. Then, it was known that $BC \approx 8000$ miles. With this information, we conclude:

$$3.5 \text{ units} \approx 8000$$

$$1 \text{ unit} \approx 2300.$$

- We already know that the distance from the moon to the earth is 108 times the diameter of the moon. If the moon has a diameter of approximately 2,300 miles, then the distance from the moon to the earth is roughly (2300×108) miles, or 248,000 miles.
- With some careful observation and measurement (that of a lunar eclipse) and basic geometry, Aristarchus was able to determine a fairly accurate measure of the distance from the earth to the moon.

Example (5 minutes)

Example

a. **If the circumference of the earth is about $25,000$ miles, what is the earth's diameter in miles?**

$\frac{25000}{\pi} \approx 8000$ *The earth's diameter is approximately $8,000$ miles.*

b. **Using part (a), what is the moon's diameter in miles?**

$\frac{2}{7} \cdot \frac{25000}{\pi} \approx 2300$ *The moon's diameter is approximately $2,300$ miles.*

c. **How far away is the moon in miles?**

$108 \cdot \frac{2}{7} \cdot \frac{25000}{\pi} \approx 246\,000$ *The moon is approximately $246,000$ miles from the earth.*

- Modern-day calculations show that the distance from the earth to the moon varies between 225,622 miles and 252,088 miles. The variation is because the orbit is actually elliptical versus circular.

Closing (1 minute)

- With some methodical observations and the use of geometry, Aristarchus was able to make a remarkable approximation of the distance from the earth to the moon using similarity.

Exit Ticket (5 minutes)

Name ____ Date ____________________

Lesson 20: How Far Away Is the Moon?

Exit Ticket

1. On Planet A, a $\frac{1}{4}$-inch diameter ball must be held at a height of 72 inches to *just* block the sun. If a moon orbiting Planet A just blocks the sun during an eclipse, approximately how many moon diameters is the moon from the planet?

2. Planet A has a circumference of $93,480$ miles. Its moon has a diameter that is approximated to be $\frac{1}{8}$ that of Planet A. Find the approximate distance of the moon from Planet A.

Exit Ticket Sample Solutions

1. On Planet A, a $\frac{1}{4}$-inch diameter ball must be held at a height of 72 inches to *just* block the sun. If a moon orbiting Planet A just blocks the sun during an eclipse, approximately how many moon diameters is the moon from the planet?

   The ratio of the diameter of the ball to the specified height is $\frac{\frac{1}{4}}{72} = \frac{1}{288}$. The moon's distance in order to just block the sun would be proportional since the shadows formed are similar triangles, so the moon would orbit approximately 288 moon diameters from Planet A.

2. Planet A has a circumference of $93,480$ miles. Its moon has a diameter that is approximately $\frac{1}{8}$ that of Planet A. Find the approximate distance of the moon from Planet A.

   To find the diameter of Planet A:

   $\frac{93480}{\pi} = d_{\text{planet A}}$ *The diameter of Planet A is approximately $29,756$ miles.*

   To find the diameter of the moon:

   $d_{\text{moon}} = \frac{1}{8} d_{\text{Planet A}}$

   $d_{\text{moon}} = \frac{1}{8}\left(\frac{93480}{\pi}\right)$

   $d_{\text{moon}} = \frac{93480}{8\pi} = \frac{11685}{\pi}$ *The diameter of the moon is approximately $3,719$ miles.*

   To find the distance of the moon from Planet A:

   $\text{distance}_{\text{moon}} = 288(d_{\text{moon}})$

   $\text{distance}_{\text{moon}} = 288\left(\frac{11685}{\pi}\right)$

   $\text{distance}_{\text{moon}} = \frac{3\,365\,280}{\pi}$ *The distance from Planet A to its moon is approximately $1,071,202$ miles.*

Problem Set Sample Solutions

1. If the sun and the moon do not have the same diameter, explain how the sun's light can be covered by the moon during a solar eclipse.

   The farther away an object is from the viewer, the smaller that object appears. The moon is closer to the earth than the sun, and it casts a shadow where it blocks some of the light from the sun. The sun is much farther away from the earth than the moon, and because of the distance, it appears much smaller in size.

2. What would a lunar eclipse look like when viewed from the moon?

   The sun would be completely blocked out by the earth for a time because the earth casts an umbra that spans a greater distance than the diameter of the moon, meaning the moon would be passing through the earth's shadow.

3. Suppose you live on a planet with a moon, where during a solar eclipse, the moon appears to be half the diameter of the sun.

a. Draw a diagram of how the moon would look against the sun during a solar eclipse.

Sample response:

b. A 1-inch diameter marble held 100 inches away on the planet barely blocks the sun. How many moon diameters away is the moon from the planet? Draw and label a diagram to support your answer.

If the diameter of the moon appears to be half the diameter of the sun as viewed from the planet, then the moon will not cause a total eclipse of the sun. In the diagram, PQ is the diameter of the moon, and TU is the diameter of the sun as seen from the planet at point I.

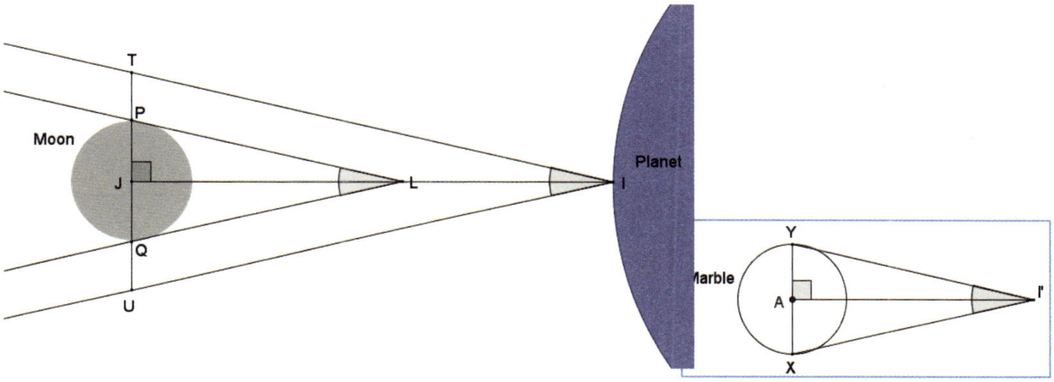

The diameter of the moon is represented by distance PQ. For me to view a total eclipse, where the sun is just blocked by the moon, the moon would have to be twice as wide, so $TU = 2(PQ)$. $\triangle TIU$ and $\triangle YI'X$ are both isosceles triangles, and their vertex angles are the same, so the triangles are similar by the SAS criterion. If the triangles are similar, then their altitudes are in the same ratio as their bases. This means $\frac{IJ}{TU} = \frac{I'A}{YX}$. By substituting values, $\frac{IJ}{2PQ} = \frac{100}{1}$, so $IJ = 200(PQ)$. The moon is approximately 200 moon diameters from the planet.

c. **If the diameter of the moon is approximately $\frac{3}{5}$ of the diameter of the planet and the circumference of the planet is $185,000$ miles, approximately how far is the moon from the planet?**

The diameter of the planet:

$$d_{\text{planet}} = \frac{185\,000}{\pi}$$

$$d_{\text{planet}} = \frac{185\,000}{\pi}$$

The diameter of the planet is approximately $58,887$ miles.

The diameter of the moon:

$$d_{\text{moon}} = \frac{3}{5} d_{\text{planet}}$$

$$d_{\text{moon}} = \frac{3}{5}\left(\frac{185\,000}{\pi}\right)$$

$$d_{\text{moon}} = \frac{111\,000}{\pi}$$

The diameter of the moon is approximately $35,332$ miles.

The distance of the moon from the planet:

$$IJ = 200(PQ)$$

$$IJ = 200\left(\frac{111\,000}{\pi}\right)$$

$$IJ = \frac{22\,200\,000}{\pi}$$

The planet's moon is approximately $7,066,479$ miles from the planet.

Name ______________________________ Date ______________

1. The coordinates of $\triangle ABC$ are shown on the coordinate plane below. $\triangle ABC$ is dilated from the origin by scale factor $r = 2$.

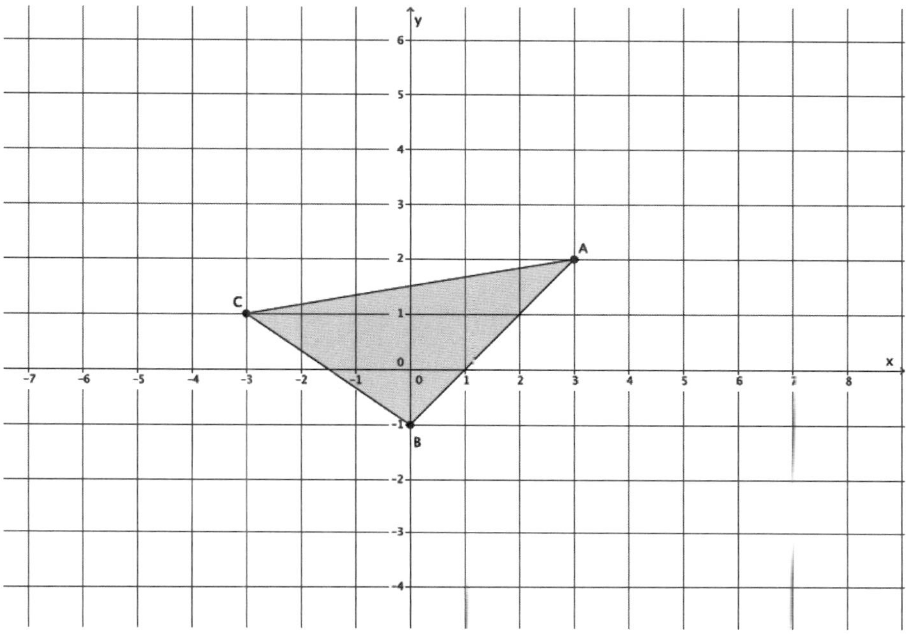

a. Identify the coordinates of the dilated $\triangle A'B'C'$.

b. Is $\triangle A'B'C' \sim \triangle ABC$? Explain.

2. Points A, B, and C are not collinear, forming $\angle BAC$. Extend $\overrightarrow{AB}$ to point P. Line ℓ passes through P and is parallel to segment BC. It meets $\overrightarrow{AC}$ at point Q.

a. Draw a diagram to represent the situation described.

b. Is $\overline{PQ}$ longer or shorter than $\overline{BC}$?

c. Prove that $\triangle ABC \sim \triangle APQ$.

d. What other pairs of segments in this figure have the same ratio of lengths that $\overline{PQ}$ has to $\overline{BC}$?

3. There is a triangular floor space $\triangle ABC$ in a restaurant. Currently, a square portion $DEFG$ is covered with tile. The owner wants to remove the existing tile and then tile the largest square possible within $\triangle ABC$, keeping one edge of the square on $\overline{AC}$.

   a. Describe a construction that uses a dilation with center A that can be used to determine the maximum square $D'E'F'G'$ within $\triangle ABC$ with one edge on $\overline{AC}$.

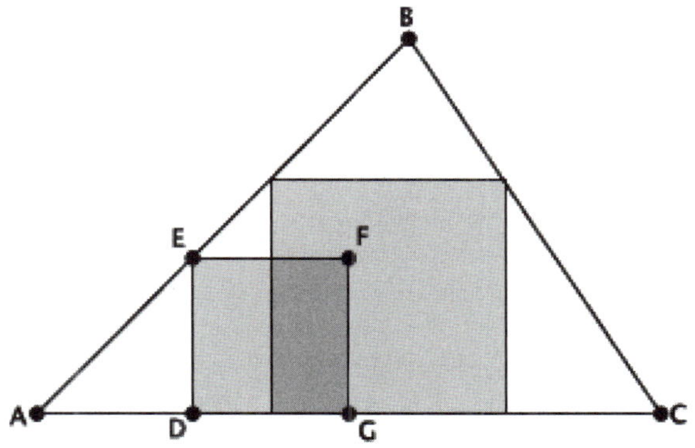

   b. What is the scale factor of $\overline{FG}$ to $\overline{F'G'}$ in terms of the distances $\overline{AF}$ and $\overline{AF'}$?

c. The owner uses the construction in part (a) to mark off where the square would be located. He measures AE to be 15 feet and EE' to be 5 feet. If the original square is 144 square feet, how many square feet of tile does he need for $D'E'F'G'$?

4. $ABCD$ is a parallelogram, with the vertices listed counterclockwise around the figure. Points $M, N, O,$ and P are the midpoints of sides $\overline{AB}$, $\overline{BC}$, $\overline{CD}$, and $\overline{DA}$, respectively. The segments MO and NP cut the parallelogram into four smaller parallelograms, with the point W in the center of $ABCD$ as a common vertex.

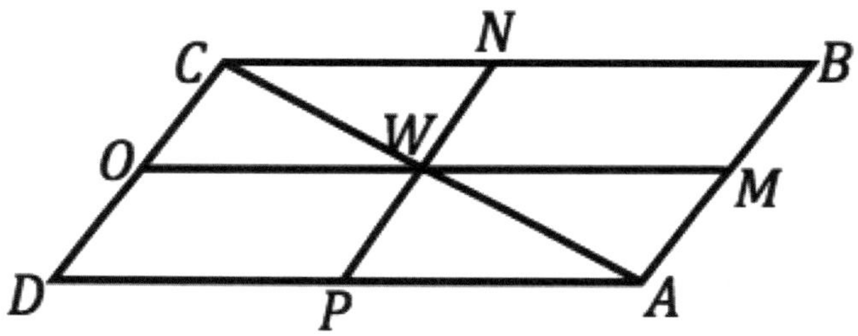

a. Exhibit a sequence of similarity transformations that takes $\triangle AMW$ to $\triangle CDA$. Be specific in describing the parameter of each transformation; for example, if describing a reflection, state the line of reflection.

b. Given the correspondence in $\triangle AMW$ similar to $\triangle CDA$, list all corresponding pairs of angles and corresponding pairs of sides. What is the ratio of the corresponding pairs of angles? What is the ratio of the corresponding pairs of sides?

5. Given two triangles, $\triangle ABC$ and $\triangle DEF$, $m\angle CAB = m\angle FDE$, and $m\angle CBA = m\angle FED$. Points A, B, D, and E lie on line l as shown. Describe a sequence of rigid motions and/or dilations to show that $\triangle ABC \sim \triangle DEF$, and sketch an image of the triangles after each transformation.

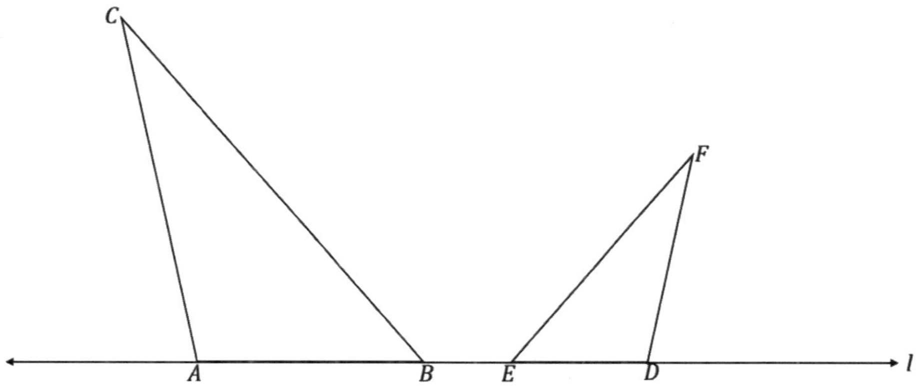

6. $\triangle JKL$ is a right triangle; $\overline{NP} \perp \overline{KL}$, $\overline{NO} \perp \overline{JK}$, $\overline{MN} \parallel \overline{OP}$.

   a. List all sets of similar triangles. Explain how you know.

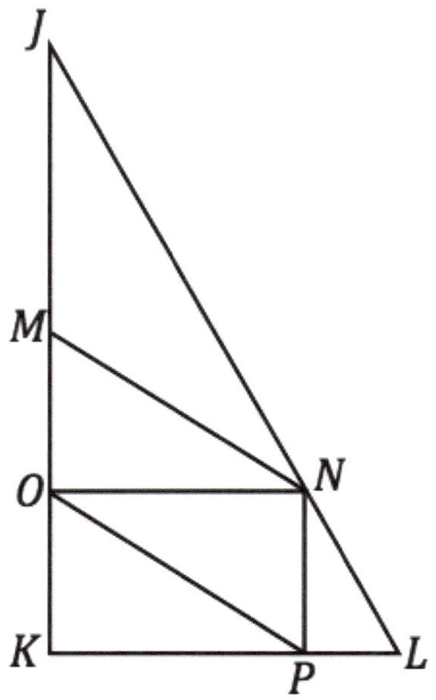

   b. Select any two similar triangles, and show why they are similar.

7.

a. The line PQ contains point O. What happens to $\overleftrightarrow{PQ}$ with a dilation about O and scale factor of $r = 2$? Explain your answer.

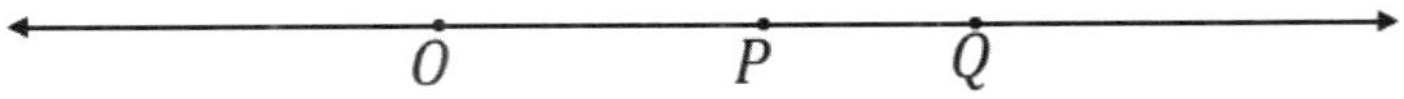

b. The line PQ does not contain point O. What happens to $\overleftrightarrow{PQ}$ with a dilation about O and scale factor of $r = 2$?

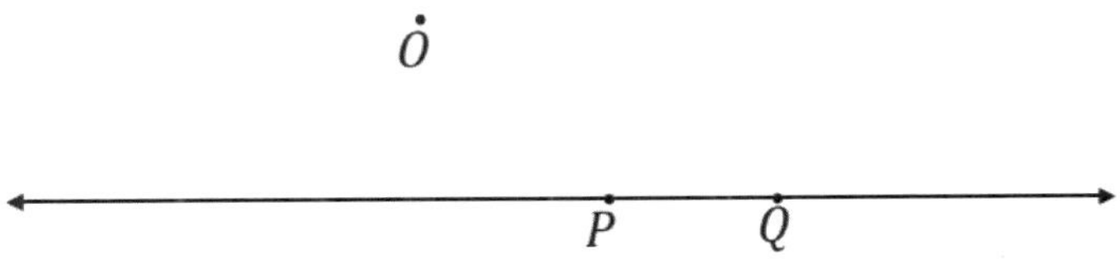

8. Use the diagram below to answer the following questions.

   a. State the pair of similar triangles. Which similarity criterion guarantees their similarity?

   b. Calculate DE to the hundredths place.

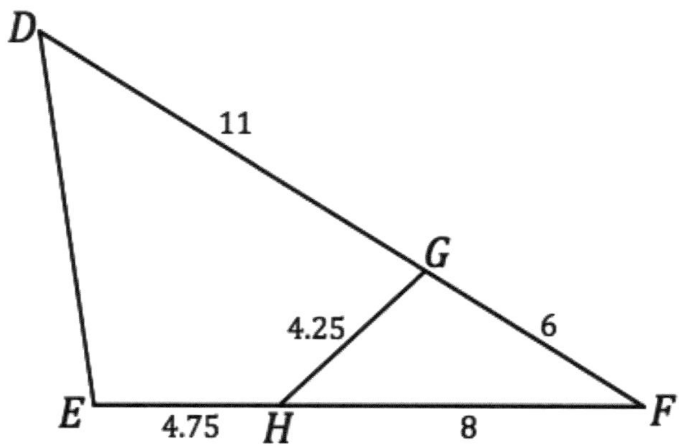

9. In $\triangle ABC$, $m\angle A$ is $40°$, $m\angle B$ is $60°$, and $m\angle C$ is $80°$. The triangle is dilated by a factor of 2 about point P to form $\triangle A'B'C'$. It is also dilated by a factor of 3 about point Q to form $\triangle A''B''C''$. What is the measure of the angle formed by line $A'B'$ and line $B''C''$? Explain how you know.

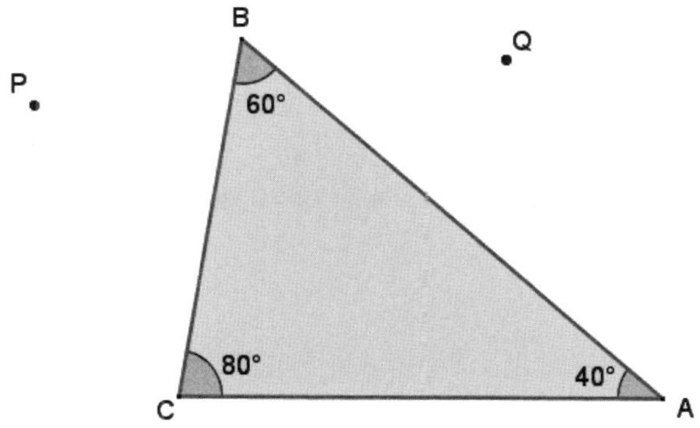

10. In the diagram below, $|AC| = |CE| = |EG|$, and $\angle BAC$, $\angle DCE$, and $\angle FEG$ are right. The two lines meet at a point to the right. Are the triangles similar? Why or why not?

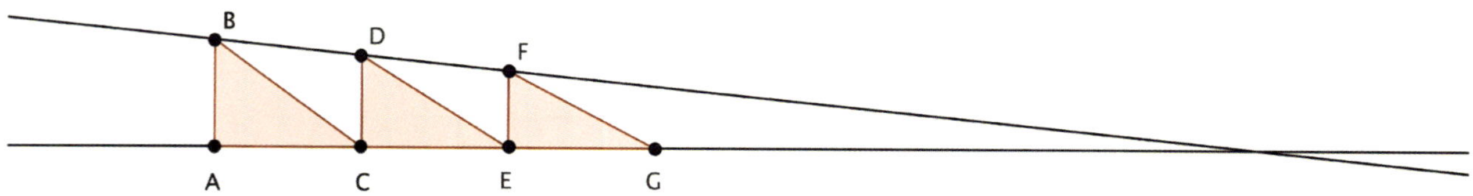

11. The side lengths of the following right triangle are 16, 30, and 34. An altitude of a right triangle from the right angle splits the hypotenuse into line segments of length x and y.

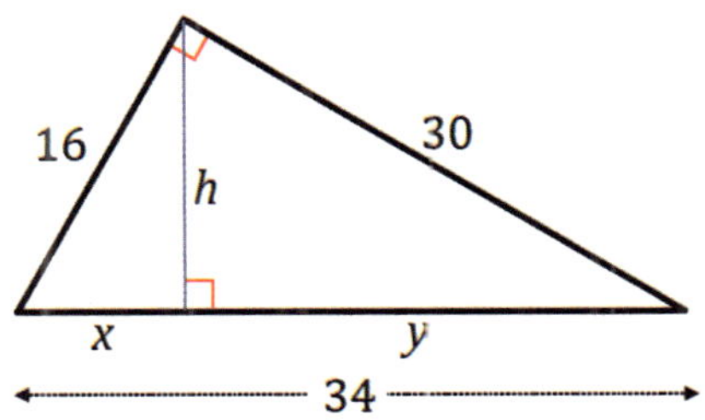

a. What is the relationship between the large triangle and the two sub-triangles? Why?

b. Solve for h, x, and y.

c. **Extension:** Find an expression that gives h in terms of x and y.

12. The sentence below, as shown, is being printed on a large banner for a birthday party. The height of the banner is 18 inches. There must be a minimum 1-inch margin on all sides of the banner. Use the dimensions in the image below to answer each question.

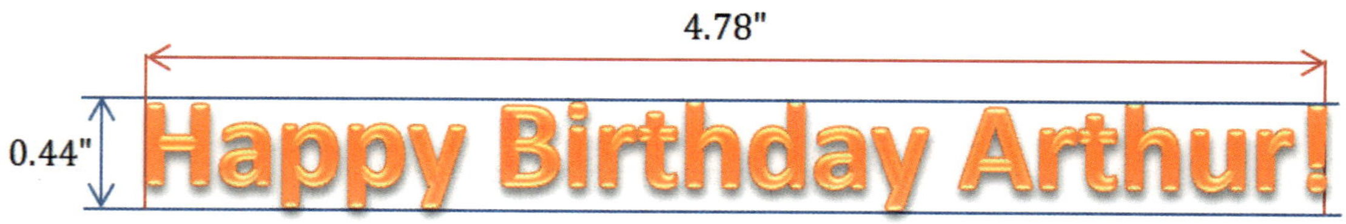

a. Describe a reasonable figure in the plane to model the printed image.

b. Find the scale factor that maximizes the size of the characters within the given constraints.

c. What is the total length of the banner based on your answer to part (a)?

A Progression Toward Mastery					
Assessment Task Item		**STEP 1** Missing or incorrect answer and little evidence of reasoning or application of mathematics to solve the problem.	**STEP 2** Missing or incorrect answer but evidence of some reasoning or application of mathematics to solve the problem.	**STEP 3** A correct answer with some evidence of reasoning or application of mathematics to solve the problem, OR an incorrect answer with substantial evidence of solid reasoning or application of mathematics to solve the problem.	**STEP 4** A correct answer supported by substantial evidence of solid reasoning or application of mathematics to solve the problem.
1	**a–b** **G-SRT.A.2**	Student provides at least two incorrect coordinates of part (a), and student does not show clear understanding of why $\triangle A'B'C' \sim \triangle ABC$ for part (b).	Student provides only partially correct and clear answers for both parts (a) and (b).	Student answers part (a) correctly, but part (b) lacks clear explanation regarding the corresponding angles and length measurements. OR Student answers part (b) correctly, and part (a) has errors in the coordinates of the dilated vertices.	Student answers both parts (a) and (b) correctly.
2	**a–d** **G-SRT.B.5**	Student provides a correct response for part (b), but student shows insufficient understanding or inaccurate answers for parts (a), (c), and (d).	Student incorrectly answers two parts.	Student incorrectly answers one part (e.g., incorrect diagram in part (a) or insufficient evidence provided for (c)).	Student correctly answers all four parts.
3	**a–c** **G-SRT.A.1**	Student provides a response for part (a) that is missing two or more construction steps, and student incorrectly answers parts (b) and (c).	Student provides a response for part (a) that is missing two or more construction steps, and student incorrectly answers part (b) or part (c).	Student provides a response for part (a) that is missing two or more construction steps but correctly answers parts (b) and (c). OR Student correctly answers part (a) but incorrectly answers part (b) or part (c).	Student correctly answers all three parts.

4	a–b G-SRT.A.2	Student provides an incomplete or otherwise inaccurate description of a similarity transformation for part (a). Student provides an incomplete list of correspondences for part (b) and makes errors in the ratios of pairs of angles or pairs of sides.	Student provides an incomplete or otherwise inaccurate description of a similarity transformation for part (a). Student provides an incomplete list of correspondences or makes errors in the ratios of pairs of angles or pairs of sides in part (b).	Student provides an incomplete or otherwise inaccurate description of a similarity transformation for part (a) but correctly answers part (b). OR Student provides a correct answer for part (a) but provides an incomplete list of correspondences or makes errors in the ratios of pairs of angles or pairs of sides in part (b).	Student correctly answers both parts (a) and (b).
5	G-SRT.A.3	Student describes an incoherent sequence of transformations or provides no description and does not explain parameters to show that one triangle maps to the other.	Student describes a sequence of transformations that maps one triangle to the other but does not provide detail regarding the parameters (e.g., a reflection is cited, but no detail regarding the line of reflection is mentioned).	Student describes a sequence of transformations that maps one triangle to the other but does not provide detail regarding the parameters of one of the transformations, or one additional transformation and its respective parameters are needed to complete a correct sequence of transformations.	Student clearly describes a sequence of appropriate transformations and provides the appropriate parameters for each transformation.
6	a–b G-SRT.B.5	Student provides a response that is missing three or more similar triangles out of both sets.	Student lists all but two of the similar triangles out of both sets.	Student lists all but one of the similar triangles out of both sets.	Student correctly lists all the similar triangles in each set.
7	a–b G-SRT.B.5	Student does not show an understanding of the properties of dilations in either part (a) or part (b).	Student correctly answers part (a) but incorrectly answers part (b).	Student correctly answers part (b) but does not provide justification as to why the line maps to itself in part (a).	Student correctly answers parts (a) and (b).

8	a–b G-SRT.B.5	Student provides incorrect similarity criteria for part (a) and makes more than one conceptual or computational error in part (b).	Student provides incorrect similarity criteria for part (a) and makes one conceptual or computational error in part (b).	Student correctly answers part (a) but makes one conceptual or computational error in part (b). OR Student provides incorrect similarity criteria for part (a) but correctly answers part (b).	Student correctly answers parts (a) and (b).
9	G-SRT.A.1	Student shows little or no understanding of why the two dilations lead to an angle of $60°$ formed by $A'B'$ and $B''C''$.	Student includes an attempted diagram but is missing multiple conclusive elements or has an error in the verbal justification.	Student includes an attempted diagram but is missing one conclusive element (e.g., missing one of the dilations or an incorrect dilation) or has an error in the verbal justification.	Student includes an accurate diagram and provides a complete and correct justification.
10	G-SRT.B.4	Student shows little or no understanding of why the triangles are not similar.	Student includes a justification that demonstrates why the triangles are not similar but has two conceptual errors.	Student includes a justification that demonstrates why the triangles are not similar but has one conceptual error.	Student includes an accurate justification that demonstrates why the triangles are not similar.
11	a–c G-SRT.B.5	Student does not provide fully correct answers for parts (a), (b), and (c).	Student provides a fully correct answer for one of the three parts.	Student provides fully correct answers for any two of the three parts.	Student provides correct answers for parts (a), (b), and (c) and clearly explains each.
12	a–c G-MG.A.1 G-MG.A.3	Student does not provide fully correct answers for parts (a), (b), and (c).	Student provides a fully correct answer for one of the three parts.	Student correctly computes scaled dimensions for the printed image but fails to consider the required margins.	Student provides correct answers for parts (a), (b), and (c) and clearly explains each.

Name ______________________________ Date ______________

1. The coordinates of $\triangle ABC$ are shown on the coordinate plane below. $\triangle ABC$ is dilated from the origin by scale factor $r = 2$.

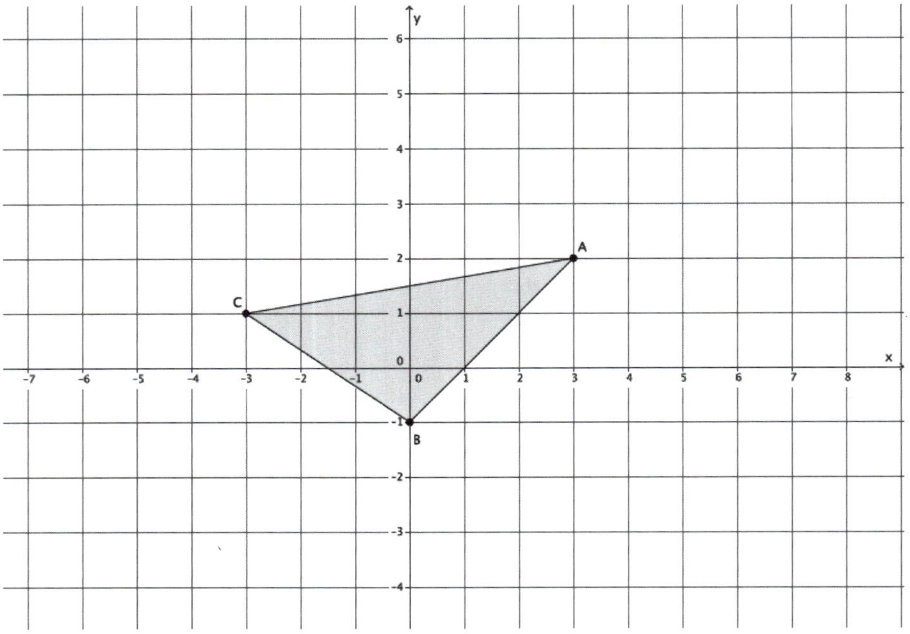

a. Identify the coordinates of the dilated $\triangle A'B'C'$.

Point A = (3, 2), then A' = (2×(3), 2×(2)) = (6, 4)

Point B = (0, −1), then B' = (2×(0), 2×(−1)) = (0, −2)

Point C = (−3, 1), then C' = (2×(−3), 2×(1)) = (−6, 2)

b. Is $\triangle A'B'C' \sim \triangle ABC$? Explain.

Yes. The side lengths of △A'B'C' are each two times the length of the sides of △ABC, and corresponding sides are proportional in length. Also, the corresponding angles are equal in measurement because dilations preserve the measurements of angles.

2. Points A, B, and C are not collinear, forming $\angle BAC$. Extend $\overrightarrow{AB}$ to point P. Line ℓ passes through P and is parallel to segment BC. It meets ray AC at point Q.

   a. Draw a diagram to represent the situation described.

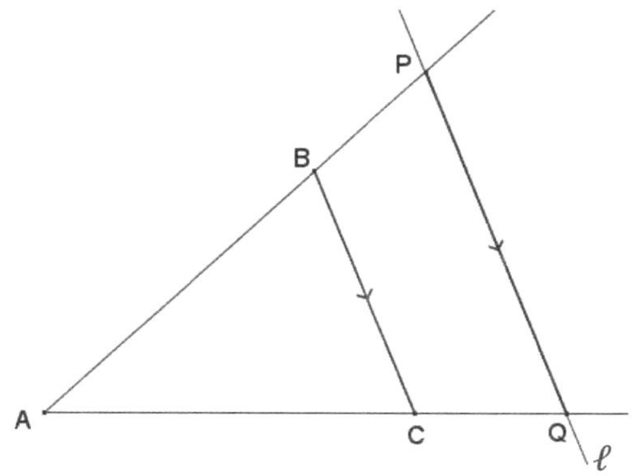

   b. Is $\overline{PQ}$ longer or shorter than $\overline{BC}$?

   $\overline{PQ}$ is longer than $\overline{BC}$.

   c. Prove that $\triangle ABC \sim \triangle APQ$.

   $\overline{PQ} \parallel \overline{BC}$. Since corresponding angles are equal in measure, then $m\angle ABC = m\angle APQ$. Additionally, $m\angle A = m\angle A$. Then $\triangle ABC \sim \triangle APQ$ by AA similarity criterion.

   d. What other pairs of segments in this figure have the same ratio of lengths that $\overline{PQ}$ has to $\overline{BC}$?

   PA:BA, QA:CA

3. There is a triangular floor space $\triangle ABC$ in a restaurant. Currently, a square portion $DEFG$ is covered with tile. The owner wants to remove the existing tile and then tile the largest square possible within $\triangle ABC$, keeping one edge of the square on AC.

   a. Describe a construction that uses a dilation with center A that can be used to determine the maximum square $D'E'F'G'$ within $\triangle ABC$ with one edge on $\overline{AC}$.

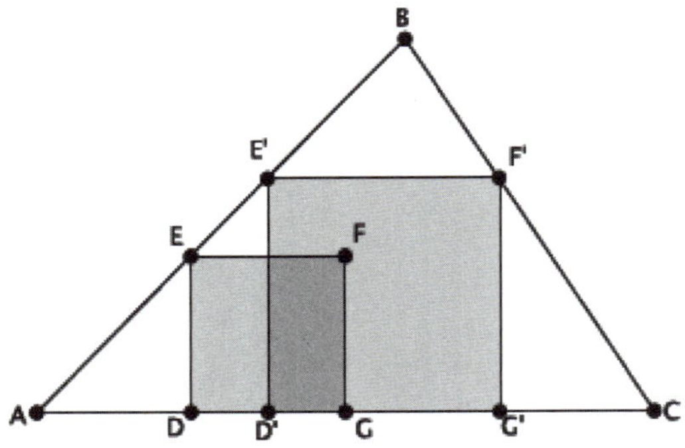

   1. Use A as a center of dilation.
   2. Draw $\overrightarrow{AF}$ through BC.
   3. Label the intersection of $\overrightarrow{AF}$ and BC as F'.
   4. Construct $\overleftrightarrow{E'F'}$ parallel to EF, where E' is the intersection of AB and the parallel line.
   5. Construct $\overleftrightarrow{F'G'}$ parallel to FG, where G' is the intersection of AC and the parallel line.
   6. Construct $\overleftrightarrow{E'D'}$ parallel to ED, where D' is the intersection of AC and the parallel line.
   7. Connect D', E', F', G'.

   b. What is the scale factor of FG to $F'G'$ in terms of the distances AF and AF'?

   $\frac{AF'}{AF}$

c. The owner uses the construction in part (a) to mark off where the square would be located. He measures AE to be 15 feet and EE' to be 5 feet. If the original square is 144 square feet, how many square feet of tile does he need for $D'E'F'G'$?

The distance $AE' = 20$ ft, so the scale factor from DEFG to D'E'F'G' is $\frac{20}{15} = \frac{4}{3}$.
The areas of similar figures are related by the square of the scale factor; therefore,

$$\text{Area}(D'E'F'G') = \left(\frac{4}{3}\right)^2 \text{Area}(DEFG)$$

$$\text{Area}(D'E'F'G') = \frac{16}{9}(144)$$

$$\text{Area}(D'E'F'G') = 256$$

The owner needs 256 square feet of tile for D'E'F'G'.

4. $ABCD$ is a parallelogram, with the vertices listed counterclockwise around the figure. Points M, N, O and P are the midpoints of sides $\overline{AB}$, $\overline{BC}$, $\overline{CD}$, and $\overline{DA}$, respectively. The segments MO and NP cut the parallelogram into four smaller parallelograms, with the point W in the center of $ABCD$ as a common vertex.

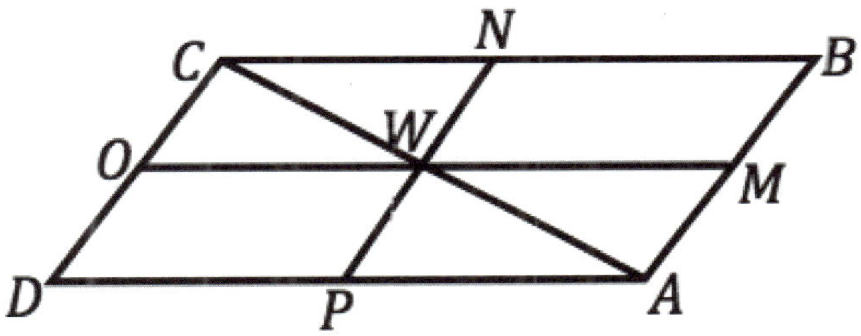

a. Exhibit a sequence of similarity transformations that takes $\triangle AMW$ to $\triangle CDA$. Be specific in describing the parameter of each transformation; for example, if describing a reflection, state the line of reflection.

Answers will vary (e.g., 180° rotation about W; dilation with center of dilation C).

b. Given the correspondence in $\triangle AMW$ similar to $\triangle CDA$, list all corresponding pairs of angles and corresponding pairs of sides. What is the ratio of the corresponding pairs of angles? What is the ratio of the corresponding pairs of sides?

$\angle MAW$ corresponds to $\angle DCA$; $\angle AMW$ corresponds to $\angle CDA$; $\angle AWM$ corresponds to $\angle CAD$. $\overline{AM}$ corresponds to $\overline{CD}$; $\overline{MW}$ corresponds to $\overline{DA}$; $\overline{WA}$ corresponds to $\overline{AC}$.
The ratio of corresponding pairs of angles is 1: 1.
The ratio of corresponding pairs of sides is 1: 2.

5. Given two triangles, ABC and DEF, $m\angle CAB = m\angle FDE$, and $m\angle CBA = m\angle FED$. Points A, B, D, and E lie on line l as shown. Describe a sequence of rigid motions and/or dilations to show that $\triangle ABC \sim \triangle DEF$, and sketch an image of the triangles after each transformation.

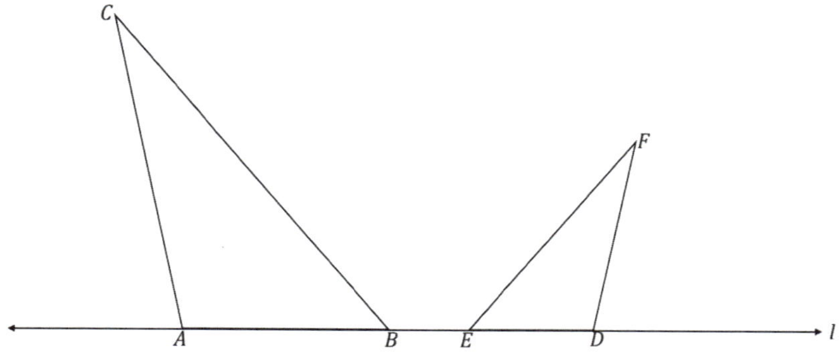

Reflect $\triangle DEF$ over the perpendicular bisector of $\overline{BE}$. The reflection takes E to B and D to a point on line AB. Since angle measures are preserved in rigid motions, F must map to a point on $\overline{BC}$.

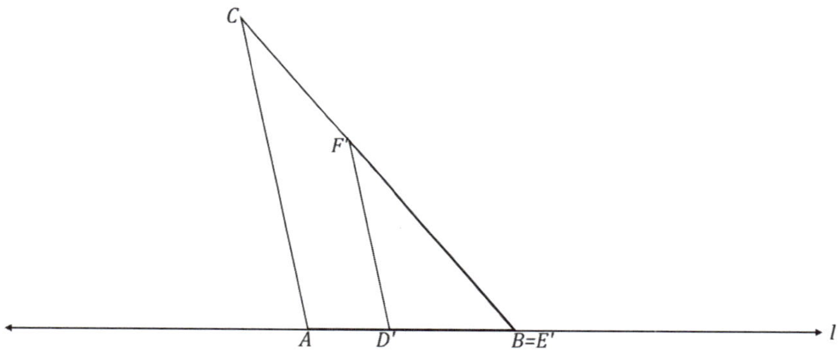

- By the hypothesis, $m\angle A = m\angle D$; therefore, $\overline{D'F'} \parallel \overline{AC}$ since corresponding angles are equal in measure.
- Since dilations map a segment to a parallel line segment, dilate $\triangle D'E'F'$ about E' and by scale factor $r = \frac{BA}{BD'}$ and that sends D' to A.
- By the dilation theorem, F' goes to C.

6. $\triangle JKL$ is a right triangle; $\overline{NP} \perp \overline{KL}$, $\overline{NO} \perp \overline{JK}$, $\overline{MN} \parallel \overline{OP}$.

a. List all sets of similar triangles. Explain how you know.

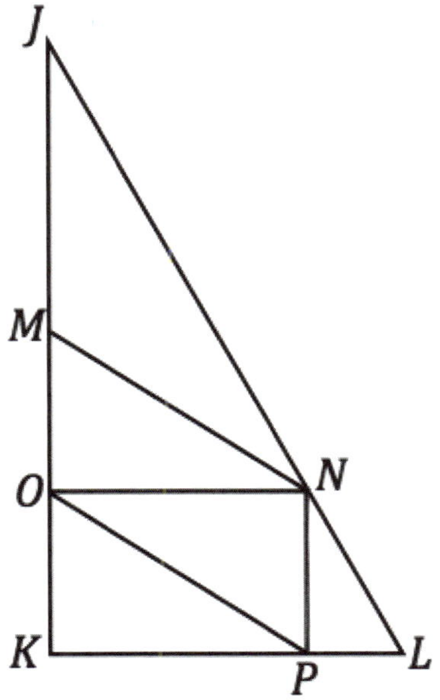

Set 1

- $\triangle MNO$
- $\triangle PON$
- $\triangle OPK$

Set 2

- $\triangle JON$
- $\triangle JKL$
- $\triangle NPL$

The triangles are similar because of the AA criterion.

b. Select any two similar triangles, and show why they are similar.

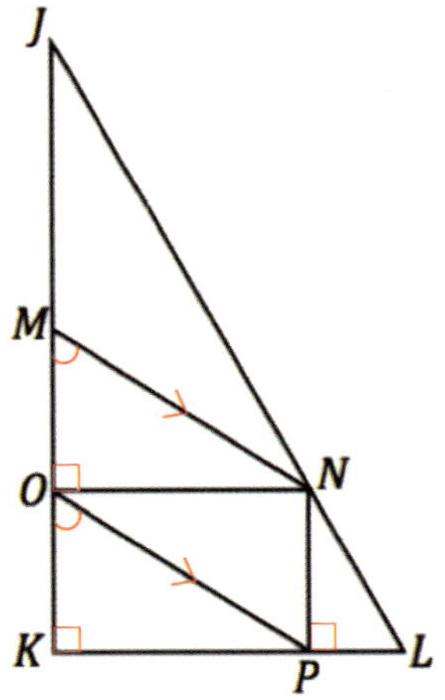

Possible response:

$\triangle MNO \sim \triangle PON$ by the AA criterion.

$\angle K$ is a right angle since $\triangle JKL$ is a right triangle.

$\angle MON$ is a right angle since $\overline{NO} \perp \overline{JK}$.

$m\angle NMO = m\angle POK$ since $\overline{MN} \parallel \overline{OP}$, and $\overline{JK}$ is a transversal that intersects $\overline{MN}$ and $\overline{OP}$; corresponding $\angle$'s are equal in measure.

Therefore, by the AA criterion, $\triangle MNO \sim \triangle PON$.

7.

a. The line PQ contains point O. What happens to $\overleftrightarrow{PQ}$ with a dilation about O and scale factor of $r = 2$? Explain your answer.

- *Since the points P and Q are collinear with the center O, then by definition of a dilation, both P′ and Q′ will also be collinear with the center O.*
- *The line PQ maps to itself.*

b. The line $\overleftrightarrow{PQ}$ does not contain point O. What happens to $\overleftrightarrow{PQ}$ with a dilation about O and scale factor of $r = 2$?

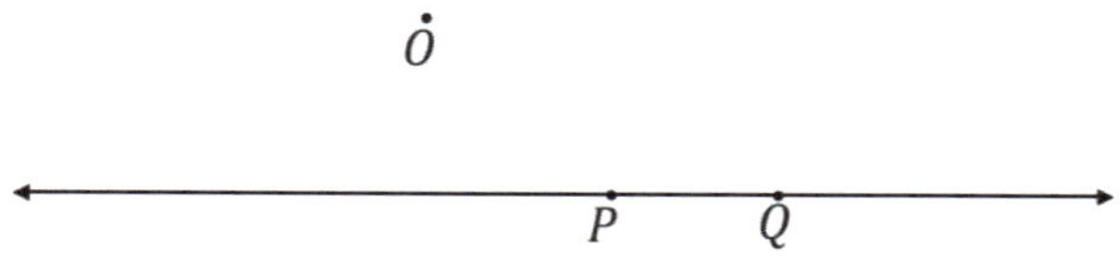

The line PQ maps to a parallel line P′Q′.

8. Use the diagram below to answer the following questions.

   a. State the pair of similar triangles. Which similarity criterion guarantees their similarity?

   $\triangle DEF \sim \triangle FGH$
   SAS

   b. Calculate DE to the hundredths place.

   $\frac{DE}{FG} = \frac{DF}{FH}$

   $\frac{DE}{6} = \frac{17}{8}$

   $8DE = 102$

   $DE \approx 12.75$

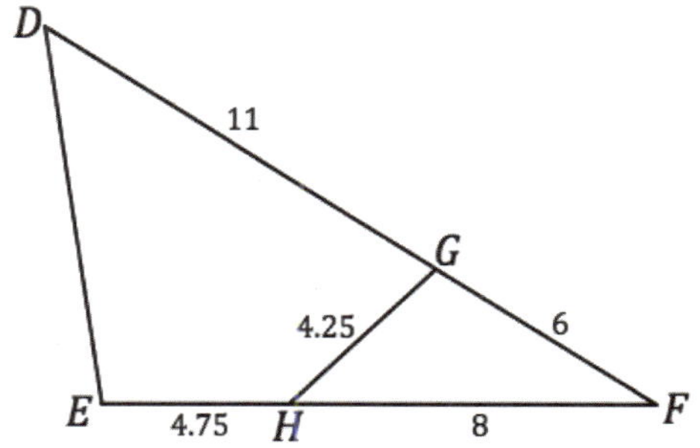

9. In $\triangle ABC$, $m\angle A$ is $40°$, $m\angle B$ is $60°$, and $m\angle C$ is $80°$. The triangle is dilated by a factor of 2 about point P to form $\triangle A'B'C'$. It is also dilated by a factor of 3 about point Q to form $\triangle A''B''C''$. What is the measure of the angle formed by line $A'B'$ and line $B''C''$? Explain how you know.

$60°$. $\overline{B'C'}$ and $\overline{A'B'}$ meet to form a $60°$ angle. Since dilations map segments to parallel segments, $\overline{B'C'} \parallel \overline{B''C''}$. Then the angle formed by lines $A'B'$ and $B''C''$ is a corresponding angle to $\angle B$ and has a measure of $60°$.

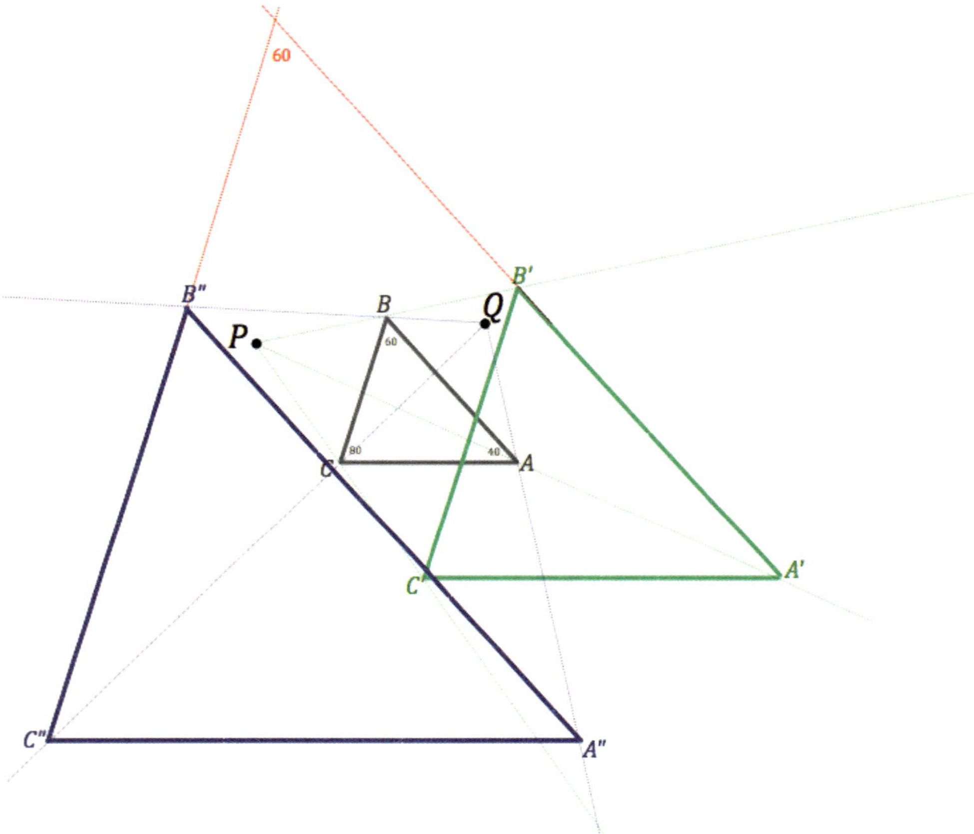

10. In the diagram below, $AC = CE = EG$, and $\angle BAC$, $\angle DCE$, and $\angle FEG$ are right. The two lines meet at a point to the right. Are the triangles similar? Why or why not?

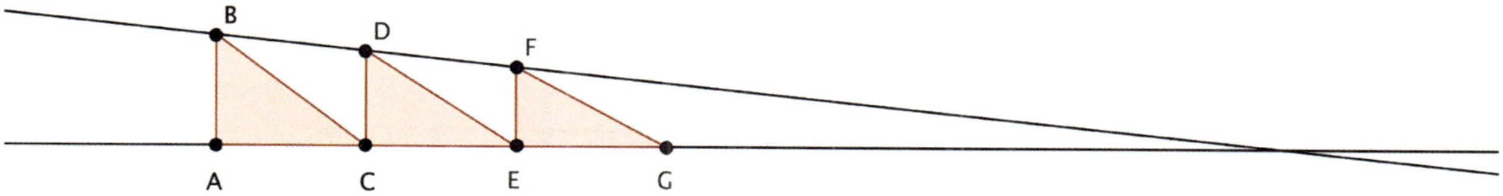

The triangles are not similar.

If they were similar, then there would have to be a similarity transformation taking A to C, B to D, and C to E. But then the dilation factor for this transformation would have to be 1, since AC = CE.

But since the dilation factor is 1 and $\overline{CD}$ is the image of $\overline{AB}$, then it must be true that AB = CD. Additionally, we know that $\overline{AB}$ is parallel to $\overline{CD}$, and since ∠BAC and ∠DCE are right, this implies ACDB is a rectangle. This implies that $\overline{BD}$ is parallel to $\overline{AC}$, but this is contrary to the given.

11. The side lengths of the following right triangle are 16, 30, and 34. An altitude of a right triangle from the right angle splits the hypotenuse into line segments of length x and y.

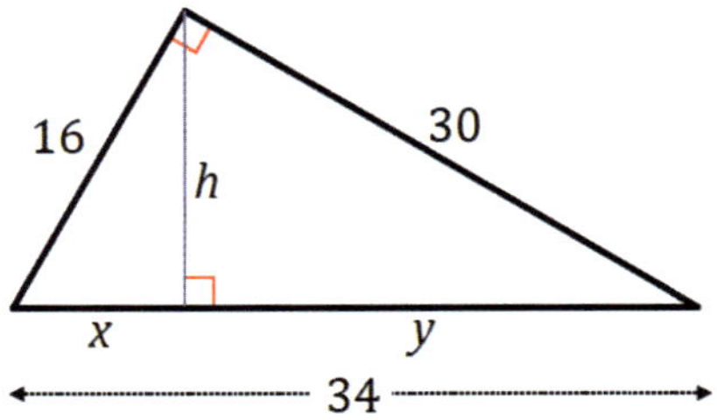

a. What is the relationship between the large triangle and the two sub-triangles? Why?

An altitude drawn from the vertex of the right angle of a right triangle to the hypotenuse divides the right triangle into two sub-triangles that are similar to the original triangle by the AA criterion.

b. Solve for h, x, and y.

$$\frac{h}{30} = \frac{16}{34} \qquad h = \frac{240}{17}$$

$$\frac{x}{16} = \frac{16}{34} \qquad x = \frac{128}{17}$$

$$\frac{y}{30} = \frac{30}{34} \qquad y = \frac{450}{17}$$

c. **Extension.** Find an expression that gives h in terms of x and y.

The large triangle is similar to both sub-triangles, and both sub-triangles are, therefore, similar. Then

$$\frac{x}{h} = \frac{h}{y}$$

$$h^2 = xy$$

$$h = \sqrt{xy}.$$

12. The sentence below, as shown, is being printed on a large banner for a birthday party. The height of the banner is 18 inches. There must be a minimum 1-inch margin on all sides of the banner. Use the dimensions in the image below to answer each question.

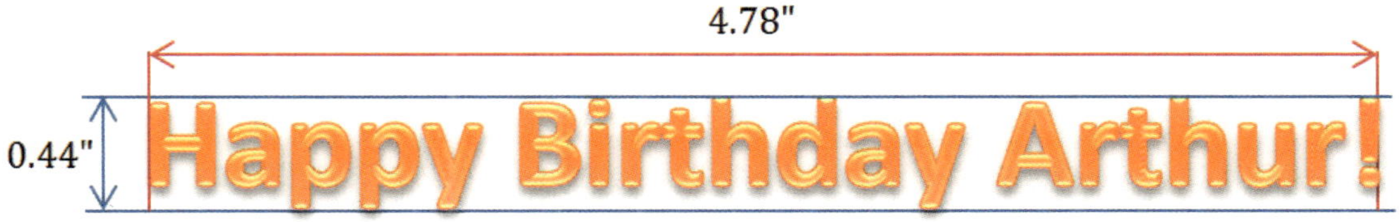

a. Describe a reasonable figure in the plane to model the printed image.

The sentence can be modeled as a rectangle with dimensions 4.78" × 0.44".

b. Find the scale factor that maximizes the size of the characters within the given constraints.

The scaled height of the rectangle cannot exceed 16" to allow for 1" margins above and below on the banner.

16 = k(0.44), where k represents the scale factor of the banner;

$$k = \frac{16}{0.44} = \frac{400}{11} = 36\frac{4}{11}$$

c. What is the total length of the banner based on your answer to part (a)?

Using the scale factor from part (a),

$$y = \left(\frac{400}{11}\right)\cdot(4.78)$$

$$y = 173\frac{9}{11}$$

The total length of the image in the banner is $173\frac{9}{11}$ inches; however, the banner must have a minimum 1" margin on all sides of the image, so the banner must be at least 2" longer. The total length of the banner must be at least $175\frac{9}{11}$ inches.

Topic D

Applying Similarity to Right Triangles

G-SRT.B.4

Focus Standard:	G-SRT.B.4	Prove theorems about triangles. *Theorems include: a line parallel to one side of a triangle divides the other two proportionally, and conversely; the Pythagorean Theorem proved using triangle similarity.*
Instructional Days:	4	
Lesson 21:	Special Relationships Within Right Triangles—Dividing into Two Similar Sub-Triangles (P)[1]	
Lesson 22:	Multiplying and Dividing Expressions with Radicals (P)	
Lesson 23:	Adding and Subtracting Expressions with Radicals (P)	
Lesson 24:	Proving the Pythagorean Theorem Using Similarity (E)	

In Topic D, students use their understanding of similarity and focus on right triangles as a lead up to trigonometry. In Lesson 21, students use the AA criterion to show how an altitude drawn from the vertex of the right angle of a right triangle to the hypotenuse creates two right triangles similar to the original right triangle. Students examine how the ratios within the three similar right triangles can be used to find unknown side lengths. Work with lengths in right triangles lends itself to expressions with radicals. In Lessons 22 and 23, students learn to rationalize fractions with radical expressions in the denominator and also to simplify, add, and subtract radical expressions. In the final lesson of Topic D, students use the relationships created by an altitude to the hypotenuse of a right triangle to prove the Pythagorean theorem.

[1]Lesson Structure Key: **P**-Problem Set Lesson, **M**-Modeling Cycle Lesson, **E**-Exploration Lesson, **S**-Socratic Lesson

Lesson 21: Special Relationships Within Right Triangles—Dividing into Two Similar Sub-Triangles

Student Outcomes

- Students understand that the altitude of a right triangle from the vertex of the right angle to the hypotenuse divides the triangle into two similar right triangles that are also similar to the original right triangle.
- Students complete a table of ratios for the corresponding sides of the similar triangles that are the result of dividing a right triangle into two similar sub-triangles.

Lesson Notes

This lesson serves as a foundational piece for understanding trigonometric ratios related to right triangles. The goal of the lesson is to show students how ratios within figures can be used to find lengths of sides in another triangle when those triangles are known to be similar.

Classwork

Opening Exercise (5 minutes)

Opening Exercise

Use the diagram below to complete parts (a)–(c).

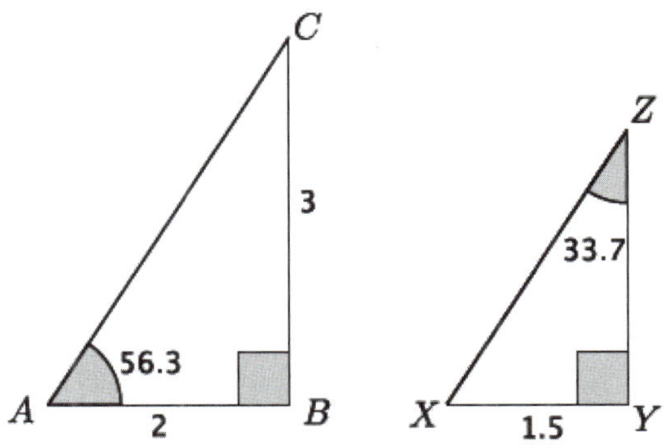

a. **Are the triangles shown above similar? Explain.**

Yes, the triangles are similar by the AA criterion. Both triangles have a right angle, and $m\angle A = m\angle X$ *and* $m\angle C = m\angle Z$.

b. **Determine the unknown lengths of the triangles.**

Let x represent the length of $\overline{YZ}$.

$$\frac{3}{2}=\frac{x}{1.5}$$
$$2x=4.5$$
$$x=2.25$$

Let y be the length of the hypotenuse of $\triangle ABC$.

$$2^2+3^2=y^2$$
$$4+9=y^2$$
$$13=y^2$$
$$\sqrt{13}=y$$

Let z be the length of the hypotenuse of $\triangle XYZ$.

$$1.5^2+2.25^2=z^2$$
$$2.25+5.0625=z^2$$
$$7.3125=z^2$$
$$\sqrt{7.3125}=z$$

c. **Explain how you found the lengths in part (a).**

Since the triangles are similar, I used the values of the ratios of the corresponding side lengths to determine the length of $\overline{YZ}$. To determine the lengths of $\overline{AC}$ and $\overline{XZ}$, I used the Pythagorean theorem.

Example 1 (15 minutes)

In Example 1, students learn that when a perpendicular is drawn from the right angle to the hypotenuse of a right triangle, the triangle is divided into two sub-triangles. Further, students show that all three of the triangles, the original one and the two formed by the perpendicular, are similar.

Scaffolding:

- A good hands-on visual that can be used here requires a 3×5 notecard. Have students draw the diagonal and then draw the perpendicular line from B to side AC.

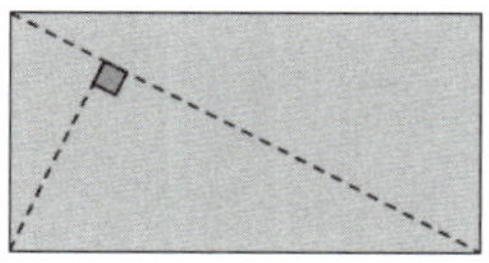

- Make sure students label all of the parts to match the triangles before they make the cuts. Next, have students cut out the three triangles. Students then have a notecard version of the three triangles shown and are better able to see the relationships among them.

Example 1

Recall that an altitude of a triangle is a perpendicular line segment from a vertex to the line determined by the opposite side. In $\triangle ABC$ below, $\overline{BD}$ is the altitude from vertex B to the line containing $\overline{AC}$.

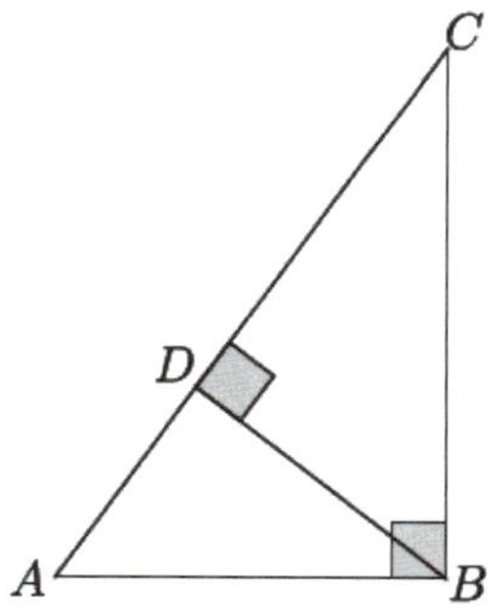

a. **How many triangles do you see in the figure?**

There are three triangles in the figure.

b. **Identify the three triangles by name.**

Note that there are many ways to name the three triangles. Ensure that the names students give show corresponding angles.

$\triangle ABC$, $\triangle ADB$, and $\triangle BDC$.

We want to consider the altitude of a right triangle from the right angle to the hypotenuse. The altitude of a right triangle splits the triangle into two right triangles, each of which shares a common acute angle with the original triangle. In $\triangle ABC$, the altitude $\overline{BD}$ divides the right triangle into two sub-triangles, $\triangle BDC$ and $\triangle ADB$.

c. **Is $\triangle ABC \sim \triangle BDC$? Is $\triangle ABC \sim \triangle ADB$? Explain.**

$\triangle ABC$ and $\triangle BDC$ are similar by the AA criterion. Each has a right angle, and each shares $\angle C$. $\triangle ABC$ and $\triangle ADB$ are similar because each has a right angle, and each shares $\angle A$, so, again, these triangles are similar by the AA criterion.

d. **Is $\triangle ABC \sim \triangle DBC$? Explain.**

$\triangle ABC$ and $\triangle DBC$ are not similar, because their corresponding angles, under the given correspondence of vertices, do not have equal measure.

e. **Since $\triangle ABC \sim \triangle BDC$ and $\triangle ABC \sim \triangle ADB$, can we conclude that $\triangle BDC \sim \triangle ADB$? Explain.**

Since similarity is transitive, $\triangle ABC \sim \triangle BDC$ and $\triangle ABC \sim \triangle ADB$ implies that $\triangle ABC \sim \triangle BDC \sim \triangle ADB$.

f. **Identify the altitude drawn in $\triangle EFG$.**

$\overline{GH}$ is the altitude from vertex G to the line containing $\overline{EF}$.

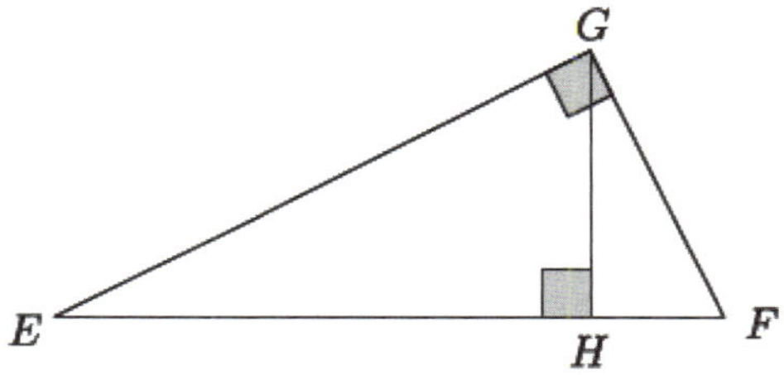

g. **As before, the altitude divides the triangle into two sub-triangles, resulting in a total of three triangles including the given triangle. Identify them by name so that the corresponding angles match up.**

$\triangle EFG$, $\triangle GFH$, and $\triangle EGH$

h. **Does the altitude divide $\triangle EFG$ into two similar sub-triangles as the altitude did with $\triangle ABC$?**

Yes.

Allow students time to investigate whether the triangles are similar. Students should conclude that $\triangle EFG \sim \triangle GFH \sim \triangle EGH$ using the same reasoning as before, that is, the AA criterion and the fact that similarity is transitive.

The fact that the altitude drawn from the right angle of a right triangle divides the triangle into two similar sub-triangles, which are also similar to the original triangle, allows us to determine the unknown lengths of right triangles.

Example 2 (15 minutes)

In this example, students use ratios within figures to determine unknown side lengths of triangles.

Example 2

Consider the right triangle $\triangle ABC$ below.

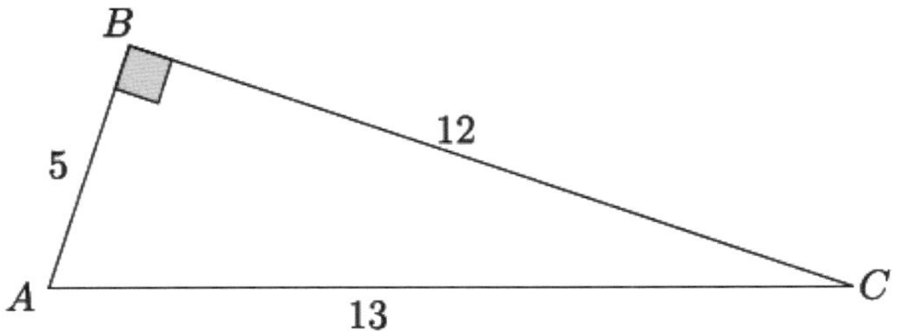

Draw the altitude $\overline{BD}$ from vertex B to the line containing $\overline{AC}$. Label $\overline{AD}$ as x, $\overline{DC}$ as y, and $\overline{BD}$ as z.

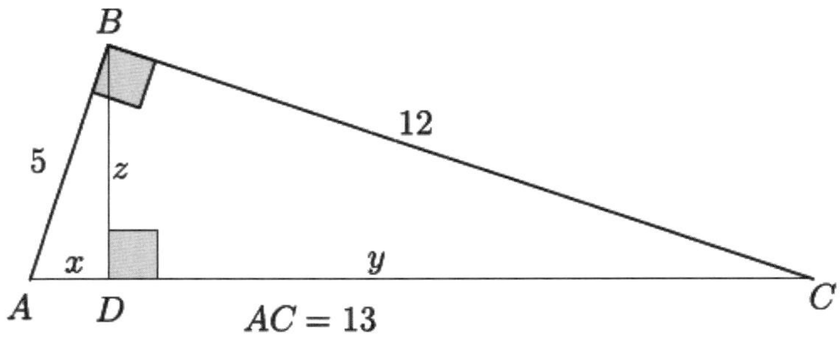

Find the values of x, y, and z.

Scaffolding:
It may be helpful for students to use the cutouts from Example 1. Have students label the side lengths on the front and back of the notecard according to this diagram and then use the cutouts to complete the table of ratios.

MP.1 Provide students time to find the values of x, y, and z. Allow students to use any reasonable strategy to complete the task. The suggested time allotment for this part of the example is 5 minutes. Next, have students briefly share their solutions and explanations for finding the lengths $x = 1\frac{12}{13}$, $y = 11\frac{1}{13}$, and $z = 4\frac{8}{13}$. For example, students may first use what they know about similar triangles and corresponding side lengths having equal ratios to determine x, then use the equation $x + y = 13$ to determine the value of y, and finally use the Pythagorean theorem to determine the length of z.

Now we will look at a different strategy for determining the lengths of x, y, and z. The strategy requires that we complete a table of ratios that compares different parts of each triangle.

Students may struggle with the initial task of finding the values of x, y, and z. Encourage them by letting them know that they have all the necessary tools to find these values. When transitioning to the use of ratios to find values, explain any method that yields acceptable answers. We are simply looking to add another tool to the toolbox of strategies that apply in this situation.

Provide students a moment to complete the ratios related to $\triangle ABC$.

Make a table of ratios for each triangle that relates the sides listed in the column headers.

	shorter leg: hypotenuse	longer leg: hypotenuse	shorter leg: longer leg
$\triangle ABC$	$5:13$	$12:13$	$5:12$

MP.2 & MP.7

Ensure that students have written the correct ratios before moving on to complete the table of ratios for $\triangle ADB$ and $\triangle CDB$.

	shorter leg: hypotenuse	longer leg: hypotenuse	shorter leg: longer leg
$\triangle ADB$	$x:5$	$z:5$	$x:z$
$\triangle CDB$	$z:12$	$y:12$	$z:y$

Our work in Example 1 showed us that $\triangle ABC \sim \triangle ADB \sim \triangle CDB$. Since the triangles are similar, the ratios of their corresponding sides are equal. For example, we can find the length of x by equating the values of shorter leg: hypotenuse ratios of $\triangle ABC$ and $\triangle ADB$.

$$\frac{x}{5} = \frac{5}{13}$$
$$13x = 25$$
$$x = \frac{25}{13} = 1\frac{12}{13}$$

Why can we use these ratios to determine the length of x?

We can use these ratios because the triangles are similar. Similar triangles have ratios of corresponding sides that are equal. We also know that we can use ratios between figures or within figures. The ratios used were within-figure ratios.

Which ratios can we use to determine the length of y?

To determine the value of y, we can equate the values longer leg: hypotenuse ratios for $\triangle CDB$ and $\triangle ABC$.

$$\frac{y}{12} = \frac{12}{13}$$
$$13y = 144$$
$$y = \frac{144}{13} = 11\frac{1}{13}$$

Use ratios to determine the length of z.

Students have several options of ratios to determine the length of z. As students work, identify those students using different ratios, and ask them to share their work with the class.

To determine the value of z, we can equate the values of longer leg: hypotenuse ratios for $\triangle ADB$ and $\triangle ABC$:

$$\frac{z}{5} = \frac{12}{13}$$
$$13z = 60$$
$$z = \frac{60}{13} = 4\frac{8}{13}$$

To determine the value of z, we can equate the values of shorter leg: hypotenuse ratios for $\triangle CDB$ and $\triangle ABC$:

$$\frac{z}{12} = \frac{5}{13}$$
$$13z = 60$$
$$z = \frac{60}{13} = 4\frac{8}{13}$$

To determine the value of z, we can equate the values of shorter leg: longer leg ratios for $\triangle ADB$ and $\triangle ABC$:

$$\frac{\frac{25}{13}}{z} = \frac{5}{12}$$
$$5z = \frac{25}{13}(12)$$
$$z = \frac{25}{13}\left(\frac{12}{5}\right)$$
$$z = \frac{60}{13} = 4\frac{8}{13}$$

To determine the value of z, we can equate the values of shorter leg: longer leg ratios for $\triangle CDB$ and $\triangle ABC$:

$$\frac{z}{\frac{144}{13}} = \frac{5}{12}$$
$$12z = \frac{144}{13}(5)$$
$$z = \frac{144}{13}\left(\frac{5}{12}\right)$$
$$z = \frac{60}{13} = 4\frac{8}{13}$$

Since corresponding ratios within similar triangles are equal, we can solve for any unknown side length by equating the values of the corresponding ratios. In the coming lessons, we will learn about more useful ratios for determining unknown side lengths of right triangles.

Closing (5 minutes)

Ask students the following questions. Students may respond in writing, to a partner, or to the whole class.

- What is an altitude, and what happens when an altitude is drawn from the right angle of a right triangle?
  - *An altitude is the perpendicular line segment from a vertex of a triangle to the line containing the opposite side. When an altitude is drawn from the right angle of a right triangle, then the triangle is divided into two similar sub-triangles.*
- What is the relationship between the original right triangle and the two similar sub-triangles?
  - *By the AA criterion and transitive property, we can show that all three triangles are similar.*
- Explain how to use the ratios of the similar right triangles to determine the unknown lengths of a triangle.

Note that we have used *shorter leg* and *longer leg* in the lesson and would expect students to do the same in responding to this prompt. It may be valuable to point out to students that an isosceles right triangle would not have a shorter or longer leg.

  - *Ratios of side lengths can be written using "shorter leg," "longer leg," and hypotenuse. The ratios of corresponding sides of similar triangles are equivalent and can be used to find unknown lengths of a triangle.*

Exit Ticket (5 minutes)

Name ____ Date ____________________

Lesson 21: Special Relationships Within Right Triangles—Dividing into Two Similar Sub-Triangles

Exit Ticket

Given $\triangle RST$, with altitude $\overline{SU}$ drawn to its hypotenuse, $ST = 15$, $RS = 36$, and $RT = 39$, answer the questions below.

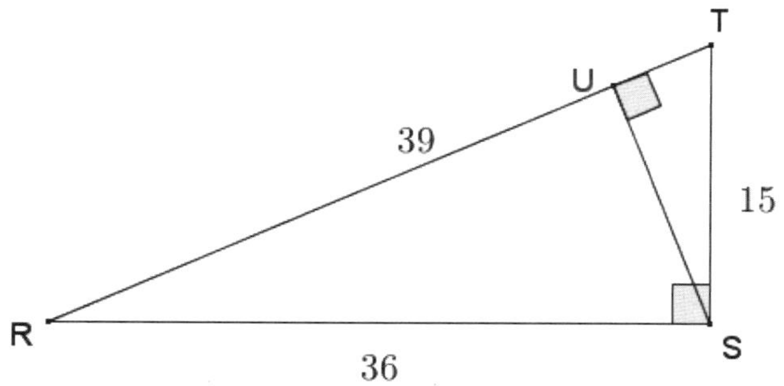

1. Complete the similarity statement relating the three triangles in the diagram.

$\triangle RST \sim \triangle$ ________ $\sim \triangle$ __________

2. Complete the table of ratios specified below.

	shorter leg: hypotenuse	**longer leg: hypotenuse**	**shorter leg: longer leg**
$\triangle \boldsymbol{RST}$			
$\triangle \boldsymbol{RSU}$			
$\triangle \boldsymbol{STU}$			

3. Use the values of the ratios you calculated to find the length of $\overline{SU}$.

Exit Ticket Sample Solutions

Given $\triangle RST$, with altitude $\overline{SU}$ drawn to its hypotenuse, $ST = 15$, $RS = 36$, and $RT = 39$, answer the questions below.

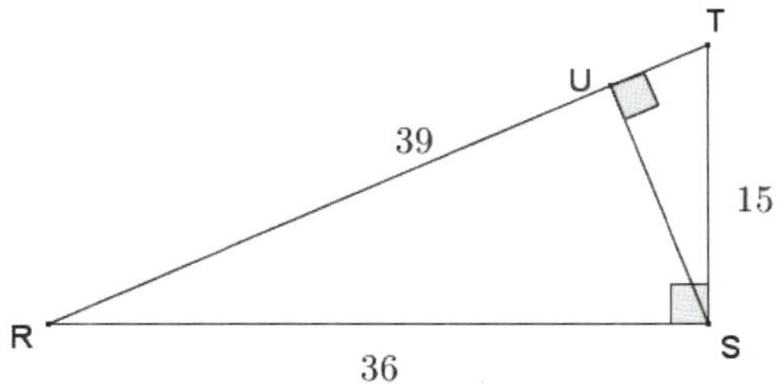

1. **Complete the similarity statement relating the three triangles in the diagram.**

$$\triangle RST \sim \triangle \underline{RUS} \sim \triangle \underline{SUT}$$

Using the right angles and shared angles, the triangles are similar by AA criterion. The transitive property may also be used.

2. **Complete the table of ratios specified below.**

	shorter leg: hypotenuse	longer leg: hypotenuse	shorter leg: longer leg
$\triangle RST$	$\frac{15}{39}$	$\frac{36}{39}$	$\frac{15}{36}$
$\triangle RSU$	$\frac{SU}{36}$	$\frac{RU}{36}$	$\frac{SU}{RU}$
$\triangle STU$	$\frac{TU}{15}$	$\frac{SU}{15}$	$\frac{TU}{SU}$

3. **Use the values of the ratios you calculated to find the length of $\overline{SU}$.**

$$\frac{15}{39} = \frac{SU}{36}$$
$$540 = 39(SU)$$
$$\frac{540}{39} = SU$$
$$13\frac{11}{13} = SU$$

Problem Set Sample Solutions

1. **Use similar triangles to find the length of the altitudes labeled with variables in each triangle below.**

   a.

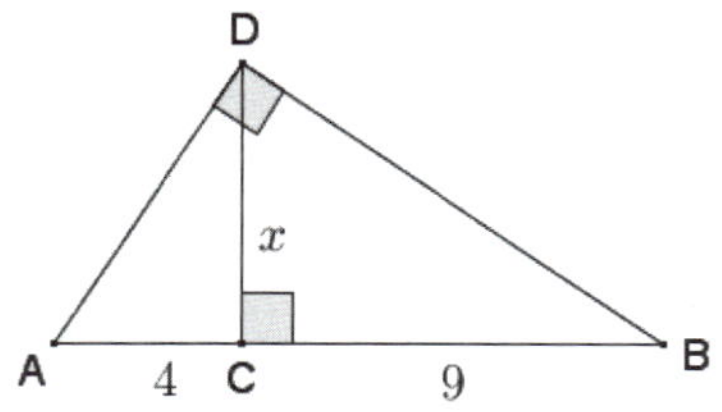

$\triangle ACD \sim \triangle DCB$ by AA criterion, so corresponding sides are proportional.

$$\frac{x}{4} = \frac{9}{x}$$
$$x^2 = 36$$
$$x = \sqrt{36} = 6$$

   b.

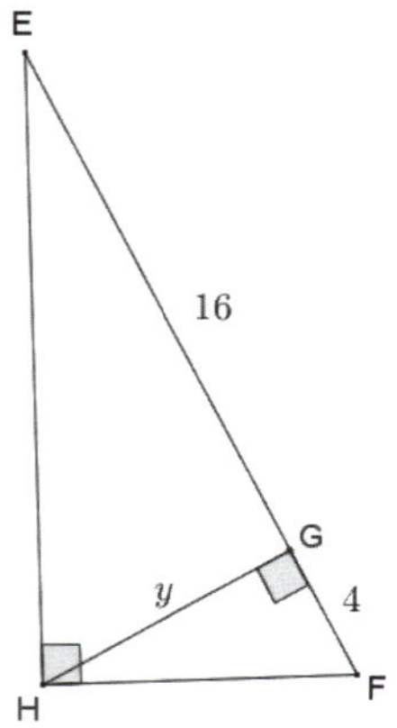

$\triangle GHE \sim \triangle GFH$ by AA criterion, so corresponding sides are proportional.

$$\frac{y}{4} = \frac{16}{y}$$
$$y^2 = 64$$
$$y = \sqrt{64} = 8$$

   c.

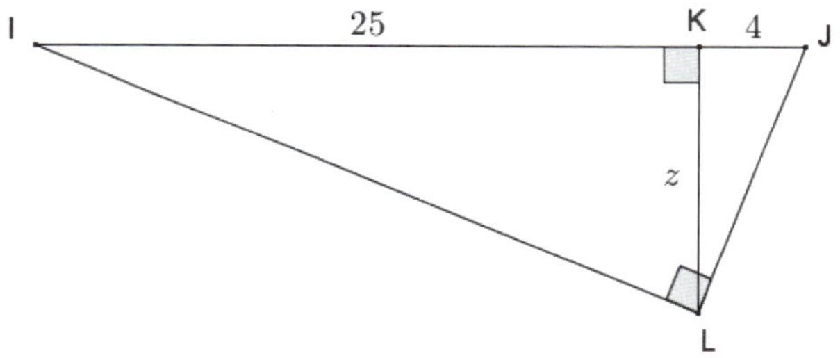

$\triangle LKI \sim \triangle JKL$ by AA criterion, so corresponding sides are proportional:

$$\frac{z}{25} = \frac{4}{z}$$
$$z^2 = 100$$
$$z = \sqrt{100} = 10$$

   d. **Describe the pattern that you see in your calculations for parts (a) through (c).**

   For each of the given right triangles, the length of the altitude drawn to its hypotenuse is equal to the square root of the product of the lengths of the pieces of the hypotenuse that it cuts.

2. **Given right triangle EFG with altitude $\overline{FH}$ drawn to the hypotenuse, find the lengths of $\overline{EH}$, $\overline{FH}$, and $\overline{GH}$.**

The altitude drawn from F to H cuts triangle EFG into two similar sub-triangles, providing the following correspondence:

$$\triangle EFG \sim \triangle EHF \sim \triangle FHG$$

Using the ratio shorter leg: hypotenuse for the similar triangles:

$$\frac{12}{20} = \frac{HF}{16}$$
$$192 = 20(HF)$$
$$\frac{192}{20} = HF$$
$$9\frac{12}{20} = 9\frac{3}{5} = HF$$

$$\frac{12}{20} = \frac{HE}{12}$$
$$144 = 20(HE)$$
$$\frac{144}{20} = HE$$
$$7\frac{4}{20} = 7\frac{1}{5} = HE$$

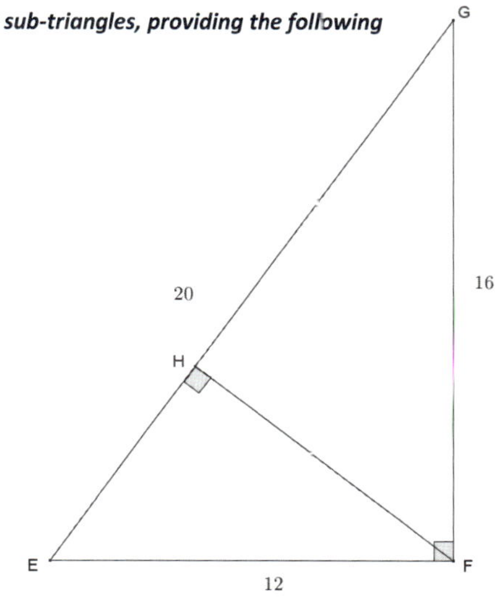

By addition:

$$EH + GH = EG$$
$$7\frac{1}{5} + GH = 20$$
$$GH = 12\frac{4}{5}$$

3. **Given triangle IMJ with altitude $\overline{JL}$, $JL = 32$, and $IL = 24$, find IJ, JM, LM, and IM.**

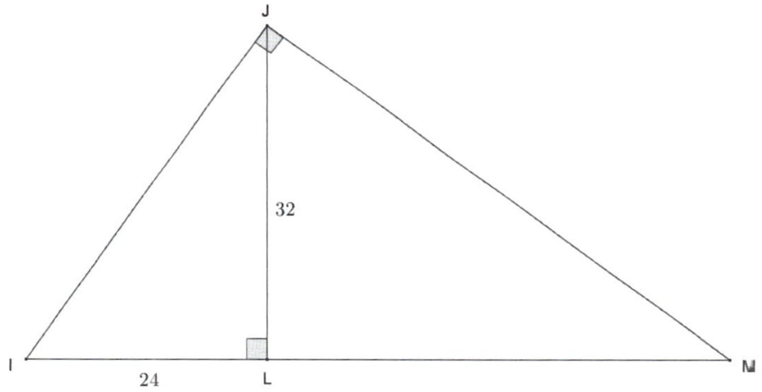

Altitude $\overline{JL}$ cuts $\triangle IMJ$ into two similar sub-triangles such that $\triangle IMJ \sim \triangle JML \sim \triangle IJL$.

By the Pythagorean theorem:

$$24^2 + 32^2 = IJ^2$$
$$576 + 1024 = IJ^2$$
$$1600 = IJ^2$$
$$\sqrt{1600} = IJ$$
$$40 = IJ$$

Using the ratio shorter leg: longer leg:

$$\frac{24}{32} = \frac{40}{JM}$$
$$24(JM) = 1280$$
$$JM = \frac{1280}{24}$$
$$JM = 53\frac{1}{3}$$

Using the ratio shorter leg: hypotenuse:

$$\frac{24}{40} = \frac{40}{IM}$$
$$24(IM) = 1600$$
$$IM = \frac{1600}{24}$$
$$IM = 66\frac{2}{3}$$

Using addition:

$$IL + LM = IM$$
$$24 + LM = 66\frac{2}{3}$$
$$LM = 42\frac{2}{3}$$

4. Given right triangle RST with altitude $\overline{RU}$ to its hypotenuse, $TU = 1\frac{24}{25}$, and $RU = 6\frac{18}{25}$, find the lengths of the sides of $\triangle RST$.

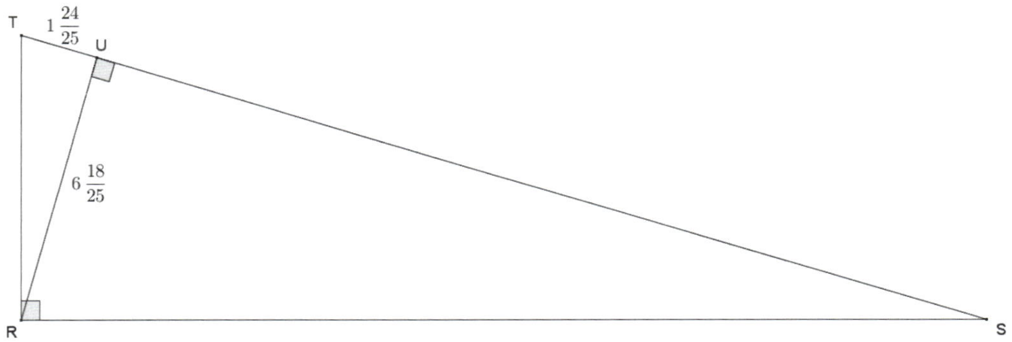

Altitude $\overline{RU}$ cuts $\triangle RST$ into similar sub-triangles: $\triangle URT \sim \triangle USR$.

Using the Pythagorean theorem:

$$TU^2 + RU^2 = TR^2$$
$$\left(1\frac{24}{25}\right)^2 + \left(6\frac{18}{25}\right)^2 = TR^2$$
$$\left(\frac{49}{25}\right)^2 + \left(\frac{168}{25}\right)^2 = TR^2$$
$$\frac{2401}{625} + \frac{28224}{625} = TR^2$$
$$\frac{30625}{625} = TR^2$$
$$49 = TR^2$$
$$\sqrt{49} = TR$$
$$7 = TR$$

Using the ratio shorter leg: hypotenuse:

$$\frac{\frac{49}{25}}{7} = \frac{7}{ST}$$
$$\frac{49}{25}(ST) = 49$$
$$ST = 25$$

Using the Pythagorean theorem:

$$RS^2 + RT^2 = ST^2$$
$$RS^2 + 7^2 = 25^2$$
$$RS^2 + 49 = 625$$
$$RS^2 = 576$$
$$RS = \sqrt{576} = 24$$

Note to the teacher: The next problem involves radical values that students have used previously; however, rationalizing is a focus of Lessons 23 and 24, so the solutions provided in this problem involve nonrationalized values.

5. **Given right triangle ABC with altitude $\overline{CD}$, find AD, BD, AB, and DC.**

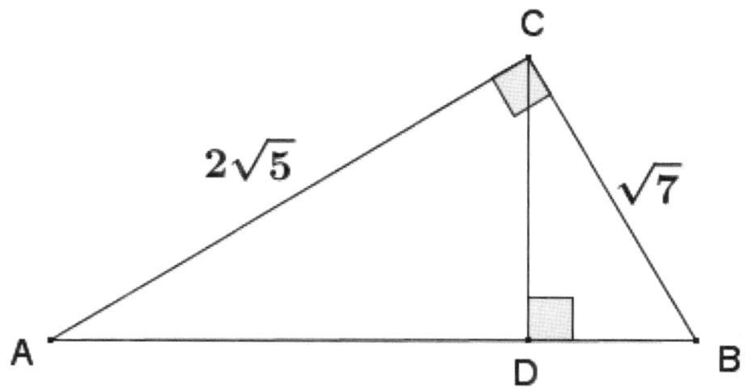

Using the Pythagorean theorem:

$$(2\sqrt{5})^2 + (\sqrt{7})^2 = AB^2$$
$$20 + 7 = AB^2$$
$$27 = AB^2$$
$$\sqrt{27} = AB$$
$$3\sqrt{3} = AB$$

An altitude from the right angle in a right triangle to the hypotenuse cuts the triangle into two similar right triangles: $\triangle ABC \sim \triangle ACD \sim \triangle CBD$.

***Using the ratio* shorter leg: hypotenuse:**

$$\frac{\sqrt{7}}{3\sqrt{3}} = \frac{DC}{2\sqrt{5}} \qquad \frac{\sqrt{7}}{3\sqrt{3}} = \frac{DB}{\sqrt{7}}$$
$$2\sqrt{35} = DC(3\sqrt{3}) \qquad 7 = DB(3\sqrt{3})$$
$$\frac{2\sqrt{35}}{3\sqrt{3}} = DC \qquad \frac{7}{3\sqrt{3}} = DB$$

Using the Pythagorean theorem:

$$AD^2 + DC^2 = AC^2$$
$$AD^2 + \left(\frac{2\sqrt{35}}{3\sqrt{3}}\right)^2 = (2\sqrt{5})^2$$
$$AD^2 + \frac{4 \cdot 35}{9 \cdot 3} = (4 \cdot 5)$$
$$AD^2 + \frac{140}{27} = 20$$
$$AD^2 = 14\frac{22}{27}$$
$$AD = \sqrt{14\frac{22}{27}}$$

6. **Right triangle DEC is inscribed in a circle with radius $AC = 5$. $\overline{DC}$ is a diameter of the circle, $\overline{EF}$ is an altitude of $\triangle DEC$, and $DE = 6$. Find the lengths x and y.**

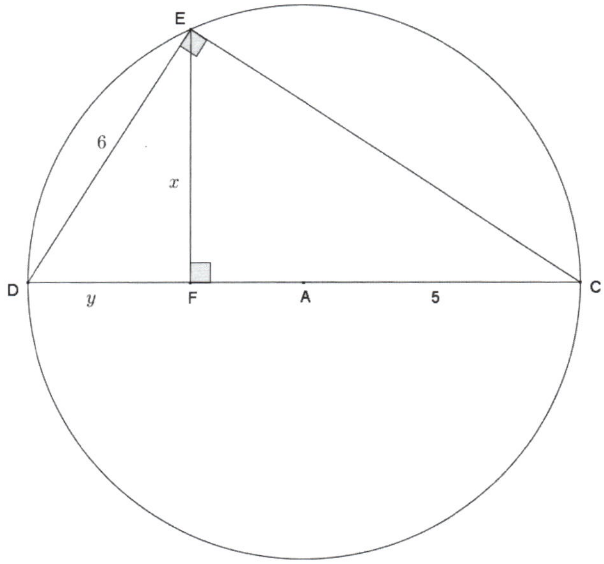

The radius of the circle is 5, and $DC = 2(5) = 10$.

By the Pythagorean theorem:

$$6^2 + EC^2 = 10^2$$
$$36 + EC^2 = 100$$
$$EC^2 = 64$$
$$EC = 8$$

($EC = -8$ is also a solution; however, since EC represents a distance, its value must be positive, so the solution $EC = -8$ is disregarded.)

We showed that an altitude from the right angle of a right triangle to the hypotenuse cuts the triangle into two similar sub-triangles, so $\triangle DEC \sim \triangle DFE \sim \triangle EFC$.

Using the ratio shorter leg: hypotenuse for similar right triangles:

$$\frac{6}{10} = \frac{y}{6} \qquad \frac{6}{10} = \frac{x}{8}$$
$$36 = 10y \qquad 48 = 10x$$
$$3.6 = y \qquad 4.8 = x$$

7. **In right triangle ABD, $AB = 53$, and altitude $DC = 14$. Find the lengths of $\overline{BC}$ and $\overline{AC}$.**

Let length $BC = x$. Then, $AC = 53 - x$.

Using the pattern from Problem 1,

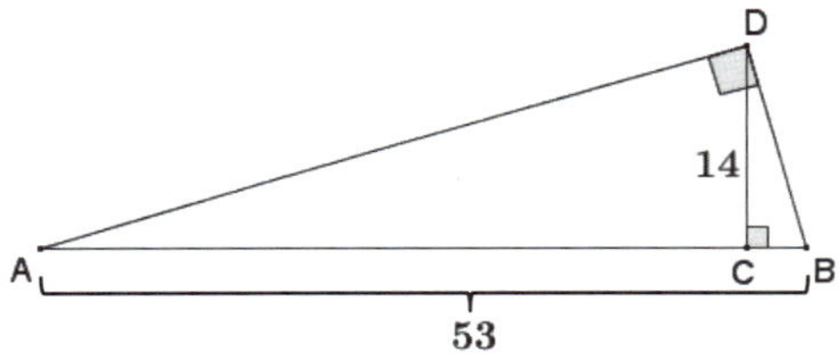

$$14^2 = x(53 - x)$$
$$196 = 53x - x^2$$
$$x^2 - 53x + 196 = 0$$
$$(x - 49)(x - 4) = 0$$
$$x = 49 \text{ or } x = 4.$$

Using the solutions from the equation and the given information, either $BC = 4$ and $AC = 49$, or $BC = 49$ and $AC = 4$.

Lesson 22: Multiplying and Dividing Expressions with Radicals

Student Outcomes

- Students multiply and divide expressions that contain radicals to simplify their answers.
- Students rationalize the denominator of a number expressed as a fraction.

Lesson Notes

Exercises 1–5 and the Discussion are meant to remind students of what they learned about roots in Grade 8 and Algebra I. In Grade 8, students learned the notation related to square roots and understood that the square root symbol automatically denotes the positive root (Grade 8, Module 7). In Algebra I, students used both the positive and negative roots of a number to find the location of the roots of a quadratic function. Because of the upcoming work with special triangles in this module, this lesson reviews what students learned about roots in Grade 8 Module 7 Lesson 4. For example, students need to understand that $\frac{1}{\sqrt{2}} \; \frac{\sqrt{2}}{2}$ when they are writing the trigonometric ratios of right triangles. To achieve this understanding, students must learn how to rationalize the denominator of numbers expressed as fractions. It is also important for students to get a sense of the value of a number. When a radical is in the denominator or is not simplified, it is more challenging to estimate its value, for example, $\sqrt{3750}$ compared to $25\sqrt{6}$.

For students who are struggling with the concepts of multiplying and dividing expressions with radicals, it may be necessary to divide the lesson so that multiplication is the focus one day and division the next. This lesson is a stepping-stone, as it moves students toward an understanding of how to rewrite expressions involving radical and rational exponents using the properties of exponents (**N.RN.A.2**), which are not mastered until Algebra II.

The lesson focuses on simplifying expressions and solving equations that contain terms with roots. By the end of the lesson, students should understand that one reason the denominator of a number expressed as a fraction is rationalized is to better estimate the value of the number. For example, students can more accurately estimate the value of $\frac{3}{\sqrt{3}}$ when written as $\frac{3\sqrt{3}}{3}$ or simply $\sqrt{3}$. Further, putting numbers in this form allows students to more easily recognize when numbers can be combined. For example, if adding $\sqrt{3}$ and $\frac{1}{\sqrt{3}}$, students may not recognize that they can be combined until $\frac{1}{\sqrt{3}}$ is rewritten as $\frac{\sqrt{3}}{3}$. Then, the sum of $\sqrt{3}$ and $\frac{\sqrt{3}}{3}$ is $\frac{4\sqrt{3}}{3}$. As a teacher, it is easier to check answers when there is an expected standard form such as a rationalized expression.

Classwork

Exercises 1–5 (8 minutes)

The first three exercises review square roots that are perfect squares. The last two exercises require students to compare the value of two radical expressions and make a conjecture about their relationship. These last two exercises exemplify what is studied in this lesson. Students may need to be reminded that the square root symbol automatically denotes the positive root of the number.

Scaffolding:

- Some students may need to review the perfect squares. A reproducible sheet for squares of numbers 1–30 is provided at the end of the lesson.
- Consider doing a fluency activity that allows students to learn their perfect squares up to 30. This may include choral recitation.
- English language learners may benefit from choral practice with the word *radical*.

Exercises 1–5

Simplify as much as possible.

1. $\sqrt{17^2} =$

   $\sqrt{17^2} = 17$

2. $\sqrt{5^{10}} =$

   $\sqrt{5^{10}} = \sqrt{5^2} \times \sqrt{5^2} \times \sqrt{5^2} \times \sqrt{5^2} \times \sqrt{5^2}$
   $= 5 \times 5 \times 5 \times 5 \times 5$
   $= 5^5$

3. $\sqrt{4x^4} =$

   $\sqrt{4x^4} = \sqrt{4} \times \sqrt{x^2} \times \sqrt{x^2}$
   $= 2 \times x \times x$
   $= 2|x|^2$

4. Complete parts (a) through (c).

   a. Compare the value of $\sqrt{36}$ to the value of $\sqrt{9} \times \sqrt{4}$.

   The value of the two expressions is equal. The square root of 36 is 6, and the product of the square roots of 9 and 4 is also 6.

   b. Make a conjecture about the validity of the following statement: For nonnegative real numbers a and b, $\sqrt{ab} = \sqrt{a} \cdot \sqrt{b}$. Explain.

   Answers will vary. Students should say that the statement $\sqrt{ab} = \sqrt{a} \cdot \sqrt{b}$ is valid because of the problem that was just completed: $\sqrt{36} = \sqrt{9} \times \sqrt{4} = 6$.

   c. Does your conjecture hold true for $a = -4$ and $b = -9$?

   No. The conjecture is not true when the numbers are negative because we cannot take the square root of a negative number. $\sqrt{(-4)(-9)} = \sqrt{36} = 6$, but we cannot calculate $\sqrt{-4} \times \sqrt{-9}$ in order to compare.

5. Complete parts (a) through (c).

   a. Compare the value of $\sqrt{\frac{100}{25}}$ to the value of $\frac{\sqrt{100}}{\sqrt{25}}$.

   The value of the two expressions is equal. The fraction $\frac{100}{25}$ simplifies to 4, and the square root of 4 is 2. The square root of 100 divided by the square root of 25 is equal to $\frac{10}{5}$, which is equal to 2.

b. **Make a conjecture about the validity of the following statement: For nonnegative real numbers a and b, when $b \neq 0$, $\sqrt{\frac{a}{b}} = \frac{\sqrt{a}}{\sqrt{b}}$. Explain.**

Answers will vary. Students should say that the statement $\sqrt{\frac{a}{b}} = \frac{\sqrt{a}}{\sqrt{b}}$ *is valid because of the problem that was just completed:* $\sqrt{\frac{100}{25}} = \frac{\sqrt{100}}{\sqrt{25}} = 2$.

c. **Does your conjecture hold true for $a = -100$ and $b = -25$?**

No. The conjecture is not true when the numbers are negative because we cannot take the square root of a negative number. $\sqrt{\frac{-100}{-25}} = \sqrt{4} = 2$, *but we cannot calculate* $\frac{\sqrt{-100}}{\sqrt{-25}}$ *in order to compare.*

Discussion (8 minutes)

Debrief Exercises 1–5 by reminding students of the definition for square root, the facts given by the definition, and the rules associated with square roots for positive radicands. Whenever possible, elicit the facts and definitions from students based on their work in Exercises 1–5. Within this Discussion is the important distinction between *a square root* of a number and *the square root* of a number. *A* square root of a number may be negative; however, *the* square root of a number always refers to the principle square root or the positive root of the number.

- Definition of *the square root*: If $x \geq 0$, then $\sqrt{x}$ is the nonnegative number p so that $p^2 = x$. This definition gives us four facts. The definition should not be confused with finding *a square root* of a number. For example, -2 is *a square root* of 4, but *the square root* of 4, that is, $\sqrt{4}$, is 2.

Consider asking students to give an example of each fact using concrete numbers. Sample responses are included below each fact.

Fact 1: $\sqrt{a^2} = a$ if $a \geq 0$

- $\sqrt{12^2} = 12$, *for any positive squared number in the radicand*

Fact 2: $\sqrt{a^2} = -a$ if $a < 0$

This may require additional explanation because students see the answer as "negative a," as opposed to the opposite of a. For this fact, it is assumed that a is a negative number; therefore, $-a$ is a positive number. It is similar to how students think about the absolute value of a number a: $|a| = a$ if $a > 0$, but $|a| = -a$ if $a < 0$. Simply put, the minus sign is just telling students they need to take the opposite of the negative number a to produce the desired result, that is, $-(-a) = a$.

- $\sqrt{(-5)^2} = 5$, *for any negative squared number in the radicand*

Fact 3: $\sqrt{a^2} = |a|$ for all real numbers a

- $\sqrt{13^2} = |13|$, *and* $\sqrt{(-13)^2} = |-13|$

Fact 4: $\sqrt{a^{2n}} = \sqrt{(a^n)^2} = a^n$ when a^n is nonnegative

- $\sqrt{7^{16}} = \sqrt{(7^8)^2} = 7^8$

Consider asking students which of the first five exercises used Rule 1.

- When $a \geq 0, b \geq 0$, and $c \geq 0$, then the following rules can be applied to square roots:

  Rule 1: $\sqrt{ab} = \sqrt{a} \cdot \sqrt{b}$

  A consequence of Rule 1 and the associative property gives us the following:

  $$\sqrt{abc} = \sqrt{a(bc)} = \sqrt{a} \cdot \sqrt{bc} = \sqrt{a} \cdot \sqrt{b} \cdot \sqrt{c}$$

  Rule 2: $\sqrt{\dfrac{a}{b}} = \dfrac{\sqrt{a}}{\sqrt{b}}$ when $b \neq 0$

  We want to show that $\sqrt{ab} = \sqrt{a} \cdot \sqrt{b}$ for $a \geq 0$ and $b \geq 0$. To do so, we use the definition of square root.

Consider allowing time for students to discuss with partners how they can prove Rule 1. Shown below are three proofs of Rule 1. Share one or all with the class.

The proof of Rule 1: Let p be the nonnegative number so that $p^2 = a$, and let q be the nonnegative number so that $q^2 = b$. Then, $\sqrt{ab} = pq$ because pq is nonnegative (it is the product of two nonnegative numbers), and $(pq)^2 = pqpq = p^2q^2 = ab$. Then, by definition, $\sqrt{ab} = pq = \sqrt{a} \cdot \sqrt{b}$. Since both sides equal pq, the equation is true.

The proof of Rule 1: Let $C = \sqrt{a} \cdot \sqrt{b}$, and let $D = \sqrt{ab}$. We need to show that $C = D$. Given positive numbers C, D, and exponent 2, if we can show that $C^2 = D^2$, then we know that $C = D$, and Rule 1 is proved.

Consider asking students why it is true that if they can show that $C^2 = D^2$, then they know that $C = D$. Students should refer to what they know about the definition of exponents. That is, since $C^2 = C \times C$ and $D^2 = D \times D$ and $C \times C = D \times D$, then C must be the same number as D.

- With that goal in mind, we take each of $C = \sqrt{a}\sqrt{b}$ and $D = \sqrt{ab}$, and by the multiplication property of equality, we raise both sides of each equation to a power of 2.

$$\begin{aligned} C^2 &= \left(\sqrt{a} \cdot \sqrt{b}\right)^2 \\ &= \left(\sqrt{a} \cdot \sqrt{b}\right) \cdot \left(\sqrt{a} \cdot \sqrt{b}\right) \\ &= \sqrt{a} \cdot \sqrt{a} \cdot \sqrt{b} \cdot \sqrt{b} \\ &= ab \end{aligned} \qquad \begin{aligned} D^2 &= \left(\sqrt{ab}\right)^2 \\ &= \sqrt{ab} \cdot \sqrt{ab} \\ &= ab \end{aligned}$$

Since $C^2 = D^2$ implies $C = D$, then $\sqrt{ab} = \sqrt{a}\sqrt{b}$.

The proof of Rule 1: Let $C, D > 0$. If $C^2 = D^2$, then $C = D$. Assume $C^2 = D^2$; then, $C^2 - D^2 = 0$. By factoring the difference of squares, we have $(C + D)(C - D) = 0$. Since both C and D are positive, then $C + D > 0$, which means that $C - D$ must be equal to zero because of the zero product property. Since $C - D = 0$, then $C = D$.

EUREKA MATH™

Example 1 (4 minutes)

- We can use Rule 1 to rationalize the denominators of fractional expressions. One reason we do this is so that we can better estimate the value of a number. For example, if we know that $\sqrt{2} \approx 1.414$, what is the value of $\frac{1}{\sqrt{2}}$? Isn't it is easier to determine the value of $\frac{\sqrt{2}}{2}$? The fractional expressions $\frac{1}{\sqrt{2}}$ and $\frac{\sqrt{2}}{2}$ are equivalent. Notice that the first expression has the irrational number $\sqrt{2}$ as its denominator, and the second expression has the rational number 2 as its denominator. What we will learn today is how to rationalize the denominator of a fractional expression using Rule 1.
- Another reason to rationalize the denominators of fractional expressions is because putting numbers in this form allows us to more easily recognize when numbers can be combined. For example, if you have to add $\sqrt{3}$ and $\frac{1}{\sqrt{3}}$, you may not recognize that they can be combined until $\frac{1}{\sqrt{3}}$ is rewritten as $\frac{\sqrt{3}}{3}$. Then, the sum of $\sqrt{3}$ and $\frac{\sqrt{3}}{3}$ is $\frac{4\sqrt{3}}{3}$.
- We want to express numbers in their *simplest radical form*. An expression is in its *simplest radical form* when the radicand (the expression under the radical sign) has no factor that can be raised to a power greater than or equal to the index (either 2 or 3), and there is no radical in the denominator.
- Using Rule 1 for square roots, we can simplify expressions that contain square roots by writing the factors of the number under the square root sign as products of perfect squares, if possible. For example, to simplify $\sqrt{75}$, we consider all of the factors of 75, focusing on those factors that are perfect squares. Which factors should we use?
  - *We should use 25 and 3 because 25 is a perfect square.*

  Then,

$$\begin{aligned}\sqrt{75} &= \sqrt{25} \cdot \sqrt{3}\\ &= 5\sqrt{3}\end{aligned}$$

> *Scaffolding:*
>
> Consider showing multiple simple examples. For example:
>
> $$\begin{aligned}\sqrt{28} &= \sqrt{4} \cdot \sqrt{7}\\ &= 2\sqrt{7}\end{aligned}$$
> $$\begin{aligned}\sqrt{45} &= \sqrt{9} \cdot \sqrt{5}\\ &= 3\sqrt{5}\end{aligned}$$
> $$\begin{aligned}\sqrt{32} &= \sqrt{16} \cdot \sqrt{2}\\ &= 4\sqrt{2}\end{aligned}$$

Example 2 (2 minutes)

In Example 2, we first use Rule 2 to rewrite a number as a rational expression and then use Rule 1 to rationalize a denominator, that is, rewrite the denominator as an integer. We have not yet proved this rule, because it is an exercise in the Problem Set. Consider mentioning this fact to students.

- Rules 1 and 2 for square roots are used to rationalize denominators of fractional expressions.

Consider asking students what it means to rationalize the denominator. Students should understand that rationalizing the denominator means expressing it as an integer.

- Consider the expression $\sqrt{\frac{3}{5}}$. By Rule 2, $\sqrt{\frac{3}{5}} = \frac{\sqrt{3}}{\sqrt{5}}$. We want to write an expression that is equivalent to $\frac{\sqrt{3}}{\sqrt{5}}$ with a rational number for the denominator.

$$\frac{\sqrt{3}}{\sqrt{5}} \times \frac{\sqrt{5}}{\sqrt{5}} = \frac{\sqrt{3}\sqrt{5}}{\sqrt{5}\sqrt{5}} \quad \text{By multiplication rule for fractional expressions}$$

$$= \frac{\sqrt{15}}{\sqrt{25}} \quad \text{By Rule 1}$$

$$= \frac{\sqrt{15}}{5}.$$

Example 3 (3 minutes)

MP.3

- Demarcus found the scale factor of a dilation to be $\frac{1}{\sqrt{2}}$. When he compared his answer to Yesenia's, which was $\frac{\sqrt{2}}{2}$, he told her that one of them must have made a mistake. Show work, and provide an explanation to Demarcus and Yesenia that proves they are both correct.
  - *Student work:*

$$\frac{1}{\sqrt{2}} \times \frac{\sqrt{2}}{\sqrt{2}} = \frac{1\sqrt{2}}{\sqrt{2}\sqrt{2}} \quad \text{By multiplication rule for fractional expressions}$$

$$= \frac{\sqrt{2}}{\sqrt{4}} \quad \text{By Rule 1}$$

$$= \frac{\sqrt{2}}{2} \quad \text{By definition of } \textit{square root}$$

If Demarcus were to rationalize the denominator of his answer, he would see that it is equal to Yesenia's answer. Therefore, they are both correct.

Example 4 (5 minutes)

- Assume $x > 0$. Rationalize the denominator of $\frac{x}{\sqrt{x^3}}$, and then simplify your answer as much as possible.

Provide time for students to work independently or in pairs. Use the question below if necessary. Seek out students who multiplied by different factors to produce an equivalent fractional expression to simplify this problem. For example, some students may have multiplied by $\frac{\sqrt{x^3}}{\sqrt{x^3}}$, while others may have used $\frac{\sqrt{x}}{\sqrt{x}}$ or some other fractional expression that would produce an exponent of x with an even number, which can be simplified. Have students share their work and compare their answers.

- We need to multiply $\sqrt{x^3}$ by a number so that it becomes a perfect square. What should we multiply by?

Students may say to multiply by $\sqrt{x^3}$ because that is what was done in the two previous examples. If so, finish the problem that way, and then show that we can multiply by $\sqrt{x}$ and get the same answer. Ask students why both methods work. They should mention equivalent expressions and the role that the number $\frac{\sqrt{x^3}}{\sqrt{x^3}}$ or $\frac{\sqrt{x}}{\sqrt{x}}$ plays in producing the equivalent expression.

- *Student work:*

$$\begin{aligned}\frac{x}{\sqrt{x^3}}\times\frac{\sqrt{x^3}}{\sqrt{x^3}}&=\frac{x\sqrt{x^3}}{\sqrt{x^3}\sqrt{x^3}}\\&=\frac{x\sqrt{x^3}}{\sqrt{x^6}}\\&=\frac{xx\sqrt{x}}{x^3}\\&=\frac{x^2\sqrt{x}}{x^3}\\&=\frac{\sqrt{x}}{x}\end{aligned}$$

$$\begin{aligned}\frac{x}{\sqrt{x^3}}\times\frac{\sqrt{x}}{\sqrt{x}}&=\frac{x\sqrt{x}}{\sqrt{x^3}\sqrt{x}}\\&=\frac{x\sqrt{x}}{\sqrt{x^4}}\\&=\frac{x\sqrt{x}}{x^2}\\&=\frac{\sqrt{x}}{x}\end{aligned}$$

Exercises 6–17 (7 minutes)

Have students work through all of the exercises in this set, or select problems for students to complete based on their level. Students who are struggling should complete Exercises 6–10. Students who are on level should complete Exercises 9–13. Students who are accelerated should complete Exercises 13–16. All students should attempt to complete Exercise 17.

Exercises 6–17

Simplify each expression as much as possible, and rationalize denominators when applicable.

6. $\sqrt{72} =$

$$\begin{aligned}\sqrt{72}&=\sqrt{36}\cdot\sqrt{2}\\&=6\sqrt{2}\end{aligned}$$

7. $\sqrt{\frac{17}{25}} =$

$$\begin{aligned}\sqrt{\frac{17}{25}}&=\frac{\sqrt{17}}{\sqrt{25}}\\&=\frac{\sqrt{17}}{5}\end{aligned}$$

8. $\sqrt{32x} =$

$\sqrt{32x} = \sqrt{16}\sqrt{2x}$
$= 4\sqrt{2x}$

9. $\sqrt{\frac{1}{3}} =$

$\sqrt{\frac{1}{3}} = \frac{\sqrt{1}}{\sqrt{3}}$
$= \frac{\sqrt{1}}{\sqrt{3}} \times \frac{\sqrt{3}}{\sqrt{3}}$
$= \frac{\sqrt{3}}{\sqrt{9}}$
$= \frac{\sqrt{3}}{3}$

10. $\sqrt{54x^2} =$

$\sqrt{54x^2} = \sqrt{9}\sqrt{6}\sqrt{x^2}$
$= 3|x|\sqrt{6}$

11. $\frac{\sqrt{36}}{\sqrt{18}} =$

$\frac{\sqrt{36}}{\sqrt{18}} = \sqrt{\frac{36}{18}}$
$= \sqrt{2}$

12. $\sqrt{\frac{4}{x^4}} =$

$\sqrt{\frac{4}{x^4}} = \frac{\sqrt{4}}{\sqrt{x^4}}$
$= \frac{2}{x^2}$

13. $\frac{4x}{\sqrt{64x^2}} =$

$\frac{4x}{\sqrt{64x^2}} = \frac{4x}{8x}$
$= \frac{1}{2}$

14. $\frac{5}{\sqrt{x^7}} =$

$\frac{5}{\sqrt{x^7}} = \frac{5}{\sqrt{x^7}} \times \frac{\sqrt{x}}{\sqrt{x}}$
$= \frac{5\sqrt{x}}{\sqrt{x^8}}$
$= \frac{5\sqrt{x}}{x^4}$

15. $\sqrt{\frac{x^5}{2}} =$

$\sqrt{\frac{x^5}{2}} = \frac{\sqrt{x^5}}{\sqrt{2}}$
$= \frac{x^2\sqrt{x}}{\sqrt{2}} \times \frac{\sqrt{2}}{\sqrt{2}}$
$= \frac{x^2\sqrt{2x}}{2}$

16. $\frac{\sqrt{18x}}{3\sqrt{x^5}} =$

$$\frac{\sqrt{18x}}{3\sqrt{x^5}} = \frac{\sqrt{9}\sqrt{2x}}{3x^2\sqrt{x}}$$
$$= \frac{3\sqrt{2x}}{3x^2\sqrt{x}}$$
$$= \frac{\sqrt{2x}}{x^2\sqrt{x}} \times \frac{\sqrt{x}}{\sqrt{x}}$$
$$= \frac{x\sqrt{2}}{x^3}$$
$$= \frac{\sqrt{2}}{x^2}$$

17. The captain of a ship recorded the ship's coordinates, then sailed north and then west, and then recorded the new coordinates. The coordinates were used to calculate the distance they traveled, $\sqrt{578}$ km. When the captain asked how far they traveled, the navigator said, "About 24 km." Is the navigator correct? Under what conditions might he need to be more precise in his answer?

Sample student responses:

The number $\sqrt{578}$ is close to the perfect square $\sqrt{576}$. The perfect square $\sqrt{576} = 24$; therefore, the navigator is correct in his estimate of the distance traveled.

When the number $\sqrt{578}$ is simplified, the result is $17\sqrt{2}$. The number 578 has factors of 289 and 2. Then:

$$\sqrt{578} = \sqrt{289}\sqrt{2}$$
$$= 17\sqrt{2}$$
$$= 24.04163 \ldots$$
$$\approx 24$$

Yes, the navigator is correct in his estimate of the distance traveled.

A more precise answer may be needed if the captain were looking for a particular location, such as the location of a shipwreck or buried treasure.

Closing (4 minutes)

Ask the following questions. Have students respond in writing, to a partner, or to the whole class.

- What are some of the basic facts and rules related to square roots?
  - *The basic facts about square roots show us how to simplify a square root when it is a perfect square. For example,* $\sqrt{5^2} = 5$*,* $\sqrt{(-5)^2} = 5$*,* $\sqrt{(-5)^2} = |-5|$*, and* $\sqrt{5^{12}} = \sqrt{(5^6)^2} = 5^6$*. The rules allow us to simplify square roots. Rule 1 shows that we can rewrite radicands as factors and simplify the factors, if possible. Rule 2 shows us that the square root of a fractional expression can be expressed as the square root of the numerator divided by the square root of a denominator.*
- What does it mean to rationalize the denominator of a fractional expression? Why might we want to do it?
  - *Rationalizing a denominator means that the fractional expression must be expressed with a rational number in the denominator. We might want to rationalize the denominator of a fractional expression to better estimate the value of the number. Another reason is to verify whether two numbers are equal or can be combined.*

Exit Ticket (4 minutes)

Name ____ Date ____________________

Lesson 22: Multiplying and Dividing Expressions with Radicals

Exit Ticket

Write each expression in its simplest radical form.

1. $\sqrt{243} =$

2. $\sqrt{\frac{7}{5}} =$

3. Teja missed class today. Explain to her how to write the length of the hypotenuse in simplest radical form.

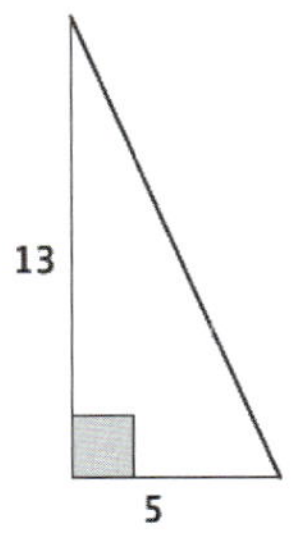

Exit Ticket Sample Solutions

Write each expression in its simplest radical form.

1. $\sqrt{243} =$

$$\sqrt{243} = \sqrt{81} \cdot \sqrt{3}$$
$$= 9\sqrt{3}$$

2. $\sqrt{\frac{7}{5}} =$

$$\sqrt{\frac{7}{5}} = \frac{\sqrt{7}}{\sqrt{5}} \times \frac{\sqrt{5}}{\sqrt{5}}$$
$$= \frac{\sqrt{7}\sqrt{5}}{\sqrt{5}\sqrt{5}}$$
$$= \frac{\sqrt{35}}{\sqrt{25}}$$
$$= \frac{\sqrt{35}}{5}$$

3. **Teja missed class today. Explain to her how to write the length of the hypotenuse in simplest radical form.**

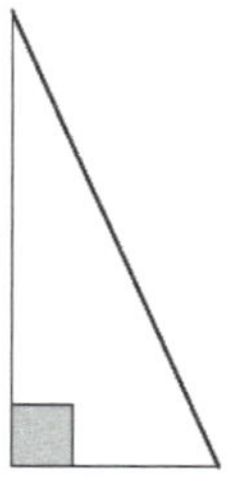

Use the Pythagorean theorem to determine the length of the hypotenuse, c:

$$5^2 + 13^2 = c^2$$
$$25 + 169 = c^2$$
$$194 = c^2$$
$$\sqrt{194} = c$$

To simplify the square root, rewrite the radicand as a product of its factors. The goal is to find a factor that is a perfect square and can then be simplified. There are no perfect square factors of the radicand; therefore, the length of the hypotenuse in simplest radical form is $\sqrt{194}$.

Problem Set Sample Solutions

Express each number in its simplest radical form.

1. $\sqrt{6} \cdot \sqrt{60} =$

$$\sqrt{6} \cdot \sqrt{60} = \sqrt{6} \cdot \sqrt{6} \cdot \sqrt{10}$$
$$= 6\sqrt{10}$$

2. $\sqrt{108} =$

$$\sqrt{108} = \sqrt{9} \cdot \sqrt{4} \cdot \sqrt{3}$$
$$= 3 \cdot 2\sqrt{3}$$
$$= 6\sqrt{3}$$

3. **Pablo found the length of the hypotenuse of a right triangle to be $\sqrt{45}$. Can the length be simplified? Explain.**

$$\sqrt{45} = \sqrt{9}\sqrt{5}$$
$$= 3\sqrt{5}$$

Yes, the length can be simplified because the number 45 has a factor that is a perfect square.

4. $\sqrt{12x^4} =$

$$\sqrt{12x^4} = \sqrt{4}\sqrt{3}\sqrt{x^4}$$
$$= 2x^2\sqrt{3}$$

5. **Sarahi found the distance between two points on a coordinate plane to be $\sqrt{74}$. Can this answer be simplified? Explain.**

The number 74 can be factored, but none of the factors are perfect squares, which are necessary to simplify. Therefore, $\sqrt{74}$ cannot be simplified.

6. $\sqrt{16x^3} =$

$$\sqrt{16x^3} = \sqrt{16}\sqrt{x^2}\sqrt{x}$$
$$= 4x\sqrt{x}$$

7. $\dfrac{\sqrt{27}}{\sqrt{3}} =$

$$\frac{\sqrt{27}}{\sqrt{3}} = \sqrt{\frac{27}{3}}$$
$$= \sqrt{9}$$
$$= 3$$

8. **Nazem and Joffrey are arguing about who got the right answer. Nazem says the answer is $\frac{1}{\sqrt{3}}$, and Joffrey says the answer is $\frac{\sqrt{3}}{3}$. Show and explain that their answers are equivalent.**

$$\frac{1}{\sqrt{3}} = \frac{1}{\sqrt{3}} \times \frac{\sqrt{3}}{\sqrt{3}}$$
$$= \frac{\sqrt{3}}{3}$$

If Nazem were to rationalize the denominator in his answer, he would see that it is equal to Joffrey's answer.

9. $\sqrt{\dfrac{5}{8}} =$

$$\sqrt{\frac{5}{8}} = \frac{\sqrt{5}}{\sqrt{8}} \times \frac{\sqrt{2}}{\sqrt{2}}$$
$$= \frac{\sqrt{10}}{\sqrt{16}}$$
$$= \frac{\sqrt{10}}{4}$$

10. **Determine the area of a square with side length $2\sqrt{7}$ in.**

$$\begin{aligned} A &= \left(2\sqrt{7}\right)^2 \\ &= 2^2\left(\sqrt{7}\right)^2 \\ &= 4(7) \\ &= 28 \end{aligned}$$

The area of the square is $28\ \text{in}^2$.

11. **Determine the exact area of the shaded region shown below.**

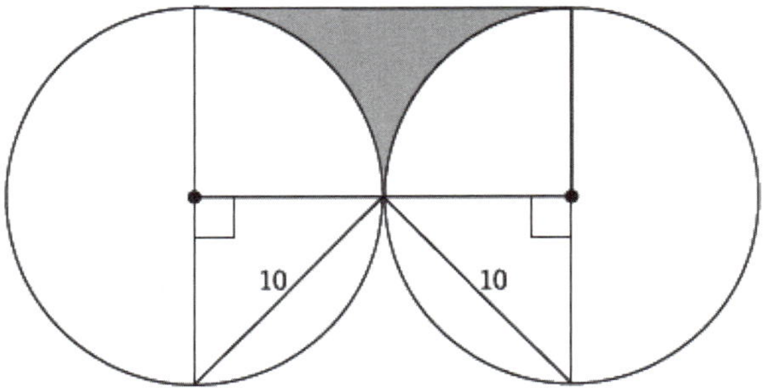

Let r be the length of the radius.

By special triangles or the Pythagorean theorem, $r = 5\sqrt{2}$.

The area of the rectangle containing the shaded region is

$$A = 2\left(5\sqrt{2}\right)\left(5\sqrt{2}\right) = 2(25)(2) = 100.$$

The sum of the two quarter circles in the rectangular region is

$$\begin{aligned} A &= \frac{1}{2}\pi\left(5\sqrt{2}\right)^2 \\ &= \frac{1}{2}\pi(25)(2) \\ &= 25\pi. \end{aligned}$$

The area of the shaded region is $100 - 25\pi$ *square units.*

12. **Determine the exact area of the shaded region shown to the right.**

The radius of each quarter circle is $\frac{1}{2}\left(\sqrt{20}\right) = \frac{1}{2}\left(2\sqrt{5}\right) = \sqrt{5}$.

The sum of the area of the four circular regions is $A = \pi\left(\sqrt{5}\right)^2 = 5\pi$.

The area of the square is $A = \left(\sqrt{20}\right)^2 = 20$.

The area of the shaded region is $20 - 5\pi$ *square units.*

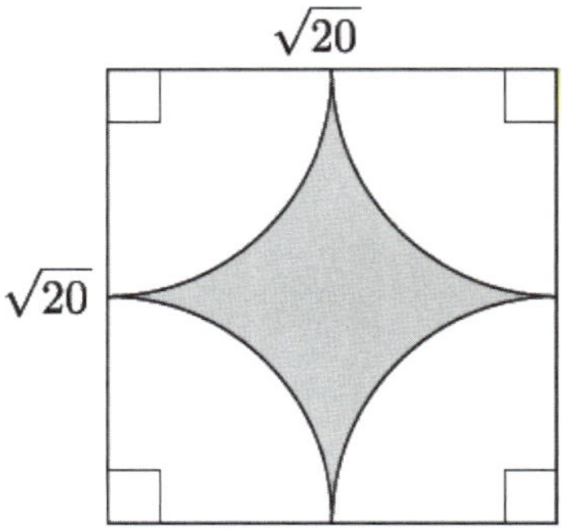

13. Calculate the area of the triangle to the right.

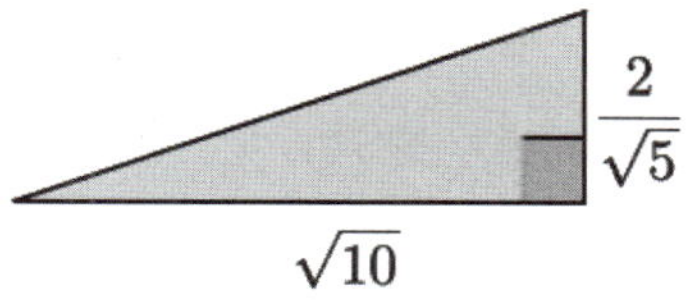

$$A = \frac{1}{2}(\sqrt{10})\left(\frac{2}{\sqrt{5}}\right)$$
$$= \frac{1}{2}\left(\frac{2\sqrt{10}}{\sqrt{5}}\right)$$
$$= \frac{\sqrt{10}}{\sqrt{5}}$$
$$= \sqrt{\frac{10}{5}}$$
$$= \sqrt{2}$$

The area of the triangle is $\sqrt{2}$ square units.

14. $\dfrac{\sqrt{2x^3}\cdot\sqrt{8x}}{\sqrt{x^3}} =$

$$\frac{\sqrt{2x^3}\cdot\sqrt{8x}}{\sqrt{x^3}} = \frac{\sqrt{16x^4}}{\sqrt{x^3}}$$
$$= \frac{4x^2}{x\sqrt{x}}$$
$$= \frac{4x}{\sqrt{x}}$$
$$= \frac{4x}{\sqrt{x}} \times \frac{\sqrt{x}}{\sqrt{x}}$$
$$= \frac{4x\sqrt{x}}{x}$$
$$= 4\sqrt{x}$$

15. Prove Rule 2 for square roots: $\sqrt{\frac{a}{b}} = \frac{\sqrt{a}}{\sqrt{b}}$ $(a \geq 0, b > 0)$

Let p be the nonnegative number so that $p^2 = a$, and let q be the nonnegative number so that $q^2 = b$. Then,

$\sqrt{\frac{a}{b}} = \sqrt{\frac{p^2}{q^2}}$ *By substitution*

$= \sqrt{\left(\frac{p}{q}\right)^2}$ *By the laws of exponents for integers*

$= \frac{p}{q}$ *By definition of square root*

$= \frac{\sqrt{a}}{\sqrt{b}}$ *By substitution*

Perfect Squares of Numbers 1–30

$1^2 = 1$	$16^2 = 256$
$2^2 = 4$	$17^2 = 289$
$3^2 = 9$	$18^2 = 324$
$4^2 = 16$	$19^2 = 361$
$5^2 = 25$	$20^2 = 400$
$6^2 = 36$	$21^2 = 441$
$7^2 = 49$	$22^2 = 484$
$8^2 = 64$	$23^2 = 529$
$9^2 = 81$	$24^2 = 576$
$10^2 = 100$	$25^2 = 625$
$11^2 = 121$	$26^2 = 676$
$12^2 = 144$	$27^2 = 729$
$13^2 = 169$	$28^2 = 784$
$14^2 = 196$	$29^2 = 841$
$15^2 = 225$	$30^2 = 900$

Lesson 23: Adding and Subtracting Expressions with Radicals

Student Outcomes

- Students use the distributive property to simplify expressions that contain radicals.

Lesson Notes

In this lesson, students add and subtract expressions with radicals, continuing their work from the previous lesson. Again, the overarching goal is for students to rewrite expressions involving radical and rational exponents using the properties of exponents (**N.RN.A.2**), which is mastered in Algebra II. Practice adding and subtracting expressions in a geometric setting is provided to achieve this goal.

Classwork

Exercises 1–5 (8 minutes)

The first three exercises are designed to informally assess students' ability to simplify square root expressions, a skill that is necessary for the topic of this lesson. The last two exercises provide a springboard for discussing how to add expressions that contain radicals. Encourage students to discuss their conjectures with a partner. The Discussion that follows debriefs Exercises 1–5.

Exercises 1–5

Simplify each expression as much as possible.

1. $\sqrt{32} =$

$\sqrt{32} = \sqrt{16}\sqrt{2}$
$= 4\sqrt{2}$

2. $\sqrt{45} =$

$\sqrt{45} = \sqrt{9}\sqrt{5}$
$= 3\sqrt{5}$

3. $\sqrt{300} =$

$\sqrt{300} = \sqrt{100}\sqrt{3}$
$= 10\sqrt{3}$

MP.3

4. **The triangle shown below has a perimeter of $6.5\sqrt{2}$ units. Make a conjecture about how this answer was reached.**

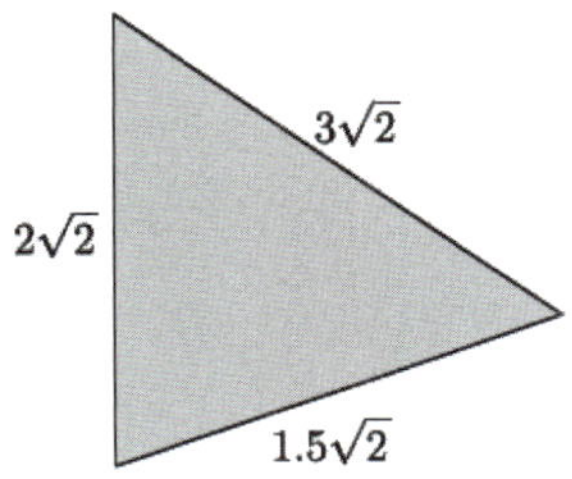

It appears that when all three sides of the triangle were added, the numbers that preceded the square roots were the only numbers that were added: $3 + 2 + 1.5 = 6.5$. The $\sqrt{2}$ shown as part of each length remained $\sqrt{2}$.

5. **The sides of a triangle are $4\sqrt{3}$, $\sqrt{12}$, and $\sqrt{75}$. Make a conjecture about how to determine the perimeter of this triangle.**

Answers will vary. The goal is for students to realize that $\sqrt{12}$ and $\sqrt{75}$ can be rewritten so that each has a factor of $\sqrt{3}$, which then strongly resembles Exercise 4. By rewriting each side length as a multiple of $\sqrt{3}$, we get

$$4\sqrt{3} + 2\sqrt{3} + 5\sqrt{3} = 11\sqrt{3}.$$

Some students may answer incorrectly by adding $3 + 12 + 75$. Show that this is incorrect using a simpler example.

$$\sqrt{9} + \sqrt{16} \neq \sqrt{25}$$

Discussion (6 minutes)

Ensure that students are correctly simplifying the expressions in Exercises 1–3 because it is a skill that is required to add and subtract the expressions in this lesson. Then, continue with the discussion that follows.

- Share your conjecture for Exercise 4. How can we explain that $1.5\sqrt{2} + 2\sqrt{2} + 3\sqrt{2} = 6.5\sqrt{2}$?

Select students to share their conjectures. The expectation is that students recognize that the rational parts of each length were added.

MP.7

- Does this remind you of anything else you have done before? Give an example.
  - *It is reminiscent of combining like terms using the distributive property. For example, $1.5x + 2x + 3x = (1.5 + 2 + 3)x = 6.5x$.*
- The distributive property is true for all real numbers. Is $\sqrt{2}$ a real number?
  - *Yes, $\sqrt{2}$ is a real number. It is irrational, but it is a real number.*
- For this reason, we can apply the distributive property to radical expressions.
- Share your conjecture for Exercise 5. How might we add $4\sqrt{3}$, $\sqrt{12}$, and $\sqrt{75}$?

Now that students know that they can apply the distributive property to radical expressions, they may need another minute or two to evaluate the conjectures they developed while working on Exercise 5. Select students to share their conjectures. The expectation is that students apply their knowledge from the previous lesson to determine a strategy for finding the perimeter of the triangle. This strategy should include simplifying each expression and combining those with the same irrational factor. Now would be a good time to point out that the $\sqrt{p}$ for any prime number p is an irrational number. This explains why many of the simplifications tend to have radicands that are either prime or a product of prime numbers.

Scaffolding:

It may be necessary to show students that $\sqrt{2}$ is a number between 1 and 2 on the number line. If they consider a unit square on the number line with a diagonal, s, as shown, then they can use a compass with a radius equal in length to the diagonal and center at 0 to show that the length of the diagonal of the square ($\sqrt{2}$) is a number.

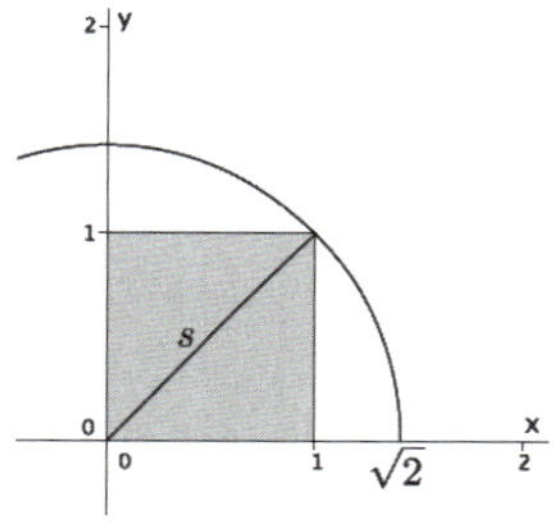

Exercise 6 (10 minutes)

To complete Exercise 6, students need to circle the expressions on the list that they think can be simplified. As students complete the task, pair them so that they can compare their lists and discuss any discrepancies. Have students who are ready for a challenge generate their own examples, possibly including expressions that contain variables.

Exercise 6

Circle the expressions that can be simplified using the distributive property. Be prepared to explain your choices.

$8.3\sqrt{2}+7.9\sqrt{2}$
$\sqrt{13}-\sqrt{6}$
$-15\sqrt{5}+\sqrt{45}$
$11\sqrt{7}-6\sqrt{7}+3\sqrt{2}$
$19\sqrt{2}+2\sqrt{8}$
$4+\sqrt{11}$
$\sqrt{7}+2\sqrt{10}$
$\sqrt{12}-\sqrt{75}$
$\sqrt{32}+\sqrt{2}$
$6\sqrt{13}+\sqrt{26}$

The expressions that can be simplified using the distributive property are noted in red.

Example 1 (5 minutes)

The expressions in this example have been taken from the list that students completed in Exercise 6. Ask students who circled this expression to explain why.

Explain how the expression $8.3\sqrt{2}+7.9\sqrt{2}$ can be simplified using the distributive property.

Each term of the expression has a common factor, $\sqrt{2}$. For that reason, the distributive property can be applied.

$$\begin{aligned} 8.3\sqrt{2}+7.9\sqrt{2} &= (8.3+7.9)\sqrt{2} && \textit{By the distributive property} \\ &= 16.2\sqrt{2} \end{aligned}$$

Explain how the expression $11\sqrt{7}-6\sqrt{7}+3\sqrt{2}$ can be simplified using the distributive property.

The expression can be simplified because the first two terms contain the expression $\sqrt{7}$. Using the distributive property, we get the following:

$$\begin{aligned} 11\sqrt{7}-6\sqrt{7}+3\sqrt{2} &= (11-6)\sqrt{7}+3\sqrt{2} && \textit{By the distributive property} \\ &= 5\sqrt{7}+3\sqrt{2} \end{aligned}$$

Example 2 (4 minutes)

MP.3

The expression in this example has been taken from the list that students completed in Exercise 6. Ask students, "Who circled this expression?" Have those students explain to small groups why they believe the expression can be simplified. Then, allow students who had not selected the expression to circle it if they have been convinced that it can be simplified. Finally, ask one of the students who changed his answer to explain how the expression can be simplified.

> **Explain how the expression $19\sqrt{2} + 2\sqrt{8}$ can be simplified using the distributive property.**
>
> ***The expression can be simplified, but first the term $2\sqrt{8}$ must be rewritten.***
>
> $$\begin{aligned} 19\sqrt{2} + 2\sqrt{8} &= 19\sqrt{2} + 2\sqrt{4}\cdot\sqrt{2} && \textit{By Rule 1} \\ &= 19\sqrt{2} + 2\cdot 2\sqrt{2} \\ &= 19\sqrt{2} + 4\sqrt{2} \\ &= (19+4)\sqrt{2} && \textit{By the distributive property} \\ &= 23\sqrt{2} \end{aligned}$$

Example 3 (6 minutes)

The expressions in this example have been taken from the list that students completed in Exercise 6. Ask students who circled this expression to explain why.

> **Can the expression $\sqrt{7} + 2\sqrt{10}$ be simplified using the distributive property?**
>
> ***No, the expression cannot be simplified because neither term can be rewritten in a way that the distributive property could be applied.***
>
> **To determine if an expression can be simplified, you must first simplify each of the terms within the expression. Then, apply the distributive property, or other properties as needed, to simplify the expression.**

MP.3 Have students return to Exercise 6 and discuss the remaining expressions in small groups. Any groups that cannot agree on an expression should present their arguments to the class.

Closing (3 minutes)

Working in pairs, have students describe to a partner how to simplify the expressions below. Once students have partner shared, ask the class how the work completed today is related to the structure of rational numbers they have observed in the past. The expected response is that the distributive property can be applied to square roots because they are numbers, too.

- Describe how to simplify the expression $3\sqrt{18} + 10\sqrt{2}$.
  - *To simplify the expression, we must first rewrite $3\sqrt{18}$ so that is has a factor of $\sqrt{2}$.*

    $$\begin{aligned} 3\sqrt{18} &= 3\sqrt{9}\sqrt{2} \\ &= 3(3)\sqrt{2} \\ &= 9\sqrt{2} \end{aligned}$$

    Now that both terms have a factor of $\sqrt{2}$, the distributive property can be applied to simplify.

    $$\begin{aligned} 9\sqrt{2} + 10\sqrt{2} &= (9+10)\sqrt{2} \\ &= 19\sqrt{2} \end{aligned}$$

- Describe how to simplify the expression $5\sqrt{3} + \sqrt{12}$.
  - *To simplify the expression, we must first rewrite* $\sqrt{12}$ *so that is has a factor of* $\sqrt{3}$*.*

$$\begin{aligned}\sqrt{12} &= \sqrt{4}\sqrt{3}\\ &= 2\sqrt{3}\end{aligned}$$

Now that both terms have a factor of $\sqrt{3}$*, the distributive property can be applied to simplify.*

$$\begin{aligned}5\sqrt{3} + 2\sqrt{3} &= (5+2)\sqrt{3}\\ &= 7\sqrt{3}\end{aligned}$$

Exit Ticket (3 minutes)

Name ____ Date ________________

Lesson 23: Adding and Subtracting Expressions with Radicals

Exit Ticket

1. Simplify $5\sqrt{11} - 17\sqrt{11}$.

2. Simplify $\sqrt{8} + 5\sqrt{2}$.

3. Write a radical addition or subtraction problem that cannot be simplified, and explain why it cannot be simplified.

Exit Ticket Sample Solutions

1. Simplify $5\sqrt{11} - 17\sqrt{11}$.

$$5\sqrt{11} - 17\sqrt{11} = (5 - 17)\sqrt{11}$$
$$= -12\sqrt{11}$$

2. Simplify $\sqrt{8} + 5\sqrt{2}$.

$$\sqrt{8} + 5\sqrt{2} = \sqrt{4}\sqrt{2} + 5\sqrt{2}$$
$$= 2\sqrt{2} + 5\sqrt{2}$$
$$= (2 + 5)\sqrt{2}$$
$$= 7\sqrt{2}$$

3. Write a radical addition or subtraction problem that cannot be simplified, and explain why it cannot be simplified.

Answers will vary. Students should state that their expression cannot be simplified because one or both terms cannot be rewritten so that each has a common factor. Therefore, the distributive property cannot be applied.

Problem Set Sample Solutions

Express each answer in simplified radical form.

1. $18\sqrt{5} - 12\sqrt{5} =$

$$18\sqrt{5} - 12\sqrt{5} = (18 - 12)\sqrt{5}$$
$$= 6\sqrt{5}$$

2. $\sqrt{24} + 4\sqrt{54} =$

$$\sqrt{24} + 4\sqrt{54} = \sqrt{4} \cdot \sqrt{6} + 4 \cdot \sqrt{9} \cdot \sqrt{6}$$
$$= 2\sqrt{6} + 4 \cdot 3\sqrt{6}$$
$$= (2 + 12)\sqrt{6}$$
$$= 14\sqrt{6}$$

3. $2\sqrt{7} + 4\sqrt{63} =$

$$2\sqrt{7} + 4\sqrt{63} = 2\sqrt{7} + 4\sqrt{9}\sqrt{7}$$
$$= 2\sqrt{7} + 4(3)\sqrt{7}$$
$$= (2 + 12)\sqrt{7}$$
$$= 14\sqrt{7}$$

4. What is the perimeter of the triangle shown below?

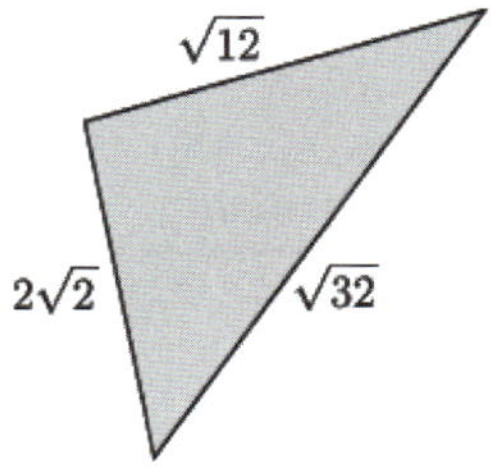

$$2\sqrt{2} + \sqrt{12} + \sqrt{32} = 2\sqrt{2} + \sqrt{4} \cdot \sqrt{3} + \sqrt{16} \cdot \sqrt{2}$$
$$= 2\sqrt{2} + 2\sqrt{3} + 4\sqrt{2}$$
$$= (2 + 4)\sqrt{2} + 2\sqrt{3}$$
$$= 6\sqrt{2} + 2\sqrt{3}$$

The perimeter of the triangle is $6\sqrt{2} + 2\sqrt{3}$ units.

5. **Determine the area and perimeter of the triangle shown. Simplify as much as possible.**

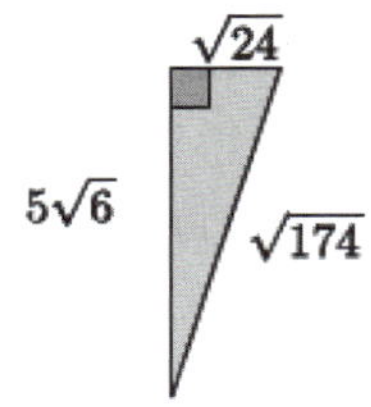

The perimeter of the triangle is

$$\sqrt{24} + 5\sqrt{6} + \sqrt{174} = \sqrt{4}\sqrt{6} + 5\sqrt{6} + \sqrt{174}$$
$$= 2\sqrt{6} + 5\sqrt{6} + \sqrt{174}$$
$$= (2+5)\sqrt{6} + \sqrt{174}$$
$$= 7\sqrt{6} + \sqrt{174}.$$

The area of the triangle is

$$\frac{\sqrt{24}(5\sqrt{6})}{2} = \frac{2\sqrt{6}(5\sqrt{6})}{2} = \frac{60}{2} = 30.$$

The perimeter is $7\sqrt{6} + \sqrt{174}$ units, and the area is 30 square units.

6. **Determine the area and perimeter of the rectangle shown. Simplify as much as possible.**

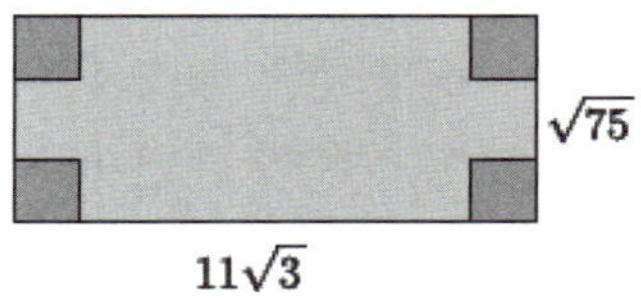

The perimeter of the rectangle is

$$11\sqrt{3} + 11\sqrt{3} + \sqrt{75} + \sqrt{75} = 2(11\sqrt{3}) + 2(\sqrt{25}\sqrt{3})$$
$$= 22\sqrt{3} + 10\sqrt{3}$$
$$= (22+10)\sqrt{3}$$
$$= 32\sqrt{3}.$$

The area of the rectangle is

$$11\sqrt{3}(5\sqrt{3}) = 55(3)$$
$$= 165.$$

The perimeter is $32\sqrt{3}$ units, and the area is 165 square units.

7. **Determine the area and perimeter of the triangle shown. Simplify as much as possible.**

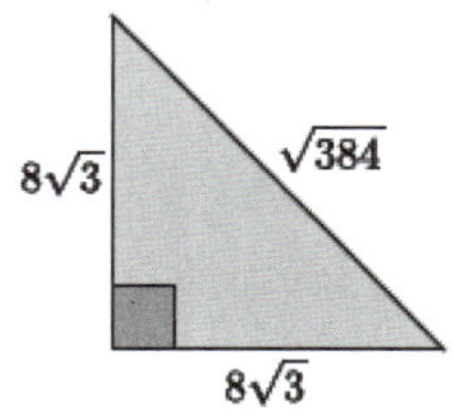

The perimeter of the triangle is

$$8\sqrt{3} + 8\sqrt{3} + \sqrt{384} = (8+8)\sqrt{3} + \sqrt{384}$$
$$= 16\sqrt{3} + \sqrt{384}$$
$$= 16\sqrt{3} + \sqrt{64}\sqrt{6}$$
$$= 16\sqrt{3} + 8\sqrt{6}.$$

The area of the triangle is

$$\frac{(8\sqrt{3})^2}{2} = \frac{8^2(\sqrt{3})^2}{2}$$
$$= \frac{64(3)}{2}$$
$$= 32(3)$$
$$= 96.$$

The perimeter of the triangle is $16\sqrt{3} + 8\sqrt{6}$ units, and the area of the triangle is 96 square units.

8. **Determine the area and perimeter of the triangle shown. Simplify as much as possible.**

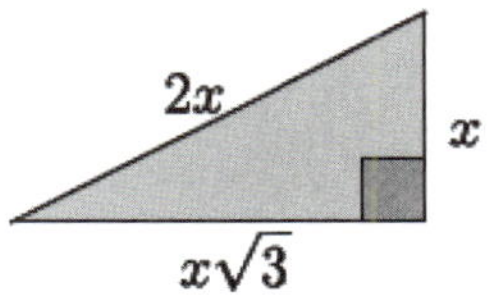

The perimeter of the triangle is

$$2x + x + x\sqrt{3} = 3x + x\sqrt{3}.$$

The area of the triangle is

$$\frac{x(x\sqrt{3})}{2} = \frac{x^2\sqrt{3}}{2}.$$

The perimeter is $3x + x\sqrt{3}$ *units, and the area is* $\frac{x^2\sqrt{3}}{2}$ *square units.*

9. **The area of the rectangle shown in the diagram below is 160 square units. Determine the area and perimeter of the shaded triangle. Write your answers in simplest radical form, and then approximate to the nearest tenth.**

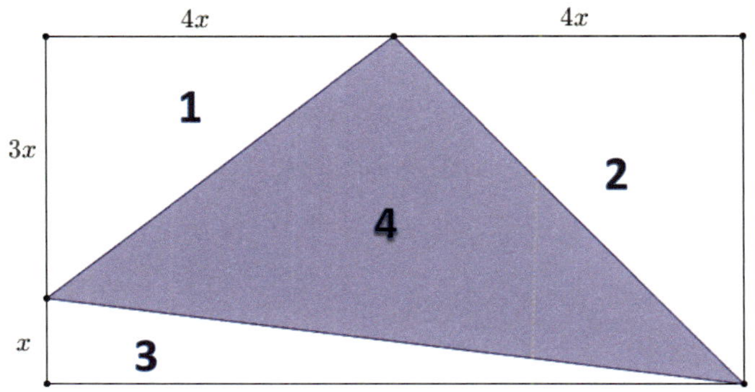

The length of the rectangle is $8x$, *and the width is* $4x$. *Using the given area of the rectangle:*

$$\text{Area} = \text{length} \times \text{width}$$

$$160 = 8x \cdot 4x$$
$$160 = 32x^2$$
$$5 = x^2$$
$$\sqrt{5} = x$$

$$\text{Area}_{\text{rectangle}} = A_1 + A_2 + A_3 + A_4$$

$\text{Area}_1 = \frac{1}{2}bh$	$\text{Area}_2 = \frac{1}{2}bh$	$\text{Area}_3 = \frac{1}{2}bh$
$\text{Area}_1 = \frac{1}{2} \cdot 3x \cdot 4x$	$\text{Area}_2 = \frac{1}{2} \cdot 4x \cdot 4x$	$\text{Area}_3 = \frac{1}{2} \cdot 8x \cdot x$
$\text{Area}_1 = 6x^2$	$\text{Area}_2 = 8x^2$	$\text{Area}_3 = 4x^2$
$\text{Area}_1 = 6(\sqrt{5})^2$	$\text{Area}_2 = 8(\sqrt{5})^2$	$\text{Area}_3 = 4(\sqrt{5})^2$
$\text{Area}_1 = 30$	$\text{Area}_2 = 40$	$\text{Area}_3 = 20$

$$160 = 30 + 40 + 20 + A_4$$
$$A_4 = 70$$

The area of the shaded triangle in the diagram is 70 *square units.*

The perimeter of the shaded triangle requires use of the Pythagorean theorem to find the hypotenuses of right triangles 1, 2, and 3. Let h_1, h_2, *and* h_3 *represent the lengths of the hypotenuses of triangles 1, 2, and 3, respectively.*

$(3\sqrt{5})^2 + (4\sqrt{5})^2 = (c_1)^2$	$(4\sqrt{5})^2 + (4\sqrt{5})^2 = (c_2)^2$	$(\sqrt{5})^2 + (8\sqrt{5})^2 = (c_3)^2$
$45 + 80 = (c_1)^2$	$80 + 80 = (c_2)^2$	$5 + 320 = (c_3)^2$
$125 = (c_1)^2$	$160 = (c_2)^2$	$325 = (c_3)^2$
$5\sqrt{5} = c_1$	$4\sqrt{10} = c_2$	$5\sqrt{13} = c_3$

$$\text{Perimeter} = c_1 + c_2 + c_3$$

$$\text{Perimeter} = 5\sqrt{5} + 4\sqrt{10} + 5\sqrt{13}$$

The perimeter of the shaded triangle is approximately 41.9 *units.*

10. Parallelogram $ABCD$ has an area of $9\sqrt{3}$ square units. $DC = 3\sqrt{3}$, and G and H are midpoints of $\overline{DE}$ and $\overline{CE}$, respectively. Find the area of the shaded region. Write your answer in simplest radical form.

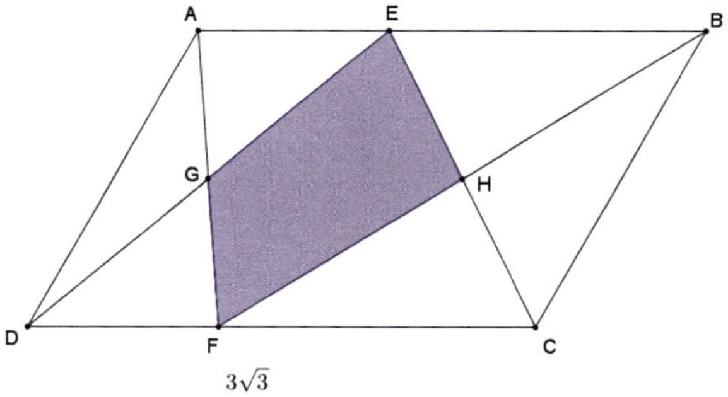

Using the area of a parallelogram:

$\text{Area}(ABCD) = bh$

$9\sqrt{3} = 3\sqrt{3} \cdot h$
$3 = h$

The height of the parallelogram is 3.

The area of the shaded region is the sum of the areas of $\triangle EGH$ *and* $\triangle FGH$.

The given points G *and* H *are midpoints of* $\overline{DE}$ *and* $\overline{CE}$; *therefore, by the triangle side splitter theorem,* $\overline{GH}$ *must be parallel to* $\overline{DC}$, *and thus, also parallel to* $\overline{AB}$. *Furthermore,* $GH = \frac{1}{2}CD = \frac{1}{2}AB = \frac{3}{2}\sqrt{3}$.

$\triangle EGH \sim \triangle EDC$ *by AA~ criterion with a scale factor of* $\frac{1}{2}$. *The areas of scale drawings are related by the square of the scale factor; therefore,* $\text{Area}(\triangle EGH) = \left(\frac{1}{2}\right)^2 \cdot \text{Area}(\triangle EDC)$.

$\text{Area}(\triangle EDC) = \frac{1}{2} \cdot 3\sqrt{3} \cdot 3$

$\text{Area}(\triangle EDC) = \frac{9}{2}\sqrt{3}$

$\text{Area}(\triangle EGH) = \left(\frac{1}{2}\right)^2 \cdot \frac{9}{2}\sqrt{3}$

$\text{Area}(\triangle EGH) = \frac{1}{4} \cdot \frac{9}{2}\sqrt{3}$

$\text{Area}(\triangle EGH) = \frac{9}{8}\sqrt{3}$

By a similar argument:

$\text{Area}(\triangle FGH) = \frac{9}{8}\sqrt{3}$

$\text{Area}(EHFG) = \text{Area}(\triangle EGH) + \text{Area}(\triangle FGH)$

$\text{Area}(EHFG) = \frac{9}{8}\sqrt{3} + \frac{9}{8}\sqrt{3}$

$\text{Area}(EHFG) = \frac{9}{4}\sqrt{3}$

The area of the shaded region is $\frac{9}{4}\sqrt{3}$ *square units.*

Lesson 24: Prove the Pythagorean Theorem Using Similarity

Student Outcomes

- Students prove the Pythagorean theorem using similarity.
- Students use similarity and the Pythagorean theorem to find the unknown side lengths of a right triangle.
- Students are familiar with the ratios of the sides of special right triangles with angle measures $45–45–90$ and $30–60–90$.

Lesson Notes

In Grade 8, students proved the Pythagorean theorem using what they knew about similar triangles. The basis of this proof is the same, but students are better prepared to understand the proof because of their work in Lesson 23. This proof differs from what students did in Grade 8 because it uses knowledge of ratios within similar triangles and more advanced algebraic skills.

The proof of the Pythagorean theorem and the Exploratory Challenge addressing special right triangles are the essential components of this lesson. Exercises 1–3 and the bullet points in the Discussion between the exercises can be moved to the Problem Set if necessary.

Classwork

Opening (5 minutes)

Show the diagram below, and then ask the three questions that follow, which increase in difficulty. Have students respond using a white board. Ask advanced students to attempt to show that $a^2 + b^2 = c^2$, without the scaffolded questions. Discuss the methods used as a class.

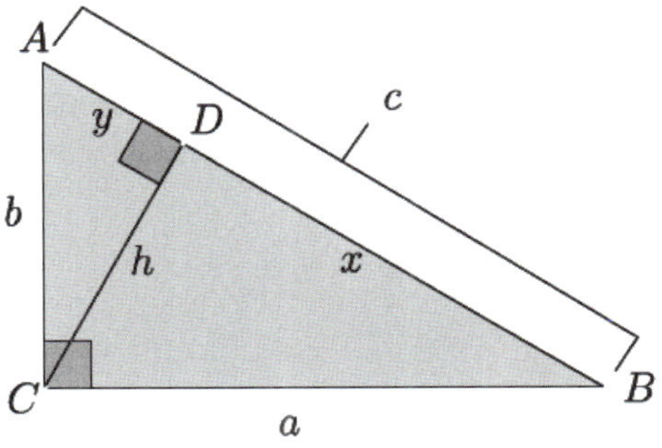

Scaffolding:

Some groups of students may respond better to a triangle with numeric values.

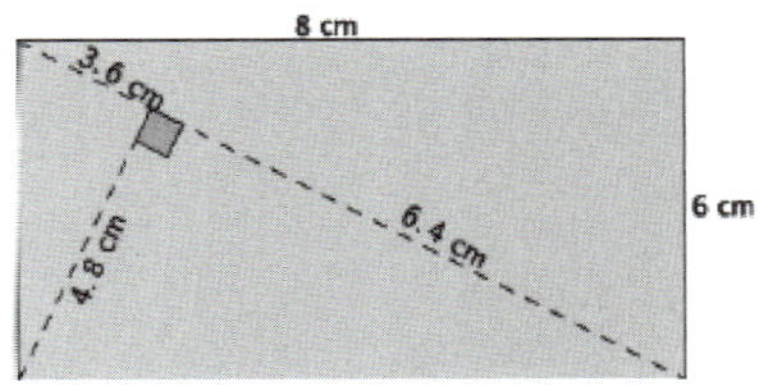

- Write an expression in terms of c for x and y.
  - $x + y = c$
- Write a similarity statement for the three right triangles in the image.
  - $\triangle ACB \sim \triangle ADC \sim \triangle CDB$

- Write a ratio based on the similarity statement from the previous question.
  - *Several answers are acceptable. Ensure that students are writing ratios as they did in Lesson 21, that is, ratios that compare shorter leg: hypotenuse, longer leg: hypotenuse, or shorter leg: longer leg.*

Discussion (10 minutes)

In the Discussion that follows, students prove the Pythagorean theorem using similarity and the converse of the Pythagorean theorem using SSS for congruent triangles.

- Our goal is to prove the Pythagorean theorem using what we know about similar triangles. Consider the right triangle $\triangle ABC$ so that $\angle C$ is a right angle. We label the side lengths $a, b,$ and c so that side length a is opposite $\angle A$, side length b is opposite $\angle B$, and hypotenuse c is opposite $\angle C$, as shown.

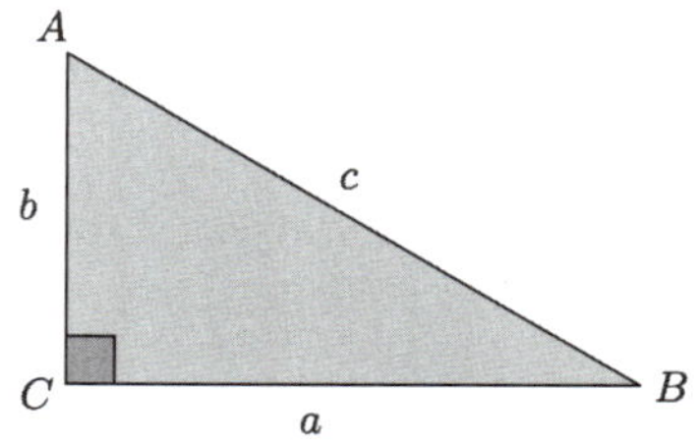

- Next, we draw the altitude, h, from the right angle to the hypotenuse so that it splits the hypotenuse at point D into lengths x and y so that $x + y = c$, as shown.

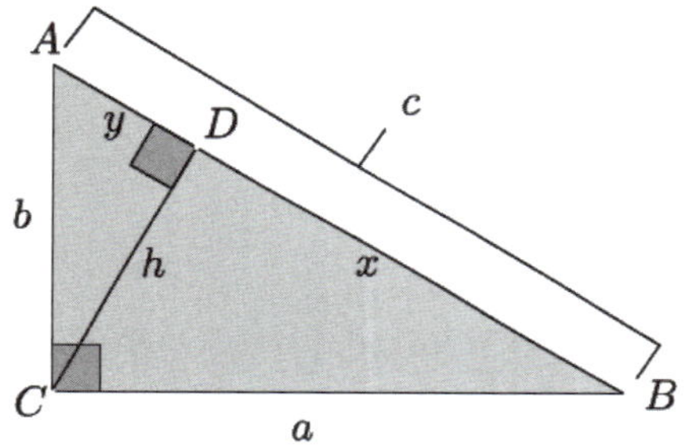

Scaffolding:
Students may want to use the cutouts that they created in Lesson 21.

- What do we know about the original triangle and the two sub-triangles?
  - *The three triangles are similar.*

If necessary, show students the set of three triangles in the diagram below. The original triangle and the two triangles formed by the altitude h have been moved via rigid motions that make their corresponding sides and angles easier to see.

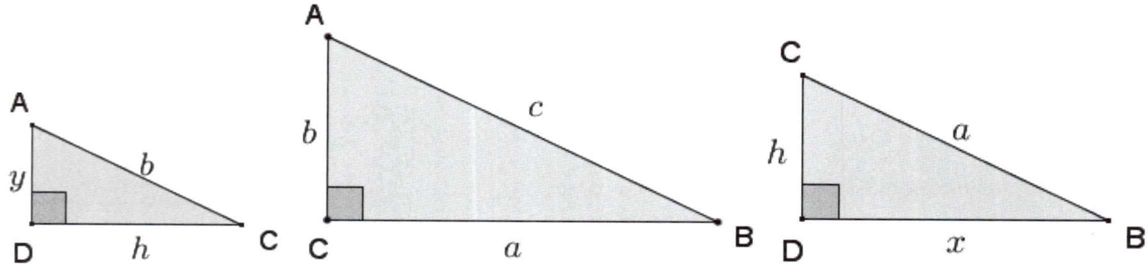

- We want to prove the Pythagorean theorem using what we know about similar triangles and the ratios we wrote in Lesson 23, that is, shorter leg: hypotenuse, longer leg: hypotenuse, or shorter leg: longer leg.
- We begin by identifying the two triangles that each have $\angle B$ as one of their angles. Which are they?
  - *The two triangles are* $\triangle ABC$ *and* $\triangle CBD$.
- Since they are similar, we can write equivalent ratios of the two similar triangles we just named in terms of longer leg: hypotenuse.

$$\frac{a}{c} = \frac{x}{a}$$
$$a^2 = xc$$

- Now, identify two right triangles in the figure that each have $\angle A$ as an acute angle.
  - *The two triangles are* $\triangle ABC$ *and* $\triangle ACD$.
- Write equivalent ratios in terms of shorter leg: hypotenuse, and simplify as we just did:

$$\frac{b}{c} = \frac{y}{b}$$
$$b^2 = yc$$

- Our goal is to show that $a^2 + b^2 = c^2$. We know that $a^2 = xc$ and $b^2 = yc$. By substitution,

$$\begin{aligned} a^2 + b^2 &= xc + yc \\ &= (x + y)c \\ &= cc \qquad \text{(because } x + y = c) \\ &= c^2. \end{aligned}$$

  Therefore, $a^2 + b^2 = c^2$, and we have proven the Pythagorean theorem using similarity.

MP.6 Ask students to summarize the steps of the proof in writing or with a partner.

- We next show that the converse of the Pythagorean theorem is true.

  CONVERSE OF THE PYTHAGOREAN THEOREM: If a triangle has side lengths a, b, and c so that $a^2 + b^2 = c^2$, then the triangle is a right triangle with c as the length of the hypotenuse (the side opposite the right angle).

  In the diagram below, we have a triangle with side lengths a, b, and c.

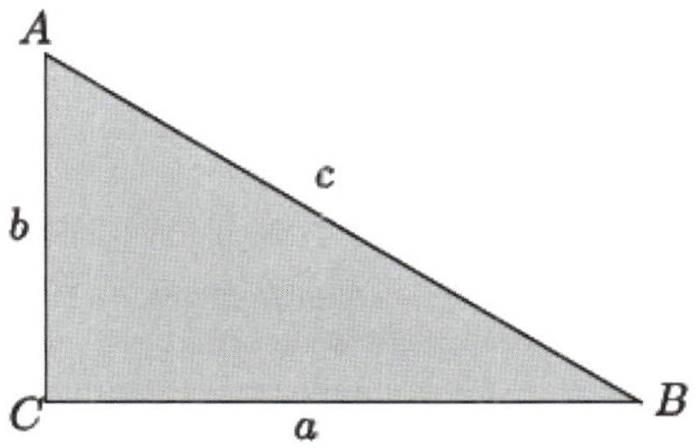

In this diagram, we have constructed a right triangle with side lengths a and b.

If time permits, actually show the construction. Draw a segment of length a, construct a line through one of the endpoints of segment a that is perpendicular to a, mark a point along the perpendicular line that is equal to the length of b, and draw the hypotenuse c.

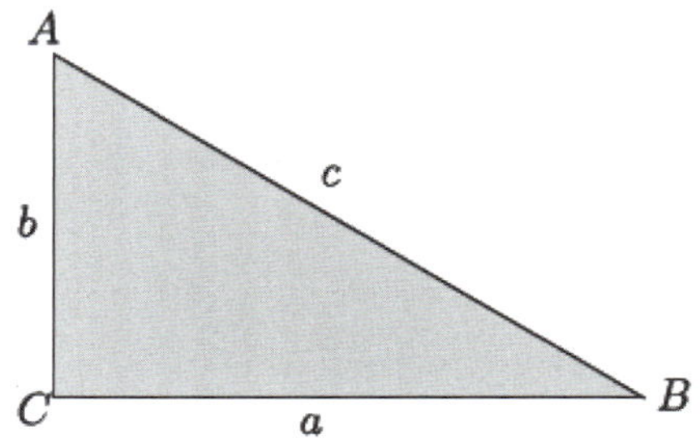

- Consider the right triangle with leg lengths of a and b. By the Pythagorean theorem, we know that $a^2 + b^2 = c^2$. Taking the square root of both sides gives the length of the hypotenuse as $\sqrt{a^2 + b^2} = c$. Then, by SSS, the two triangles are congruent because both triangles have side lengths of a, b, and c. Therefore, the given triangle is a right triangle with a right angle that is opposite the side of length c.

Exercises 1–2 (5 minutes)

Students complete Exercise 1 independently. Exercise 2 is designed so that early finishers of Exercise 1 can begin thinking about the topic of the next discussion. If time allows, consider asking students why they do not need to consider the negative solution for $c^2 = 12500$, which is $c = \pm 50\sqrt{5}$; they only consider $50\sqrt{5}$ in our solution. The desired response is that the context is length; therefore, there is no need to consider a negative length.

Exercises 1–2

1. **Find the length of the hypotenuse of a right triangle whose legs have lengths 50 and 100.**

$$c^2 = 50^2 + 100^2$$
$$c^2 = 2500 + 10000$$
$$c^2 = 12500$$
$$\sqrt{c^2} = \sqrt{12500}$$
$$c = \sqrt{2^2 \cdot 5^5}$$
$$c = 2 \cdot 5^2\sqrt{5}$$
$$c = 50\sqrt{5}$$

2. **Can you think of a simpler method for finding the length of the hypotenuse in Exercise 1? Explain.**

   Accept any reasonable methods. Students may recall from Grade 8 that they can use what they know about similar triangles and scale factors to make their computations easier.

Scaffolding:

Guide students' thinking by asking, "How might similar triangles make the work in Exercise 1 easier?"

Discussion (4 minutes)

In the Discussion that follows, students use what they know about similar triangles to simplify their work from Exercise 1.

- To simplify our work with large numbers, as in the leg lengths of 50 and 100 from Exercise 1, we can find the greatest common factor (GCF) of 50 and 100 and then consider a similar triangle with those smaller side lengths. Since $\text{GCF}(50, 100) = 50$, we can use that GCF to determine the side lengths of a dilated triangle.

- Specifically, we can consider a triangle that has been dilated by a scale factor of $\frac{1}{50}$, which produces a similar triangle with leg lengths 1 and 2.
- A triangle with leg lengths 1 and 2 has a hypotenuse length of $\sqrt{5}$. The original triangle has side lengths that are 50 times longer than this one; therefore, the length of the hypotenuse of the original triangle is $50\sqrt{5}$.

Exercise 3 (2 minutes)

Students complete Exercise 3 independently.

Exercise 3

3. Find the length of the hypotenuse of a right triangle whose legs have lengths 75 and 225.

A right triangle with leg lengths 75 and 225 has leg lengths that are 75 times longer than a triangle with leg lengths 1 and 3. A triangle with leg lengths 1 and 3 has a hypotenuse of length $\sqrt{10}$. Therefore, the length of the hypotenuse of a triangle with leg lengths 75 and 225 is $75\sqrt{10}$.

Exploratory Challenge/Exercises 4–5 (10 minutes)

This challenge reveals the relationships of special right triangles with angle measures 30–60–90 and 45–45–90. Divide the class so that one half investigates the 30–60–90 triangle and the other half investigates the 45–45–90 triangle. Consider having pairs of students from each half become one small group to share their results. Another option is to discuss the results as a whole class using the closing questions below.

Exploratory Challenge/Exercises 4–5

4. An equilateral triangle has sides of length 2 and angle measures of $60°$, as shown below. The altitude from one vertex to the opposite side divides the triangle into two right triangles.

a. Are those triangles congruent? Explain.

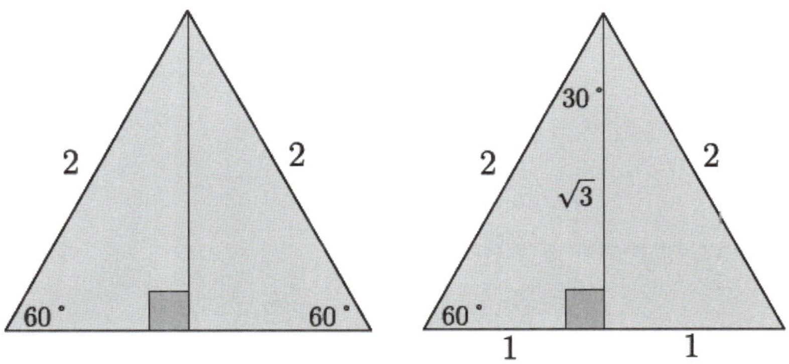

Yes, the two right triangles are congruent by ASA. Since the altitude is perpendicular to the base, then each of the right triangles has angles of measure $90°$ and $60°$. By the triangle sum theorem, the third angle has a measure of $30°$. Then, each of the right triangles has corresponding angle measures of $30°$ and $60°$, and the included side length is 2.

b. What is the length of the shorter leg of each of the right triangles? Explain.

Since the total length of the base of the equilateral triangle is 2, and the two right triangles formed are congruent, then the bases of each must be equal in length. Therefore, the length of the base of one right triangle is 1.

c. Use the Pythagorean theorem to determine the length of the altitude.

Let h represent the length of the altitude.

$$1^2 + h^2 = 2^2$$
$$h^2 = 2^2 - 1^2$$
$$h^2 = 3$$
$$h = \sqrt{3}$$

d. Write the ratio that represents shorter leg: hypotenuse.

$$1:2$$

e. Write the ratio that represents longer leg: hypotenuse.

$$\sqrt{3}:2$$

f. Write the ratio that represents shorter leg: longer leg.

$$1:\sqrt{3}$$

g. By the AA criterion, any triangles with measures 30–60–90 will be similar to this triangle. If a 30–60–90 triangle has a hypotenuse of length 16, what are the lengths of the legs?

Consider providing the following picture for students:

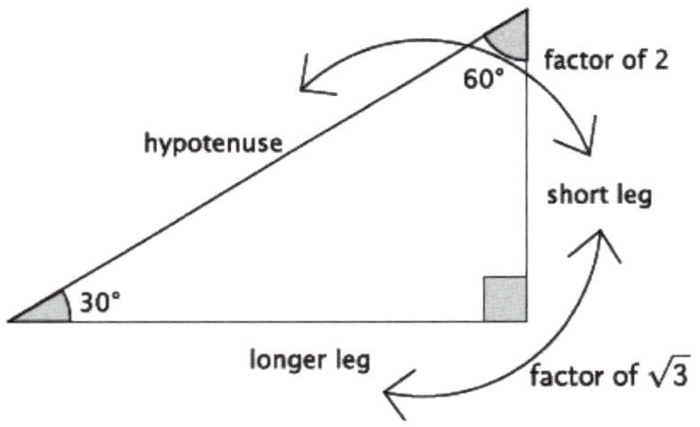

Let a represent the length of the shorter leg.

$$\frac{a}{16} = \frac{1}{2}$$
$$a = 8$$

Let b represent the length of the longer leg.

$$\frac{b}{16} = \frac{\sqrt{3}}{2}$$
$$2b = 16\sqrt{3}$$
$$b = 8\sqrt{3}$$

The length $a = 8$, and the length $b = 8\sqrt{3}$.

Note: After finding the length of one of the legs, some students may have used the ratio shorter leg: longer leg to determine the length of the other leg.

5. An isosceles right triangle has leg lengths of 1, as shown.

a. What are the measures of the other two angles? Explain.

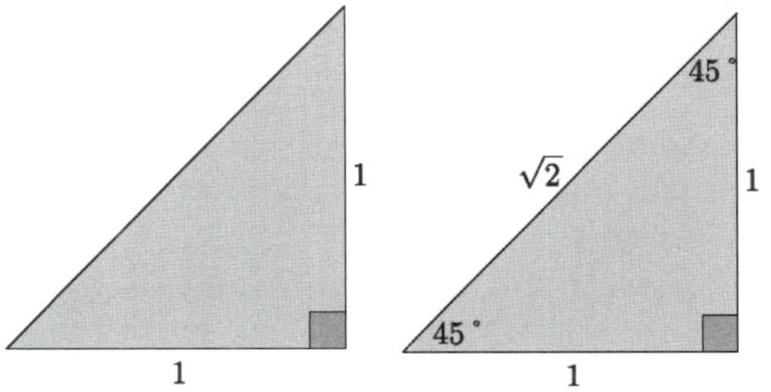

Base angles of an isosceles triangle are equal; therefore, the other two angles have a measure of $45°$.

b. Use the Pythagorean theorem to determine the length of the hypotenuse of the right triangle.

Let c represent the length of the hypotenuse.

$$1^2 + 1^2 = c^2$$
$$2 = c^2$$
$$\sqrt{2} = c$$

c. Is it necessary to write all three ratios: shorter leg: hypotenuse, longer leg: hypotenuse, and shorter leg: longer leg? Explain.

No, it is not necessary to write all three ratios. The reason is that the shorter leg and the longer leg are the same length. Therefore, the ratios shorter leg: hypotenuse and longer leg: hypotenuse will be the same. Additionally, the shorter leg: longer leg ratio would be $1:1$*, which is not useful since we are given that the right triangle is an isosceles right triangle.*

d. Write the ratio that represents leg: hypotenuse.

$$1:\sqrt{2}$$

e. By the AA criterion, any triangles with measures 45–45–90 will be similar to this triangle. If a 45–45–90 triangle has a hypotenuse of length 20, what are the lengths of the legs?

Let a represent the length of the leg.

$$\frac{a}{20} = \frac{1}{\sqrt{2}}$$
$$a\sqrt{2} = 20$$
$$a = \frac{20}{\sqrt{2}}$$
$$a = \frac{20}{\sqrt{2}}\left(\frac{\sqrt{2}}{\sqrt{2}}\right)$$
$$a = \frac{20\sqrt{2}}{2}$$
$$a = 10\sqrt{2}$$

Closing (4 minutes)

Ask students to summarize the key points of the lesson. Additionally, consider asking students the following questions. Have students respond in writing, to a partner, or to the whole class.

- Explain in your own words the proof of the Pythagorean theorem using similarity.
- Triangles with angle measures 30–60–90 are a result of drawing an altitude from one angle of an equilateral triangle to the opposite side. Explain how to use ratios of legs and the hypotenuse to find the lengths of any 30–60–90 triangle. Why does it work?
- Triangles with angle measures 45–45–90 are isosceles right triangles. Explain how to use ratios of legs and the hypotenuse to find the lengths of any 45–45–90 triangle. Why does it work?

Exit Ticket (5 minutes)

Name ____ Date ________________

Lesson 24: Prove the Pythagorean Theorem Using Similarity

Exit Ticket

A right triangle has a leg with a length of 18 and a hypotenuse with a length of 36. Bernie notices that the hypotenuse is twice the length of the given leg, which means it is a 30–60–90 triangle. If Bernie is right, what should the length of the remaining leg be? Explain your answer. Confirm your answer using the Pythagorean theorem.

Exit Ticket Sample Solutions

A right triangle has a leg with a length of 18 and a hypotenuse with a length of 36. Bernie notices that the hypotenuse is twice the length of the given leg, which means it is a 30–60–90 triangle. If Bernie is right, what should the length of the remaining leg be? Explain your answer. Confirm your answer using the Pythagorean theorem.

A right angle and two given sides of a triangle determine a unique triangle. All 30–60–90 triangles are similar by AA criterion, and the lengths of their sides are $1c$, $2c$, and $c\sqrt{3}$, for some positive number c. The given hypotenuse is twice the length of the given leg of the right triangle, so Bernie's conclusion is accurate. The ratio of the length of the short leg to the length of the longer leg of any 30–60–90 triangle is $1:\sqrt{3}$. The given leg has a length of 18, which is $\frac{1}{2}$ of the hypotenuse and, therefore, must be the shorter leg of the triangle.

$$\frac{1}{\sqrt{3}} = \frac{18}{\text{leg}_2}$$
$$\text{leg}_2 = 18\sqrt{3}$$

$$18^2 + \text{leg}^2 = 36^2$$
$$324 + \text{leg}^2 = 1296$$
$$\text{leg}^2 = 972$$
$$\text{leg} = \sqrt{972}$$
$$\text{leg} = \sqrt{324}\sqrt{3}$$
$$\text{leg} = 18\sqrt{3}$$

Problem Set Sample Solutions

1. In each row of the table below are the lengths of the legs and hypotenuses of different right triangles. Find the missing side lengths in each row, in simplest radical form.

Leg_1	Leg_2	Hypotenuse
15	20	25
15	36	39
3	$2\sqrt{10}$	7
100	200	$100\sqrt{5}$

Answers provided in table

By the Pythagorean theorem:

$$\text{leg}_1^2 + \text{leg}_1^2 = \text{hyp}^2$$
$$15^2 + y^2 = 25^2$$
$$225 + y^2 = 625$$
$$y^2 = 400$$
$$y = 20 \text{ or } y = -20$$

The case where $y = -20$ does not make sense since it represents a length of a side of a triangle and is, therefore, disregarded.

Alternative strategy:

Divide each side length by the greatest common factor to get the side lengths of a similar right triangle. Find the missing side length for the similar triangle, and multiply by the GCF.

Leg: 15, Hypotenuse: 25, $\text{GCF}(15, 25) = 5$

Consider the triangle with Leg: 3 *and* Hypotenuse: 5. *This is a 3–4–5 right triangle. The missing leg length is $5 \cdot 4 = 20$.*

Legs: 100 *and* 200: $\text{GCF}(100, 200) = 100$

Consider the right triangle with Legs: 1 *and* 2. Hypotenuse $= \sqrt{5}$. *The missing hypotenuse length is $100\sqrt{5}$.*

2. **Claude sailed his boat due south for 38 miles and then due west for 25 miles. Approximately how far is Claude from where he began?**

Claude's path forms a right triangle since south and west are perpendicular to each other. His distance from where he began is a straight line segment represented by the hypotenuse of the triangle.

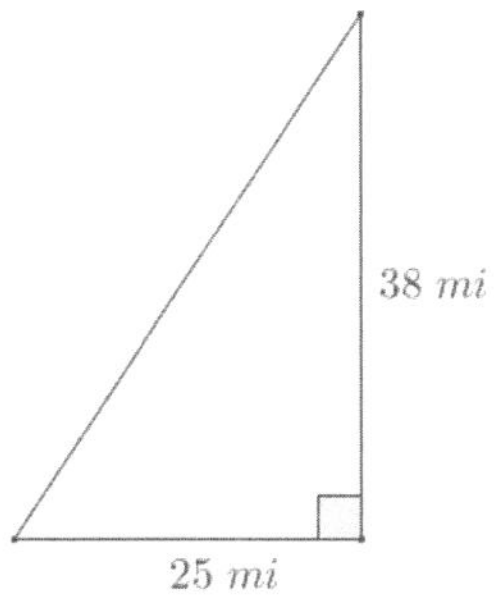

$$\begin{aligned} 38^2 + 25^2 &= \text{hyp}^2 \\ 1444 + 625 &= \text{hyp}^2 \\ 2069 &= \text{hyp}^2 \\ \sqrt{2069} &= \text{hyp} \end{aligned}$$

Claude is approximately 45.5 miles from where he began.

3. **Find the lengths of the legs in the triangle given the hypotenuse with length 100.**

By the Pythagorean theorem:

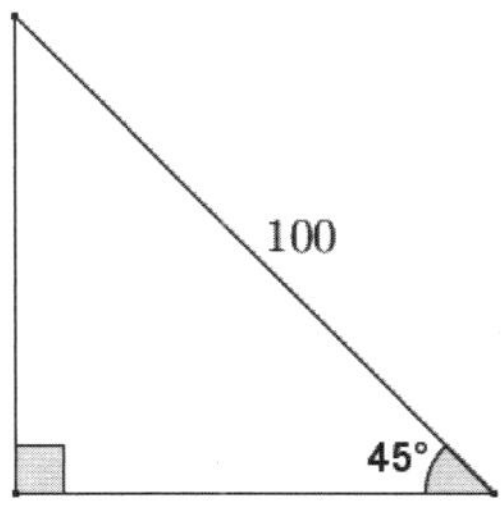

$$\begin{aligned} \text{leg}^2 + \text{leg}^2 &= 100^2 \\ 2\ \text{leg}^2 &= 10000 \\ \text{leg}^2 &= 5000 \\ \text{leg} &= \sqrt{5000} \\ \text{leg} &= \sqrt{2500}\sqrt{2} \\ \text{leg} &= 50\sqrt{2} \end{aligned}$$

Alternative strategy:

The right triangle is an isosceles right triangle, so the leg lengths are equal. The hypotenuse of an isosceles right triangle can be calculated as follows:

$$\begin{aligned} \text{hyp} &= \text{leg} \cdot \sqrt{2} \\ 100 &= \text{leg}\sqrt{2} \\ \frac{100}{\sqrt{2}} &= \text{leg} \\ \frac{100\sqrt{2}}{2} &= 50\sqrt{2} = \text{leg} \end{aligned}$$

The legs of the 45–45–90 right triangle with a hypotenuse of 100 are $50\sqrt{2}$.

4. **Find the length of the hypotenuse in the right triangle given that the legs have lengths of 100.**

By the Pythagorean theorem:

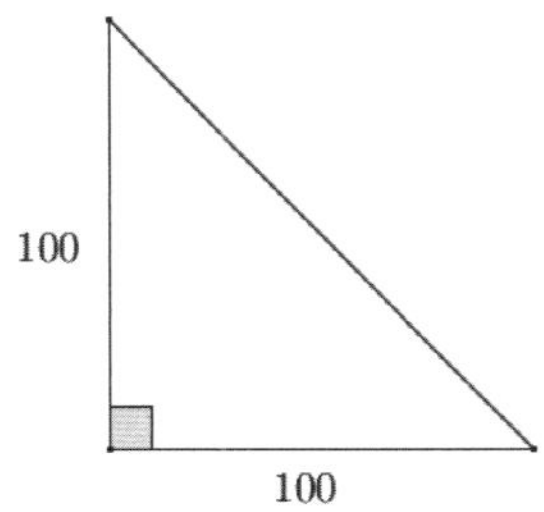

$$\begin{aligned} 100^2 + 100^2 &= \text{hyp}^2 \\ 10000 + 10000 &= \text{hyp}^2 \\ 20000 &= \text{hyp}^2 \\ \sqrt{20000} &= \text{hyp} \\ \sqrt{10000}\sqrt{2} &= \text{hyp}^2 \\ 100\sqrt{2} &= \text{hyp} \end{aligned}$$

Alternative strategy:

The given right triangle is a 45–45–90 triangle. Therefore, the ratio of the length of its legs to the length of its hypotenuse is $1 : \sqrt{2}$.

$$\begin{aligned} \text{hyp} &= \text{leg} \cdot \sqrt{2} \\ \text{hyp} &= 100\sqrt{2} \end{aligned}$$

The hypotenuse of the right triangle with legs of length 100 is $100\sqrt{2}$.

Each row in the table below shows the side lengths of a different 30–60–90 right triangle. Complete the table with the missing side lengths in simplest radical form. Use the relationships of the values in the first three rows to complete the last row. How could the expressions in the last row be used?

Shorter Leg	Longer Leg	Hypotenuse
25	$25\sqrt{3}$	50
15	$15\sqrt{3}$	30
$\sqrt{3}$	3	$2\sqrt{3}$
x	$x\sqrt{3}$	$2x$

The last row of the table shows that the sides of a 30–60–90 right triangle are multiples of 1, 2, and $\sqrt{3}$ by some constant x, with $2x$ being the longest and, therefore, the hypotenuse. The expressions could be used to find two unknown sides of a 30–60–90 triangle where only one of the sides is known.

5. **In right triangle ABC with $\angle C$ a right angle, an altitude of length h is dropped to side $\overline{AB}$ that splits the side $\overline{AB}$ into segments of length x and y. Use the Pythagorean theorem to show $h^2 = xy$.**

By the Pythagorean theorem, $a^2 + b^2 = c^2$.

Since $c = x + y$, we have

$a^2 + b^2 = (x + y)^2$

$a^2 + b^2 = x^2 + 2xy + y^2$.

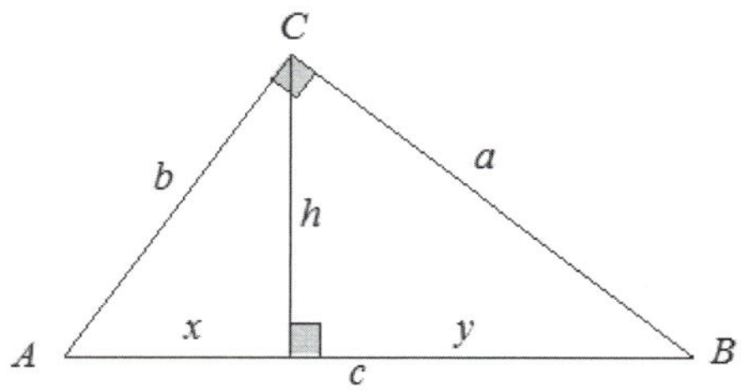

Also by the Pythagorean theorem,

$a^2 = h^2 + y^2$, *and* $b^2 = h^2 + x^2$, *so*

$a^2 + b^2 = h^2 + y^2 + h^2 + x^2$

$a^2 + b^2 = 2h^2 + x^2 + y^2$.

Thus, by substitution,

$x^2 + 2xy + y^2 = 2h^2 + x^2 + y^2$

$2xy = 2h^2$

$xy = h^2$.

6. **In triangle ABC, the altitude from $\angle C$ splits side $\overline{AB}$ into two segments of lengths x and y. If h denotes the length of the altitude and $h^2 = xy$, use the Pythagorean theorem and its converse to show that triangle ABC is a right triangle with $\angle C$ a right angle.**

Let a, b, and c be the lengths of the sides of the triangle opposite $\angle A$, $\angle B$, and $\angle C$, respectively. By the Pythagorean theorem:

$a^2 = h^2 + y^2$ *and* $b^2 = h^2 + x^2$, *so*

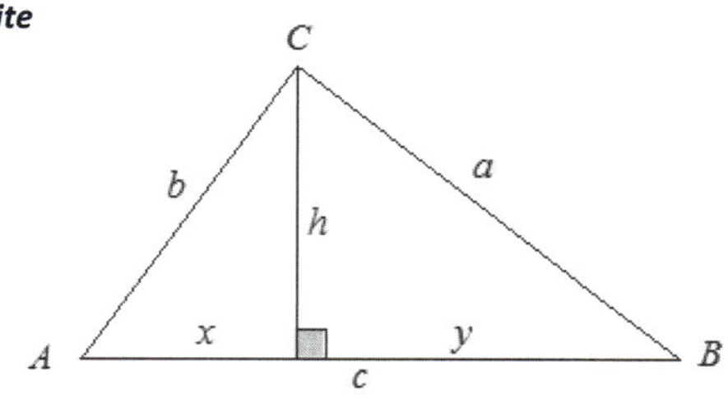

$a^2 + b^2 = h^2 + y^2 + h^2 + x^2$

$a^2 + b^2 = x^2 + 2h^2 + y^2$

$a^2 + b^2 = x^2 + 2xy + y^2$

$a^2 + b^2 = (x + y)^2$

$a^2 + b^2 = c^2$.

So, by the converse of the Pythagorean theorem, $\triangle ABC$ is a right triangle, a and b are the lengths of legs of the triangle, and c is the hypotenuse, which lies opposite the right angle. Therefore, $\angle C$ is the right angle of the right triangle.

Topic E

Trigonometry

G-SRT.C.6, G-SRT.C.7, G-SRT.C.8

Focus Standards:	G-SRT.C.6	Understand that by similarity, side ratios in right triangles are properties of the angles in the triangle, leading to definitions of trigonometric ratios for acute angles.
	G-SRT.C.7	Explain and use the relationship between the sine and cosine of complementary angles.
	G-SRT.C.8	Use trigonometric ratios and the Pythagorean Theorem to solve right triangles in applied problems.★
Instructional Days:	10	
Lesson 25:	Incredibly Useful Ratios (E)[1]	
Lesson 26:	The Definition of Sine, Cosine, and Tangent (P)	
Lesson 27:	Sine and Cosine of Complementary Angles and Special Angles (P)	
Lesson 28:	Solving Problems Using Sine and Cosine (P)	
Lesson 29:	Applying Tangents (P)	
Lesson 30:	Trigonometry and the Pythagorean Theorem (P)	
Lesson 31:	Using Trigonometry to Determine Area (P)	
Lesson 32:	Using Trigonometry to Find Side Lengths of an Acute Triangle (S)	
Lesson 33:	Applying the Laws of Sines and Cosines (P)	
Lesson 34:	Unknown Angles (P)	

Students begin the study of trigonometry in the final topic of the module. The emphasis in the module on side length relationships within similar triangles (Topic C) and the specific emphasis on right triangles (Topic D) help set the foundation for trigonometry. Lesson 25 is a last highlight of the side length ratios within and between right triangles. Students are guided to the idea that the values of the ratios depend solely on a given acute angle in the right triangle before the basic trigonometric ratios are explicitly defined in Lesson 26 (**G.SRT.C.6**). After practice with ratios labeled as $\text{shorter leg}: \text{hypotenuse}$ (Lesson 21) and $\text{opp}: \text{hyp}$ (Lesson 25), students are introduced to the trigonometric ratios sine, cosine, and tangent (**G-SRT.C.6**) in Lesson 26. Students examine the relationship between sine and cosine in Lesson 27, discovering that the sine and cosine of complementary angles are equal (**G-SRT.C.7**). They are also

[1]Lesson Structure Key: **P**-Problem Set Lesson, **M**-Modeling Cycle Lesson, **E**-Exploration Lesson, **S**-Socratic Lesson

introduced to the common sine and cosine values of angle measures frequently seen in trigonometry. Students apply the trigonometric ratios to solve for unknown lengths in Lessons 28 and 29; students also learn about the relationship between tangent and slope in Lesson 29 (**G-SRT.C.8**). In Lesson 30, students use the Pythagorean theorem to prove the identity $\sin^2\theta + \cos^2\theta = 1$ and also show why $\tan\theta = \frac{\sin\theta}{\cos\theta}$. In Lessons 31–33, students study the application of trigonometry to determine area and solve for unknown lengths using the laws of sines and cosines (**G-SRT.9**, **G-SRT.10**, **G-SRT.11**). Finally, in Lesson 34, students learn how to determine the unknown measure of an angle of a right triangle. Students are introduced to the trigonometric functions $\arcsin$, $\arccos$, and $\arctan$. These inverse functions are taught formally in Algebra II. For now, students should understand the meaning of and how to use $\arcsin$, $\arccos$, and $\arctan$ to determine unknown measures of angles.

Lesson 25: Incredibly Useful Ratios

Student Outcomes

- For a given acute angle of a right triangle, students identify the opposite side, adjacent side, and hypotenuse.
- Students understand that the values of the $\frac{\text{adj}}{\text{hyp}}$ and $\frac{\text{opp}}{\text{hyp}}$ ratios for a given acute angle are constant.

Lesson Notes

In Lesson 25, students discover that the values of the $\frac{\text{adj}}{\text{hyp}}$ and $\frac{\text{opp}}{\text{hyp}}$ ratios in a right triangle depend solely on the measure of the acute angle by which the adjacent, opposite, and hypotenuse sides are identified. To do this, students first learn how to identify these reference labels. Then, two groups take measurements and make calculations of the values of the $\frac{\text{adj}}{\text{hyp}}$ and $\frac{\text{opp}}{\text{hyp}}$ ratios for two sets of triangles, where each triangle in one set is similar to a triangle in the other. This exploration leads to the conclusion regarding the "incredibly useful ratios."

Classwork

Exercises 1–3 (4 minutes)

MP.7

Exercises 1–3

Use the right triangle $\triangle ABC$ to answer Exercises 1–3.

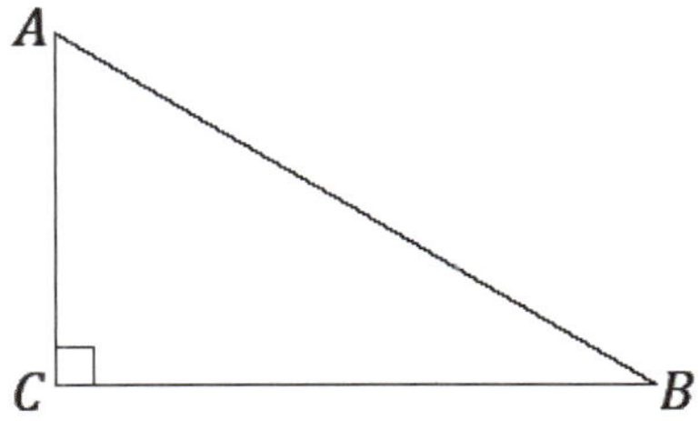

1. **Name the side of the triangle opposite $\angle A$.**

   $\overline{BC}$

2. **Name the side of the triangle opposite $\angle B$.**

   $\overline{AC}$

3. **Name the side of the triangle opposite $\angle C$.**

   $\overline{AB}$

- Another common way of labeling the sides of triangles is to make use of the relationship between a vertex and the side opposite the vertex and the lowercase letter of each vertex angle.

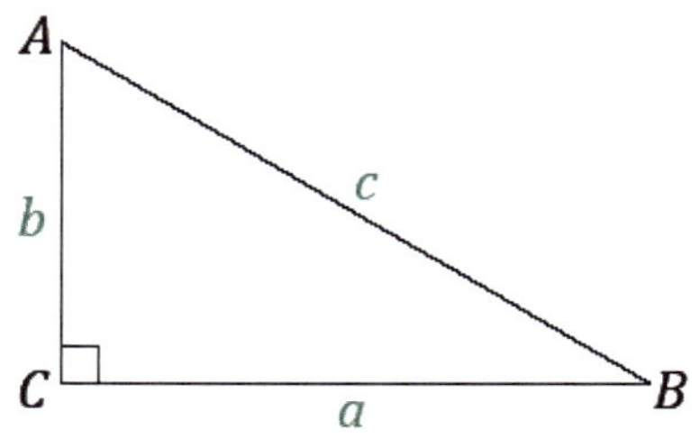

- In the diagram above, label the side opposite $\angle A$ as a. Similarly, label the side opposite $\angle B$ as b and the side opposite $\angle C$ as c.
- Why is this way of labeling the triangle useful?
  - *Using a lowercase letter to represent the side* opposite *a vertex immediately describes a relationship within a triangle. It might be useful when we are sketching diagrams for problems that provide partial information about a triangle.*

Discussion (8 minutes)

- Now we will discuss another set of labels. These labels are specific to right triangles.
- The triangle below is right triangle $\triangle ABC$. Denote $\angle B$ as the right angle of the triangle.
- Of the two acute angles, we are going to focus our attention on labeling $\angle A$. Mark $\angle A$ in the triangle with a single arc.
- With respect to acute angle $\angle A$ of a right triangle $\triangle ABC$, we identify the opposite side, the adjacent side, and the hypotenuse as follows:
  - With respect to $\angle A$, the opposite side, denoted opp, is side $\overline{BC}$,
  - With respect to $\angle A$, the adjacent side, denoted adj, is side $\overline{AB}$,
  - The hypotenuse, denoted hyp, is side $\overline{AC}$ and is opposite from the $90°$ angle.
- With respect to $\angle A$, label the sides of the triangle as opp, adj, and hyp.

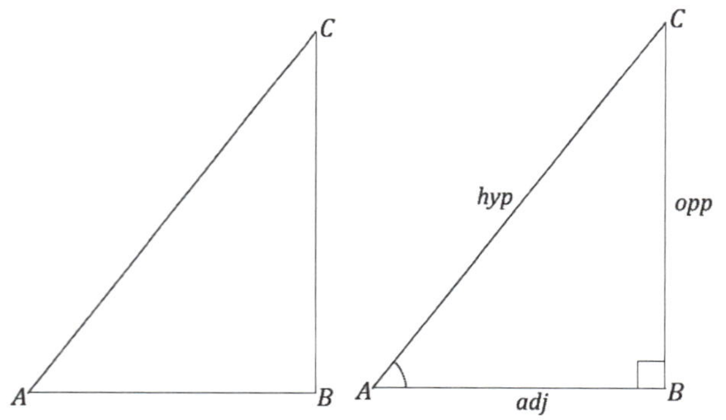

> *Scaffolding:*
> - Consider asking students to create a poster of this diagram to display on the classroom wall (alternatively, a sample is provided in this lesson before the Exit Ticket).
> - Consider color coding the sides and the definitions. For example, write the word *hypotenuse* in red, and then trace that side of the triangle in red.
> - Rehearse the various terms by pointing to each and asking for choral responses of *opposite, adjacent,* and *hypotenuse*.

- Note that the sides are identified with respect to $\angle A$, or using $\angle A$ as a reference. The labels change if the acute angle selected changes. How would the labels of the sides change if we were looking at acute angle $\angle C$, instead of $\angle A$, as a reference?
  - *The hypotenuse would remain the same, but the opposite side of $\angle C$ is $\overline{AB}$, and the adjacent side of $\angle C$ is $\overline{BC}$.*

- The label *hypotenuse* will always be the hypotenuse of the right triangle, that is, the side opposite from the right angle, but the *adjacent* and *opposite* side labels are dependent upon which acute angle is used as a reference.

Exercises 4–6 (4 minutes)

In Exercises 4–6, students practice identifying the adjacent side, the opposite side, and the hypotenuse of a right triangle, with respect to a given acute angle.

Exercises 4–6

For each exercise, label the appropriate sides as *adjacent, opposite,* and *hypotenuse,* with respect to the marked acute angle.

4.

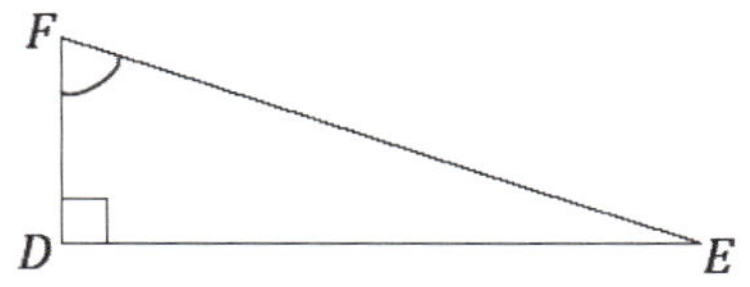

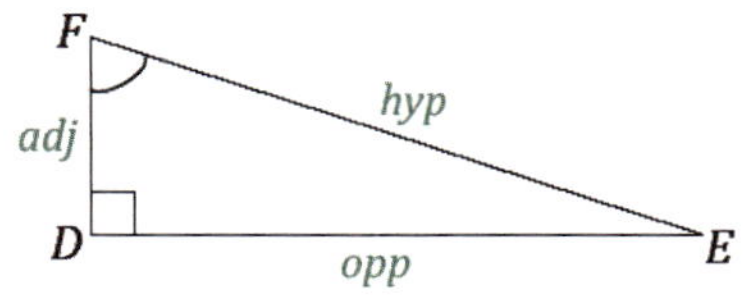

5.

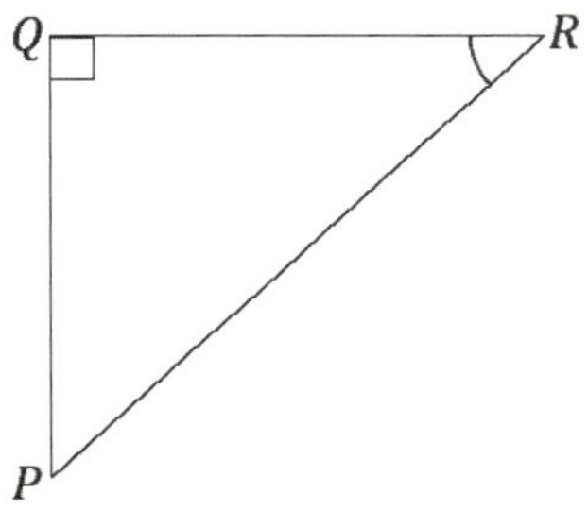

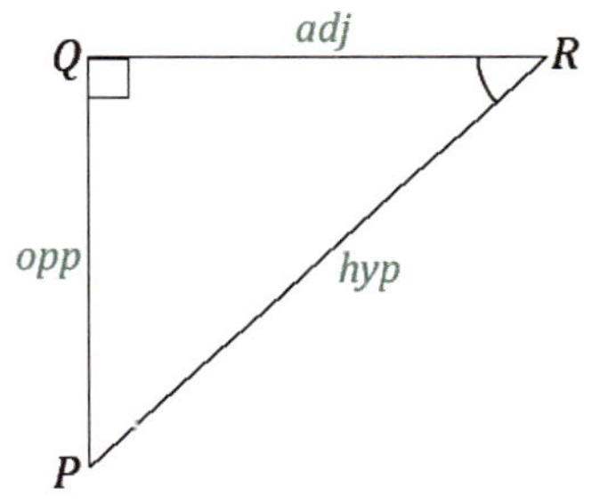

6.

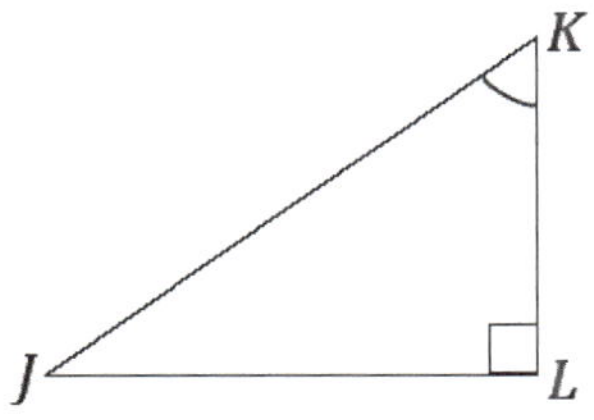

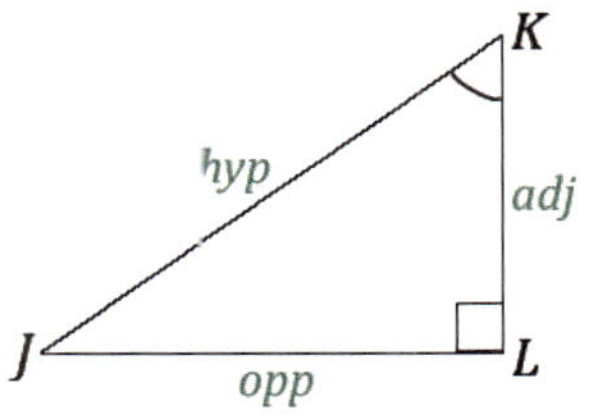

Exploratory Challenge (10 minutes)

In the following Exploratory Challenge, assign one set of triangles to each half of the class. Every triangle in each set has a corresponding, similar triangle in the other set. Students complete the table for missing angle and side length measurements and values of the ratios $\frac{\text{adj}}{\text{hyp}}$ and $\frac{\text{opp}}{\text{hyp}}$.

Exploratory Challenge

Note: Angle measures are approximations.

For each triangle in your set, determine missing angle measurements and side lengths. Side lengths should be measured to one decimal place. Make sure that each of the $\frac{\text{adj}}{\text{hyp}}$ and $\frac{\text{opp}}{\text{hyp}}$ ratios are set up and missing values are calculated and rounded appropriately.

Group 1					
	Triangle	Angle Measures	Length Measures	$\frac{\text{opp}}{\text{hyp}}$	$\frac{\text{adj}}{\text{hyp}}$
1.	$\triangle ABC$	$m\angle A \approx 67°$	$AB = 5$ cm $BC = 12$ cm $AC = 13$ cm	$\frac{12}{13} \approx 0.92$	$\frac{5}{13} \approx 0.38$
2.	$\triangle DEF$	$m\angle D \approx 53°$	$DE = 3$ cm $EF = 4$ cm $DF = 5$ cm	$\frac{4}{5} = 0.8$	$\frac{3}{5} = 0.6$
3.	$\triangle GHI$	$m\angle I \approx 41°$	$GH = 5.3$ cm $HI = 6$ cm $IG = 8$ cm	$\frac{5.3}{8} \approx 0.66$	$6/8 = 0.75$
4.	$\triangle JKL$	$m\angle L \approx 30°$	$JK = 4$ cm $KL = 6.93$ cm $JL = 8$ cm	$\frac{4}{8} = 0.5$	$\frac{6.93}{8} \approx 0.87$
5.	$\triangle MNO$	$m\angle M \approx 28°$	$MN = 7.5$ cm $NO = 4$ cm $OM = 8.5$ cm	$\frac{4}{8.5} \approx 0.47$	$\frac{7.5}{8.5} \approx 0.88$

Group 2					
	Triangle	Angle Measures	Length Measures	$\frac{\text{opp}}{\text{hyp}}$	$\frac{\text{adj}}{\text{hyp}}$
1.	$\triangle A'B'C'$	$m\angle A' \approx 67°$	$A'B' = 2.5$ cm $B'C' = 6$ cm $A'C' = 6.5$ cm	$\frac{6}{6.5} \approx 0.92$	$\frac{2.5}{6.5} \approx 0.38$
2.	$\triangle D'E'F'$	$m\angle D' \approx 53°$	$D'E' = 6$ cm $E'F' = 8$ cm $D'F' = 10$ cm	$\frac{8}{10} = 0.8$	$\frac{6}{10} = 0.6$
3.	$\triangle G'H'I'$	$m\angle I' \approx 41°$	$G'H' = 7.9$ cm $H'I' = 9$ cm $I'G' = 12$ cm	$\frac{7.9}{12} \approx 0.66$	$\frac{9}{12} = 0.75$
4.	$\triangle J'K'L'$	$m\angle L' \approx 30°$	$J'K' = 6$ cm $K'L' = 10.4$ cm $J'L' = 12$ cm	$\frac{6}{12} = 0.5$	$\frac{10.4}{12} \approx 0.87$
5.	$\triangle M'N'O'$	$m\angle M' \approx 28°$	$M'N' = 15$ cm $N'O' = 8$ cm $O'M' = 17$ cm	$\frac{8}{17} \approx 0.47$	$\frac{15}{17} \approx 0.88$

Have each half of the class share all of the actual measurements and values of the ratios for both sets of triangles (on a poster or on the board).

MP.7

With a partner, discuss what you can conclude about each pair of triangles between the two sets.

Each pair of triangles is similar by the AA criterion. The marked acute angle for each pair of corresponding triangles is the same. The values of the $\frac{\text{adj}}{\text{hyp}}$ ***and*** $\frac{\text{opp}}{\text{hyp}}$ ***ratios are the same.***

- Since the triangles are similar, it is no surprise that corresponding angles are equal in measure and that the values of the ratios of lengths are in constant proportion.
- Use these observations to guide you through solving for unknown lengths in Exercises 7–10. Consider why it is that you are able to actually find the missing lengths.

Exercises 7–10 (10 minutes)

Students approximate unknown lengths to one decimal place. The answers are approximations because the acute angles are really only approximations, not exact measurements.

Exercises 7–10

For each question, round the unknown lengths appropriately. Refer back to your completed chart from the Exploratory Challenge; each indicated acute angle is the same approximated acute angle measure as in the chart. Set up and label the appropriate length ratios, using the terms opp, adj, and hyp in the setup of each ratio.

7.

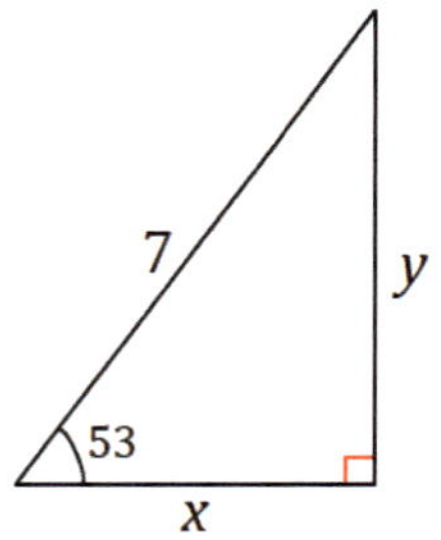

$\frac{\text{opp}}{\text{hyp}} = 0.8$

$\frac{y}{7} = 0.8$

$y = 5.6$

$\frac{\text{adj}}{\text{hyp}} = 0.6$

$\frac{x}{7} = 0.6$

$x = 4.2$

8.

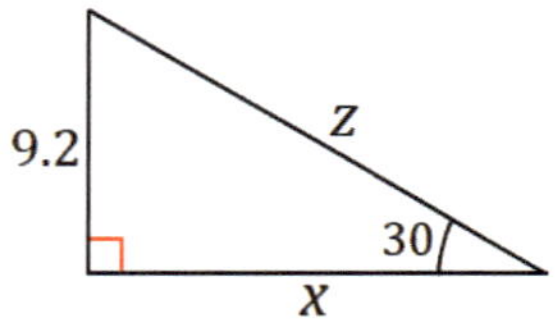

$\frac{\text{opp}}{\text{hyp}} = 0.5$

$\frac{9.2}{z} = 0.5$

$z = 18.4$

$\frac{\text{adj}}{\text{hyp}} = 0.87$

$\frac{x}{18.4} = 0.87$

$x = 16.0$

9.

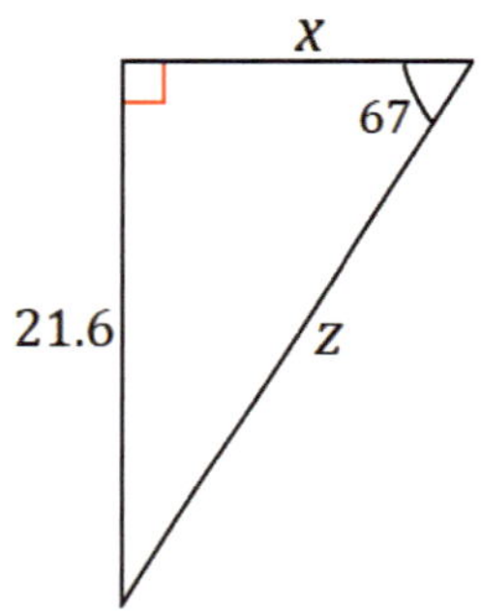

$\frac{\text{opp}}{\text{hyp}} = 0.92$

$\frac{21.6}{z} = 0.92$

$z = 23.48$

$\frac{\text{adj}}{\text{hyp}} = 0.38$

$\frac{x}{z} = 0.38$

$\frac{x}{23.48} = 0.38$

$x = 8.9$

10. **From a point 120 m away from a building, Serena measures the angle between the ground and the top of a building and finds it measures $41°$.**

What is the height of the building? Round to the nearest meter.

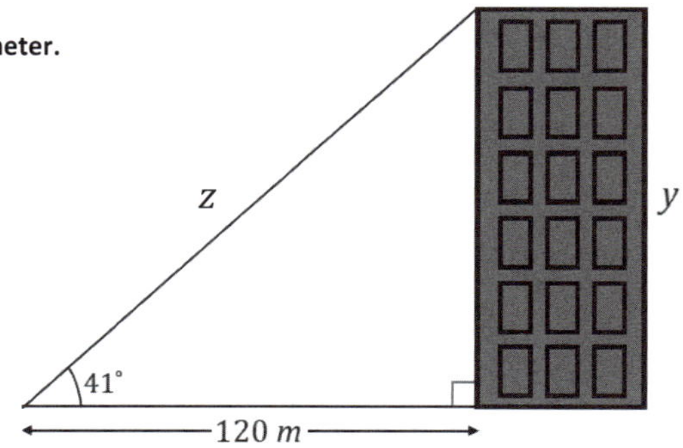

$\frac{\text{adj}}{\text{hyp}} = 0.75$

$\frac{120}{z} = 0.75$

$z = 160$

$\frac{\text{opp}}{\text{hyp}} = 0.66$

$\frac{y}{160} = 0.66$

$y = 105.6$

The height of the building is approximately 106 *meters.*

Closing (4 minutes)

- What did you notice about the triangles in Exercises 7–10 and the exercises in the Exploratory Challenge? Explain.
  - *The triangles in the exercises were similar to those used in the Exploratory Challenge. We can be sure of this because the triangles in the exercises and the Exploratory Challenge each had two angle measures in common: each had a right angle and an acute angle equal in measure.*
- Why were you able to rely on the Exploratory Challenge chart to determine the unknown lengths in Exercises 7–10?
  - *Because the triangles in Exercises 7–10 are similar to those triangles in the Exploratory Challenge, we were able to use the values of the* $\frac{\text{opp}}{\text{hyp}}$ *and* $\frac{\text{adj}}{\text{hyp}}$ *ratios to set up an equation for each question.*
- Since all the triangles used in the lesson are right triangles in general, what determines when two right triangles have $\frac{\text{opp}}{\text{hyp}}$ *and* $\frac{\text{adj}}{\text{hyp}}$ *ratios equal in value*?
  - *Both right triangles must have corresponding acute angles of equal measure, and the same acute angle must be referenced in both triangles for the values of the* $\frac{\text{opp}}{\text{hyp}}$ *and* $\frac{\text{adj}}{\text{hyp}}$ *ratios to be equal.*
- In this lesson, you were provided with just a handful of ratio values, each of which corresponded to a given acute angle measure in a right triangle. Knowing these ratio values for the given acute angle measure allowed us to solve for unknown lengths in triangles similar to right triangles; this is what makes these ratios so incredible. In the next lesson, we learn that we have easy access to these ratios for *any* angle measure.

Relevant Vocabulary

SIDES OF A RIGHT TRIANGLE: The *hypotenuse* of a right triangle is the side opposite the right angle; the other two sides of the right triangle are called the *legs*. Let θ be the angle measure of an acute angle of the right triangle. The *opposite side* is the leg opposite that angle. The *adjacent side* is the leg that is contained in one of the two rays of that angle (the hypotenuse is contained in the other ray of the angle).

Exit Ticket (5 minutes)

Name ________________________________ Date ______________

Lesson 25: Incredibly Useful Ratios

Exit Ticket

1. Use the chart from the Exploratory Challenge to approximate the unknown lengths y and z to one decimal place.

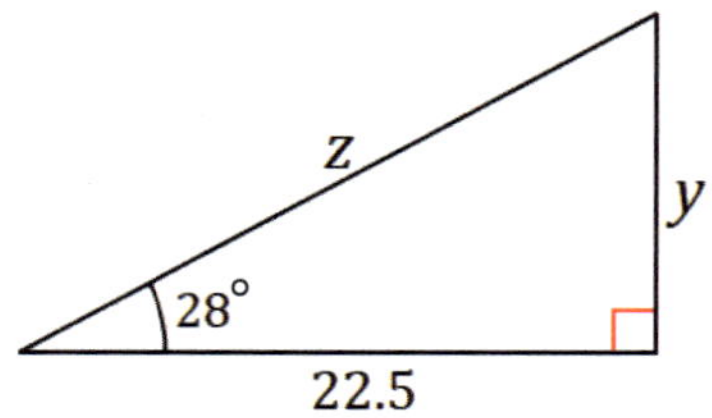

2. Why can we use the chart from the Exploratory Challenge to approximate the unknown lengths?

Exit Ticket Sample Solutions

1. Use the chart from the Exploratory Challenge to approximate the unknown lengths y and z to one decimal place.

$\frac{\text{adj}}{\text{hyp}} = 0.88$

$\frac{22.5}{z} = 0.88$

$z = 25.6$

$\frac{\text{opp}}{\text{hyp}} = 0.47$

$\frac{y}{25.6} = 0.47$

$y \approx 12$

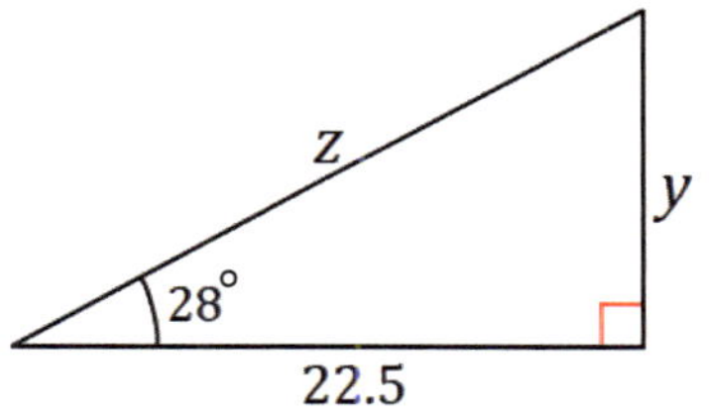

2. Why can we use the chart from the Exploratory Challenge to approximate the unknown lengths?

The triangle in Problem 1 is similar to a triangle in the chart from the Exploratory Challenge. Since the triangles are similar, the values of the $\frac{\text{opp}}{\text{hyp}}$ *and* $\frac{\text{adj}}{\text{hyp}}$ *ratios in reference to the acute angle of* $28°$ *can be used in the equations needed to solve for unknown lengths.*

Problem Set Sample Solutions

The table below contains the values of the ratios $\frac{\text{opp}}{\text{hyp}}$ and $\frac{\text{adj}}{\text{hyp}}$ for a variety of right triangles based on a given acute angle, θ, from each triangle. Use the table and the diagram of the right triangle below to complete each problem.

θ (degrees)	0	10	20	30	40	45	50	60	70	80	90
$\frac{\text{opp}}{\text{hyp}}$	0	0.1736	0.3420	$\frac{1}{2} = 0.5$	0.6428	0.7071	0.7660	0.8660	0.9397	0.9848	1
$\frac{\text{adj}}{\text{hyp}}$	1	0.9848	0.9397	0.8660	0.7660	0.7071	0.6428	$\frac{1}{2} = 0.5$	0.3420	0.1736	0

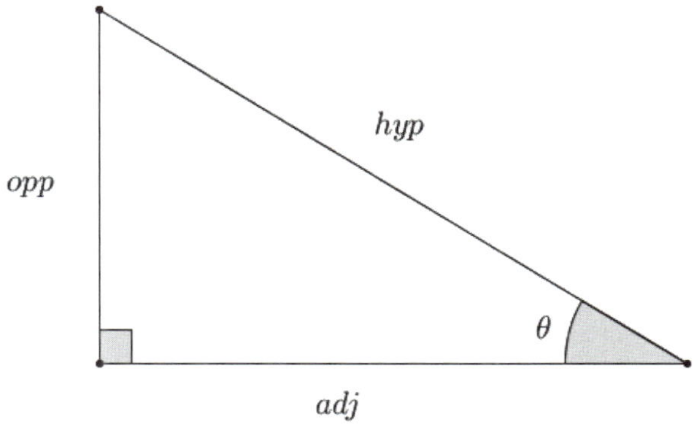

NOT DRAWN TO SCALE

For each problem, approximate the unknown lengths to one decimal place. Write the appropriate length ratios using the terms opp, adj, and hyp in the setup of each ratio.

1. **Find the approximate length of the leg opposite the $80°$ angle.**

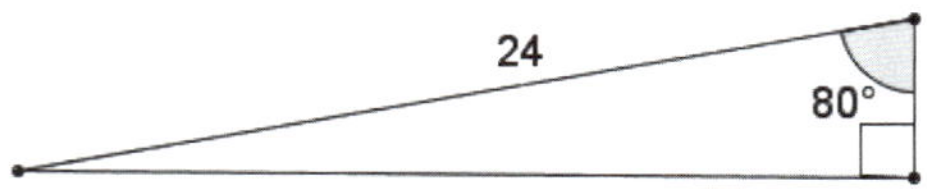

Using the value of the ratio $\frac{\text{opp}}{\text{hyp}}$ *for an* $80°$ *angle:*

$$\frac{\text{opp}}{24} = 0.9848$$
$$\text{opp} = 23.6352$$

The length of the leg opposite the $80°$ *angle is approximately* 23.6.

2. **Find the approximate length of the hypotenuse.**

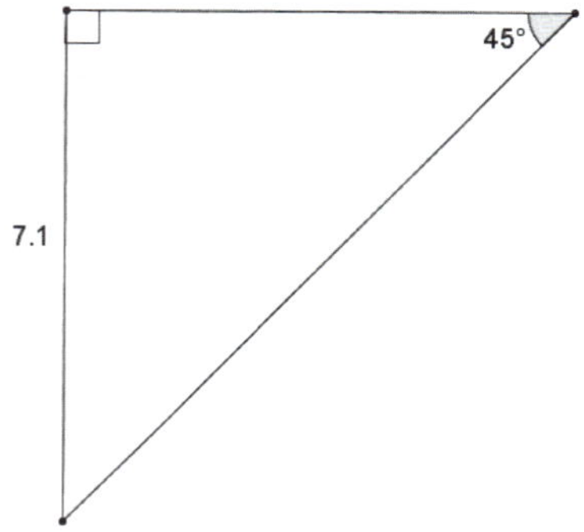

Using the value of the ratio $\frac{\text{opp}}{\text{hyp}}$ *for a* $45°$ *angle:*

$$\frac{7.1}{\text{hyp}} = 0.7071$$
$$\frac{7.1}{0.7071} = \text{hyp}$$
$$10.0410 \approx \text{hyp}$$

The length of the hypotenuse is approximately 10.0.

3. **Find the approximate length of the hypotenuse.**

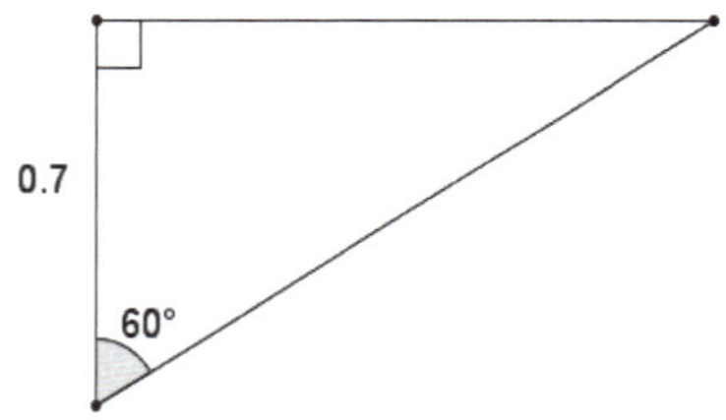

Using the value of the ratio $\frac{\text{adj}}{\text{hyp}}$ *for a* $60°$ *angle:*

$$\frac{0.7}{\text{hyp}} = \frac{1}{2}$$
$$\frac{0.7}{\frac{1}{2}} = \text{hyp}$$
$$1.4 = \text{hyp}$$

The length of the hypotenuse is 1.4.

4. **Find the approximate length of the leg adjacent to the $40°$ angle.**

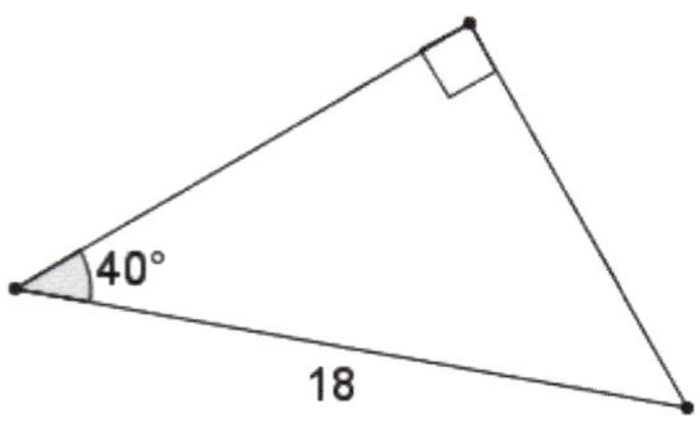

Using the value of the ratio $\frac{\text{adj}}{\text{hyp}}$ *for a* $40°$ *angle:*

$$\frac{\text{adj}}{18} = 0.7660$$
$$\text{adj} = 13.788$$

The length of the leg adjacent to the given $40°$ *angle is approximately* 13.8.

5. **Find the length of both legs of the right triangle below. Indicate which leg is adjacent and which is opposite the given angle of $30°$.**

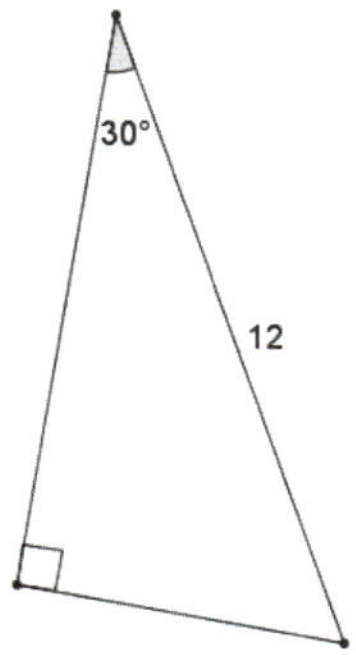

Using the value of the ratio $\frac{\text{opp}}{\text{hyp}}$ for a $30°$ angle:

$$\frac{\text{opp}}{12} = \frac{1}{2}$$
$$\text{opp} = 6$$

The length of the leg that is opposite the given $30°$ angle is 6.

Using the value of the ratio $\frac{\text{adj}}{\text{hyp}}$ for a $30°$ angle:

$$\frac{\text{adj}}{12} = 0.8660$$
$$\text{adj} = 10.392$$

The length of the leg that is adjacent to the given $30°$ angle is approximately 10.4.

6. **Three city streets form a right triangle. Main Street and State Street are perpendicular. Laura Street and State Street intersect at a $50°$ angle. The distance along Laura Street to Main Street is 0.8 mile. If Laura Street is closed between Main Street and State Street for a festival, approximately how far (to the nearest tenth) will someone have to travel to get around the festival if they take only Main Street and State Street?**

The leg opposite the $50°$ angle represents the distance along Main Street, the leg adjacent to the $50°$ angle represents the distance along State Street, and the hypotenuse of the triangle represents the 0.8 mile distance along Laura Street.

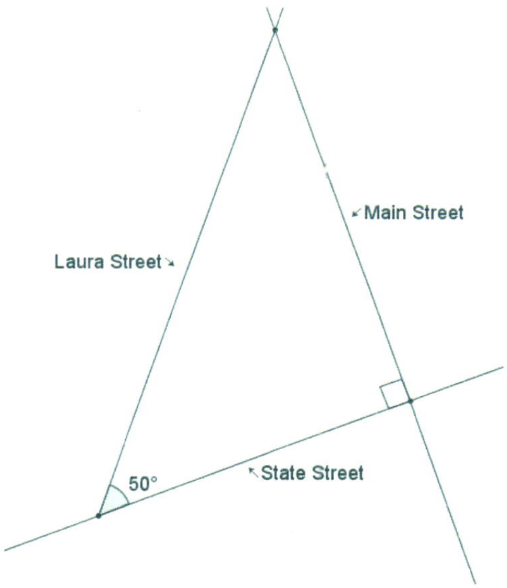

Using the ratio $\frac{\text{opp}}{\text{hyp}}$ for $50°$:

$$0.7660 = \frac{\text{opp}}{0.8}$$
$$0.6128 = \text{opp}$$

Using the ratio $\frac{\text{adj}}{\text{hyp}}$ for $50°$:

$$0.6428 = \frac{\text{adj}}{0.8}$$
$$0.51424 = \text{adj}$$

The total distance of the detour:

$$0.6128 + 0.51424 = 1.12704$$

The total distance to travel around the festival along State Street and Main Street is approximately 1.1 miles.

7. A cable anchors a utility pole to the ground as shown in the picture. The cable forms an angle of $70°$ with the ground. The distance from the base of the utility pole to the anchor point on the ground is 3.8 meters. Approximately how long is the support cable?

The hypotenuse of the triangle represents the length of the support cable in the diagram, and the leg adjacent to the given $70°$ angle represents the distance between the anchor point and the base of the utility pole. Using the value of the ratio $\frac{\text{adj}}{\text{hyp}}$ for $70°$:

$$0.3420 = \frac{3.8}{\text{hyp}}$$
$$\text{hyp} = \frac{3.8}{0.3420}$$
$$\text{hyp} = 11.1111$$

The length of the support cable is approximately 11.1 meters long.

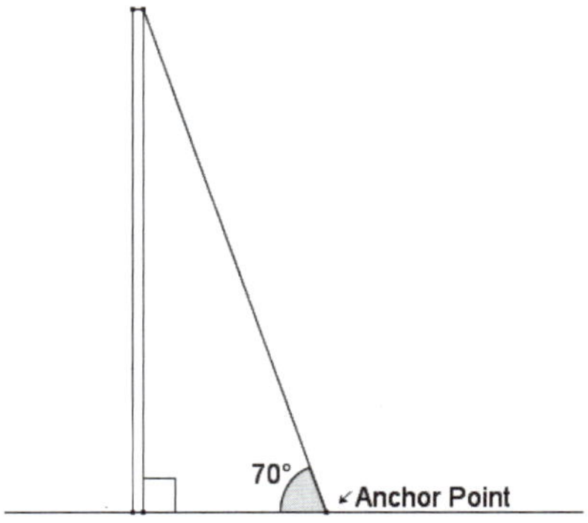

8. Indy says that the ratio of $\frac{\text{opp}}{\text{adj}}$ for an angle of $0°$ has a value of 0 because the opposite side of the triangle has a length of 0. What does she mean?

Indy's triangle is not actually a triangle since the opposite side does not have length, which means that it does not exist. As the degree measure of the angle considered gets closer to $0°$, the value of the ratio gets closer to 0.

Group 1

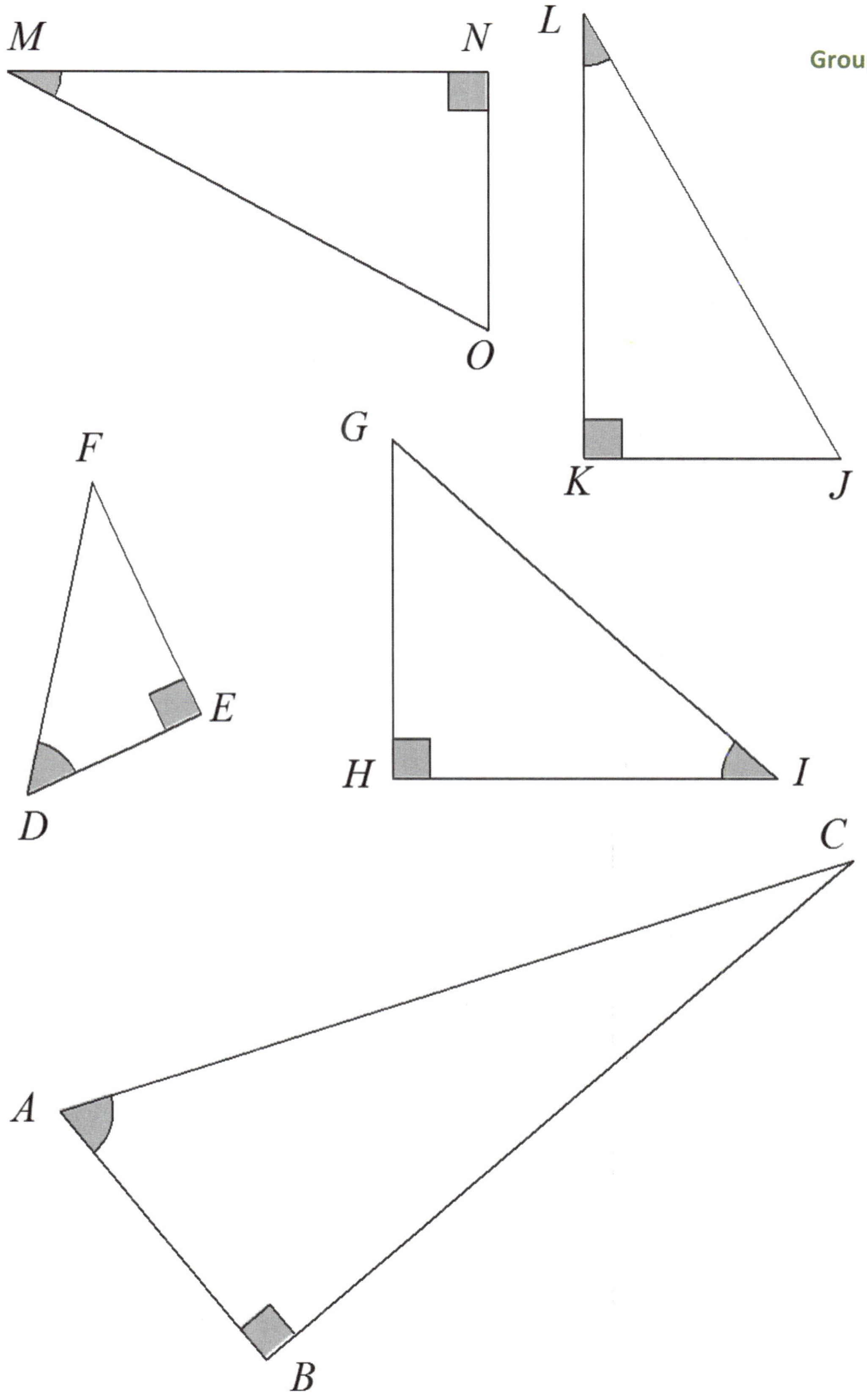

Group 2

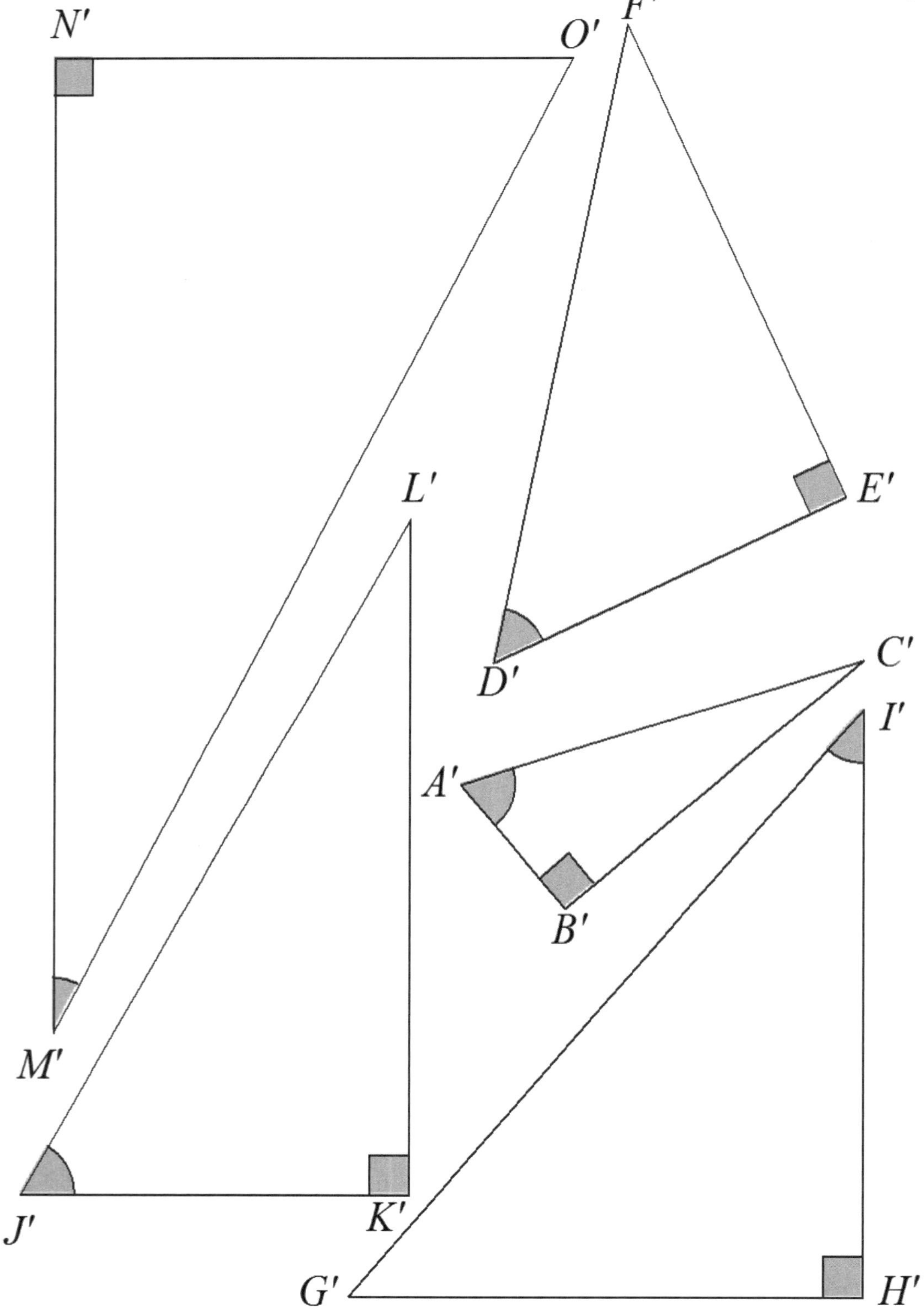

Identifying Sides of a Right Triangle with Respect to a Given Acute Angle

Poster

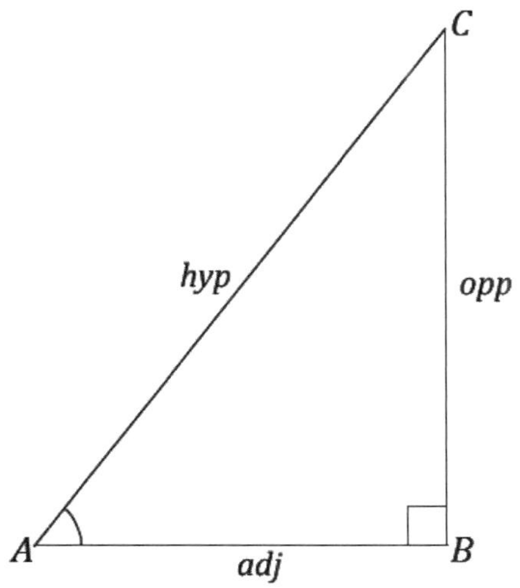

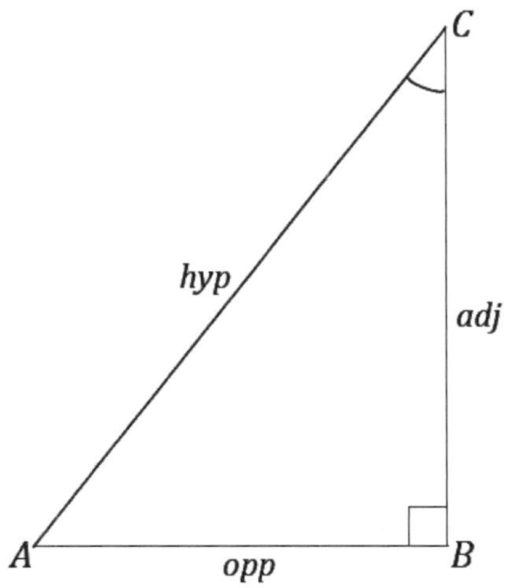

• With respect to $\angle A$, the opposite side, opp, is side $\overline{BC}$. • With respect to $\angle A$, the adjacent side, adj, is side $\overline{AB}$. • The hypotenuse, hyp, is side $\overline{AC}$ and is always opposite from the $90°$ angle.	• With respect to $\angle C$, the opposite side, opp, is side $\overline{AB}$. • With respect to $\angle C$, the adjacent side, adj, is side $\overline{BC}$. • The hypotenuse, hyp, is side $\overline{AC}$ and is always opposite from the $90°$ angle.

Lesson 26: The Definition of Sine, Cosine, and Tangent

Student Outcomes

- Students define sine, cosine, and tangent of θ, where θ is the angle measure of an acute angle of a right triangle. Students denote sine, cosine, and tangent as sin, cos, and tan, respectively.
- If $\angle A$ is an acute angle whose measure in degrees is θ, then $\sin \angle A = \sin\theta$, $\cos \angle A = \cos\theta$, and $\tan \angle A = \tan\theta$.
- Given the side lengths of a right triangle with acute angles, students find sine, cosine, and tangent of each acute angle.

Lesson Notes

It is convenient, as adults, to use the notation "$\sin^2 x$" to refer to the value of the square of the sine function. However, rushing too fast to this abbreviated notation for trigonometric functions leads to incorrect understandings of how functions are manipulated, which can lead students to think that $\sin x$ is short for "$\sin \cdot x$" and to incorrectly divide out the variable $\frac{\sin x}{x} = \sin$.

To reduce these types of common notation-driven errors later, this curriculum is very deliberate about how and when to use abbreviated function notation for sine, cosine, and tangent:

1. In Geometry, sine, cosine, and tangent are thought of as the value of *ratios* of triangles, not as *functions.* No attempt is made to describe the trigonometric ratios as functions of the real number line. Therefore, the notation is just an abbreviation for the sine of an angle ($\sin \angle A$) or sine of an angle measure ($\sin \theta$). Parentheses are used more for grouping and clarity reasons than as symbols used to represent a function.
2. In Algebra II, to distinguish between the ratio version of sine in geometry, all sine functions are notated as functions: $\sin(x)$ is the value of the sine function for the real number x, just like $f(x)$ is the value of the function f for the real number x. In this grade, students maintain function notation integrity and strictly maintain parentheses as part of function notation, writing, for example, $\sin\left(\frac{\pi}{2} - \theta\right) = cos(\theta)$ instead of $\sin\frac{\pi}{2} - \theta = \cos\theta$.
3. By Precalculus and Advanced Topics, students have had two full years of working with sine, cosine, and tangent as both ratios and functions. It is finally in this year that the distinction between ratio and function notations begins to blur and students write, for example, $\sin^2(\theta)$ as the value of the square of the sine function for the real number θ, which is how most calculus textbooks notate these functions.

Classwork

Exercises 1–3 (6 minutes)

The following exercises provide students with practice identifying specific ratios of sides based on the labels learned in Lesson 25. The third exercise leads into an understanding of the relationship between sine and cosine, which is observed in this lesson and formalized in Lesson 27.

Exercises 1–3

1. **Identify the $\frac{\text{opp}}{\text{hyp}}$ ratios for $\angle A$ and $\angle B$.**

   For $\angle A$: $\frac{12}{13}$

   For $\angle B$: $\frac{5}{13}$

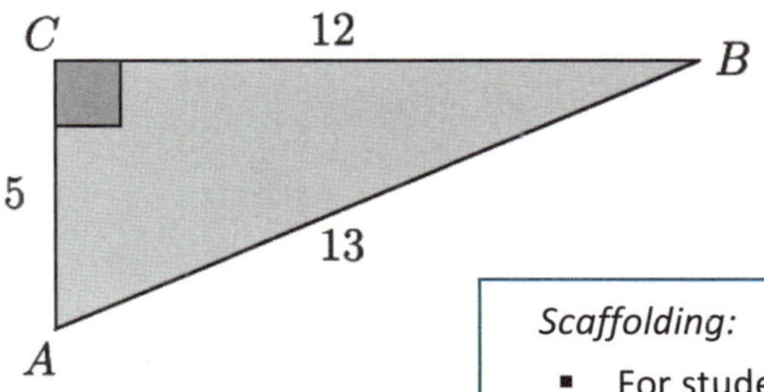

2. **Identify the $\frac{\text{adj}}{\text{hyp}}$ ratios for $\angle A$ and $\angle B$.**

   For $\angle A$: $\frac{5}{13}$

   For $\angle B$: $\frac{12}{13}$

3. **Describe the relationship between the ratios for $\angle A$ and $\angle B$.**

   The $\frac{\text{opp}}{\text{hyp}}$ *ratio for* $\angle A$ *is equal to the* $\frac{\text{adj}}{\text{hyp}}$ *ratio for* $\angle B$.

   The $\frac{\text{opp}}{\text{hyp}}$ *ratio for* $\angle B$ *is equal to the* $\frac{\text{adj}}{\text{hyp}}$ *ratio for* $\angle A$.

Scaffolding:

- For students who are struggling, model labeling opp, hyp, and adj for $\angle A$.
- Allow students to use the reference chart provided with Lesson 25.
- For advanced students, include an extension: Draw another triangle ABC with different side lengths that have the same ratios for $\angle A$ and $\angle B$. Explain how you know the ratios is the same, and justify your answer.

Discussion (6 minutes)

The Discussion defines sine, cosine, and tangent. As the names opp, adj, and hyp are relatively new to students, it is important that students have a visual of the triangle to reference throughout the Discussion. Following the Discussion is an exercise that can be used to informally assess students' understanding of these definitions.

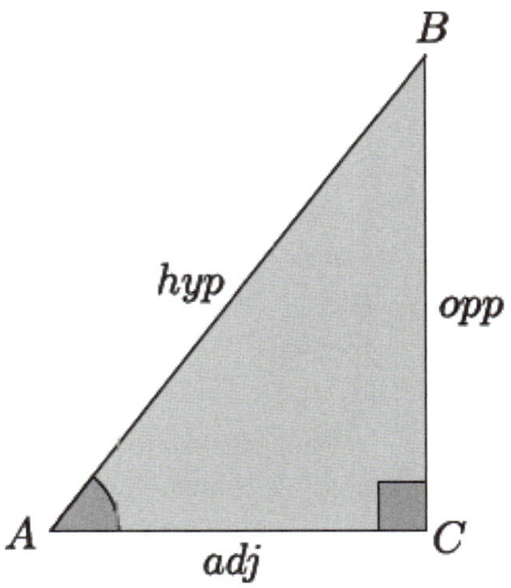

- In everyday life, we reference objects and people by name, especially when we use the object or see the person frequently. Imagine always calling your friend *hey* or *him* or *her* or *friend!* We want to be able to easily distinguish and identify a person, so we use a name. The same reasoning can be applied to the fractional expressions that we have been investigating: $\frac{\text{opp}}{\text{hyp}}$, $\frac{\text{adj}}{\text{hyp}}$, and $\frac{\text{opp}}{\text{adj}}$ need names.

Normally, these fractional expressions are said to be values of the ratios. However, to avoid saying "value of the ratio opp: hyp as $\frac{\text{opp}}{\text{hyp}}$," all of the time, these fractional expressions are called, collectively, the *trigonometric ratios,* based upon historical precedence, even though they really are not defined as ratios in the standards. Consider sharing this fact with students.

- These incredibly useful ratios were discovered long ago and have had several names. The names we currently use are translations of Latin words. You will learn more about the history behind these ratios in Algebra II.
- Presently, mathematicians have agreed upon the names *sine, cosine,* and *tangent*.

- If θ is the angle measure of $\angle A$, as shown, then we define:

  The *sine* of θ is the value of the ratio of the length of the opposite side to the length of the hypotenuse. As a formula,

  $$\sin\theta = \frac{\text{opp}}{\text{hyp}}.$$

  We also say $\sin\angle A = \sin\theta$. Then, $\sin\angle A = \frac{BC}{AB}$.

  The *cosine* of θ is the value of the ratio of the length of the adjacent side to the length of the hypotenuse. As a formula,

  $$\cos\theta = \frac{\text{adj}}{\text{hyp}}.$$

  We also say $\cos\angle A = \cos\theta$. Then, $\cos\angle A = \frac{AC}{AB}$.

  The *tangent* of θ is the value of the ratio of the length of the opposite side to the length of the adjacent side. As a formula,

  $$\tan\theta = \frac{\text{opp}}{\text{adj}}.$$

  We also say $\tan\angle A = \tan\theta$. Then, $\tan\angle A = \frac{BC}{AC}$.

- There are still three possible combinations of quotients of the side lengths; we briefly introduce them here.

  The *secant* of θ is the value of the ratio of the length of the hypotenuse to the length of the adjacent side. As a formula,

  $$\sec\theta = \frac{\text{hyp}}{\text{adj}}.$$

  The *cosecant* of θ is the value of the ratio of the length of the hypotenuse to the length of the opposite side. As a formula,

  $$\csc\theta = \frac{\text{hyp}}{\text{opp}}.$$

  The *cotangent* of θ is the value of the ratio of the length of the adjacent side to the length of the opposite side. As a formula,

  $$\cot\theta = \frac{\text{adj}}{\text{opp}}.$$

- We have little need in this course for secant, cosecant, and cotangent because, given any two sides, it is possible to write the quotient so as to get sine, cosine, and tangent. In more advanced courses, secant, cosecant, and cotangent are more useful.

Exercises 4–9 (15 minutes)

Have students complete Exercises 4–6 independently, or divide the work among students, and have them share their results. Once Exercises 4–6 have been completed, encourage students to discuss in small groups the relationships they notice between the sine of the angle and the cosine of its complement. Also, encourage them to discuss the relationship they notice about the tangent of both angles. Finally, have students share their observations with the whole class and then complete Exercises 7–8. Note that because students are not being asked to rationalize denominators, the relationships are clearer.

Scaffolding:

- Consider having students draw two right triangles and then color-code and/or label each with opp, adj, and hyp, relative to the angle they are looking at.
- Consider having advanced students draw two right triangles, $\triangle DEF$ and $\triangle GHI$, such that they are not congruent but so that $\sin E = \sin H$. Then, have students explain how they know.

Exercises 4–9

4. In $\triangle PQR$, $m\angle P = 53.2°$ and $m\angle Q = 36.8°$. Complete the following table.

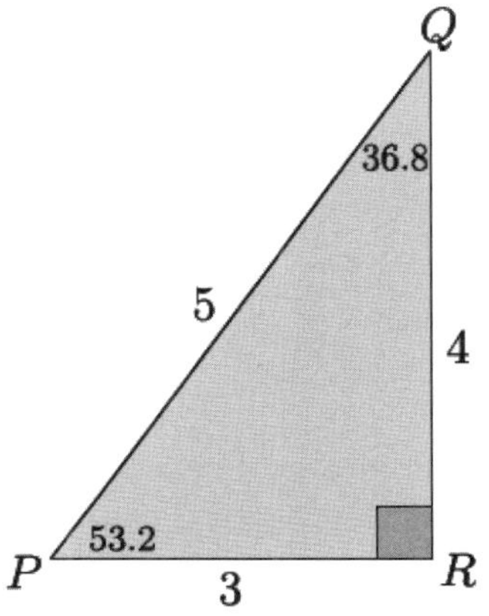

Measure of Angle	Sine $\left(\frac{\text{opp}}{\text{hyp}}\right)$	Cosine $\left(\frac{\text{adj}}{\text{hyp}}\right)$	Tangent $\left(\frac{\text{opp}}{\text{adj}}\right)$
53.2	$\sin 53.2 = \frac{4}{5}$	$\cos 53.2 = \frac{3}{5}$	$\tan 53.2 = \frac{4}{3}$
36.8	$\sin 36.8 = \frac{3}{5}$	$\cos 36.8 = \frac{4}{5}$	$\tan 36.8 = \frac{3}{4}$

5. In the triangle below, $m\angle A = 33.7°$ and $m\angle B = 56.3°$. Complete the following table.

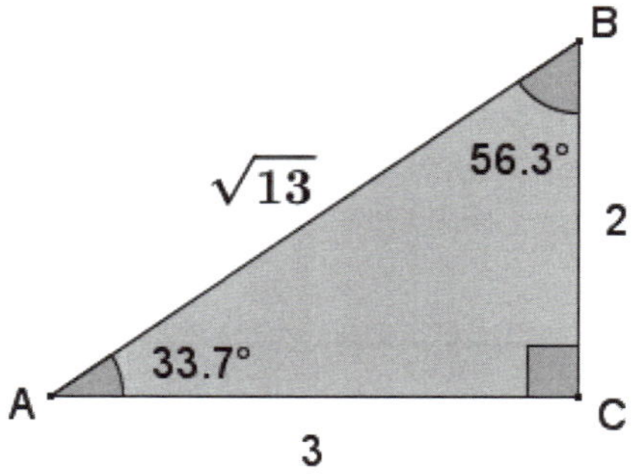

Measure of Angle	Sine	Cosine	Tangent
33.7	$\sin 33.7 = \frac{2}{\sqrt{13}}$	$\cos 33.7 = \frac{3}{\sqrt{13}}$	$\tan 33.7 = \frac{2}{3}$
56.3	$\sin 56.3 = \frac{3}{\sqrt{13}}$	$\cos 56.3 = \frac{2}{\sqrt{13}}$	$\tan 56.3 = \frac{3}{2}$

6. In the triangle below, let e be the measure of $\angle E$ and d be the measure of $\angle D$. Complete the following table.

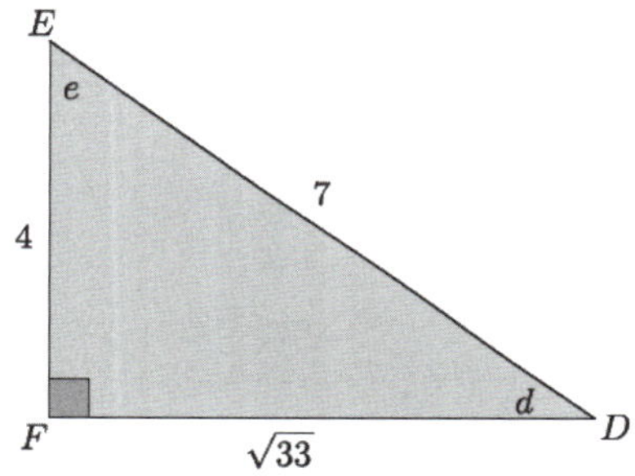

Measure of Angle	Sine	Cosine	Tangent
d	$\sin d = \frac{4}{7}$	$\cos d = \frac{\sqrt{33}}{7}$	$\tan d = \frac{4}{\sqrt{33}}$
e	$\sin e = \frac{\sqrt{33}}{7}$	$\cos e = \frac{4}{7}$	$\tan e = \frac{\sqrt{33}}{4}$

7. In the triangle below, let x be the measure of $\angle X$ and y be the measure of $\angle Y$. Complete the following table.

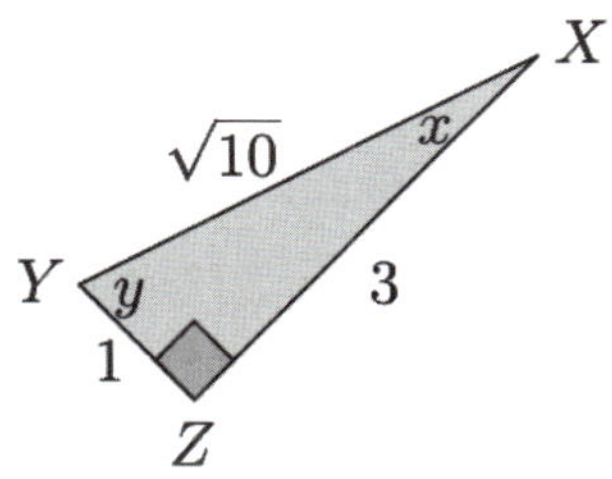

Measure of Angle	Sine	Cosine	Tangent
x	$\sin x = \frac{1}{\sqrt{10}}$	$\cos x = \frac{3}{\sqrt{10}}$	$\tan x = \frac{1}{3}$
y	$\sin y = \frac{3}{\sqrt{10}}$	$\cos y = \frac{1}{\sqrt{10}}$	$\tan y = \frac{3}{1}$

MP.8

8. Tamer did not finish completing the table below for a diagram similar to the previous problems that the teacher had on the board where p was the measure of $\angle P$ and q was the measure of $\angle Q$. Use any patterns you notice from Exercises 1–4 to complete the table for Tamer.

Measure of Angle	Sine	Cosine	Tangent
p	$\sin p = \frac{11}{\sqrt{157}}$	$\cos p = \frac{6}{\sqrt{157}}$	$\tan p = \frac{11}{6}$
q	$\sin q = \frac{6}{\sqrt{157}}$	$\cos q = \frac{11}{\sqrt{157}}$	$\tan q = \frac{6}{11}$

9. Explain how you were able to determine the sine, cosine, and tangent of $\angle Q$ in Exercise 8.

I was able to complete the table for Tamer by observing the patterns of previous problems. For example, I noticed that the sine of one angle was always equal to the ratio that represented the cosine of the other angle. Since I was given $\sin p$*, I knew the ratio* $\frac{11}{\sqrt{157}}$ *would be the* $\cos q$*. Similarly,* $\cos p = \sin q = \frac{6}{\sqrt{157}}$ *Finally, I noticed that the tangents of the angles were always reciprocals of each other. Since I was given the* $\tan p = \frac{11}{6}$*, I knew that the* $\tan q$ *must be equal to* $\frac{6}{11}$.

Discussion (8 minutes)

The sine, cosine, and tangent of an angle can be used to find unknown lengths of other triangles using *within*-figure ratios of similar triangles. The discussion that follows begins by posing a question to students. Provide time for students to discuss the answer in pairs or small groups, and then have them share their thoughts with the class.

- If $0 < \theta < 90$, we can define sine, cosine, and tangent of θ: Take a right triangle that has an acute angle with angle degree measure θ, and use the appropriate side lengths.
- If we use different right triangles, why will we get the same value for $\sin \theta$, $\cos \theta$, and $\tan \theta$?

MP.3 Provide time for students to talk to a partner or small group. If necessary, use the questions and diagrams below to guide students' thinking. The decimal values for the side lengths are used to make less obvious the fact that the ratios of the side lengths are equal.

- For example, consider the following two triangles.

The triangles below contain approximations for all lengths. If the Pythagorean theorem is used to verify that the triangles are right triangles, then you will notice that the values are slightly off. For example, the length of the hypotenuse of $\triangle DEF$ contains 11 decimal digits, not the 5 shown.

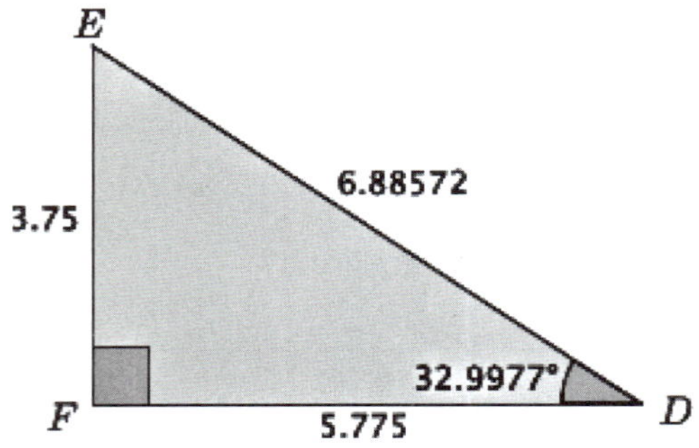

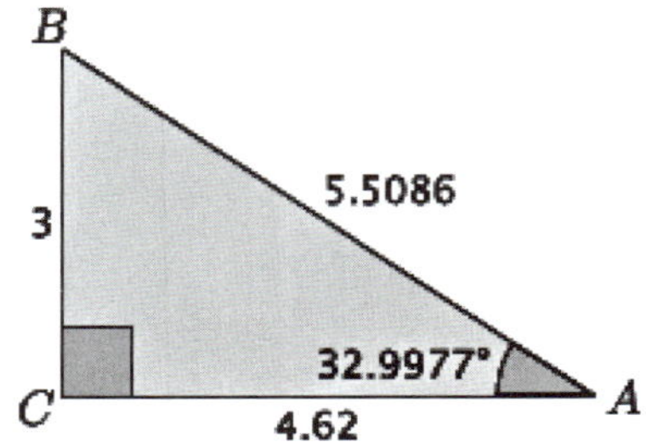

- Find the $\sin A$, $\cos A$, and $\tan A$. Compare those ratios to the $\sin D$, $\cos D$, and $\tan D$. What do you notice?

The task of finding the ratios can be divided among students and shared with the group.

  - *Students should notice that* $\sin A = \sin D$, $\cos A = \cos D$, *and* $\tan A = \tan D$.

- Under what circumstances have we observed ratios within one figure being equal to ratios within another figure?
  - *Within-figure ratios are equal when the figures are similar.*
- Two right triangles, each having an acute angle of angle measure θ, are similar by the AA criterion. So, we know that the values of corresponding ratios of side lengths are equal. That means $\sin\theta$, $\cos\theta$, and $\tan\theta$ do not depend on which right triangle we use.
- The ratios we write for the sine, cosine, and tangent of an angle are useful because they allow us to solve for two sides of a triangle when we know only the length of one side.

Closing (5 minutes)

Ask students the following questions. Have students respond in writing, to a partner, or to the whole class.

- Describe the ratios that we used to calculate sine, cosine, and tangent.
  - *Given an angle,* θ, $\sin\theta = \frac{opp}{hyp}$, $\cos\theta = \frac{adj}{hyp}$, *and* $\tan\theta = \frac{opp}{adj}$.
- Given any two right triangles that each have an acute angle with measure θ, why would we get the same value for $\sin\theta$, $\cos\theta$, and $\tan\theta$ using either triangle?
  - *Since the two right triangles each have an acute angle with measure θ, they are similar by the AA criterion. Similar triangles have corresponding side lengths that are equal in ratio. Additionally, based on our investigations in Lesson 25, we know that the value of the ratios of corresponding sides for a particular angle size are equal to the same constant.*

- Given a right triangle, describe the relationship between the sine of one acute angle and the cosine of the other acute angle.
  - *The sine of one acute angle of a right triangle is equal to the cosine of the other acute angle in the triangle.*

Exit Ticket (5 minutes)

Name ______________________________ Date ______________

Lesson 26: The Definition of Sine, Cosine, and Tangent

Exit Ticket

1. Given the diagram of the triangle, complete the following table.

Angle Measure	$\sin\theta$	$\cos\theta$	$\tan\theta$
s			
t			

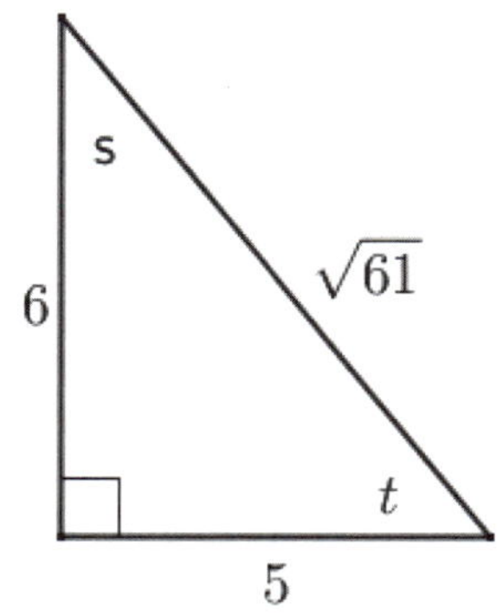

a. Which values are equal?

b. How are $\tan s$ and $\tan t$ related?

2. If u and v are the measures of complementary angles such that $\sin u = \frac{2}{5}$ and $\tan v = \frac{\sqrt{21}}{2}$, label the sides and angles of the right triangle in the diagram below with possible side lengths.

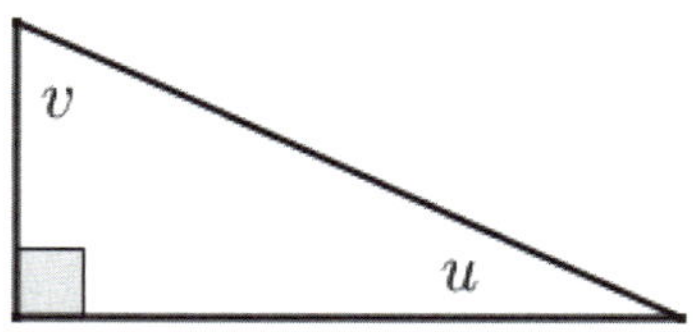

Exit Ticket Sample Solutions

1. Given the diagram of the triangle, complete the following table.

Angle Measure	$\sin\theta$	$\cos\theta$	$\tan\theta$
s	$\frac{5}{\sqrt{61}}=\frac{5\sqrt{61}}{61}$	$\frac{6}{\sqrt{61}}=\frac{6\sqrt{61}}{61}$	$\frac{5}{6}$
t	$\frac{6}{\sqrt{61}}=\frac{6\sqrt{61}}{61}$	$\frac{5}{\sqrt{61}}=\frac{5\sqrt{61}}{61}$	$\frac{6}{5}$

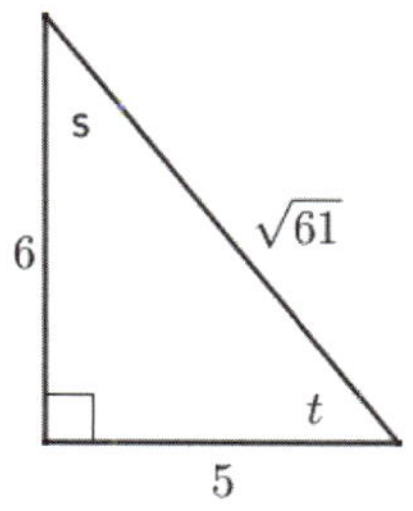

a. Which values are equal?

$\sin s=\cos t$ and $\cos s=\sin t$

b. How are $\tan s$ and $\tan t$ related?

They are reciprocals: $\frac{5}{6}\cdot\frac{6}{5}=1$.

2. If u and v are the measures of complementary angles such that $\sin u=\frac{2}{5}$ and $\tan v=\frac{\sqrt{21}}{2}$, label the sides and angles of the right triangle in the diagram below with possible side lengths:

A possible solution is shown below; however, any similar triangle having a shorter leg with length of $2x$, longer leg with length of $x\sqrt{21}$, and hypotenuse with length of $5x$, for some positive number x, is also correct.

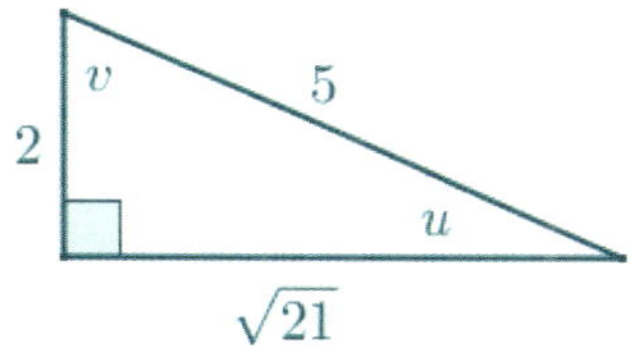

Problem Set Sample Solutions

1. Given the triangle in the diagram, complete the following table.

Angle Measure	sin	cos	tan
α	$\frac{2\sqrt{5}}{6}$	$\frac{4}{6}=\frac{2}{3}$	$\frac{2\sqrt{5}}{4}=\frac{\sqrt{5}}{2}$
β	$\frac{4}{6}=\frac{2}{3}$	$\frac{2\sqrt{5}}{6}$	$\frac{4}{2\sqrt{5}}=\frac{2}{\sqrt{5}}$

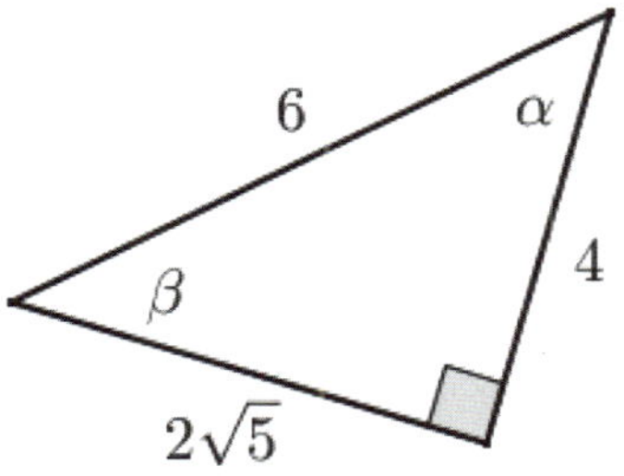

2. **Given the table of values below (not in simplest radical form), label the sides and angles in the right triangle.**

Angle Measure	sin	cos	tan
α	$\frac{4}{2\sqrt{10}}$	$\frac{2\sqrt{6}}{2\sqrt{10}}$	$\frac{4}{2\sqrt{6}}$
β	$\frac{2\sqrt{6}}{2\sqrt{10}}$	$\frac{4}{2\sqrt{10}}$	$\frac{2\sqrt{6}}{4}$

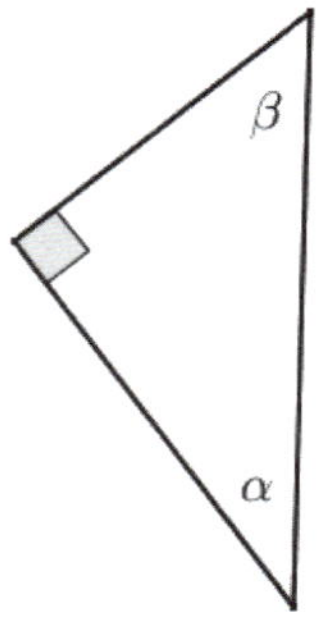

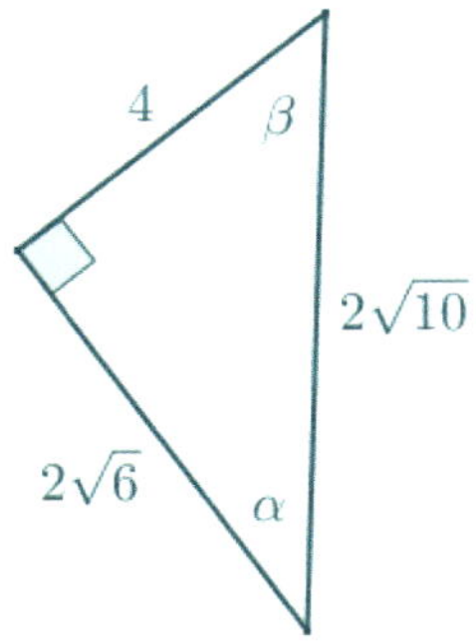

3. **Given $\sin\alpha$ and $\sin\beta$, complete the missing values in the table. You may draw a diagram to help you.**

Angle Measure	sin	cos	tan
α	$\frac{\sqrt{2}}{3\sqrt{3}}$	$\frac{5}{3\sqrt{3}}$	$\frac{\sqrt{2}}{5}$
β	$\frac{5}{3\sqrt{3}}$	$\frac{\sqrt{2}}{3\sqrt{3}}$	$\frac{5}{\sqrt{2}}$

4. **Given the triangle shown to the right, fill in the missing values in the table.**

Using the Pythagorean theorem:

$$\text{hyp}^2 = 2^2 + 6^2$$
$$\text{hyp}^2 = 4 + 36$$
$$\text{hyp}^2 = 40$$
$$\text{hyp} = \sqrt{40}$$
$$\text{hyp} = 2\sqrt{10}$$

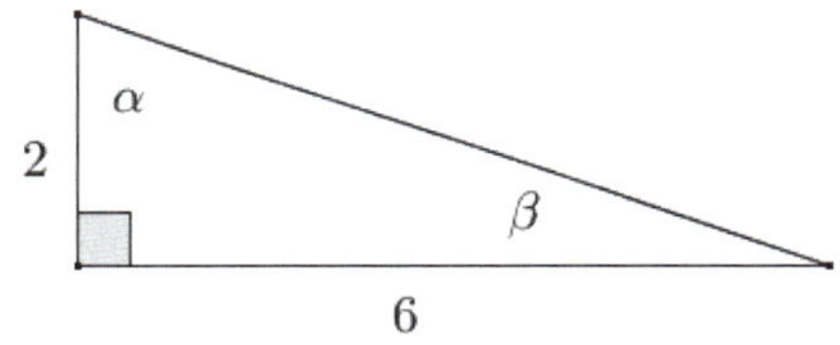

Angle Measure	sin	cos	tan
α	$\frac{6}{2\sqrt{10}} = \frac{3\sqrt{10}}{10}$	$\frac{2}{2\sqrt{10}} = \frac{\sqrt{10}}{10}$	$\frac{6}{2} = 3$
β	$\frac{2}{2\sqrt{10}} = \frac{\sqrt{10}}{10}$	$\frac{6}{2\sqrt{10}} = \frac{3\sqrt{10}}{10}$	$\frac{2}{6} = \frac{1}{3}$

5. **Jules thinks that if α and β are two different acute angle measures, then $\sin\alpha \neq \sin\beta$. Do you agree or disagree? Explain.**

I agree. If α and β are different acute angle measures, then either $\alpha > \beta$ or $\beta > \alpha$. A right triangle with acute angle α cannot be similar to a right triangle with acute angle β (unless $\alpha + \beta = 90$) because the triangles fail the AA criterion. If the triangles are not similar, then their corresponding sides are not in proportion, meaning their within-figure ratios are not in proportion; therefore, $\sin\alpha \neq \sin\beta$. In the case where $\alpha + \beta = 90$, the given right triangles are similar; however, α and β must be alternate acute angles, meaning $\sin\alpha = \cos\beta$, and $\sin\beta = \cos\alpha$, but $\sin\alpha \neq \sin\beta$.

6. **Given the triangle in the diagram, complete the following table.**

Angle Measure	sin	cos	tan
α	$\frac{3\sqrt{6}}{9}$	$\frac{3\sqrt{3}}{9}$	$\frac{3\sqrt{6}}{3\sqrt{3}}$
β	$\frac{3\sqrt{3}}{9}$	$\frac{3\sqrt{6}}{9}$	$\frac{3\sqrt{3}}{3\sqrt{6}}$

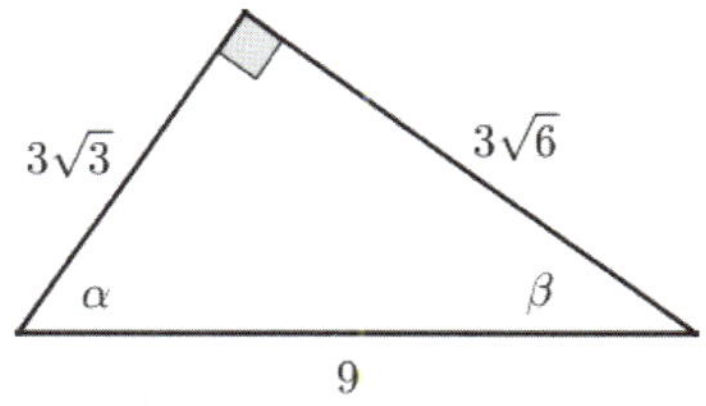

Rewrite the values from the table in simplest terms.

Angle Measure	sin	cos	tan
α	$\frac{\sqrt{6}}{3}$	$\frac{\sqrt{3}}{3}$	$\frac{\sqrt{6}}{\sqrt{3}} = \sqrt{2}$
β	$\frac{\sqrt{3}}{3}$	$\frac{\sqrt{6}}{3}$	$\frac{\sqrt{3}}{\sqrt{6}} = \frac{\sqrt{2}}{2}$

Draw and label the sides and angles of a right triangle using the simplified values of the ratios sin and cos. How is the new triangle related to the original triangle?

The triangles are similar by SSS criterion because the new triangle has sides that are $\frac{1}{3}$ of the length of their corresponding sides in the original triangle.

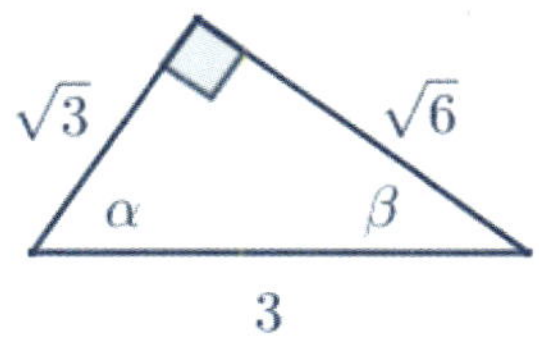

7. **Given $\tan \alpha$ and $\cos \beta$, in simplest terms, find the missing side lengths of the right triangle if one leg of the triangle has a length of 4. Draw and label the sides and angles of the right triangle.**

Angle Measure	$\sin \theta$	$\cos \theta$	$\tan \theta$
α	$\frac{\sqrt{3}}{2}$	$\frac{1}{2}$	$\sqrt{3}$
β	$\frac{1}{2}$	$\frac{\sqrt{3}}{2}$	$\frac{\sqrt{3}}{3}$

The problem does not specify which leg is 4, so there are two possible solutions to this problem. The values given in the table do not represent the actual lengths of the sides of the triangles; however, they do represent the lengths of the sides of a similar triangle, which is a 30–60–90 right triangle with side lengths 1, 2, and $\sqrt{3}$.

Case 1: The short leg of the right triangle is 4:

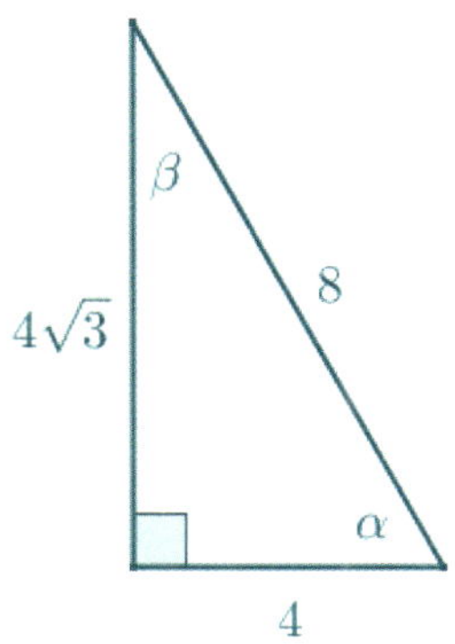

Case 2: The long leg of the right triangle is 4:

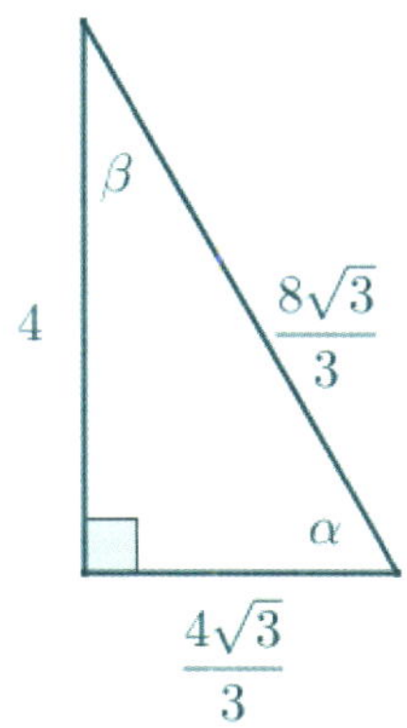

8. Eric wants to hang a rope bridge over a small ravine so that it is easier to cross. To hang the bridge, he needs to know how much rope is needed to span the distance between two trees that are directly across from each other on either side of the ravine. Help Eric devise a plan using sine, cosine, and tangent to determine the approximate distance from tree A to tree B without having to cross the ravine.

Student solutions will vary. Possible solution:

If Eric walks a path parallel to the ravine to a point P at a convenient distance from A, he could measure the angle formed by his line of sight to both trees. Using the measured angle and distance, he could use the value of the tangent ratio of the angle to determine the length of the opposite leg of the triangle. The length of the opposite leg of the triangle represents the distance between the two trees.

9. A fisherman is at point F on the open sea and has three favorite fishing locations. The locations are indicated by points A, B, and C. The fisherman plans to sail from F to A, then to B, then to C, and then back to F. If the fisherman is 14 miles from $\overline{AC}$, find the total distance that he will sail.

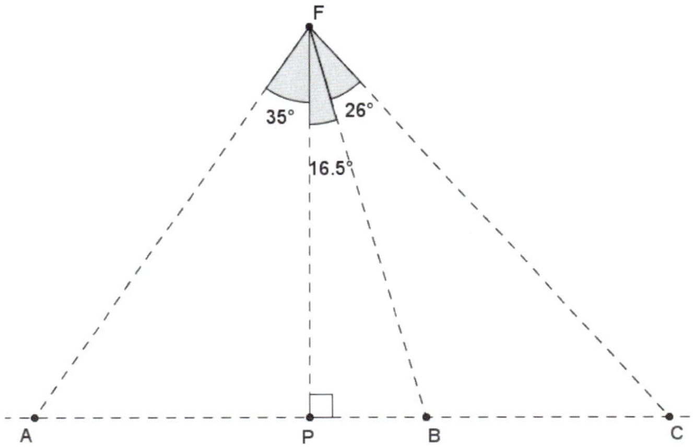

$FP = 14$ and can be considered the adjacent side to the $35°$ angle shown in triangle APF.

Using cosine:

$$\cos 35 = \frac{14}{AF}$$

$$AF = \frac{14}{\cos 35}$$

$$AF \approx 17.09$$

Using tangent:

$$\tan 35 = \frac{AP}{14}$$

$$AP = 14 \tan 35$$

$$AP \approx 9.8029$$

$\overline{PC}$ is the leg opposite angle PFC in triangle PFC and has a degree measure of 42.5.

Using tangent:

$$\tan 42.5 = \frac{PC}{14}$$

$$PC = 14 \tan 42.5$$

$$PC \approx 12.8286$$

Using cosine:

$$\cos 42.5 = \frac{14}{FC}$$

$$FC = \frac{14}{\cos 42.5}$$

$$FC \approx 18.9888$$

The total distance that the fisherman will sail:

$$\text{distance} = AF + AP + PC + FC$$

$$\text{distance} = \frac{14}{\cos 35} + 14 \tan 35 + 14 \tan 42.5 + \frac{14}{\cos 42.5}$$

$$\text{distance} \approx 58.7$$

The total distance that the fisherman will sail is approximately 58.7 miles.

Lesson 27: Sine and Cosine of Complementary Angles and Special Angles

Student Outcomes

- Students understand that if α and β are the measurements of complementary angles, then $\sin \alpha = \cos \beta$.
- Students solve triangle problems using special angles.

Lesson Notes

Students examine the sine and cosine relationship more closely and find that the sine and cosine of complementary angles are equal. Students become familiar with the values associated with sine and cosine of special angles. Once familiar with these common values, students use them to find unknown values in problems.

Scaffolding:

- If students are struggling to see the connection, use a right triangle with side lengths 3, 4, and 5 to help make the values of the ratios more apparent.
- Use the cutouts from Lesson 21.
- Ask students to calculate values of sine and cosine for the acute angles (by measuring), and then ask them, "What do you notice?"
- As an extension, ask students to write a letter to a middle school student explaining why the sine of an angle is equal to the cosine of its complementary angle.

Classwork

Example 1 (8 minutes)

MP.2

Students discover why cosine has the prefix *co-*. It may be necessary to remind students why we know alpha and beta are complementary.

Example 1

If α and β are the measurements of complementary angles, then we are going to show that $\sin \alpha = \cos \beta$.

In right triangle ABC, the measurement of acute angle $\angle A$ is denoted by α, and the measurement of acute angle $\angle B$ is denoted by β.

Determine the following values in the table:

$\sin \alpha$	$\sin \beta$	$\cos \alpha$	$\cos \beta$
$\sin \alpha = \frac{\text{opp}}{\text{hyp}} = \frac{a}{c}$	$\sin \beta = \frac{\text{opp}}{\text{hyp}} = \frac{b}{c}$	$\cos \alpha = \frac{\text{adj}}{\text{hyp}} = \frac{b}{c}$	$\cos \beta = \frac{\text{adj}}{\text{hyp}} = \frac{a}{c}$

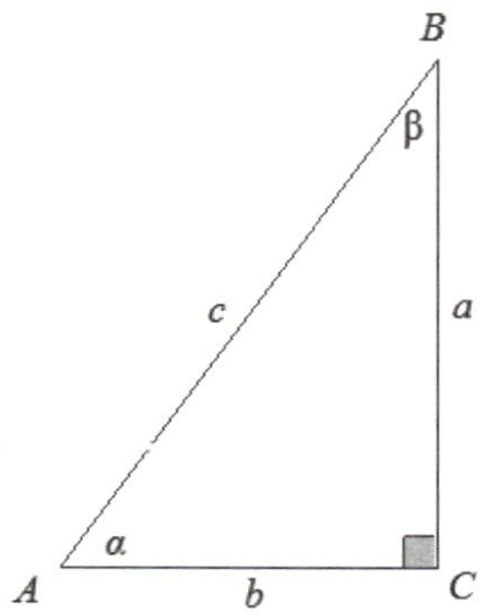

What can you conclude from the results?

Since the ratios for $\sin \alpha$ and $\cos \beta$ are the same, $\sin \alpha = \cos \beta$, and the ratios for $\cos \alpha$ and $\sin \beta$ are the same; additionally, $\cos \alpha = \sin \beta$. The sine of an angle is equal to the cosine of its complementary angle, and the cosine of an angle is equal to the sine of its complementary angle.

- Therefore, we conclude for complementary angles α and β that $\sin \alpha = \cos \beta$, or in other words, when $0 < \theta < 90$ that $\sin(90 - \theta) = \cos \theta$, and $\sin \theta = \cos(90 - \theta)$. Any two angles that are complementary can be realized as the acute angles in a right triangle. Hence, the *co-* prefix in *cosine* is a reference to the fact that the cosine of an angle is the sine of its complement.

Exercises 1–3 (7 minutes)

Students apply what they know about the sine and cosine of complementary angles to solve for unknown angle values.

Exercises 1–3

1. **Consider the right triangle ABC so that $\angle C$ is a right angle, and the degree measures of $\angle A$ and $\angle B$ are α and β, respectively.**

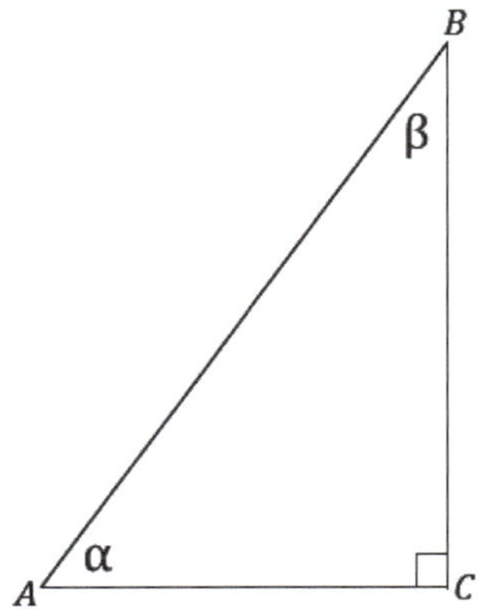

   a. **Find $\alpha + \beta$.**

      $90°$

   b. **Use trigonometric ratios to describe $\frac{BC}{AB}$ two different ways.**

      $\sin \angle A = \frac{BC}{AB}$, $\cos \angle B = \frac{BC}{AB}$

   c. **Use trigonometric ratios to describe $\frac{AC}{AB}$ two different ways.**

      $\sin \angle B = \frac{AC}{AB}$, $\cos \angle A = \frac{AC}{AB}$

   d. **What can you conclude about $\sin \alpha$ and $\cos \beta$?**

      $\sin \alpha = \cos \beta$

   e. **What can you conclude about $\cos \alpha$ and $\sin \beta$?**

      $\cos \alpha = \sin \beta$

2. **Find values for θ that make each statement true.**

   a. $\sin \theta = \cos(25)$

      $\theta = 65$

   b. $\sin 80 = \cos \theta$

      $\theta = 10$

   c. $\sin \theta = \cos(\theta + 10)$

      $\theta = 40$

   d. $\sin(\theta - 45) = \cos(\theta)$

      $\theta = 67.5$

3. **For what angle measurement must sine and cosine have the same value? Explain how you know.**

   Sine and cosine have the same value for $\theta = 45$. The sine of an angle is equal to the cosine of its complement. Since the complement of 45 is 45, $\sin 45 = \cos 45$.

Example 2 (8 minutes)

Students begin to examine special angles associated with sine and cosine, starting with the angle measurements of $0°$ and $90°$. Consider modeling this on the board by drawing a sketch of the following figure and using a meter stick to represent c.

Example 2

What is happening to a and b as θ changes? What happens to $\sin\theta$ and $\cos\theta$?

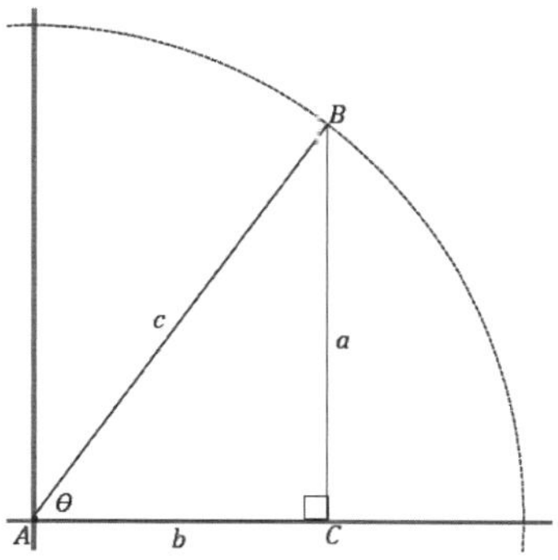

- There are values for sine and cosine commonly known for certain angle measurements. Two such angle measurements are when $\theta = 0°$ and $\theta = 90°$.
- To better understand sine and cosine values, imagine a right triangle whose hypotenuse has a fixed length c of 1 unit. We illustrate this by imagining the hypotenuse as the radius of a circle, as in the image.
- What happens to the value of the sine ratio as θ approaches $0°$? Consider what is happening to the opposite side, a.

With one end of the meter stick fixed at A, rotate it like the hands of a clock, and show how a decreases as θ decreases. Demonstrate the change in the triangle for each case.

  - *As θ gets closer to $0°$, a decreases. Since $\sin\theta = \frac{a}{1}$, the value of $\sin\theta$ is also approaching 0.*

- Similarly, what happens to the value of the cosine ratio as θ approaches $0°$? Consider what is happening to the adjacent side, b.
  - *As θ gets closer to $0°$, b increases and becomes closer to 1. Since $\cos\theta = \frac{b}{1}$, the value of $\cos\theta$ is approaching 1.*
- Now, consider what happens to the value of the sine ratio as θ approaches $90°$. Consider what is happening to the opposite side, a.
  - *As θ gets closer to $90°$, a increases and becomes closer to 1. Since $\sin\theta = \frac{a}{1}$, the value of $\sin\theta$ is also approaching 1.*
- What happens to the value of the cosine ratio as θ approaches $90°$? Consider what is happening to the adjacent side, b.
  - *As θ gets closer to $90°$, b decreases and becomes closer to 0. Since $\cos\theta = \frac{b}{1}$, the value of $\cos\theta$ is approaching 0.*
- Remember, because there are no right triangles with an acute angle of $0°$ or of $90°$, in the above thought experiment, we are really defining $\sin 0 = 0$ and $\cos 0 = 1$.

- Similarly, we define $\sin 90 = 1$ and $\cos 90 = 0$; notice that this falls in line with our conclusion that the sine of an angle is equal to the cosine of its complementary angle.

Example 3 (10 minutes)

Students examine the remaining special angles associated with sine and cosine in Example 3. Consider assigning parts (b) and (c) to two halves of the class and having students present a share-out of their findings.

Example 3

There are certain special angles where it is possible to give the exact value of sine and cosine. These are the angles that measure $0°$, $30°$, $45°$, $60°$, and $90°$; these angle measures are frequently seen.

You should memorize the sine and cosine of these angles with quick recall just as you did your arithmetic facts.

a. **Learn the following sine and cosine values of the key angle measurements.**

θ	$0°$	$30°$	$45°$	$60°$	$90°$
Sine	0	$\frac{1}{2}$	$\frac{\sqrt{2}}{2}$	$\frac{\sqrt{3}}{2}$	1
Cosine	1	$\frac{\sqrt{3}}{2}$	$\frac{\sqrt{2}}{2}$	$\frac{1}{2}$	0

MP.7

We focus on an easy way to remember the entries in the table. What do you notice about the table values?

The entries for cosine are the same as the entries for sine but in the reverse order.

This is easily explained because the pairs $(0, 90)$, $(30, 60)$, and $(45, 45)$ are the measures of complementary angles. So, for instance, $\sin 30 = \cos 60$.

The sequence $0, \frac{1}{2}, \frac{\sqrt{2}}{2}, \frac{\sqrt{3}}{2}, 1$ may be easier to remember as the sequence $\frac{\sqrt{0}}{2}, \frac{\sqrt{1}}{2}, \frac{\sqrt{2}}{2}, \frac{\sqrt{3}}{2}, \frac{\sqrt{4}}{2}$.

b. **$\triangle ABC$ is equilateral, with side length 2; D is the midpoint of side $\overline{AC}$. Label all side lengths and angle measurements for $\triangle ABD$. Use your figure to determine the sine and cosine of 30 and 60.**

Provide students with a hint, if necessary, by suggesting they construct the angle bisector of $\angle B$, which is also the altitude to $\overline{AC}$.

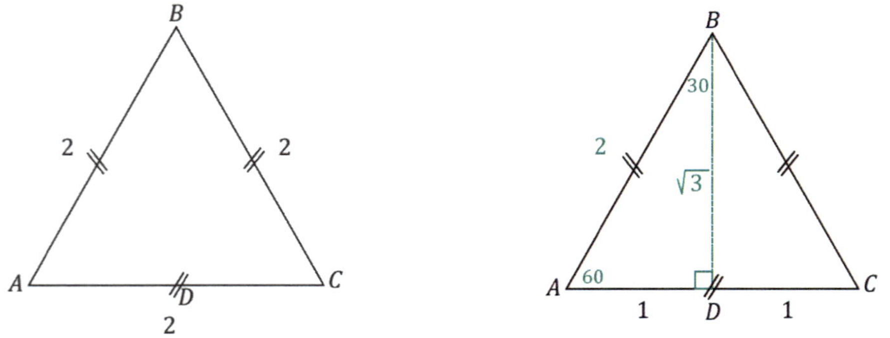

$\sin(30) = \frac{AD}{AB} = \frac{1}{2}$, $\cos(30) = \frac{BD}{AB} = \frac{\sqrt{3}}{2}$, $\sin(60) = \frac{BD}{AB} = \frac{\sqrt{3}}{2}$, $\cos(60) = \frac{AD}{AB} = \frac{1}{2}$

c. **Draw an isosceles right triangle with legs of length 1. What are the measures of the acute angles of the triangle? What is the length of the hypotenuse? Use your triangle to determine sine and cosine of the acute angles.**

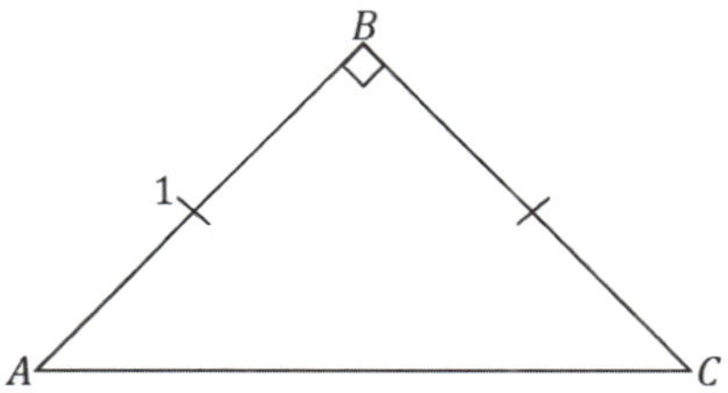

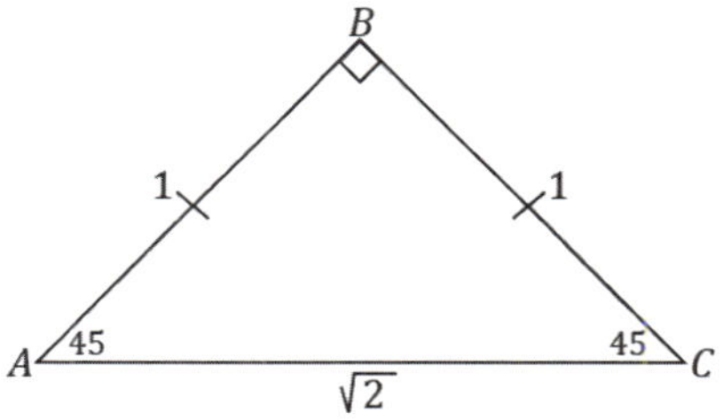

$$\sin(45) = \frac{AB}{AC} = \frac{1}{\sqrt{2}},\ \cos(45) = \frac{BC}{AC} = \frac{1}{\sqrt{2}}$$

Parts (b) and (c) demonstrate how the sine and cosine values of the mentioned special angles can be found. These triangles are common to trigonometry; we refer to the triangle in part (b) as a $30–60–90$ triangle and the triangle in part (c) as a $45–45–90$ triangle.

- Remind students that the values of the sine and cosine ratios of triangles similar to each of these are the same.

Highlight the length ratios for $30–60–90$ and $45–45–90$ triangles. Consider using a setup like the table below to begin the conversation. Ask students to determine side lengths of three different triangles similar to each of the triangles provided above. Remind them that the scale factor determines side length. Then, have them generalize the length relationships.

$30–60–90$ Triangle, side length ratio $1: 2: \sqrt{3}$	**$45–45–90$ Triangle, side length ratio $1: 1: \sqrt{2}$**
$2: 4: 2\sqrt{3}$	$2: 2: 2\sqrt{2}$
$3: 6: 3\sqrt{3}$	$3: 3: 3\sqrt{2}$
$4: 8: 4\sqrt{3}$	$4: 4: 4\sqrt{2}$
$x: 2x: x\sqrt{3}$	$x: x: x\sqrt{2}$

Scaffolding:

- For the $1: 2: \sqrt{3}$ triangle, students may develop the misconception that the last value is the length of the hypotenuse, the longest side of the right triangle. Help students correct this misconception by comparing $\sqrt{3}$ and $\sqrt{4}$ to show that $\sqrt{4} > \sqrt{3}$, and $\sqrt{4} = 2$, so $2 > \sqrt{3}$.
- The ratio $1: 2: \sqrt{3}$ is easier to remember because of the numbers $1, 2, 3$.

Exercises 4–5 (5 minutes)

Exercises 4–5

4. **Find the missing side lengths in the triangle.**

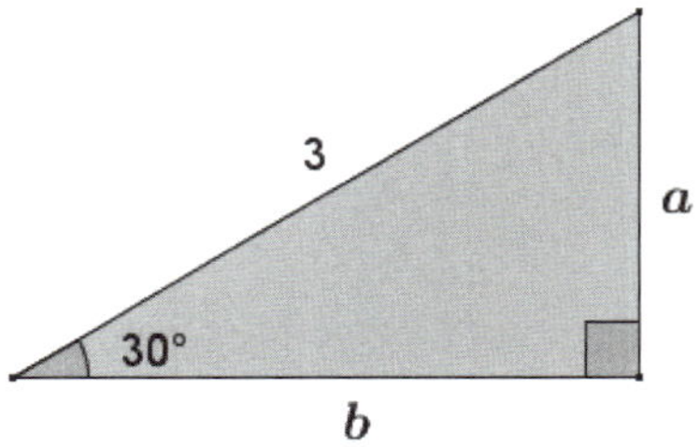

$\sin 30 = \frac{a}{3} = \frac{1}{2}, a = \frac{3}{2}$

$\cos 30 = \frac{b}{3} = \frac{\sqrt{3}}{2}, b = \frac{3\sqrt{3}}{2}$

5. **Find the missing side lengths in the triangle.**

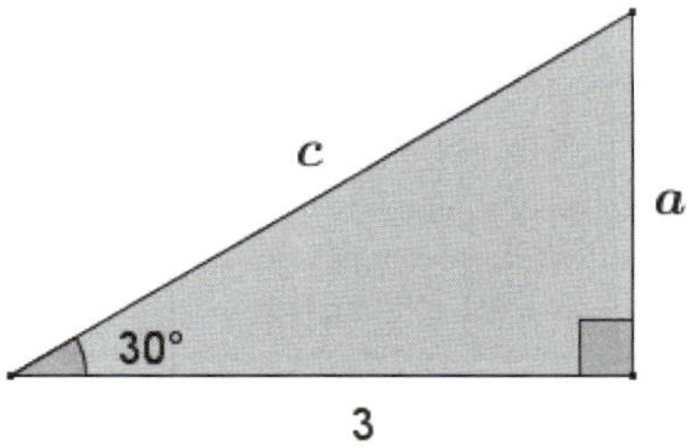

$\cos 30 = \frac{3}{c} = \frac{\sqrt{3}}{2}, c = \frac{6}{\sqrt{3}} = 2\sqrt{3}$

$\sin 30 = \frac{a}{2\sqrt{3}} = \frac{1}{2}, a = \sqrt{3}$

Closing (2 minutes)

Ask students to respond to these questions about the key ideas of the lesson independently in writing, with a partner, or as a class.

- What is remarkable about the sine and cosine of a pair of angles that are complementary?
  - *The sine of an angle is equal to the cosine of its complementary angle, and the cosine of an angle is equal to the sine of its complementary angle.*
- Why is $\sin 90 = 1$? Similarly, why is $\sin 0 = 0$, $\cos 90 = 0$, and $\cos 0 = 1$?
  - *We can see that $\sin \theta$ approaches 1 as θ approaches 90. The same is true for the other sine and cosine values for 0 and 90.*
- What do you notice about the sine and cosine of the following special angle values?
  - *The entries for cosine are the same as the entries for sine, but the values are in reverse order. This is explained by the fact the special angles can be paired up as complements, and we already know that the sine and cosine values of complementary angles are equal.*

θ	$0°$	$30°$	$45°$	$60°$	$90°$
Sine	0	$\frac{1}{2}$	$\frac{\sqrt{2}}{2}$	$\frac{\sqrt{3}}{2}$	1
Cosine	1	$\frac{\sqrt{3}}{2}$	$\frac{\sqrt{2}}{2}$	$\frac{1}{2}$	0

Exit Ticket (5 minutes)

Name ______________________________ Date ______________

Lesson 27: Sine and Cosine of Complementary Angles and Special Angles

Exit Ticket

1. Find the values for θ that make each statement true.
   a. $\sin\theta = \cos 32$

   b. $\cos\theta = \sin(\theta + 20)$

2. $\triangle LMN$ is a 30–60–90 right triangle. Find the unknown lengths x and y.

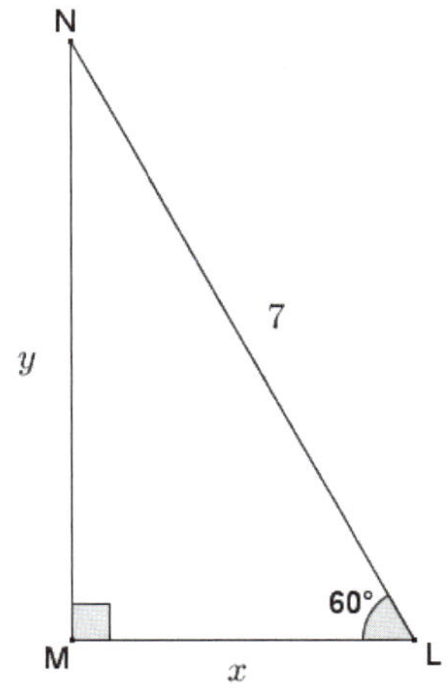

Exit Ticket Sample Solutions

1. **Find the values for θ that make each statement true.**

   a. $\sin\theta = \cos 32$

   $\theta = 90 - 32$

   $\theta = 58$

   b. $\cos\theta = \sin(\theta + 20)$

   $\sin(90 - \theta) = \sin(\theta + 20)$

   $90 - \theta = \theta + 20$

   $70 = 2\theta$

   $35 = \theta$

2. **$\triangle LMN$ is a 30–60–90 right triangle. Find the unknown lengths x and y.**

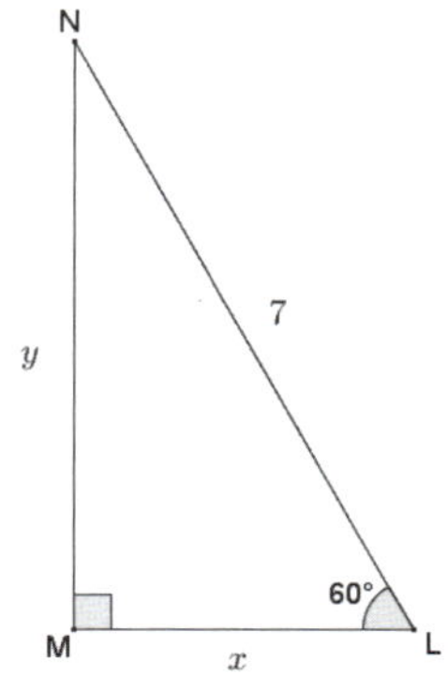

$$\sin 60 = \frac{\sqrt{3}}{2}$$
$$\frac{\sqrt{3}}{2} = \frac{y}{7}$$
$$7\sqrt{3} = 2y$$
$$y = \frac{7\sqrt{3}}{2}$$

$$\cos 60 = \frac{1}{2}$$
$$\frac{1}{2} = \frac{x}{7}$$
$$7 = 2x$$
$$\frac{7}{2} = x$$

Problem Set Sample Solutions

1. **Find the value of θ that makes each statement true.**

   a. $\sin\theta = \cos(\theta + 38)$

   $\cos(90 - \theta) = \cos(\theta + 38)$

   $90 - \theta = \theta + 38$

   $52 = 2\theta$

   $26 = \theta$

   b. $\cos\theta = \sin(\theta - 30)$

   $\sin(90 - \theta) = \sin(\theta - 30)$

   $90 - \theta = \theta - 30$

   $120 = 2\theta$

   $60 = \theta$

c. $\sin\theta = \cos(3\theta + 20)$

$\cos(90 - \theta) = \cos(3\theta + 20)$

$90 - \theta = 3\theta + 20$

$70 = 4\theta$

$17.5 = \theta$

d. $\sin\left(\frac{\theta}{3} + 10\right) = \cos\theta$

$\sin\left(\frac{\theta}{3} + 10\right) = \sin(90 - \theta)$

$\frac{\theta}{3} + 10 = 90 - \theta$

$\frac{4\theta}{3} = 80$

$\theta = 60$

2.

a. Make a prediction about how the sum $\sin 30 + \cos 60$ will relate to the sum $\sin 60 + \cos 30$.

Answers will vary; however, some students may believe that the sums will be equal. This is explored in Problems 3 through 5.

b. Use the sine and cosine values of special angles to find the sum: $\sin 30 + \cos 60$.

$\sin 30 = \frac{1}{2}$ *and* $\cos 60 = \frac{1}{2}$ *Therefore,* $\sin 30 + \cos 60 = \frac{1}{2} + \frac{1}{2} = 1.$

Alternative strategy:

$\cos 60° = \sin(90 - 60)° = \sin 30°$

$\sin 30° + \cos 60° = \sin 30° + \sin 30° = 2(\sin 30°) = 2\left(\frac{1}{2}\right) = 1$

c. Find the sum: $\sin 60 + \cos 30$.

$\sin 60 = \frac{\sqrt{3}}{2}$ *and* $\cos 30 = \frac{\sqrt{3}}{2}$ *Therefore,* $\sin 60 + \cos 30 = \frac{\sqrt{3}}{2} + \frac{\sqrt{3}}{2} = \sqrt{3}.$

d. Was your prediction a valid prediction? Explain why or why not.

Answers will vary.

3. Langdon thinks that the sum $\sin 30 + \sin 30$ is equal to $\sin 60$. Do you agree with Langdon? Explain what this means about the sum of the sines of angles.

I disagree. Explanations may vary. It was shown in the solution to Problem 3 that $\sin 30 + \sin 30 = 1$*, and it is known that* $\sin 60 = \frac{\sqrt{3}}{2} \neq 1$*. This shows that the sum of the sines of angles is not equal to the sine of the sum of the angles.*

4. A square has side lengths of $7\sqrt{2}$. Use sine or cosine to find the length of the diagonal of the square. Confirm your answer using the Pythagorean theorem.

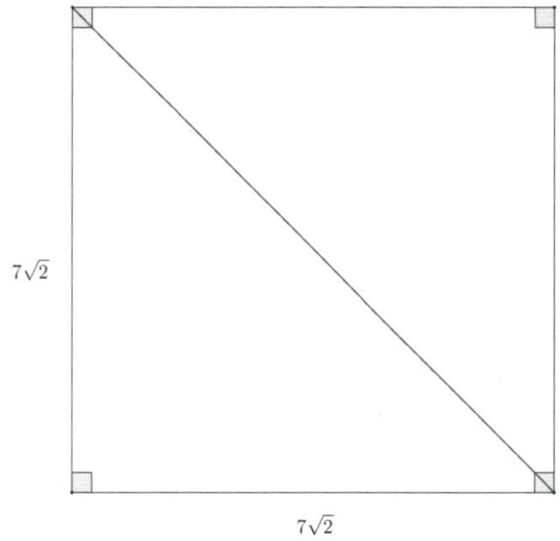

The diagonal of a square cuts the square into two congruent 45–45–90 right triangles. Let d represent the length of the diagonal of the square:

$$\cos 45 = \frac{\sqrt{2}}{2}$$
$$\frac{\sqrt{2}}{2} = \frac{7\sqrt{2}}{d}$$
$$d\sqrt{2} = 14\sqrt{2}$$
$$d = 14$$

Confirmation using Pythagorean theorem:

$$(7\sqrt{2})^2 + (7\sqrt{2})^2 = \text{hyp}^2$$
$$98 + 98 = \text{hyp}^2$$
$$196 = \text{hyp}^2$$
$$\sqrt{196} = \text{hyp}$$
$$14 = \text{hyp}$$

5. Given an equilateral triangle with sides of length 9, find the length of the altitude. Confirm your answer using the Pythagorean theorem.

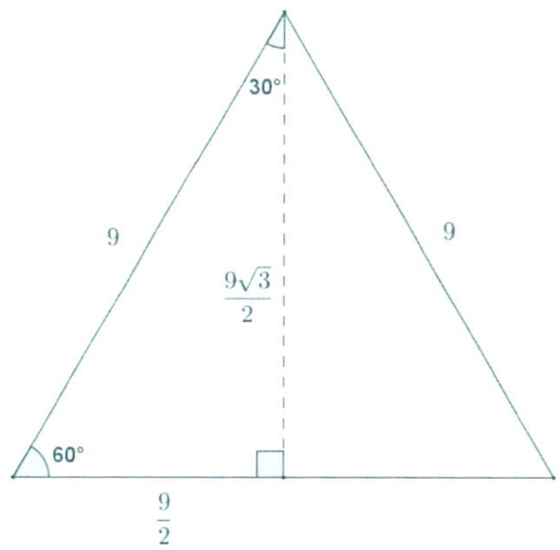

An altitude drawn within an equilateral triangle cuts the equilateral triangle into two congruent 30–60–90 right triangles. Let h represent the length of the altitude:

$$\sin 60 = \frac{\sqrt{3}}{2}$$
$$\frac{\sqrt{3}}{2} = \frac{h}{9}$$
$$9\sqrt{3} = 2h$$
$$\frac{9\sqrt{3}}{2} = h$$

The altitude of the triangle has a length of $\frac{9\sqrt{3}}{2}$.

Confirmation using Pythagorean theorem:

$$\left(\frac{9}{2}\right)^2 + \text{leg}^2 = 9^2$$
$$\frac{81}{4} + \text{leg}^2 = 81$$
$$\text{leg}^2 = \frac{243}{4}$$
$$\text{leg} = \sqrt{\frac{243}{4}}$$
$$\text{leg} = \frac{\sqrt{243}}{2}$$
$$\text{leg} = \frac{9\sqrt{3}}{2}$$

Lesson 28: Solving Problems Using Sine and Cosine

Student Outcomes

- Students use graphing calculators to find the values of $\sin\theta$ and $\cos\theta$ for θ between $0°$ and $90°$.
- Students solve for missing sides of a right triangle given the length of one side and the measure of one of the acute angles.
- Students find the length of the base of a triangle with acute base angles given the lengths of the other two sides and the measure of each of the base angles.

Lesson Notes

Students need access to a graphing calculator to calculate the sine and cosine of given angle measures. It is likely necessary to show students how to set the calculator in degree mode and to perform these operations. Encourage students to make one computation on the calculator and then approximate their answer as opposed to making intermediate approximations throughout the solution process. Intermediate approximations lead to a less accurate answer than doing the approximation once.

Classwork

Exercises 1–4 (12 minutes)

Allow students to work in pairs to complete Exercise 1. It may be necessary to demonstrate how to use a graphing calculator to perform the following calculations. Ensure that all calculators are in degree mode, not radian. Consider telling students that radian is a measure they encounter in Module 5 and use in Algebra II. For now, the unit of angle measure is degree. After completing the exercises, debrief by having students share their explanations in Exercise 4.

Exercises 1–4

1.

a. **The bus drops you off at the corner of H Street and 1st Street, approximately 300 ft. from school. You plan to walk to your friend Janneth's house after school to work on a project. Approximately how many feet will you have to walk from school to Janneth's house? Round your answer to the nearest foot. (Hint: Use the ratios you developed in Lesson 25.)**

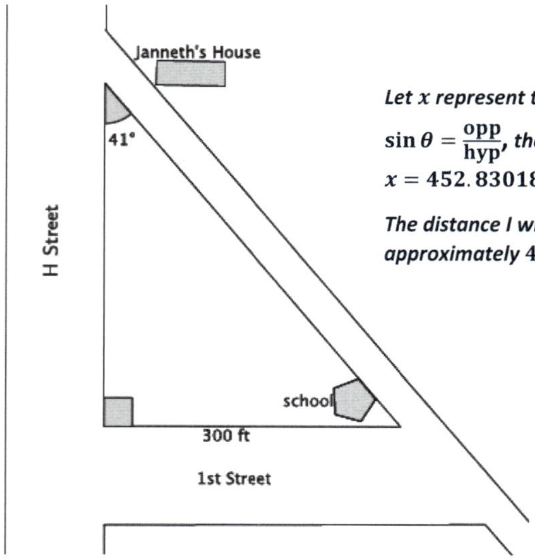

Let x represent the distance from school to Janneth's house.

$\sin\theta = \frac{\text{opp}}{\text{hyp}}$, *then* $\sin 41 = \frac{5.3}{8}$. *Then,* $\frac{300}{x} = \frac{5.3}{8}$ *and* $x = 452.8301887\ldots$

The distance I will have to walk from school to Janneth's house is approximately $453\ ft$.

b. **In real life, it is unlikely that you would calculate the distance between school and Janneth's house in this manner. Describe a similar situation in which you might actually want to determine the distance between two points using a trigonometric ratio.**

Accept any reasonable responses. Some may include needing to calculate the distance to determine if a vehicle has enough fuel to make the trip or the need to determine the length prior to attempting the walk because a friend is on crutches and cannot easily get from one location to the next when the distance is too long.

2. Use a calculator to find the sine and cosine of θ. Give your answer rounded to the ten-thousandth place.

θ	0	10	20	30	40	50	60	70	80	90
$\sin\theta$	0	0.1736	0.3420	$\frac{1}{2} = 0.5$	0.6428	0.7660	0.8660	0.9397	0.9848	1
$\cos\theta$	1	0.9848	0.9397	0.8660	0.7660	0.6428	$\frac{1}{2} = 0.5$	0.3420	0.1736	0

MP.8

3. What do you notice about the numbers in the row $\sin\theta$ compared with the numbers in the row $\cos\theta$?

The numbers are the same but reversed in order.

4. Provide an explanation for what you noticed in Exercise 2.

The pattern exists because the sine and cosine of complementary angles are equal.

Example 1 (8 minutes)

Students find the missing side length of a right triangle using sine and cosine.

Example 1

Find the values of a and b.

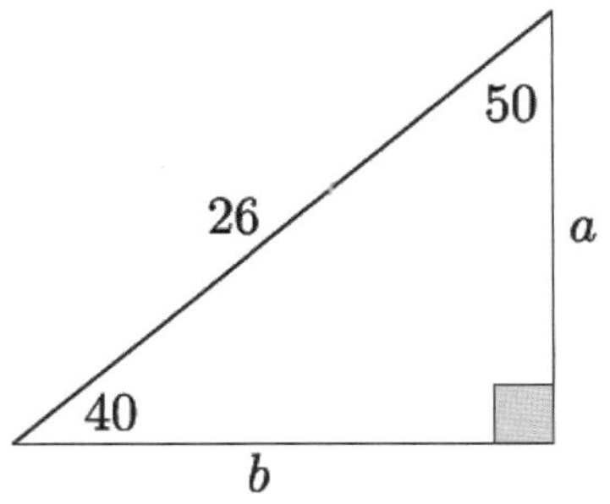

- Now that we can calculate the sine and cosine of a given angle using a calculator, we can use the decimal value of the ratio to determine the unknown side length of a triangle.

Consider the following triangle.

- What can we do to find the length of side a?
  - *We can find the* $\sin 40$ *or* $\cos 50$.
- Let's begin by using the $\sin 40$. We expect $\sin 40 = \frac{a}{26}$. Why?
  - *By definition of sine;* $\sin \theta = \frac{\text{opp}}{\text{hyp}}$.
- To calculate the length of a, we must determine the value of $26 \sin 40$ because $a = 26 \sin 40$. We will round our answer to two decimal places.
- Using the decimal approximation of $\sin 40 \approx 0.6428$, we can write

$$26(0.6428) \approx a$$
$$16.71 \approx a$$

- Now let's use $\cos 50$, which is approximately 0.6428. What do you expect the result to be? Explain.
  - *I expect the result to be the same. Since the approximation of* $\sin 40$ *is equal to the approximation of* $\cos 50$, *the computation should be the same.*

Note that students may say that $\sin 40 = \cos 50$. Ensure that students know that once decimal approximations are used in place of the functions, they are no longer looking at two quantities that are equal because the decimals are approximations. To this end, ask students to recall that in Exercise 1 they were only taking the first four decimal digits of the number; that is, they are using approximations of those values. Therefore, they cannot explicitly claim that $\sin 40 = \cos 50$, rather that their approximations are extremely close in value to one another.

If necessary, show the computation below that verifies the claim made above.

$$\cos 50 = \frac{a}{26}$$
$$26 \cos 50 = a$$
$$26(0.6428) \approx a$$
$$16.71 \approx a$$

- Now, calculate the length of side b.
  - *Side b can be determined using* $\sin 50$ *or* $\cos 40$.

$$26(0.7660) \approx b$$
$$19.92 \approx b$$

- Could we have used another method to determine the length of side b?
  - *Yes. Because this is a right triangle and two sides are known, we could use the Pythagorean theorem to determine the length of the third side.*

The points below are to make clear that the calculator gives approximations of the ratios we desire when using trigonometric functions.

- When we use a calculator to compute, what we get is a decimal approximation of the ratio $\frac{a}{26}$. Our calculators are programmed to know which number a is needed, relative to 26, so that the value of the ratio $\frac{a}{26}$ is equal to the value of $\sin 40$. For example, $\sin 40 = \frac{a}{26}$ and $\sin 40 \approx 0.6428$. Our calculators give us the number a that, when divided by 26, is closest to the actual value of $\sin 40$.
- Here is a simpler example illustrating this fact. Consider a right triangle with an acute angle of $30°$ and hypotenuse length of 9 units. Then, $\sin 30 = \frac{a}{9}$. We know that $\sin 30 = \frac{1}{2} = 0.5$. What our calculators do is find the number a so that $\frac{a}{9} = \frac{1}{2} = 0.5$, which is $a = 4.5$.

Exercise 5 (5 minutes)

Students complete Exercise 5 independently. All students should be able to complete part (a) in the allotted time. Consider assigning part (b) to only those students who finish part (a) quickly. Once completed, have students share their solutions with the class.

Scaffolding:

- Read the problem aloud, and ask students to summarize the situation with a partner.
- English language learners may benefit from labeling the horizontal distance E for east and the vertical distance S for south.
- Consider simplifying the problem by drawing only the triangle and labeling the measures of the angle and the hypotenuse and then asking students to find the unknown lengths.

Exercise 5

5. **A shipmate set a boat to sail exactly $27°$ NE from the dock. After traveling 120 miles, the shipmate realized he had misunderstood the instructions from the captain; he was supposed to set sail going directly east!**

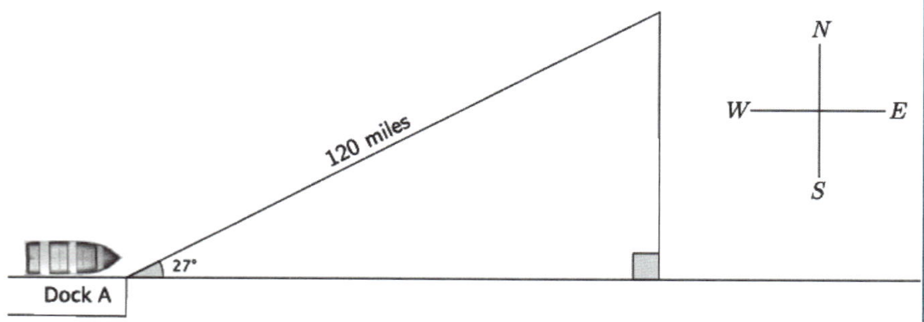

EUREKA MATH

a. **How many miles will the shipmate have to travel directly south before he is directly east of the dock? Round your answer to the nearest mile.**

Let S represent the distance they traveled directly south.

$$\sin 27 = \frac{S}{120}$$
$$120 \sin 27 = S$$
$$54.47885997\ldots = S$$

He traveled approximately 54 mi. south.

b. **How many extra miles does the shipmate travel by going the wrong direction compared to going directly east? Round your answer to the nearest mile.**

Solutions may vary. Some students may use the Pythagorean theorem while others may use the cosine function. Both are acceptable strategies. If students use different strategies, make sure to share them with the class, and discuss the benefits of each.

Let E represent the distance the boat is when it is directly east of the dock.

$$\cos 27 = \frac{E}{120}$$
$$120 \cos 27 = E$$
$$106.9207829\ldots = E$$
$$107 \approx E$$

The total distance traveled by the boat is $120 + 54 = 174$. They ended up exactly 107 miles east of the dock. $174 - 107 = 67$, so they traveled an extra 67 miles.

Example 2 (8 minutes)

Students find the missing side length of a triangle using sine and cosine.

Example 2

Johanna borrowed some tools from a friend so that she could precisely, but not exactly, measure the corner space in her backyard to plant some vegetables. She wants to build a fence to prevent her dog from digging up the seeds that she plants. Johanna returned the tools to her friend before making the most important measurement: the one that would give the length of the fence!

Johanna decided that she could just use the Pythagorean theorem to find the length of the fence she would need. Is the Pythagorean theorem applicable in this situation? Explain.

MP.1

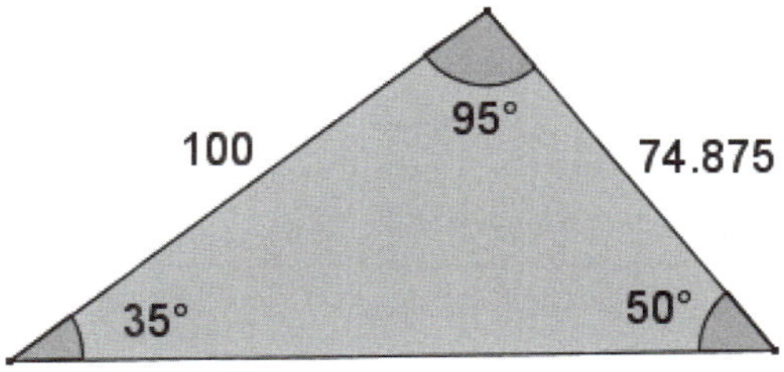

No The corner of her backyard is not a $90°$ angle; therefore, the Pythagorean theorem cannot be applied in this situation. The Pythagorean theorem will, however, provide an approximation since the given angle has a measure that is close to $90°$.

MP.1

- What can we do to help Johanna figure out the length of fence she needs?

Provide time for students to discuss this in pairs or small groups. Allow them to make sense of the problem and persevere in solving it. It may be necessary to guide their thinking using the prompts below.

- If we dropped an altitude from the angle with measure $95°$, could that help? How?
- Would we be able to use the Pythagorean theorem now? Explain.
- If we denote the side opposite the $95°$ angle as x and y, as shown, can we use what we know about sine and cosine? Explain.

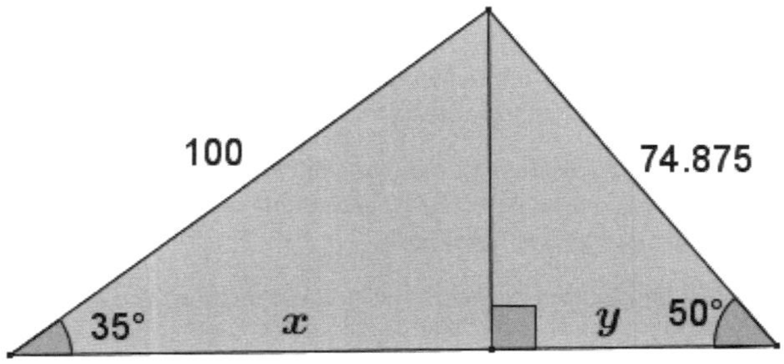

  - *The missing side length is equal to* $x + y$. *The length* x *is equal to* $100 \cos 35$, *and the length* y *is equal to* $74.875 \cos 50$. *Therefore, the length of*
$$x + y = 100 \cos 35 + 74.875 \cos 50 \approx 81.92 + 48.12872 \approx 130.05.$$

Note: The Pythagorean theorem provides a reasonable approximation of 124.93.

Exercise 6 (4 minutes)

Students complete Exercise 6 independently.

Exercise 6

6. **The measurements of the triangle shown below are rounded to the nearest hundredth. Calculate the missing side length to the nearest hundredth.**

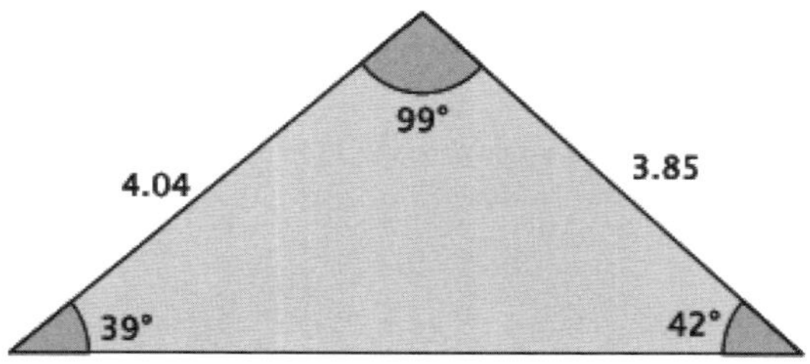

Drop an altitude from the angle that measures $99°$.

EUREKA MATH™

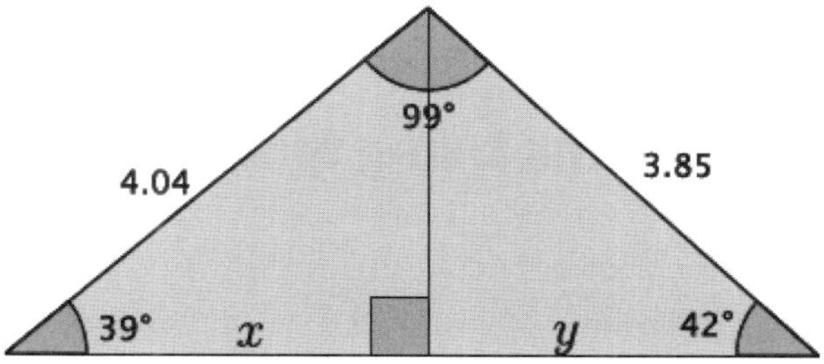

Then, the length of the missing side is $x + y$, which can be found by
$\mathbf{4.04\cos 39 + 3.85\cos 42 \approx 3.139669 + 2.861107 = 6.000776 \approx 6.00}$.

Closing (3 minutes)

Ask students to discuss the answers to the following questions with a partner, and then select students to share with the class. For the first question, elicit as many acceptable responses as possible.

- Explain how to find the unknown length of a side of a right triangle.
  - *If two sides are known, then the Pythagorean theorem can be used to determine the length of the third side.*
  - *If one side is known and the measure of one of the acute angles is known, then sine, cosine, or tangent can be used.*
  - *If the triangle is known to be similar to another triangle where the side lengths are given, then corresponding ratios or knowledge of the scale factor can be used to determine the unknown length.*
  - *Direct measurement can be used.*
- Explain when and how you can find the unknown length of a side of a triangle that does not have a right angle.
  - *You can find the length of an unknown side length of a triangle when you know two of the side lengths and the missing side is between two acute angles. Split the triangle into two right triangles, and find the lengths of two pieces of the missing side.*

Exit Ticket (5 minutes)

Name ______________________________ Date ______________

Lesson 28: Solving Problems Using Sine and Cosine

Exit Ticket

1. Given right triangle ABC with hypotenuse $AB = 8.5$ and $m\angle A = 55°$, find AC and BC to the nearest hundredth.

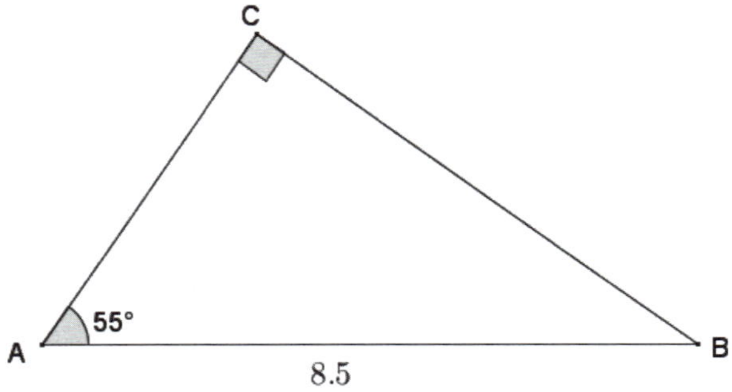

2. Given triangle DEF, $m\angle D = 22°$, $m\angle F = 91°$, $DF = 16.55$, and $EF = 6.74$, find DE to the nearest hundredth.

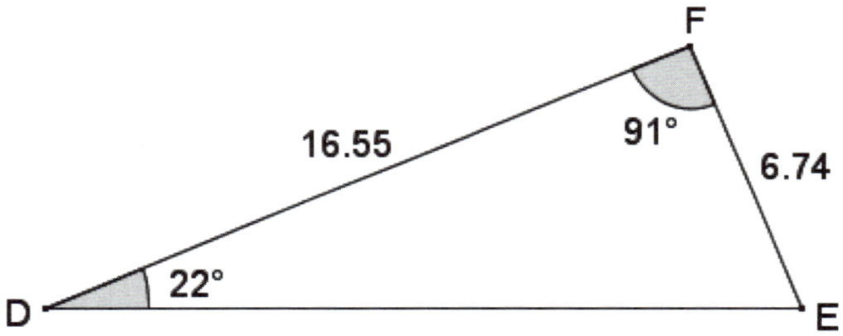

Exit Ticket Sample Solutions

1. Given right triangle ABC with hypotenuse $AB = 8.5$ and $m\angle A = 55°$, find AC and BC to the nearest hundredth.

$BC = 8.5(\sin 55)$
$BC \approx 6.96$

$AC = 8.5\,(\cos 55)$
$AC \approx 4.88$

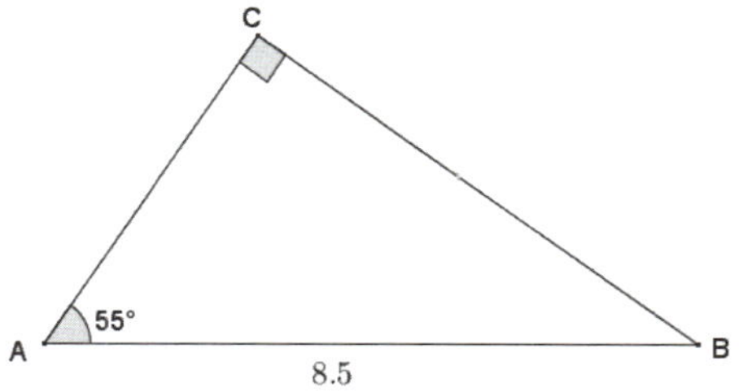

2. Given triangle DEF, $m\angle D = 22°$, $m\angle F = 91°$, $DF = 16.55$, and $EF = 6.74$, find DE to the nearest hundredth.

Draw altitude from F to $\overline{DE}$ at point P. Cosines can be used on angles D and E to determine the lengths of $\overline{DP}$ and $\overline{PE}$, which together compose $\overline{DE}$.

$PE = 6.74(\cos 67)$
$PE \approx 2.6335$

$DP = 16.55(\cos 22)$
$DP \approx 15.3449$

$DE = DP + PE$
$DE \approx 15.3449 + 2.6335$
$DE \approx 17.98$

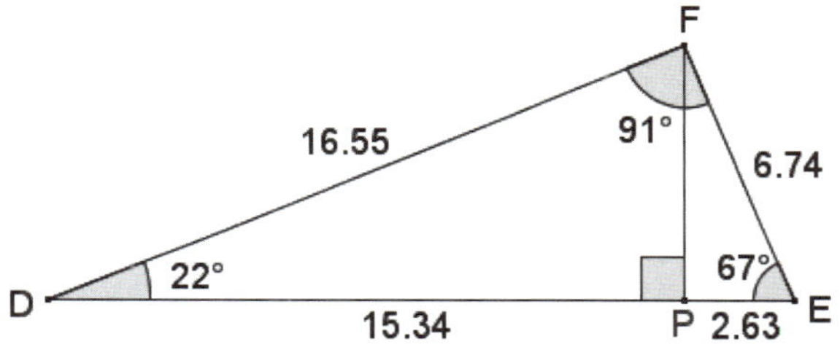

Note to teacher: Answers of $DE \approx 17.97$ result from rounding to the nearest hundredth too early in the problem.

Problem Set Sample Solutions

1. Given right triangle GHI, with right angle at H, $GH = 12.2$, and $m\angle G = 28°$, find the measures of the remaining sides and angle to the nearest tenth.

$\cos 28 = \dfrac{12.2}{GI}$

$GI = \dfrac{12.2}{\cos 28}$

$GI \approx 13.8$

$\tan 28 = \dfrac{IH}{12.2}$

$IH = 12.2 \tan 28$

$IH \approx 6.5$

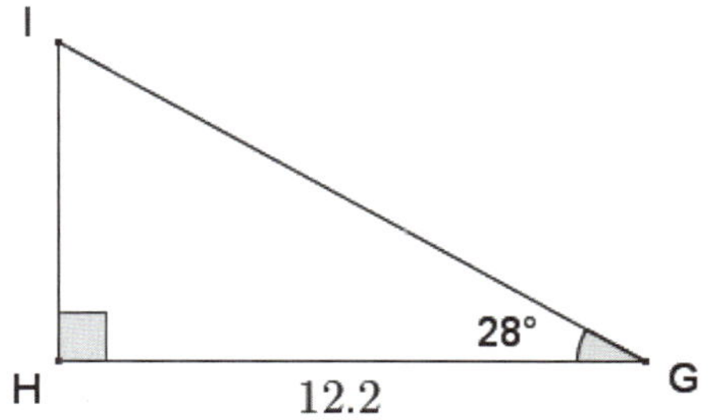

$28° + m\angle I = 90°$
$m\angle I = 62°$

2. The Occupational Safety and Health Administration (OSHA) provides standards for safety at the workplace. A ladder is leaned against a vertical wall according to OSHA standards and forms an angle of approximately $75°$ with the floor.

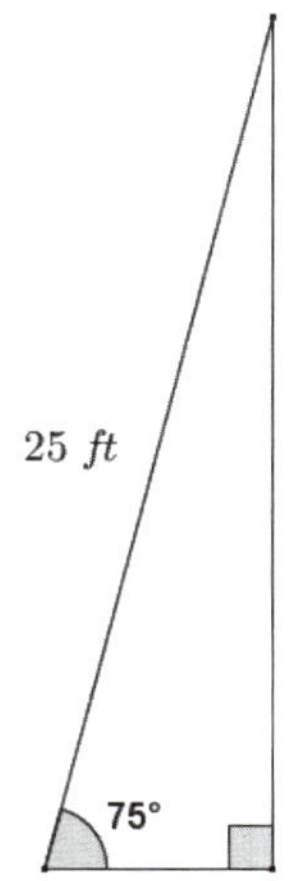

a. If the ladder is 25 ft. long, what is the distance from the base of the ladder to the base of the wall?

Let b represent the distance of the base of the ladder from the wall in feet.

$$b = 25(\cos 75)$$
$$b \approx 6.5$$

The base of the ladder is approximately 6 ft. 6 in. from the wall.

b. How high on the wall does the ladder make contact?

Let h represent the height on the wall where the ladder makes contact in feet.

$$h = 25(\sin 75)$$
$$h \approx 24.1$$

The ladder contacts the wall just over 24 ft. above the ground.

c. Describe how to safely set a ladder according to OSHA standards without using a protractor.

Answers will vary. Possible description:

The horizontal distance of the base of the ladder to the point of contact of the ladder should be approximately $\frac{1}{4}$ of the length of the ladder.

3. A regular pentagon with side lengths of 14 cm is inscribed in a circle. What is the radius of the circle?

Draw radii from center C of the circle to two consecutive vertices of the pentagon, A and B, and draw an altitude from the center C to D on $\overline{AB}$.

The interior angles of a regular pentagon have measure of $108°$, and $\overline{AC}$ and $\overline{BC}$ bisect the interior angles at A and B.

$$AD = BD = 7 \text{ cm}$$

Let x represent the lengths of $\overline{AC}$ in centimeters.

Using cosine, $\cos 54 = \frac{7}{x}$, and thus:

$$x = \frac{7}{\cos 54}$$
$$x \approx 11.9.$$

$\overline{AC}$ is a radius of the circle and has a length of approximately 11.9 cm.

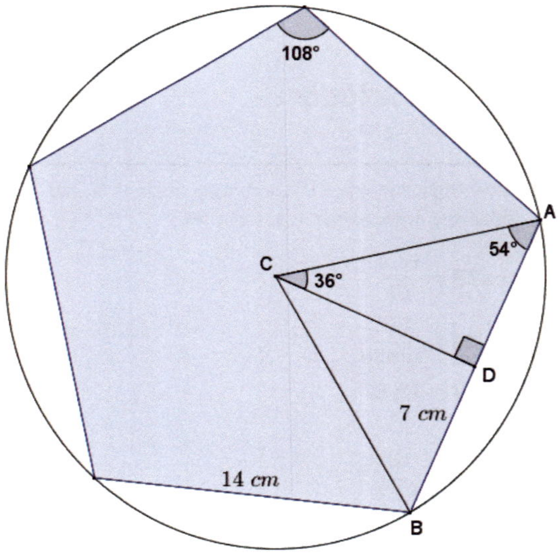

4. The circular frame of a Ferris wheel is suspended so that it sits 4 ft. above the ground and has a radius of 30 ft. A segment joins center C to point S on the circle. If $\overline{CS}$ makes an angle of $48°$ with the horizon, what is the distance of point S to the ground?

Note to teacher: There are two correct answers to this problem since the segment can make an angle of $48°$ above or below the horizon in four distinct locations, providing two different heights above the ground.

There are four locations at which the segment makes an angle of $48°$ with the horizon. In each case, $\overline{CS}$ is the hypotenuse of a right triangle with acute angles with measures of $48°$ and $42°$.

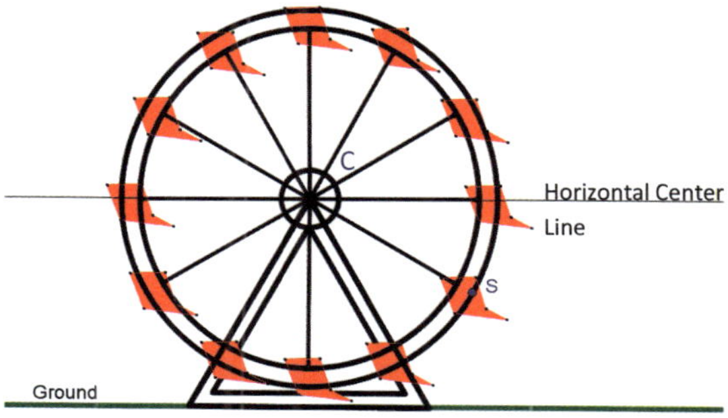

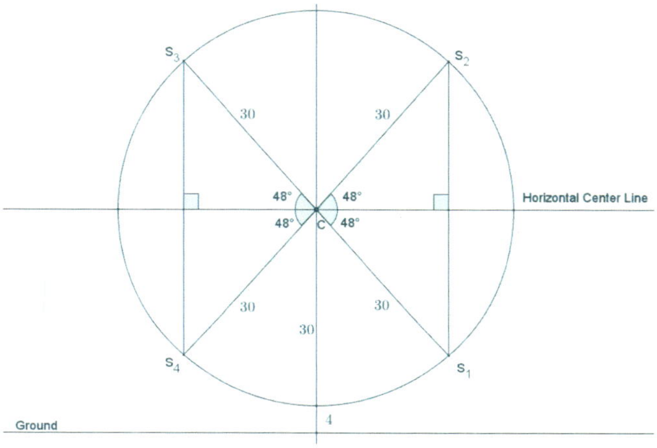

Let d represent the distance in feet from point S to the horizon (applies to each case):

$$\sin 48 = \frac{d}{30}$$
$$30\,(\sin 48) = d$$
$$22.3 \approx d$$

The center of the Ferris wheel is 34 ft. above the ground; therefore, the distance from points S_1 and S_4 to the ground in feet is

$34 - 22.3 = 11.7.$

Points S_2 and S_3 are approximately 22.3 ft. above the center of the Ferris wheel, so the distance from S_2 and S_3 to the ground in feet is

$34 + 22.3 = 56.3.$

When $\overline{CS}$ forms a $48°$ angle with the horizon, point S is either approximately 11.7 ft. above the ground or approximately 56.3 ft. above the ground.

5. Tim is a contractor who is designing a wheelchair ramp for handicapped access to a business. According to the Americans with Disabilities Act (ADA), the maximum slope allowed for a public wheelchair ramp forms an angle of approximately $4.76°$ to level ground. The length of a ramp's surface cannot exceed 30 ft. without including a flat 5 ft.$\times$ 5 ft. platform (minimum dimensions) on which a person can rest, and such a platform must be included at the bottom and top of any ramp.

Tim designs a ramp that forms an angle of $4°$ to the level ground to reach the entrance of the building. The entrance of the building is 2 ft. 9 in. above the ground. Let x and y as shown in Tim's initial design below be the indicated distances in feet.

a. Assuming that the ground in front of the building's entrance is flat, use Tim's measurements and the ADA requirements to complete and/or revise his wheelchair ramp design.

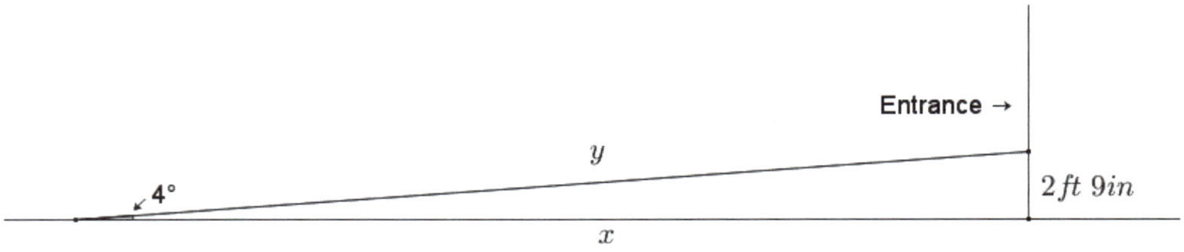

(For more information, see section 405 of the 2010 ADA Standards for Accessible Design at the following link: http://www.ada.gov/regs2010/2010ADAStandards/2010ADAstandards.htm#pgfId-1006877.)

Note to teacher: Student designs will vary; however, the length of the ramp's surface is greater than 30 ft.*, which requires at least one resting platform along the ramp. Furthermore, Tim's design does not include a platform at the top of the ramp as required by the guidelines, rendering his design incorrect.*

Possible student solution:

2 ft. 9 in. $= 2.75$ ft.

Using tangent, $\tan 4 = \frac{2.75}{x}$*, and thus*

$$x = \frac{2.75}{\tan 4}$$
$$x \approx 39.33.$$

The ramp begins approximately 39 ft. 4 in. *from the building; thus, the ramp's surface is greater than* 30 ft. *in length. The hypotenuse of the triangle represents the sloped surface of the ramp and must be longer than the legs. Tim's design will not meet the ADA guidelines because it does not include a flat resting section along the ramp's slope, nor does it include a platform at the top of the ramp. (The bottom of the ramp is flat ground. The student's design may or may not include a platform at the bottom.)*

The vertical distance from the ground to the entrance is 2.75 ft. *Using sine,* $\sin 4 = \frac{2.75}{y}$*, and thus,*

$$y = \frac{2.75}{\sin 4}$$
$$y \approx 39.42.$$

The total length of the ramp surface is approximately 39 ft. 5 in.*; however, because of its length, it requires a resting platform somewhere in the first* 30 ft. *and another platform at the top.*

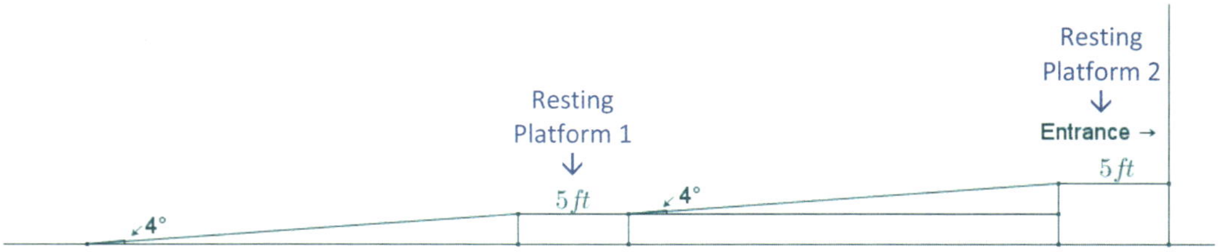

b. **What is the total distance from the start of the ramp to the entrance of the building in your design?**

If each platform is 5 ft. *in length, then the total distance along the ramp from the ground to the building is approximately* 49 ft. 5 in.

6. **Tim is designing a roof truss in the shape of an isosceles triangle. The design shows the base angles of the truss to have measures of $18.5°$. If the horizontal base of the roof truss is 36 ft. across, what is the height of the truss?**

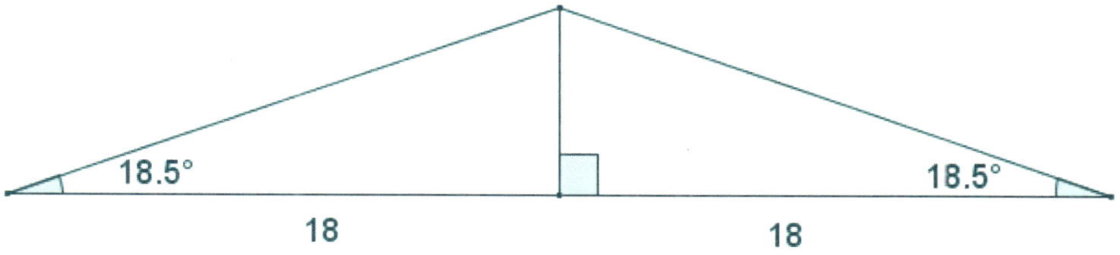

Let h represent the height of the truss in feet. Using tangent, $\tan 18.5 = \frac{h}{18}$, *and thus,*

$$h = 18(\tan 18.5)$$
$$h \approx 6.$$

The height of the truss is approximately 6 ft.

Lesson 29: Applying Tangents

Student Outcomes

- Students understand that the value of the tangent ratio of the angle of elevation or depression of a line is equal to the slope of the line.
- Students use the value of the tangent ratio of the angle of elevation or depression to solve real-world problems.

Lesson Notes

Lesson 26 introduced students to the tangent of θ as the ratio of the length of the opposite side to the length of the adjacent side. In Lesson 29, students use $\tan \theta$ in two different contexts: (1) the value of the ratio as it has to do with slope and (2) its use in solving real-world problems involving angles of elevation and depression. The focus of this lesson is modeling with mathematics, MP.4, as students work on a series of real-world applications with tangent.

Classwork

Opening Exercise (7 minutes)

For these exercises, consider asking half the class to work on part (a) and the other half to go directly to part (b), and then share results. Alternately, have students work in small groups. Students can work in pairs (or fours to split the work a little more); a student divides the values while the partner finds the tangent values, and they compare their results. Then, the class can debrief as a whole.

MP.8

Opening Exercise

a. **Use a calculator to find the tangent of θ. Enter the values, correct to four decimal places, in the last row of the table.**

θ	0	10	20	30	40	50	60	70	80	90
$\sin \theta$	0	0.1736	0.3420	0.5	0.6428	0.7660	0.8660	0.9397	0.9848	1
$\cos \theta$	1	0.9848	0.9397	0.8660	0.7660	0.6428	0.5	0.3420	0.1736	0
$\frac{\sin \theta}{\cos \theta}$	0	0.1763	0.3639	0.5774	0.8392	1.1917	1.7320	2.7477	5.6728	*undefined*
$\tan \theta$	0	0.1763	0.3640	0.5774	0.8391	1.1918	1.7321	2.7475	5.6713	*undefined*

Note to the teacher: Dividing the values in the first two rows provides a different answer than using the full decimal readout of the calculator.

MP.8

b. The table from Lesson 29 is provided here for you. In the row labeled $\frac{\sin\theta}{\cos\theta}$, divide the sine values by the cosine values. What do you notice?

For each of the listed degree values, dividing the value of $\sin\theta$ by the corresponding value of $\cos\theta$ yields a value very close to the value of $\tan\theta$. The values divided to obtain $\frac{\sin\theta}{\cos\theta}$ were approximations, so the actual values might be exactly the same.

Note to the teacher: Consider showing the calculation of $\frac{\sin\theta}{\cos\theta}$ using the calculator to reveal that the quotients are not just close to the value of $\tan\theta$, but are, in fact, equal to the value of $\tan\theta$.

Depending on student readiness, consider asking students to speculate what the relationship between sine, cosine, and tangent is. The relationship is summarized as $\frac{\frac{\text{opp}}{\text{adj}}}{\frac{\text{adj}}{\text{hyp}}} = \frac{\text{opp}}{\text{hyp}} \times \frac{\text{hyp}}{\text{adj}} = \frac{\text{opp}}{\text{adj}}$ but is expanded upon in Lesson 30. This level of explanation makes it clear why $\tan 0 = 0$ and why $\tan 90$ is undefined.

Discussion (5 minutes)

Lead students through a discussion that ties together the concepts of angle of elevation/depression in a real-world sense, and then, using the coordinate plane, tying it to tangent and slope.

- Consider the image below. How would you describe the angle of elevation? How would you describe the angle of depression?

Allow students a moment to come up with their own definitions and then share out responses before presenting the following definition:

- The angle of elevation or depression is the angle between the horizontal (parallel with the earth's surface) and the line of sight.
- In a case where two viewers can observe each other, such as in the diagram below, what do you notice about the measure of the angle of depression and the angle of elevation? Why?
  - *They have the same measure since the horizontal from each viewer forms a pair of parallel lines, and the line of sight acts as a transversal; the angle of depression and angle of elevation are alternate interior angles of equal measure.*

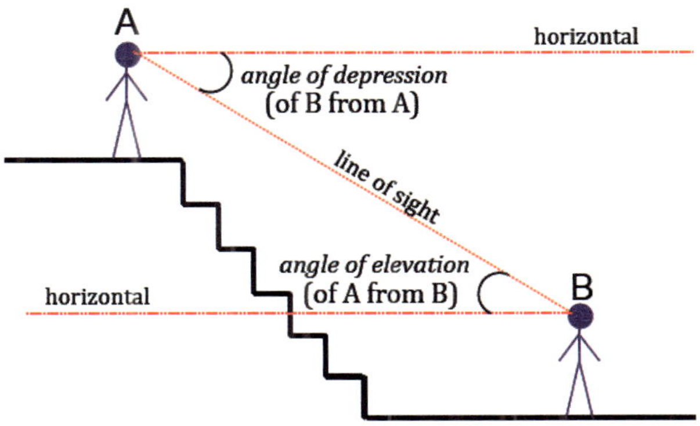

Scaffolding:

- Consider placing a poster of the following in a prominent place in the classroom:

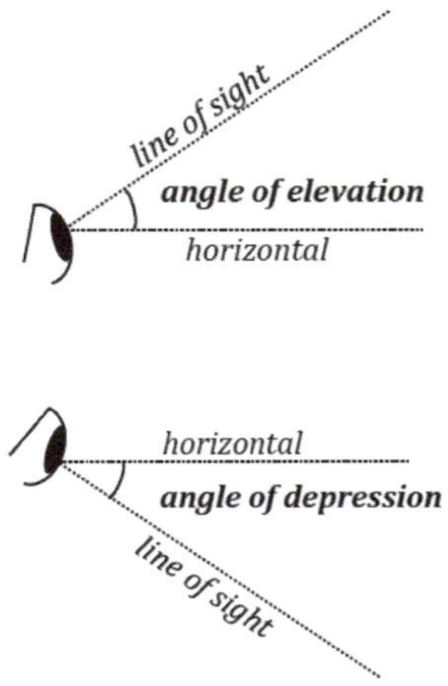

- Students may benefit from choral recitation of the description of each term.

Example 1 (7 minutes)

Example 1 is a real-world application of the angle of elevation and tangent. Encourage students to label the diagram as they read through the problem; consider using the same labels for the key vertices as a whole class. If students struggle to begin the problem, consider drawing the triangle for the class and then asking students to write or speak about how the triangle models the situation.

- Consider standing near a tree and observing a bird at the top of the tree. We can use measurements we know and the angle of elevation to help determine how high off the ground the bird is.

Please encourage students to attempt to solve first on their own or with a partner.

Example 1

Scott, whose eye level is 1.5 m above the ground, stands 30 m from a tree. The angle of elevation of a bird at the top of the tree is $36°$. How far above ground is the bird?

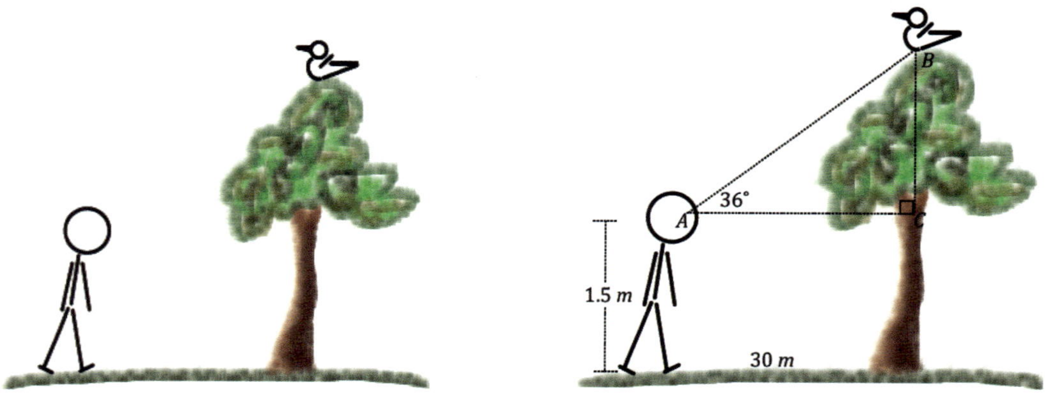

MP.1

- With respect to your diagram, think of the measurement you are looking for.
  - *In our diagram, we are looking for BC.*
- How will you find BC?
  - *I can use the tangent to determine BC in meters.*

$$\tan 36 = \frac{BC}{AC}$$

$$\tan 36 = \frac{BC}{30}$$

$$30 \tan 36 = BC$$

$$BC \approx 21.8$$

- Have we found the height at which the bird is off the ground?
  - *No. The full height must be measured from the ground, so the distance from the ground to Scott's eye level must be added to BC.*
  - *The height of the bird off of the ground is* $1.5\text{ m} + 21.8\text{ m} = 23.3\text{ m}$.

MP.1

- So, between the provided measurements, including the angle of elevation, and the use of the tangent ratio, we were able to find the height of the bird.

Example 2 (5 minutes)

Example 2 is a real-world application of the angle of depression and tangent. Encourage students to label the diagram as they read through the problem; consider having the whole class use common labels for the key vertices. Consider having students work with a partner for this example.

> **Example 2**
>
> **From an angle of depression of $40°$, John watches his friend approach his building while standing on the rooftop. The rooftop is 16 m from the ground, and John's eye level is at about 1.8 m from the rooftop. What is the distance between John's friend and the building?**

Make sure to point out the angle of depression in the diagram below. Emphasize that the $40°$ angle of depression is the angle between the line of sight and the line horizontal (to the ground) from the eye.

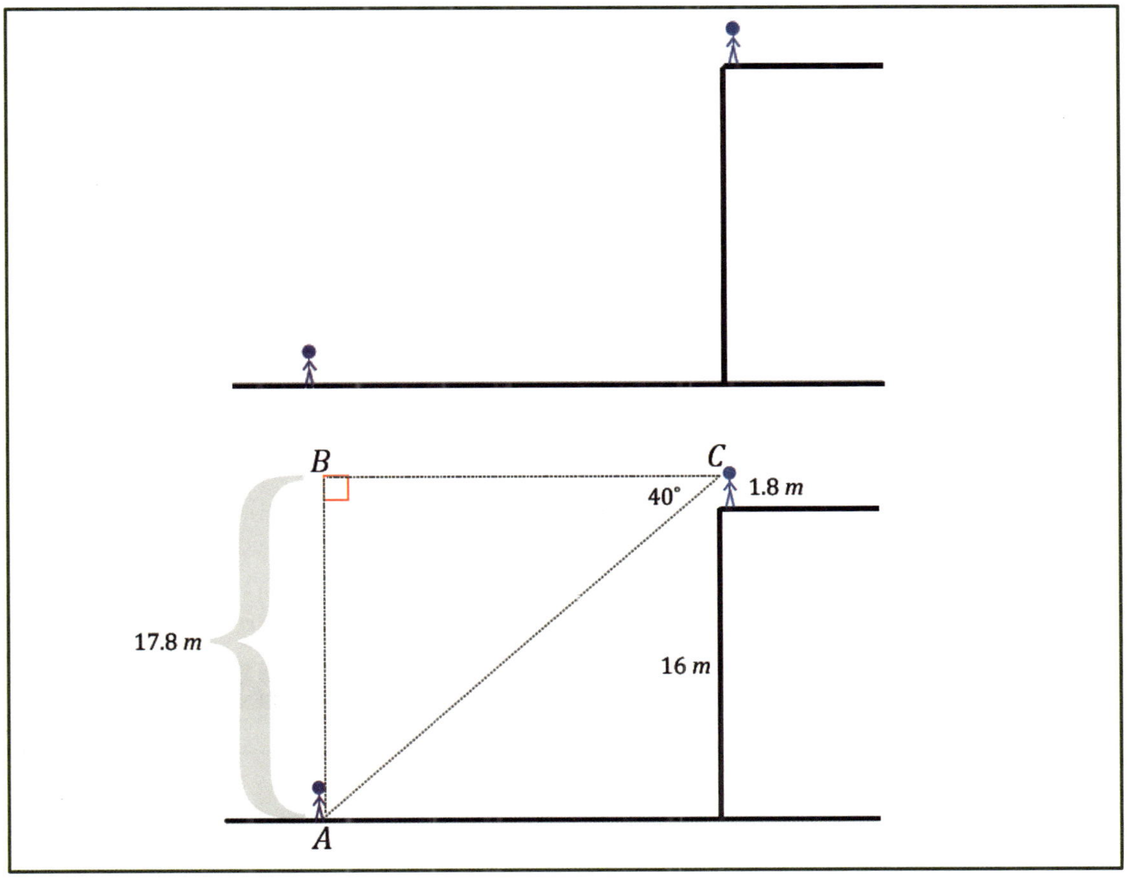

- Use the diagram to describe the distance we must determine.
  - *We are going to find BC in meters.*

- How will we find BC?
  - *We can use the tangent:* $\tan 40 = \frac{AB}{BC}$.

$$\tan 40 = \frac{AB}{BC}$$

$$\tan 40 = \frac{17.8}{BC}$$

$$BC = \frac{17.8}{\tan 40}$$

$$BC \approx 21.2$$

- Again, with the assistance of a few measurements, including the angle of depression, we were able to determine the distance between John's friend and the building, which is 21.2 meters.

Exercise 1 (4 minutes)

In the following problem, the height of the person is not provided as a piece of information. This is because the relative height of the person does not matter with respect to the other heights mentioned.

Exercise 1

Standing on the gallery of a lighthouse (the deck at the top of a lighthouse), a person spots a ship at an angle of depression of $20°$. The lighthouse is 28 m tall and sits on a cliff 45 m tall as measured from sea level. What is the horizontal distance between the lighthouse and the ship? Sketch a diagram to support your answer.

Approximately 201 m

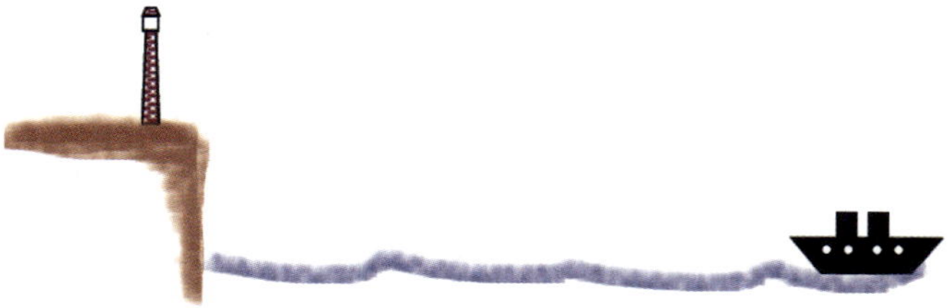

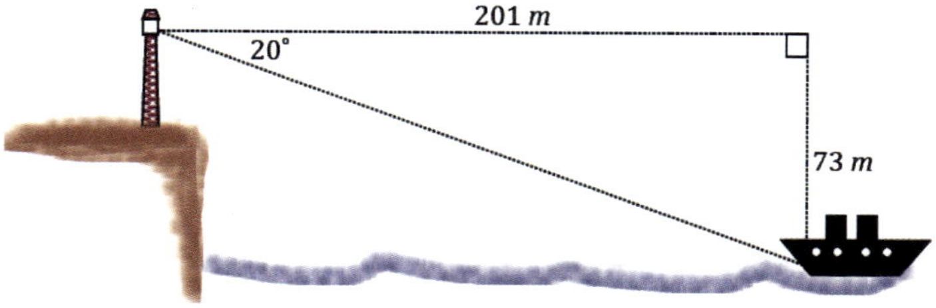

Discussion (6 minutes)

- Now, let's examine the angle of elevation and depression on the coordinate plane.
- Observe the following line, which has a positive slope and crosses the x-axis.
- The indicated angle is the angle of elevation.

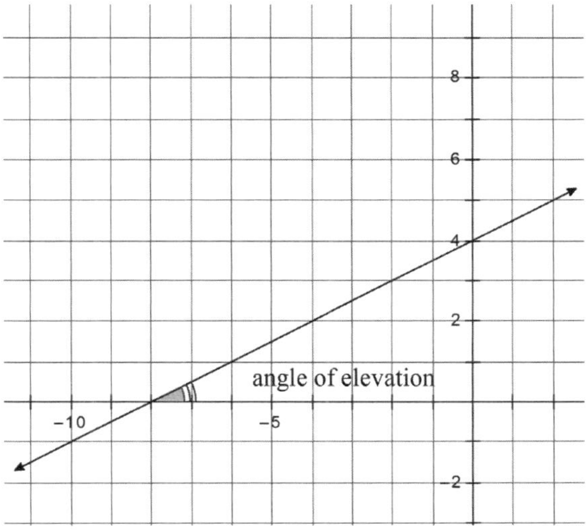

- Notice that the slope of the line is the tangent of the angle of elevation. Why is this true?
  - *By building a right triangle (so that one leg is parallel to the x-axis and the other leg is parallel to the y-axis) around any two points selected on the line, we can determine the slope by finding the value of $\frac{\text{rise}}{\text{run}}$. This is equal to the tangent of the angle of elevation because, with respect to the angle of elevation, the tangent ratio is $\frac{\text{opp}}{\text{adj}}$, which is equivalent to the value of $\frac{\text{rise}}{\text{run}}$.*
- Similar to the angle of elevation, there is also an angle of depression. The indicated angle between the line declining to the right and the x-axis is the angle of depression.

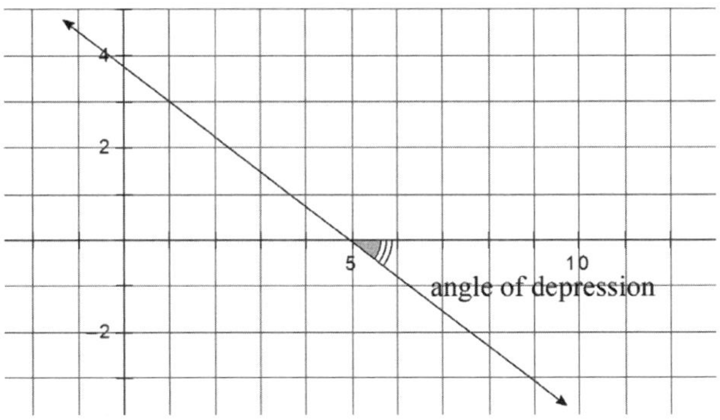

- Notice that the tangent of the angle of depression will be positive; to accurately capture the slope of the line, we must take the negative of the tangent.

Exercise 2 (4 minutes)

Exercise 2

A line on the coordinate plane makes an angle of depression of $36°$. Find the slope of the line correct to four decimal places.

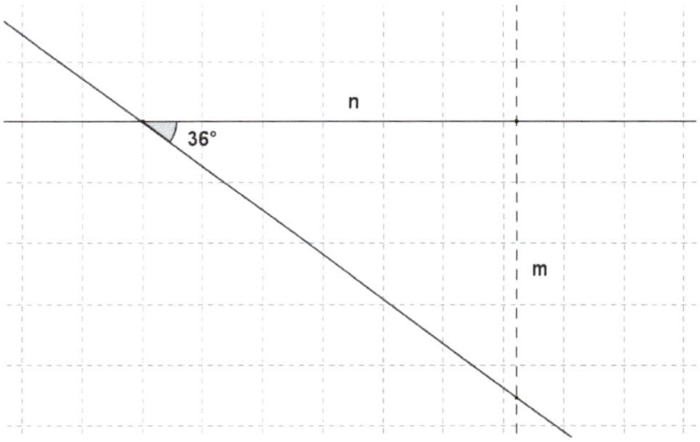

Choose a segment on the line, and construct legs of a right triangle parallel to the x- and y-axes, as shown. If m is the length of the vertical leg and n is the length of the horizontal leg, then $\tan 36 = \frac{m}{n}$. The line decreases to the right, so the value of the slope must be negative. Therefore,

$$\text{slope} = -\frac{m}{n} = -\tan 36 \approx -0.7265.$$

Closing (2 minutes)

Ask students to summarize the key points of the lesson. Additionally, consider asking students the following questions independently in writing, to a partner, or to the whole class.

- Review the relationship between tangent and slope.
  - *The slope of a line increasing to the right is the same as the tangent of the angle of elevation of the line. The slope of a line decreasing to the right is the same as the negative tangent of the angle of depression of the line.*
- Review the relationship between tangent and slope.
- Review what the angle of elevation and depression means in a real-world context.
  - *The angle of elevation or depression is the angle between the horizontal and the line of sight.*

Exit Ticket (5 minutes)

Name ___________________________ Date ______________

Lesson 29: Applying Tangents

Exit Ticket

1. The line on the coordinate plane makes an angle of depression of $24°$. Find the slope of the line correct to four decimal places.

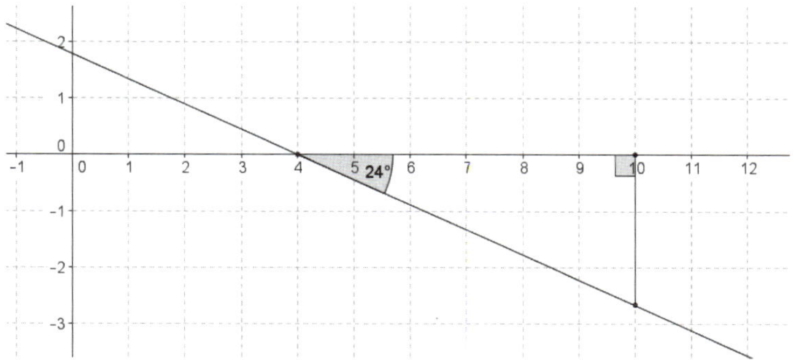

2. Samuel is at the top of a tower and will ride a trolley down a zip line to a lower tower. The total vertical drop of the zip line is 40 ft. The zip line's angle of elevation from the lower tower is $11.5°$. What is the horizontal distance between the towers?

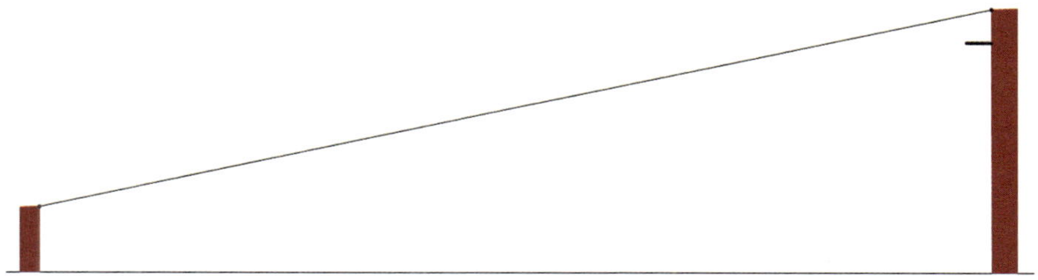

Exit Ticket Sample Solutions

1. **A line on the coordinate plane makes an angle of depression of $24°$. Find the slope of the line correct to four decimal places.**

Choose a segment on the line, and construct legs of a right triangle parallel to the x- and y-axes, as shown. If m is the length of the vertical leg and n is the length of the horizontal leg, then $\tan 24 = \frac{m}{n}$. The line decreases to the right, so the value of the slope must be negative. Therefore,

$$\text{slope} = -\frac{m}{n} = -\tan 24 \approx -0.4452.$$

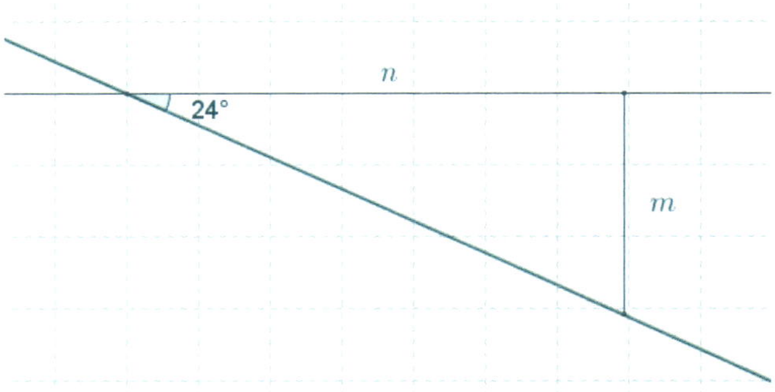

2. **Samuel is at the top of a tower and will ride a trolley down a zip line to a lower tower. The total vertical drop of the zip line is 40 ft. The zip line's angle of elevation from the lower tower is $11.5°$. What is the horizontal distance between the towers?**

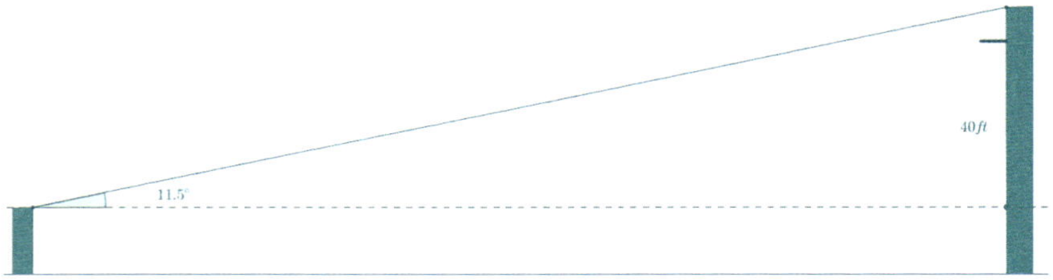

A right triangle is formed by the zip line's path, the vertical drop along the upper tower, and the horizontal distance between the towers. Let x represent the horizontal distance between the towers in feet. Using tangent:

$$\tan 11.5 = \frac{40}{x}$$

$$x = \frac{40}{\tan 11.5}$$

$$x \approx 196.6$$

The horizontal distance between the towers is approximately 196.6 ft.

Problem Set Sample Solutions

1. **A line in the coordinate plane has an angle of elevation of $53°$. Find the slope of the line correct to four decimal places.**

Since parallel lines have the same slope, we can consider the line that passes through the origin with an angle of inclination of $53°$. Draw a vertex at $(5, 0)$ on the x-axis, and draw a segment from $(5, 0)$ parallel to the y-axis to the intersection with the line to form a right triangle.

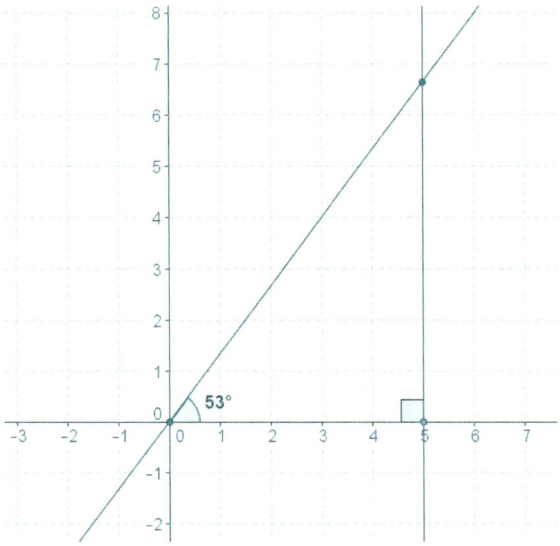

If the base of the right triangle is 5-units long, let the height of the triangle be represented by y.

$$\tan 53 = \frac{x}{5}$$
$$5(\tan 53) = x$$

The slope of the line is

$$\text{slope} = \frac{\text{rise}}{\text{run}}$$
$$\text{slope} = \frac{5(\tan 53)}{5}$$
$$\text{slope} = \tan 53 \approx 1.3270.$$

2. **A line in the coordinate plane has an angle of depression of $25°$. Find the slope of the line correct to four decimal places.**

A line crosses the x-axis, decreasing to the right with an angle of depression of $25°$. Using a similar method as in Problem 1, a right triangle is formed. If the leg along the x-axis represents the base of the triangle, let h represent the height of the triangle.

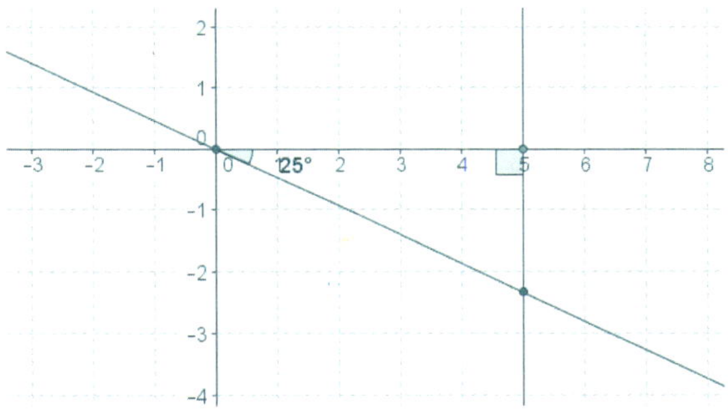

Using tangent:

$$\tan 25 = \frac{h}{5}$$
$$5(\tan 25) = h$$

The slope of the line is negative since the line decreases to the right:

$$\text{slope} = \frac{\text{rise}}{\text{run}}$$
$$\text{slope} = -\frac{h}{5}$$
$$\text{slope} = -\frac{5(\tan 25)}{5}$$
$$\text{slope} = -\tan 25 \approx -0.4663$$

3. **In Problems 1 and 2, why do the lengths of the legs of the right triangles formed not affect the slope of the line?**

When using the tangent ratio, the length of one leg of a right triangle can be determined in terms of the other leg. Let x represent the length of the horizontal leg of a slope triangle. The vertical leg of the triangle is then $x \tan \theta$, where θ is the measure of the angle of inclination or depression. The slope of the line is $\frac{\text{rise}}{\text{run}} = \frac{x \tan \theta}{x} = \tan \theta$.

4. Given the angles of depression below, determine the slope of the line with the indicated angle correct to four decimal places.

a. $35°$ angle of depression

$\tan 35 \approx 0.7002$

slope ≈ -0.7002

b. $49°$ angle of depression

$\tan 49 \approx 1.1504$

slope ≈ -1.1504

c. $80°$ angle of depression

$\tan 80 \approx 5.6713$

slope ≈ -5.6713

d. $87°$ angle of depression

$\tan 87 \approx 19.0811$

slope ≈ -19.0811

e. $89°$ angle of depression

$\tan 89 \approx 57.2900$

slope ≈ -57.2900

f. $89.9°$ angle of depression

$\tan 89.9 \approx 572.9572$

slope ≈ -572.9572

g. What appears to be happening to the slopes (and tangent values) as the angles of depression get closer to $90°$?

As the angles get closer to $90°$, their slopes (and tangent values) get much farther from zero.

h. Find the slopes of angles of depression that are even closer to $90°$ than $89.9°$. Can the value of the tangent of $90°$ be defined? Why or why not?

Choices of angles will vary. The closer an angle is in measure to $90°$, the greater the tangent value of that angle and the farther the slope of the line determined by that angle is from zero. An angle of depression of $90°$ would be a vertical line, and vertical lines have 0 run; therefore, the value of the ratio $rise: run$ is undefined. The value of the tangent of $90°$ would have a similar outcome because the adjacent leg of the "triangle" would have a length of 0, so the ratio $\frac{\text{opp}}{\text{adj}} = \frac{\text{opp}}{0}$, which is undefined.

5. For the indicated angle, express the quotient in terms of sine, cosine, or tangent. Then, write the quotient in simplest terms.

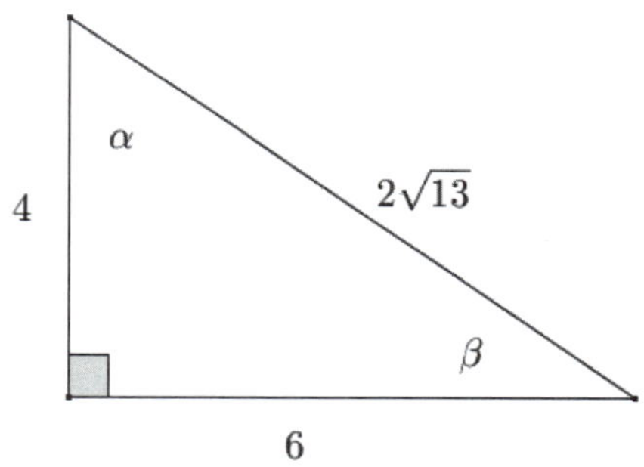

a. $\frac{4}{2\sqrt{13}}$; α

$\cos\alpha = \frac{4}{2\sqrt{13}} = \frac{2}{\sqrt{13}} = \frac{2\sqrt{13}}{13}$

b. $\frac{6}{4}$; α

$\tan\alpha = \frac{6}{4} = \frac{3}{2}$

c. $\frac{4}{2\sqrt{13}}$; β

$\sin\beta = \frac{4}{2\sqrt{13}} = \frac{2\sqrt{13}}{13}$

d. $\frac{4}{6}$; β

$\tan\beta = \frac{4}{6} = \frac{2}{3}$

6. The pitch of a roof on a home is expressed as a ratio of $\text{vertical rise}:\text{horizontal run}$ where the run has a length of 12 units. If a given roof design includes an angle of elevation of $22.5°$ and the roof spans 36 ft. as shown in the diagram, determine the pitch of the roof. Then, determine the distance along one of the two sloped surfaces of the roof.

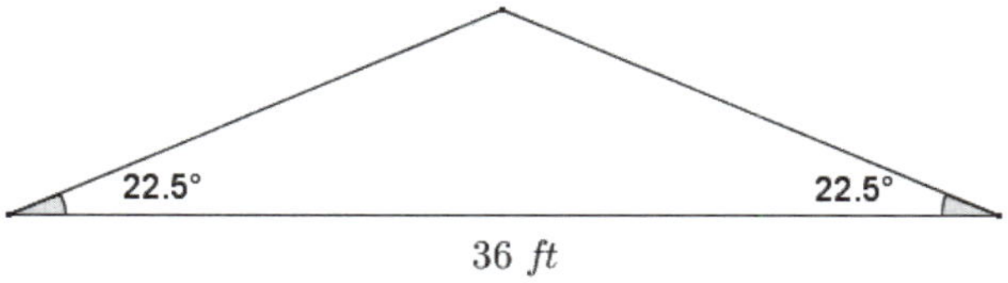

The diagram as shown is an isosceles triangle since the base angles have equal measure. The altitude, a, of the triangle is the vertical rise of the roof.

The right triangles formed by drawing the altitude of the given isosceles triangle have a leg of length 18 ft.

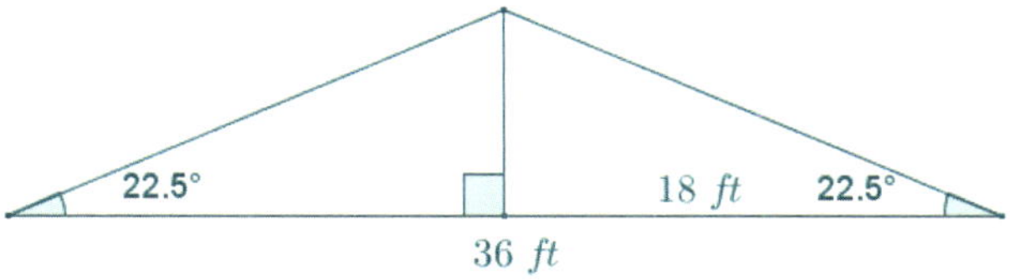

$$\tan 22.5 = \frac{a}{18}$$
$$a = 18 \tan 22.5$$
$$a \approx 7.5$$

Roof pitch:

$$\frac{7.5}{18} = \frac{h}{12}$$
$$18h = 12 \cdot 7.5$$
$$h = 5$$

$$\cos 22.5 = \frac{18}{s}$$
$$s = \frac{18}{\cos 22.5}$$
$$s \approx 19.5$$

The pitch of the roof is $5:12$.

The sloped surface of the roof has a distance of approximately 19.5 ft.

7. An anchor cable supports a vertical utility pole forming a $51°$ angle with the ground. The cable is attached to the top of the pole. If the distance from the base of the pole to the base of the cable is 5 meters, how tall is the pole?

Let h represent the height of the pole in meters.

Using tangent:

$$\tan 51 = \frac{h}{5}$$
$$5(\tan 51) = h$$
$$6.17 \approx h$$

The height of the utility pole is approximately 6.17 meters.

8. A winch is a tool that rotates a cylinder, around which a cable is wound. When the winch rotates in one direction, it draws the cable in. Joey is using a winch and a pulley (as shown in the diagram) to raise a heavy box off the floor and onto a cart. The box is 2 ft. tall, and the winch is 14 ft. horizontally from where cable drops down vertically from the pulley. The angle of elevation to the pulley is $42°$. What is the approximate length of cable required to connect the winch and the box?

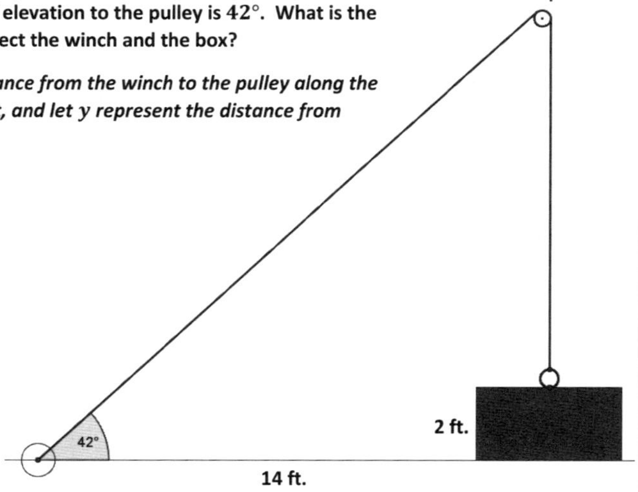

Let h represent the length of cable in the distance from the winch to the pulley along the hypotenuse of the right triangle shown in feet, and let y represent the distance from the pulley to the floor in feet.

Using tangent:

$$\tan 42 = \frac{y}{14}$$
$$14(\tan 42) = y$$
$$12.61 \approx y$$

Using cosine:

$$\cos 42 = \frac{14}{h}$$
$$h = \frac{14}{\cos 42}$$
$$h \approx 18.84$$

Let t represent the total amount of cable from the winch to the box in feet:

$$t \approx 12.61 + 18.84 - 2$$
$$t \approx 29.45$$

The total length of cable from the winch to the box is approximately 29.45 ft.

Lesson 30: Trigonometry and the Pythagorean Theorem

Student Outcomes

- Students rewrite the Pythagorean theorem in terms of sine and cosine ratios and use it in this form to solve problems.
- Students write tangent as an identity in terms of sine and cosine and use it in this form to solve problems.

Lesson Notes

Students discover the Pythagorean theorem in terms of sine and cosine ratios and demonstrate why $\tan\theta = \frac{\sin\theta}{\cos\theta}$. They begin solving problems where any one of the values of $\sin\theta$, $\cos\theta$, $\tan\theta$ are provided.

In this Geometry course, trigonometry is anchored in right triangle trigonometry as evidenced by standards **G-SRT.C.6–8** in the cluster that states: Define trigonometric ratios and solve problems involving right triangles. The focus is on the values of ratios for a given right triangle. This is an important distinction from trigonometry studied in Algebra II, which is studied from the perspective of functions; sine, cosine, and tangent are functions of any real number, not just those of right triangles. The language in **G-SRT.C.8** juxtaposes trigonometric ratios and the Pythagorean theorem in the same standard, which leads directly to the Pythagorean identity for right triangles. In Algebra II, students prove that the identity holds for all real numbers. Presently, this lesson offers an opportunity for students to develop deep understanding of the relationship between the trigonometric ratios and the Pythagorean theorem.

Classwork

Exercises 1–2 (4 minutes)

- In this lesson, we will use a fact well known to us, the Pythagorean theorem, and tie it to trigonometry.

Exercises 1–2

1. **In a right triangle with acute angle of measure θ, $\sin\theta = \frac{1}{2}$. What is the value of $\cos\theta$? Draw a diagram as part of your response.**

   $\cos\theta = \frac{\sqrt{3}}{2}$

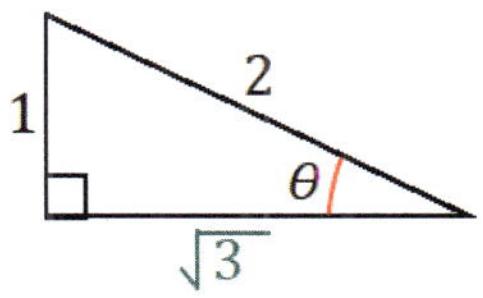

2. **In a right triangle with acute angle of measure θ, $\sin\theta = \frac{7}{9}$. What is the value of $\tan\theta$? Draw a diagram as part of your response.**

   $\tan\theta = \frac{7}{4\sqrt{2}} = \frac{7\sqrt{2}}{8}$

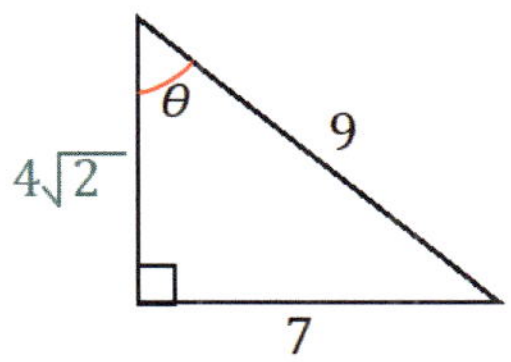

Scaffolding:

For students who are struggling, begin with an image of a right triangle with one acute angle marked and the opp, adj, and hyp sides labeled. Then, ask: "If the value of $\text{opp}:\text{hypotenuse}$ ratio is $3:5$, what is the value of $\text{adj}:\text{hypotenuse}$?" $(4:5)$

MP.3

- How did you apply the Pythagorean theorem to answer Exercises 1–2?
  - *Since the triangles are right triangles, we used the relationship between side lengths, $a^2 + b^2 = c^2$, to solve for the missing side length and then used the missing side length to determine the value of the appropriate ratio.*

Example 1 (13 minutes)

- The Great Pyramid of Giza in Egypt was constructed around 2600 B.C.E. out of limestone blocks weighing several tons each. The angle measure between the base and each of the four triangular faces of the pyramid is roughly 53°.
- Observe Figure 1, a model of the Great Pyramid, and Figure 2, which isolates the right triangle formed by the height, the slant height, and the segment that joins the center of the base to the bottom of the slant height.

Example 1

a. **What common right triangle was probably modeled in the construction of the triangle in Figure 2? Use $\sin 53° \approx 0.8$.**

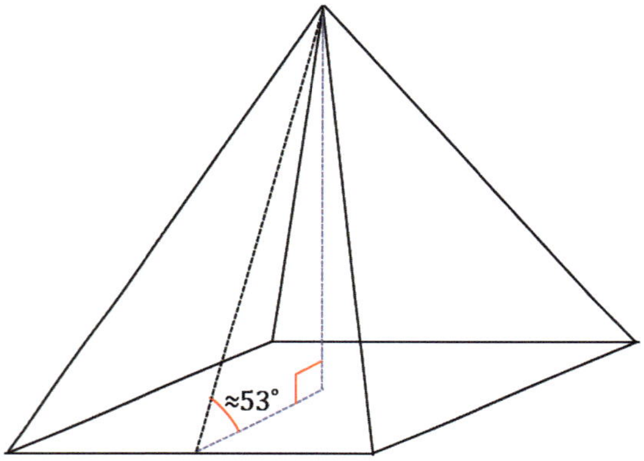

Figure 1

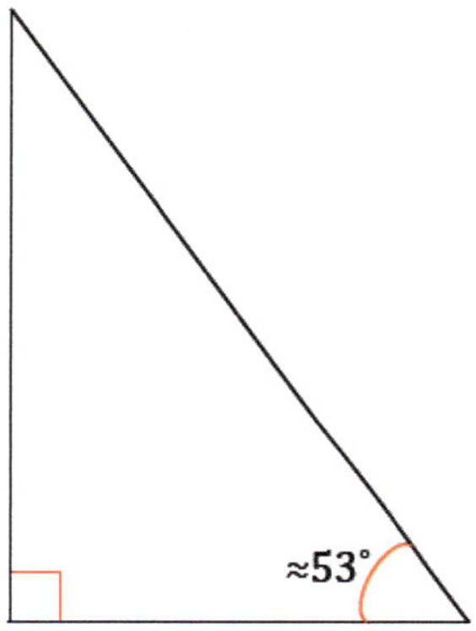

Figure 2

Right triangle with side lengths 3, 4, and 5, since $0.8 = \frac{4}{5}$.

- What common right triangle was probably modeled in the construction of the triangle in Figure 2?

Though it may not be immediately obvious to students, part (a) is the same type of question as they completed in Exercises 1–2. The difference is the visual appearance of the value of $\sin 53$ in decimal form versus in fraction form. Allow students time to sort through what they must do to answer part (a). Offer guiding questions and comments as needed such as the following:

- Revisit Exercises 1–2. What similarities and differences do you notice between Example 1, part (a), and Exercises 1–2?
- What other representation of 0.8 may be useful in this question?

Alternatively, students should also see that the value of $\sin 53°$ can be thought of as $\frac{\text{opp}}{\text{hyp}} = \frac{0.8}{1}$. We proceed to answer part (a) using this fraction. If students have responses to share, share them as a whole class, or proceed to share the following solution:

- To determine the common right triangle that was probably modeled in the construction of the triangle in Figure 2, using the approximation $\sin 53° \approx 0.8$ means we are looking for a right triangle with side-length relationships that are well known.
- Label the triangle with given acute angle measure of approximately $53°$ as is labeled in the following figure. The hypotenuse has length 1, the opposite side has length 0.8, and the side adjacent to the marked angle is labeled as x.

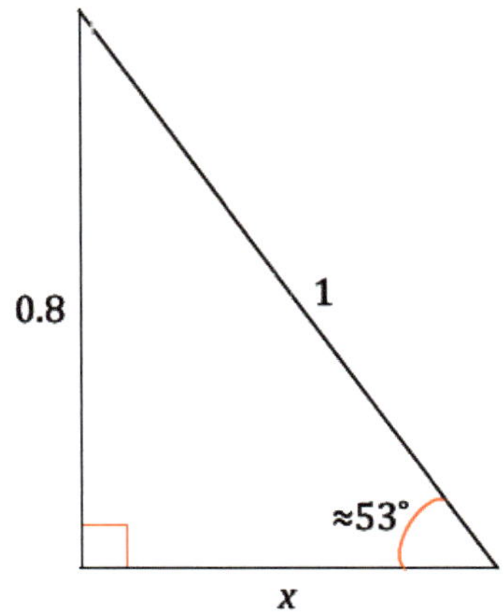

- How can we determine the value of x?
  - *We can apply the Pythagorean theorem.*
- Solve for x.
  - $(0.8)^2 + x^2 = (1)^2$
    $0.64 + x^2 = 1$
    $x^2 = 0.36$
    $x = 0.6$
- The side lengths of this triangle are 0.6, 0.8, and 1. What well-known right triangle matches these lengths?

Even though the calculations to determine the lengths of the triangle have been made, determining that this triangle is a 3–4–5 triangle is still a jump. Allow time for students to persevere after the answer. Offer guiding questions and comments as needed such as the following:

- Sketch a right triangle with side lengths 6, 8, and 10, and ask how that triangle is related to the one in the problem.
- List other triangle side lengths that are in the same ratio as a 6–8–10 triangle.

Students should conclude part (a) with the understanding that a triangle with an acute angle measure of approximately $53°$ is a 3–4–5 triangle.

b. The actual angle between the base and lateral faces of the pyramid is actually closer to $52°$. Considering the age of the pyramid, what could account for the difference between the angle measure in part (a) and the actual measure?

The Great Pyramid is approximately $4,500$ years old, and the weight of each block is several tons. It is conceivable that over time, the great weight of the blocks caused the pyramids to settle and shift the lateral faces enough so that the angle is closer to $52°$ than to $53°$.

c. Why do you think the architects chose to use a 3–4–5 as a model for the triangle?

Answers may vary. Perhaps they used it (1) because it is the right triangle with the shortest whole-number side lengths to satisfy the converse of the Pythagorean theorem and (2) because of the aesthetic it offers.

Discussion (10 minutes)

Lead students through a discussion.

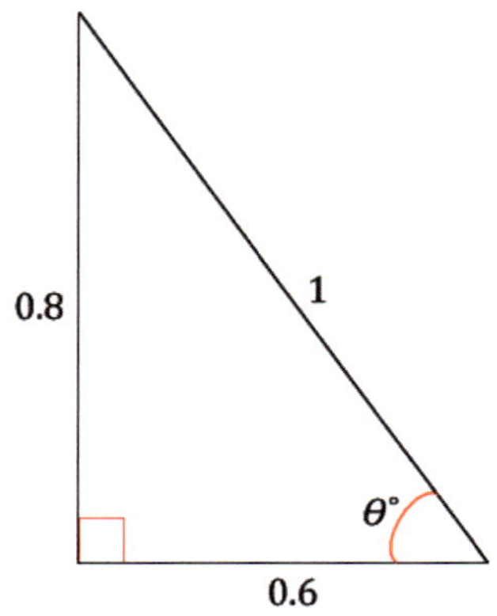

- Let θ be the angle such that $\sin\theta = 0.8$. Before, we were using an approximation. We do not know exactly what the angle measure is, but we will assign the angle measure that results in the sine of the angle as 0.8 the label θ. Then, we know (draw the following image):
- In the diagram, rewrite the leg lengths in terms of $\sin\theta$ and $\cos\theta$.

Allow students a few moments to struggle with this connection. If needed, prompt them, and ask what the values of $\sin\theta$ and $\cos\theta$ are in this right triangle.

  - *Since the value of* $\sin\theta$ *is* 0.8 *and the value of* $\cos\theta = 0.6$*, the leg lengths can be rewritten as*:

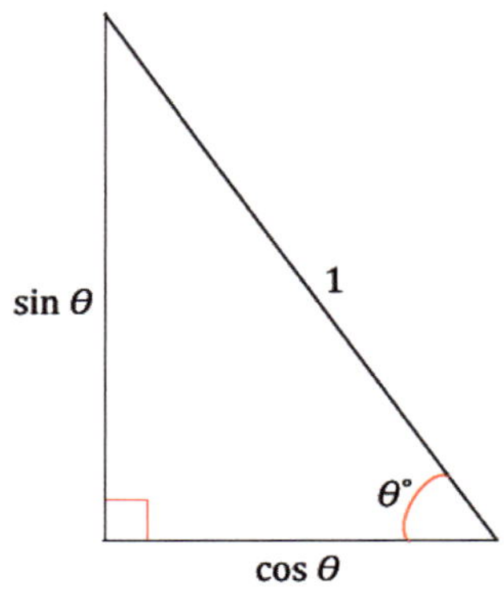

- The Pythagorean theorem states that for a right triangle with lengths a, b, and c, where c is the hypotenuse, the relationship between the side lengths is $a^2 + b^2 = c^2$. Apply the Pythagorean theorem to this triangle.
  - $(\sin\theta)^2 + (\cos\theta)^2 = 1$
- This statement is called the Pythagorean identity. This relationship is easy to show in general.

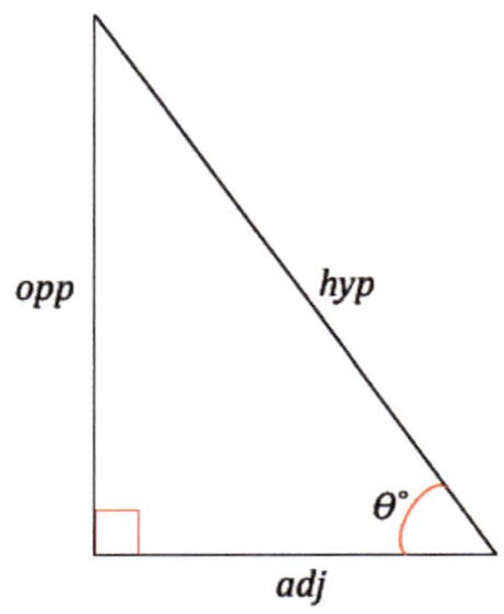

- Referencing the diagram above, we can say

$$\text{opp}^2 + \text{adj}^2 = \text{hyp}^2.$$

- Divide both sides by hyp^2.

$$\frac{\text{opp}^2}{\text{hyp}^2} + \frac{\text{adj}^2}{\text{hyp}^2} = \frac{\text{hyp}^2}{\text{hyp}^2}$$

$$(\sin\theta)^2 + (\cos\theta)^2 = 1$$

Explain to students that they might see the Pythagorean identity written in the following way on the Web and in other texts:

$$\sin^2\theta + \cos^2\theta = 1$$

Let them know that in Precalculus and Advanced Topics, they use this notation but not to worry about such notation now.

Example 2 (7 minutes)

Students discover a second trigonometric identity; this identity describes a relationship between sine, cosine, and tangent.

- Recall Opening Exercise part (b) from Lesson 29. We found that the tangent values of the listed angle measurements were equal to dividing sine by cosine for the same angle measurements. We discover why this is true in this example.

MP.7

- Use the provided diagram to reason why the trigonometric identity $\tan\theta = \frac{\sin\theta}{\cos\theta}$.

Example 2

Show why $\tan\theta = \frac{\sin\theta}{\cos\theta}$.

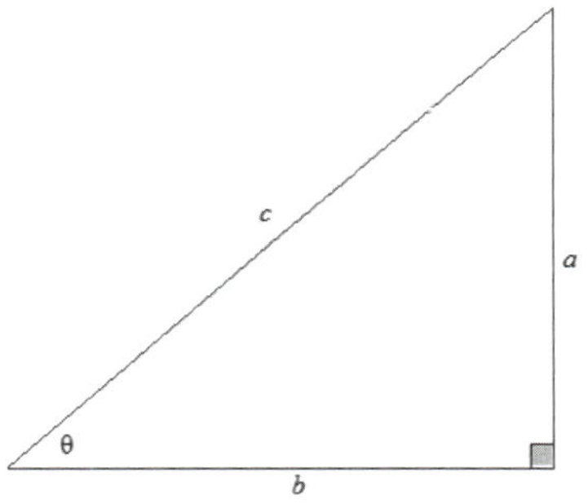

Allow students time to work through the reasoning independently before guiding them through an explanation. To provide more support, consider having the diagram on the board and then writing the following to start students off:

$$\sin\theta = \qquad\qquad \sin\theta = \frac{a}{c}$$

$$\cos\theta = \qquad\qquad \cos\theta = \frac{b}{c}$$

$$\tan\theta = \qquad\qquad \tan\theta = \frac{a}{b}$$

MP.7

- $\tan\theta = \dfrac{\sin\theta}{\cos\theta}$ *because*

$$\sin\theta = \frac{a}{c} \text{ and } \cos\theta = \frac{b}{c}.$$

Then,

$$\tan\theta = \frac{\sin\theta}{\cos\theta}$$

$\tan\theta = \dfrac{a}{b}$, *which is what we found earlier.*

- If you are given one of the values $\sin\theta$, $\cos\theta$, or $\tan\theta$, we can find the other two values using the identities $\sin^2\theta + \cos^2\theta = 1$ and $\tan\theta = \dfrac{\sin\theta}{\cos\theta}$ or by using the Pythagorean theorem.

Exercises 3–4 (5 minutes)

Exercises 3–4 are the same as Exercise 1–2; however, students answer them now by applying the Pythagorean identity.

Exercises 3–4

3. **In a right triangle with acute angle of measure θ, $\sin\theta = \frac{1}{2}$, use the Pythagorean identity to determine the value of $\cos\theta$.**

$$\sin^2\theta + \cos^2\theta = 1$$
$$\left(\frac{1}{2}\right)^2 + \cos^2\theta = 1$$
$$\frac{1}{4} + \cos^2\theta = 1$$
$$\cos^2\theta = \frac{3}{4}$$
$$\cos\theta = \frac{\sqrt{3}}{2}$$

4. **Given a right triangle with acute angle of measure θ, $\sin\theta = \frac{7}{9}$, use the Pythagorean identity to determine the value of $\tan\theta$.**

$$\sin^2\theta + \cos^2\theta = 1$$
$$\left(\frac{7}{9}\right)^2 + \cos^2\theta = 1$$
$$\frac{49}{81} + \cos^2\theta = 1$$
$$\cos^2\theta = \frac{32}{81}$$
$$\cos\theta = \frac{4\sqrt{2}}{9}$$
$$\tan\theta = \frac{\sin\theta}{\cos\theta} = \frac{\frac{7}{9}}{\frac{4\sqrt{2}}{9}} = \frac{7}{4\sqrt{2}}$$
$$\tan\theta = \frac{7\sqrt{2}}{8}$$

Closing (1 minute)

Ask students to summarize the key points of the lesson. Additionally, consider asking students the following questions independently in writing, to a partner, or to the whole class.

- What is the Pythagorean identity?
  - $\sin^2\theta + \cos^2\theta = 1$
- What are the ways the tangent can be represented?
  - $\tan\theta = \frac{\text{opp}}{\text{adj}}$
  - $\tan\theta = \frac{\sin\theta}{\cos\theta}$
- If one of the values $\sin\theta$, $\cos\theta$, or $\tan\theta$ is provided to us, we can find the other two values by using the identities $\sin^2\theta + \cos^2\theta = 1$, $\tan\theta = \frac{\sin\theta}{\cos\theta}$, or the Pythagorean theorem.

Exit Ticket (5 minutes)

Name ______________________________ Date ______________

Lesson 30: Trigonometry and the Pythagorean Theorem

Exit Ticket

1. If $\sin\beta = \frac{4\sqrt{29}}{29}$, use trigonometric identities to find $\cos\beta$ and $\tan\beta$.

2. Find the missing side lengths of the following triangle using sine, cosine, and/or tangent. Round your answer to four decimal places.

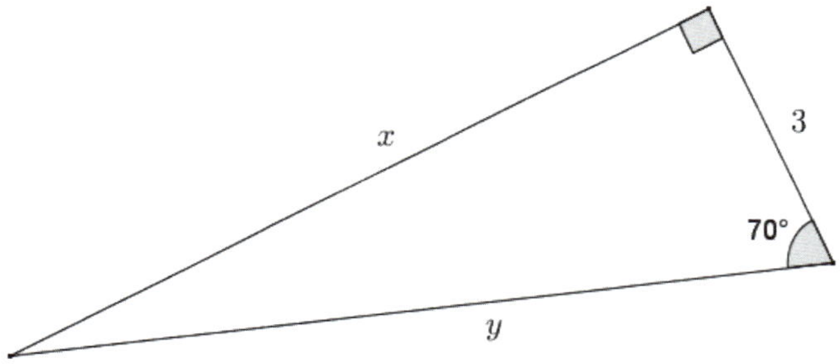

Exit Ticket Sample Solutions

1. If $\sin\beta = \frac{4\sqrt{29}}{29}$, use trigonometric identities to find $\cos\beta$ and $\tan\beta$.

$$\sin^2\beta + \cos^2\beta = 1$$
$$\left(\frac{4\sqrt{29}}{29}\right)^2 + \cos^2\beta = 1$$
$$\frac{16}{29} + \cos^2\beta = 1$$
$$\cos^2\beta = \frac{13}{29}$$
$$\cos\beta = \sqrt{\frac{13}{29}}$$
$$\cos\beta = \frac{\sqrt{13}}{\sqrt{29}} = \frac{\sqrt{377}}{29}$$

$$\tan\beta = \frac{\sin\beta}{\cos\beta}$$
$$\tan\beta = \frac{\frac{4\sqrt{29}}{29}}{\frac{\sqrt{377}}{29}}$$
$$\tan\beta = \frac{4\sqrt{29}}{\sqrt{377}}$$
$$\tan\beta = \frac{4}{\sqrt{13}}$$
$$\tan\beta = \frac{4\sqrt{13}}{13}$$

2. Find the missing side lengths of the following triangle using sine, cosine, and/or tangent. Round your answer to four decimal places.

$$\cos 70 = \frac{3}{y}$$
$$y = \frac{3}{\cos 70} \approx 8.7714$$
$$\tan 70 = \frac{x}{3}$$
$$x = 3(\tan 70) \approx 8.2424$$

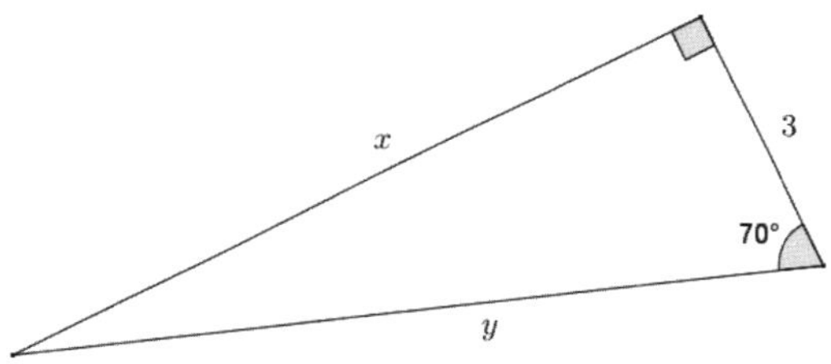

Problem Set Sample Solutions

1. If $\cos\theta = \frac{4}{5}$, find $\sin\theta$ and $\tan\theta$.

Using the identity $\sin^2\theta + \cos^2\theta = 1$:

$$\sin^2\theta + \left(\frac{4}{5}\right)^2 = 1$$
$$\sin^2\theta = 1 - \left(\frac{4}{5}\right)^2$$
$$\sin^2\theta = 1 - \left(\frac{16}{25}\right)$$
$$\sin^2\theta = \frac{9}{25}$$
$$\sin\theta = \sqrt{\frac{9}{25}}$$
$$\sin\theta = \frac{3}{5}$$

Using the identity $\tan\theta = \frac{\sin\theta}{\cos\theta}$:

$$\tan\theta = \frac{\frac{3}{5}}{\frac{4}{5}} = \frac{3}{4}$$

2. **If $\sin\theta = \frac{44}{125}$, find $\cos\theta$ and $\tan\theta$.**

$$\sin^2\theta + \cos^2\theta = 1$$
$$\left(\frac{44}{125}\right)^2 + \cos^2\theta = 1$$
$$\cos^2\theta = 1 - \frac{1936}{15625} = \frac{13689}{15625}$$
$$\cos\theta = \sqrt{\frac{13689}{15625}} = \frac{117}{125}$$

$$\tan\theta = \frac{44}{125} \div \frac{117}{125} = \frac{44}{117}$$

3. **If $\tan\theta = 5$, find $\sin\theta$ and $\cos\theta$.**

$\tan\theta = 5 = \frac{5}{1}$, *so the legs of a right triangle can be considered to have lengths of* 5 *and* 1. *Using the Pythagorean theorem:*

$$5^2 + 1^2 = \text{hyp}^2$$
$$26 = \text{hyp}^2$$
$$\sqrt{26} = \text{hyp}$$

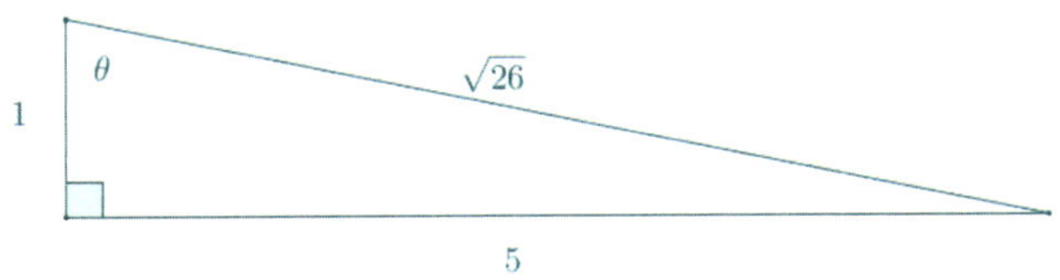

$$\sin\theta = \frac{5}{\sqrt{26}} = \frac{5\sqrt{26}}{26};\ \cos\theta = \frac{1}{\sqrt{26}} = \frac{\sqrt{26}}{26}$$

4. **If $\sin\theta = \frac{\sqrt{5}}{5}$, find $\cos\theta$ and $\tan\theta$.**

Using the identity $\sin^2\theta + \cos^2\theta = 1$*:*

$$\left(\frac{\sqrt{5}}{5}\right)^2 + \cos^2\theta = 1$$
$$\cos^2\theta = 1 - \left(\frac{\sqrt{5}}{5}\right)^2$$
$$\cos^2\theta = 1 - \left(\frac{5}{25}\right)$$
$$\cos^2\theta = 1 - \frac{1}{5} = \frac{4}{5}$$
$$\cos\theta = \sqrt{\frac{4}{5}} = \frac{\sqrt{4}}{\sqrt{5}} = \frac{2}{\sqrt{5}}$$
$$\cos\theta = \frac{2\sqrt{5}}{5}$$

Using the identity $\tan\theta = \frac{\sin\theta}{\cos\theta}$*:*

$$\tan\theta = \frac{\frac{\sqrt{5}}{5}}{\frac{2\sqrt{5}}{5}} = \frac{\sqrt{5}}{2\sqrt{5}} = \frac{1}{2}$$

5. Find the missing side lengths of the following triangle using sine, cosine, and/or tangent. Round your answer to four decimal places.

$$\frac{x}{12} = \tan 27$$
$$x = 12\tan 27 \approx 6.1143$$
$$\frac{12}{y} = \sin 63$$
$$y = \frac{12}{\sin 63} \approx 13.4679$$

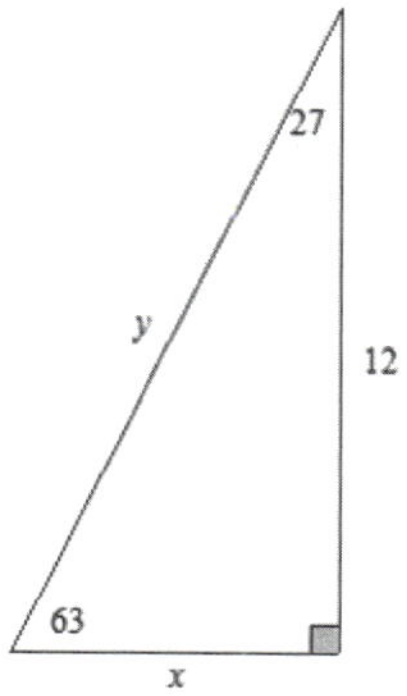

6. A surveying crew has two points A and B marked along a roadside at a distance of 400 yd. A third point C is marked at the back corner of a property along a perpendicular to the road at B. A straight path joining C to A forms a $28°$ angle with the road. Find the distance from the road to point C at the back of the property and the distance from A to C using sine, cosine, and/or tangent. Round your answer to three decimal places.

$$\tan 28 = \frac{BC}{400}$$
$$BC = 400(\tan 28)$$
$$BC \approx 212.684$$

The distance from the road to the back of the property is approximately 212.684 yd.

$$\cos 28 = \frac{400}{AC}$$
$$AC = \frac{400}{\cos 28}$$
$$AC \approx 453.028$$

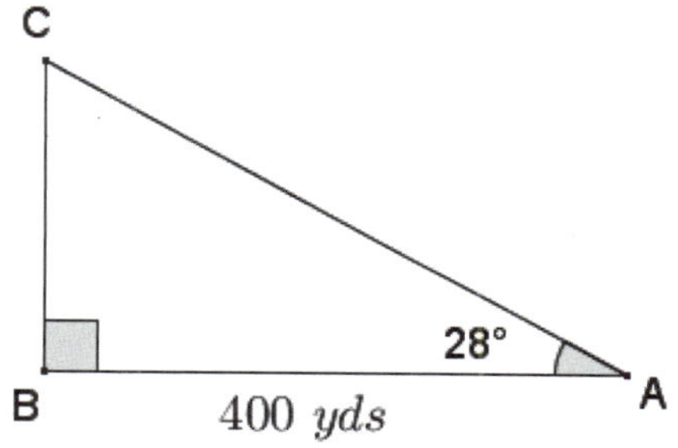

The distance from point C to point A is approximately 453.028 yd.

7. The right triangle shown is taken from a slice of a right rectangular pyramid with a square base.

a. Find the height of the pyramid (to the nearest tenth).

$$\sin 66 = \frac{h}{9}$$
$$h = 9(\sin 66)$$
$$h \approx 8.2$$

The height of the pyramid is approximately 8.2 *units.*

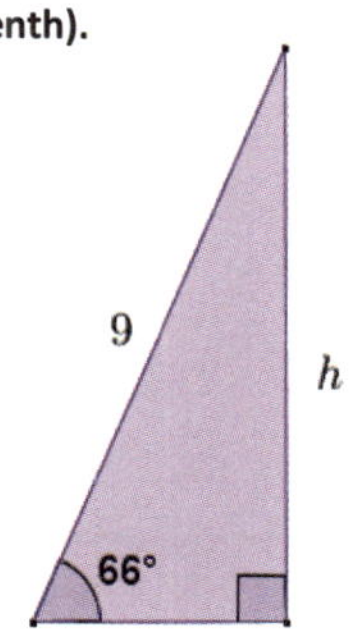

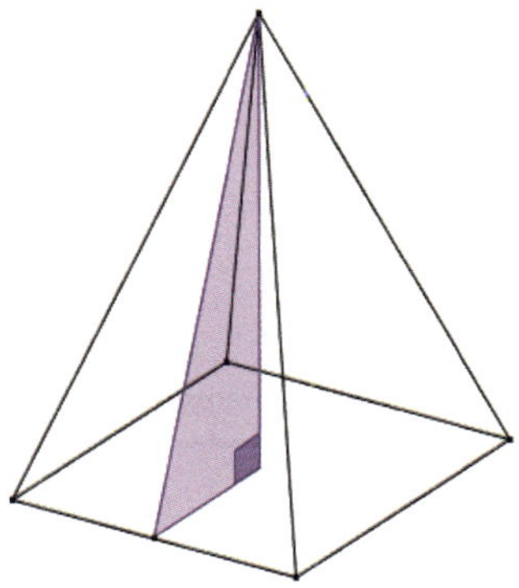

b. **Find the lengths of the sides of the base of the pyramid (to the nearest tenth).**

The lengths of the sides of the base of the pyramid are twice the length of the short leg of the right triangle shown.

$$\cos 66 = \frac{n}{9}$$
$$n = 9(\cos 66)$$

$$\text{length} = 2(9\cos 66)$$
$$\text{length} = 18\cos 66$$
$$\text{length} \approx 7.3$$

The lengths of the sides of the base are approximately 7.3 *units.*

c. **Find the lateral surface area of the right rectangular pyramid.**

The faces of the prism are congruent isosceles triangles having bases of $18\cos 66$ *and height of* 9.

$$\text{Area} = \frac{1}{2}bh$$
$$\text{Area} = \frac{1}{2}(18\cos 66)(9)$$
$$\text{Area} = 81\cos 66$$
$$\text{Area} \approx 32.9$$

The lateral surface area of the right rectangular pyramid is approximately 32.9 *square units.*

8. **A machinist is fabricating a wedge in the shape of a right triangular prism. One acute angle of the right triangular base is $33°$, and the opposite side is 6.5 cm. Find the length of the edges labeled l and m using sine, cosine, and/or tangent. Round your answer to the nearest thousandth of a centimeter.**

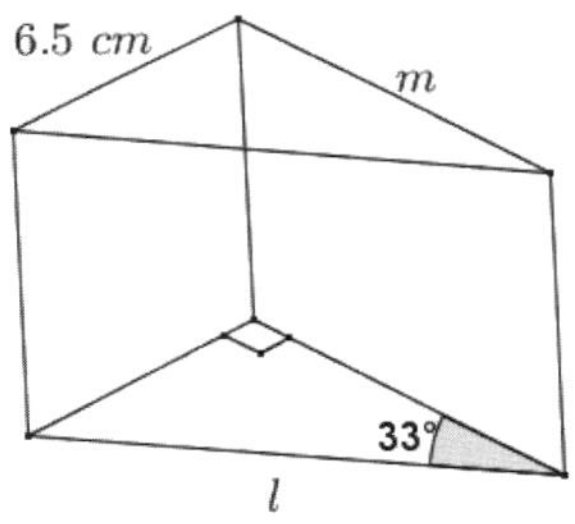

$$\sin 33 = \frac{6.5}{l}$$
$$l = \frac{6.5}{\sin 33}$$
$$l \approx 11.935$$

Distance l is approximately 11.935 cm.

$$\tan 33 = \frac{6.5}{m}$$
$$m = \frac{6.5}{\tan 33}$$
$$m \approx 10.009$$

Distance m is approximately 10.009 cm.

9. Let $\sin\theta = \frac{l}{m}$, where $l, m > 0$. Express $\tan\theta$ and $\cos\theta$ in terms of l and m.

$$\sin^2\theta + \cos^2\theta = 1$$
$$\left(\frac{l}{m}\right)^2 + \cos^2\theta = 1$$
$$\cos^2\theta = 1 - \left(\frac{l}{m}\right)^2$$
$$\cos^2\theta = \frac{m^2}{m^2} - \frac{l^2}{m^2}$$
$$\cos^2\theta = \frac{m^2 - l^2}{m^2}$$
$$\cos\theta = \sqrt{\frac{m^2 - l^2}{m^2}}$$
$$\cos\theta = \frac{\sqrt{m^2 - l^2}}{m}$$

$$\tan\theta = \frac{\sin\theta}{\cos\theta}$$

By substituting the previous result for $\cos\theta$,

$$\tan\theta = \frac{\frac{l}{m}}{\frac{\sqrt{m^2 - l^2}}{m}}$$
$$\tan\theta = \frac{l}{\sqrt{m^2 - l^2}}$$

Lesson 31: Using Trigonometry to Determine Area

Student Outcomes

- Students prove that the area of a triangle is one-half times the product of two side lengths times the sine of the included angle and solve problems using this formula.
- Students find the area of an isosceles triangle given the base length and the measure of one angle.

Lesson Notes

Students discover how trigonometric ratios can help with area calculations in cases where the measurement of the height is not provided. In order to determine the height in these cases, students must draw an altitude to create right triangles within the larger triangle. With the creation of the right triangles, students can then set up the necessary trigonometric ratios to express the height of the triangle (**G-SRT.C.8**). Students carefully connect the meanings of formulas to the diagrams they represent (MP.2 and 7). In addition, this lesson introduces the formula $\text{Area} = \frac{1}{2}ab \sin C$ as described by **G-SRT.D.9**.

Classwork

Opening Exercise (5 minutes)

Opening Exercise

Three triangles are presented below. Determine the areas for each triangle, if possible. If it is not possible to find the area with the provided information, describe what is needed in order to determine the area.

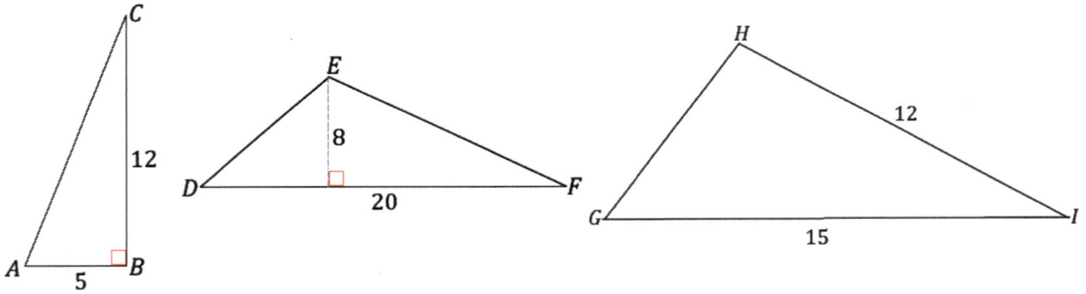

The area of $\triangle ABC$ is $\frac{1}{2}(5)(12)$, or 30 square units, and the area of $\triangle DEF$ is $\frac{1}{2}(8)(20)$, or 80 square units. There is not enough information to find the height of $\triangle GHI$ and, therefore, the area of the triangle.

Is there a way to find the missing information?

Without further information, there is no way to calculate the area.

Example 1 (13 minutes)

- What if the third side length of the triangle were provided? Is it possible to determine the area of the triangle now?

Example 1

Find the area of $\triangle GHI$.

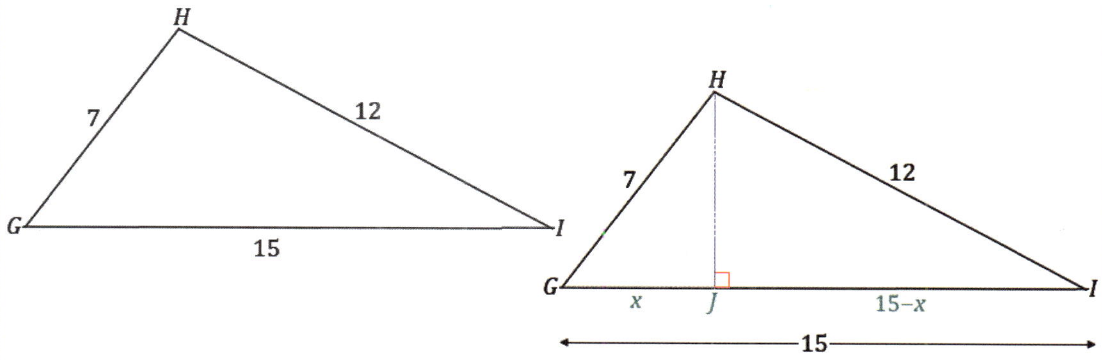

MP.1 Allow students the opportunity and the time to determine what they must find (the height) and how to locate it (one option is to drop an altitude from vertex H to side $\overline{GI}$). For students who are struggling, consider showing just the altitude and allowing them to label the newly divided segment lengths and the height.

- How can the height be calculated?
  - *By applying the Pythagorean theorem to both of the created right triangles to find x,*

$$h^2 = 49 - x^2 \qquad\qquad h^2 = 144 - (15 - x)^2$$

$$49 - x^2 = 144 - (15 - x)^2$$
$$49 - x^2 = 144 - 225 + 30x - x^2$$
$$130 = 30x$$
$$x = \frac{13}{3}$$
$$HJ = \frac{13}{3},\ IJ = \frac{32}{3}$$

  - *The value of x can then be substituted into either of the expressions equivalent to h^2 to find h.*

$$h^2 = 49 - \left(\frac{13}{3}\right)^2$$
$$h^2 = 49 - \frac{169}{9}$$
$$h = \frac{4\sqrt{17}}{3}$$

- What is the area of the triangle?

$$\text{Area} = \left(\frac{1}{2}\right)(15)\left(\frac{4\sqrt{17}}{3}\right)$$
$$\text{Area} = 10\sqrt{17}$$

Discussion (10 minutes)

Scaffolding:

- For students who are ready for a challenge, pose the unstructured question, "How could you write a formula for area that uses trigonometry, if we didn't know h?"
- For students who are struggling, use a numerical example, such as the 6–8–10 triangle in Lesson 24, to demonstrate how h can be found using $\sin\theta$.

Now, consider $\triangle ABC$, which is set up similarly to the triangle in Example 1:

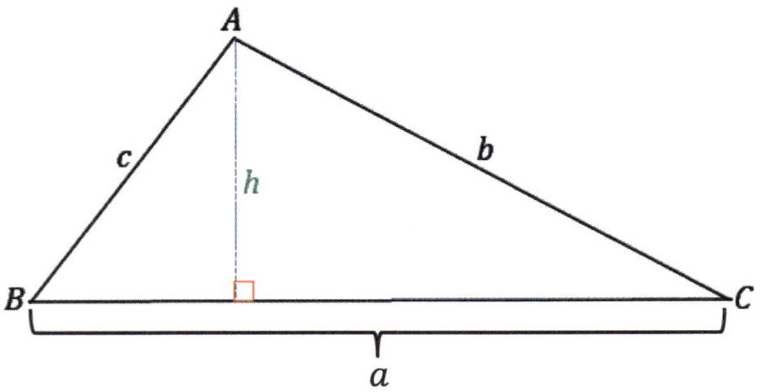

- Write an equation that describes the area of this triangle.
  - $\text{Area} = \frac{1}{2}ah$

Write the left-hand column on the board, and elicit the anticipated student response on the right-hand side after writing the second line; then, elicit the anticipated student response after writing the third line.

MP.2 & MP.7

- We will rewrite this equation. Describe what you see.

$$\text{Area} = \frac{1}{2}ah$$

$$\text{Area} = \frac{1}{2}ah\left(\frac{b}{b}\right)$$ *The statement is multiplied by* 1.

$$\text{Area} = \frac{1}{2}ab\left(\frac{h}{b}\right)$$ *The last statement is rearranged, but the value remains the same.*

Create a discussion around the third line of the newly written formula. Modify $\triangle ABC$: Add in an arc mark at vertex C, and label it θ.

- What do you notice about the structure of $\frac{h}{b}$? Can we think of this newly written area formula in a different way using trigonometry?
  - *The value of* $\frac{h}{b}$ *is equivalent to* $\sin\theta$; *the newly written formula can be written as* $\text{Area} = \frac{1}{2}ab\sin\theta$.

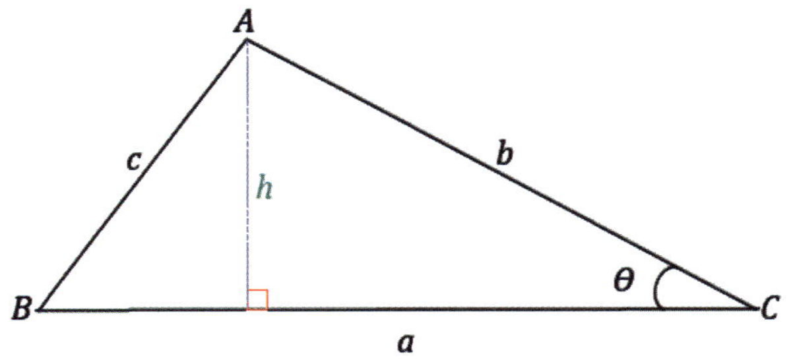

- If the area can be expressed as $\frac{1}{2}ab\sin\theta$, which part of the expression represents the height?
  - $h = b\sin\theta$

MP.2 & MP.7

- Compare the information provided to find the area of $\triangle GHI$ in the Opening Exercise, how the information changed in Example 1, and then changed again in the triangle above, $\triangle$ ABC.
  - *Only two side lengths were provided in the Opening Exercise, and the area could not be found because there was not enough information to determine the height. In Example 1, all three side lengths were provided, and then the area could be determined because the height could be determined by applying the Pythagorean theorem in two layers. In the most recent figure, we only needed two sides and the included angle to find the area.*
- If you had to determine the area of a triangle and were given the option to have three side lengths of a triangle or two side lengths and the included measure, which set of measurements would you rather have? Why?

The response to this question is a matter of opinion. Considering the amount of work needed to find the area when provided three side lengths, students should opt for the briefer, trigonometric solution!

Example 2 (5 minutes)

Example 2

A farmer is planning how to divide his land for planting next year's crops. A triangular plot of land is left with two known side lengths measuring 500 m and $1,700$ m.

What could the farmer do next in order to find the area of the plot?

- With just two side lengths known of the plot of land, what are the farmer's options to determine the area of his plot of land?
  - *He can either measure the third side length, apply the Pythagorean theorem to find the height of the triangle, and then calculate the area, or he can find the measure of the included angle between the known side lengths and use trigonometry to express the height of the triangle and then determine the area of the triangle.*
- Suppose the included angle measure between the known side lengths is 30°. What is the area of the plot of land? Sketch a diagram of the plot of land.
  - $\text{Area} = \frac{1}{2}(1700)(500)\sin 30$

    $\text{Area} = 212\,500$

    The area of the plot of land is 212,500 *square meters.*

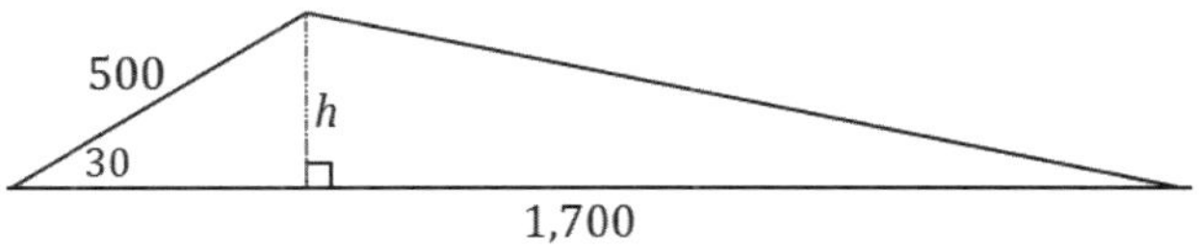

Exercise 1 (5 minutes)

Exercise 1

A real estate developer and her surveyor are searching for their next piece of land to build on. They each examine a plot of land in the shape of $\triangle ABC$. The real estate developer measures the length of $\overline{AB}$ and $\overline{AC}$ and finds them to both be approximately $4,000$ feet, and the included angle has a measure of approximately $50°$. The surveyor measures the length of $\overline{AC}$ and $\overline{BC}$ and finds the lengths to be approximately $4,000$ feet and $3,400$ feet, respectively, and measures the angle between the two sides to be approximately $65°$.

a. **Draw a diagram that models the situation, labeling all lengths and angle measures.**

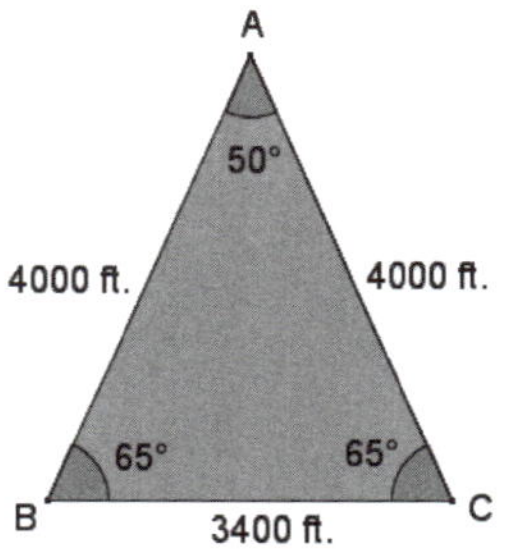

b. **The real estate developer and surveyor each calculate the area of the plot of land and both find roughly the same area. Show how each person calculated the area; round to the nearest hundred. Redraw the diagram with only the relevant labels for both the real estate agent and surveyor.**

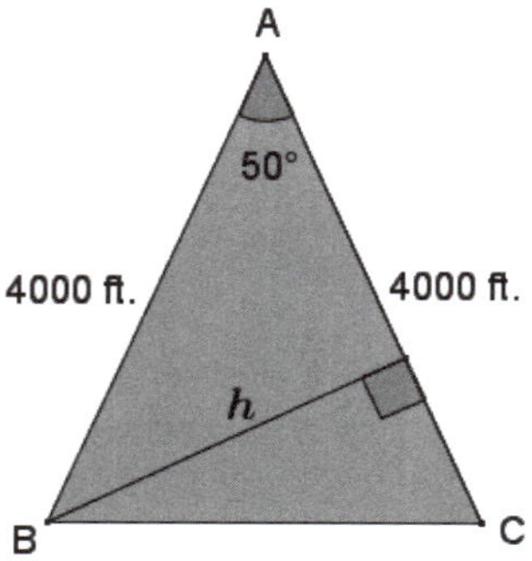

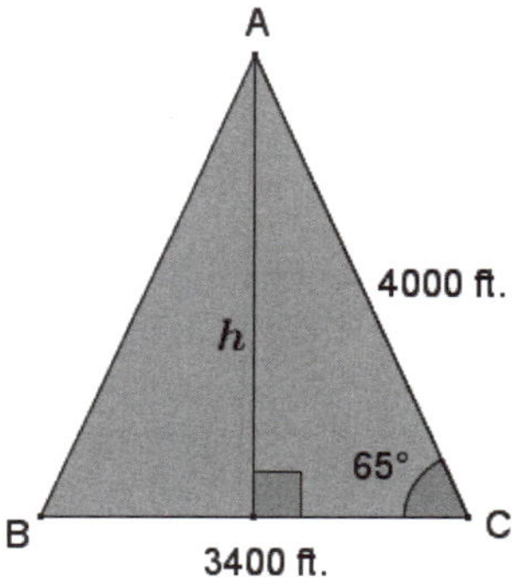

$$A = \frac{1}{2}(4000)(4000)\sin 50$$
$$A \approx 6\,128\,356$$

The area is approximately $6{,}128{,}356$ *square feet.*

$$A = \frac{1}{2}(3400)(4000)\sin 65$$
$$A \approx 6\,162\,893$$

The area is approximately $6{,}162{,}893$ *square feet.*

c. **What could possibly explain the difference between the real estate agent's and surveyor's calculated areas?**

The difference in the area of measurements can be accounted for by the approximations of the measurements taken instead of exact measurements.

Closing (2 minutes)

Ask students to summarize the key points of the lesson. Additionally, consider asking students to respond to the following questions independently in writing, to a partner, or to the whole class.

- For a triangle with side lengths a and b and included angle of measure θ, when will we need to use the area formula $\text{Area} = \frac{1}{2}ab\sin\theta$?
  - *We will need it when we are determining the area of a triangle and are not provided a height.*
- Recall how to transform the equation $\text{Area} = \frac{1}{2}bh$ to $\text{Area} = \frac{1}{2}ab\sin\theta$.

Exit Ticket (5 minutes)

Name ______________________________ Date ______________

Lesson 31: Using Trigonometry to Determine Area

Exit Ticket

1. Given two sides of the triangle shown, having lengths of 3 and 7 and their included angle of $49°$, find the area of the triangle to the nearest tenth.

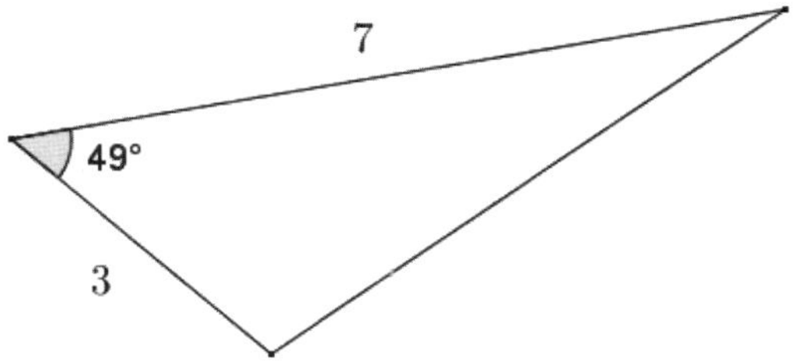

2. In isosceles triangle PQR, the base $QR = 11$, and the base angles have measures of $71.45°$. Find the area of $\triangle PQR$ to the nearest tenth.

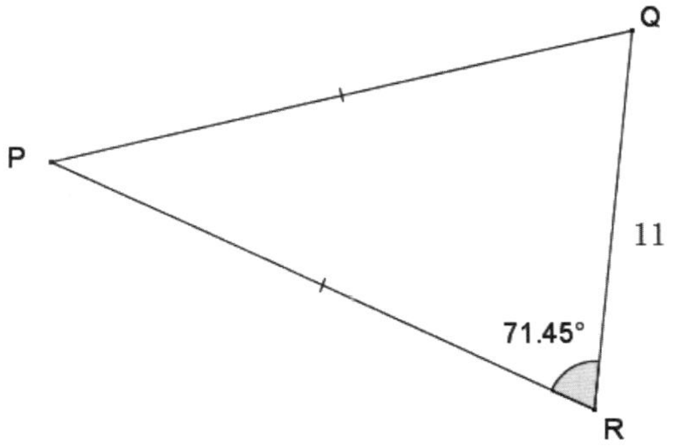

Exit Ticket Sample Solutions

1. Given two sides of the triangle shown, having lengths of 3 and 7 and their included angle of $49°$, find the area of the triangle to the nearest tenth.

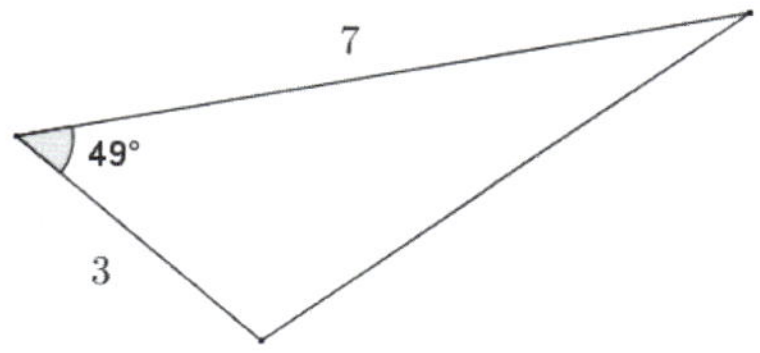

$$\text{Area} = \frac{1}{2}(3)(7)(\sin 49)$$

$$\text{Area} = 10.5(\sin 49) \approx 7.9$$

The area of the triangle is approximately 7.9 square units.

1. In isosceles triangle PQR, the base $QR = 11$, and the base angles have measures of $71.45°$. Find the area of $\triangle PQR$ to the nearest tenth.

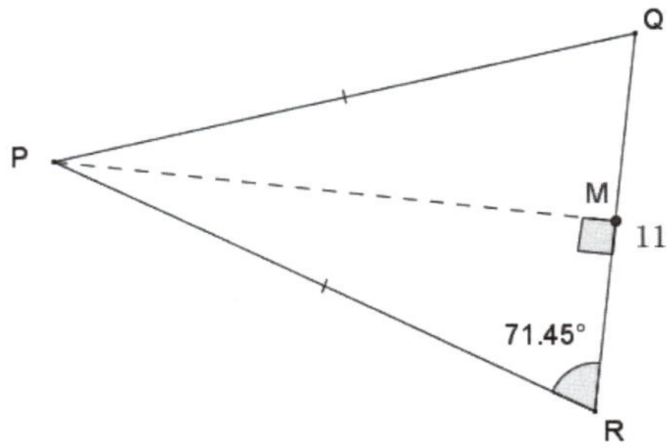

Drawing an altitude from P to midpoint M on $\overline{QR}$ cuts the isosceles triangle into two right triangles with $QM = MR = 5.5$. Using tangent, solve the following:

$$\tan 71.45 = \frac{PM}{5.5}$$

$$PM = 5.5(\tan 71.45)$$

$$\text{Area} = \frac{1}{2}bh$$

$$\text{Area} = \frac{1}{2}(11)\big(5.5(\tan 71.45)\big)$$

$$\text{Area} = 30.25(\tan 71.45) \approx 90.1$$

The area of the isosceles triangle is approximately 90.1 square units.

Problem Set Sample Solutions

Find the area of each triangle. Round each answer to the nearest tenth.

1.

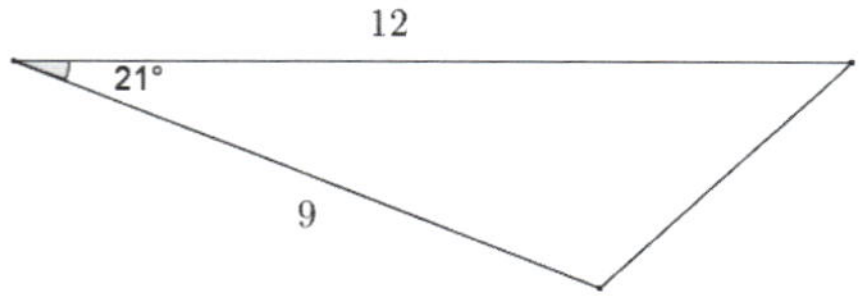

$$\text{Area} = \frac{1}{2}(12)(9)(\sin 21)$$

$$\text{Area} = 54(\sin 21) \approx 19.4$$

The area of the triangle is approximately 19.4 square units.

2.

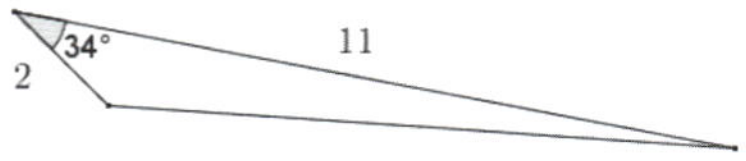

$$\text{Area} = \frac{1}{2}(2)(11)(\sin 34)$$

$$\text{Area} = 11(\sin 34) \approx 6.2$$

The area of the triangle is approximately 6.2 square units.

3.

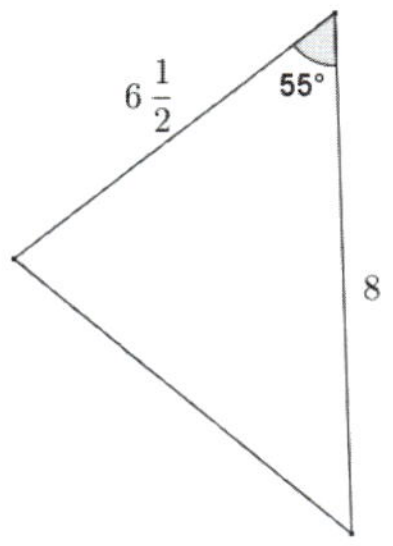

$$\text{Area} = \frac{1}{2}(8)\left(6\frac{1}{2}\right)(\sin 55)$$

$$\text{Area} = 26(\sin 55) \approx 21.3$$

The area of the triangle is approximately 21.3 square units.

4.

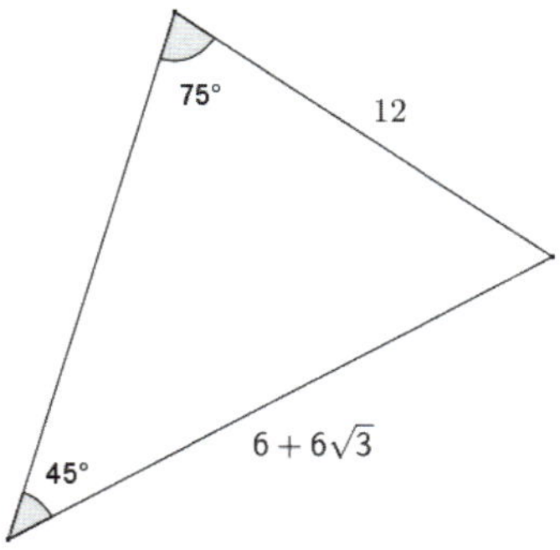

The included angle is $60°$ by the angle sum of a triangle.

$$\text{Area} = \frac{1}{2}(12)(6+6\sqrt{3})\sin 60$$

$$\text{Area} = 6(6+6\sqrt{3})\left(\frac{\sqrt{3}}{2}\right)$$

$$\text{Area} = (36+36\sqrt{3})\left(\frac{\sqrt{3}}{2}\right)$$

$$\text{Area} = 18\sqrt{3} + 18(3)$$

$$\text{Area} = 18\sqrt{3} + 54 \approx 85.2$$

The area of the triangle is approximately 85.2 square units.

5. In $\triangle DEF$, $EF = 15$, $DF = 20$, and $m\angle F = 63°$. Determine the area of the triangle. Round to the nearest tenth.

$$\text{Area} = \frac{1}{2}(20)(15)\sin(63) \approx 133.7$$

The area of $\triangle DEF$ is 133.7 units².

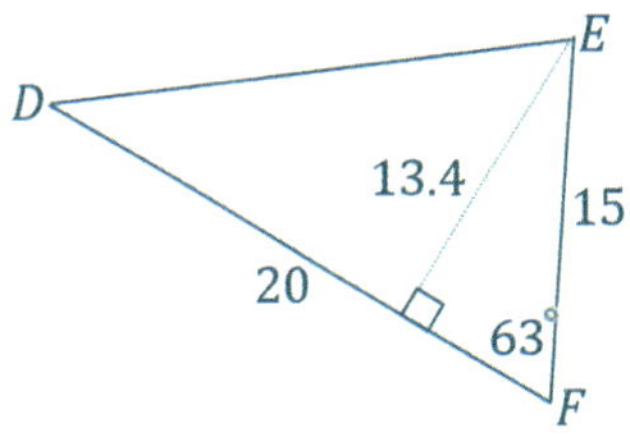

6. A landscape designer is designing a flower garden for a triangular area that is bounded on two sides by the client's house and driveway. The length of the edges of the garden along the house and driveway are 18 ft. and 8 ft., respectively, and the edges come together at an angle of $80°$. Draw a diagram, and then find the area of the garden to the nearest square foot.

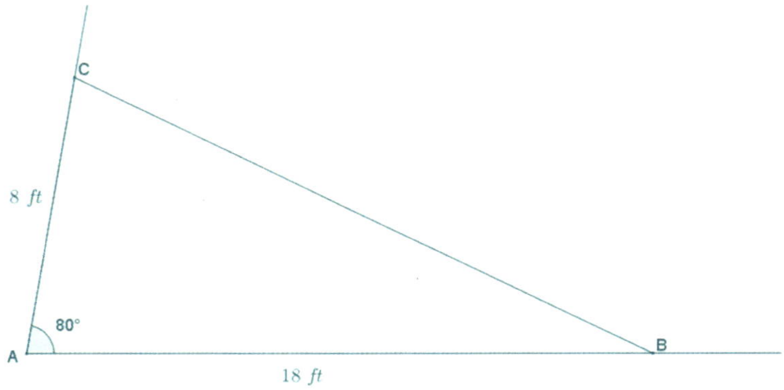

The garden is in the shape of a triangle in which the lengths of two sides and the included angle have been provided.

$$\text{Area}(ABC) = \frac{1}{2}(8\text{ ft.})(18\text{ ft.})\sin 80$$
$$\text{Area}(ABC) = (72\ \sin 80)\ \text{ft}^2$$
$$\text{Area}(ABC) \approx 71\ \text{ft}^2$$

7. A right rectangular pyramid has a square base with sides of length 5. Each lateral face of the pyramid is an isosceles triangle. The angle on each lateral face between the base of the triangle and the adjacent edge is $75°$. Find the surface area of the pyramid to the nearest tenth.

Using tangent, the altitude of the triangle to the base of length 5 is equal to $2.5\tan 75$.

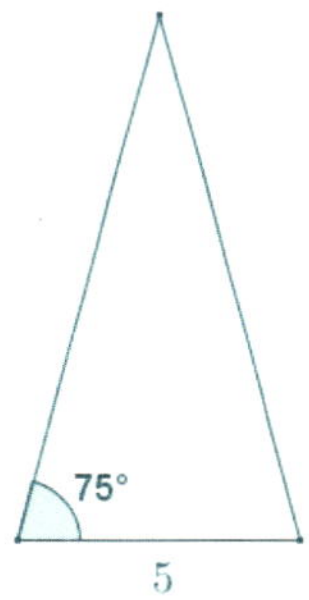

$$\text{Area} = \frac{1}{2}bh$$
$$\text{Area} = \frac{1}{2}(5)(2.5\sin 75)$$
$$\text{Area} = 6.25(\sin 75)$$

The total surface area of the pyramid is the sum of the four lateral faces and the area of the square base:

$$\text{SA} = 4\big(6.25(\sin 75)\big) + 5^2$$
$$\text{SA} = 25\sin 75 + 25$$
$$\text{SA} \approx 49.1$$

The surface area of the right rectangular pyramid is approximately 49.1 square units.

8. The Pentagon building in Washington, DC, is built in the shape of a regular pentagon. Each side of the pentagon measures 921 ft. in length. The building has a pentagonal courtyard with the same center. Each wall of the center courtyard has a length of 356 ft. What is the approximate area of the roof of the Pentagon building?

Let A_1 represent the area within the outer perimeter of the Pentagon building in square feet.

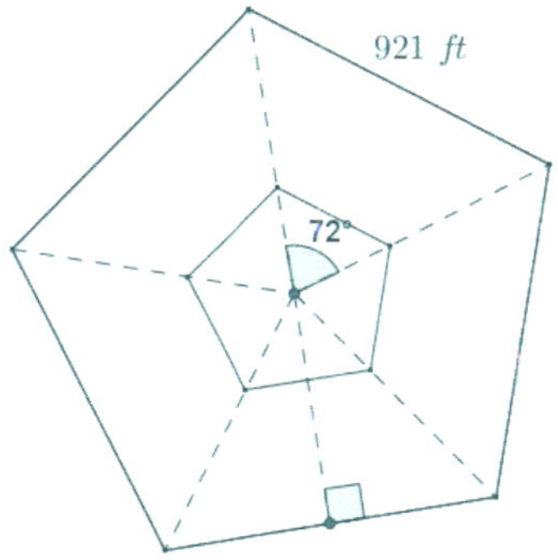

$$A_1 = \frac{nb^2}{4\tan\left(\frac{180}{n}\right)}$$

$$A_1 = \frac{5 \cdot (921)^2}{4\tan\left(\frac{180}{5}\right)}$$

$$A_1 = \frac{4\,241\,205}{4\tan(36)} \approx 1\,459\,379$$

The area within the outer perimeter of the Pentagon building is approximately $1{,}459{,}379$ ft^2.

Let A_2 represent the area within the perimeter of the courtyard of the Pentagon building in square feet.

$$A_2 = \frac{nb^2}{4\tan 36}$$

$$A_2 = \frac{5(356)^2}{4\tan 36}$$

$$A_2 = \frac{633\,680}{4\tan 36}$$

$$A_2 = \frac{158\,420}{\tan 36} \approx 218\,046$$

Let A_T represent the total area of the roof of the Pentagon building in square feet.

$$A_T = A_1 - A_2$$

$$A_T = \frac{4\,241\,205}{4\tan(36)} - \frac{158\,420}{\tan 36}$$

$$A_T = \frac{4\,241\,205}{4\tan 36} - \frac{633\,680}{4\tan 36}$$

$$A_T = \frac{3\,607\,525}{4\tan 36} \approx 1\,241\,333$$

The area of the roof of the Pentagon building is approximately $1{,}241{,}333$ ft^2.

9. A regular hexagon is inscribed in a circle with a radius of 7. Find the perimeter and area of the hexagon.

The regular hexagon can be divided into six equilateral triangular regions with each side of the triangles having a length of 7. To find the perimeter of the hexagon, solve the following:

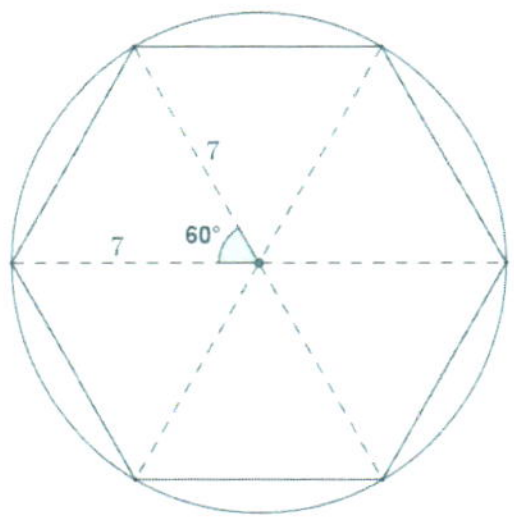

$6 \cdot 7 = 42$*, so the perimeter of the hexagon is 42 units.*

To find the area of one equilateral triangle:

$$\text{Area} = \frac{1}{2}(7)(7)\sin 60$$

$$\text{Area} = \frac{49}{2}\left(\frac{\sqrt{3}}{2}\right)$$

$$\text{Area} = \frac{49\sqrt{3}}{4}$$

The area of the hexagon is six times the area of the equilateral triangle.

$$\text{Total Area} = 6\left(\frac{49\sqrt{3}}{4}\right)$$

$$\text{Total Area} = \frac{147\sqrt{3}}{2} \approx 127.3$$

The total area of the regular hexagon is approximately 127.3 square units.

10. In the figure below, $\angle AEB$ is acute. Show that Area $(\triangle ABC) = \frac{1}{2}AC \cdot BE \cdot \sin\angle AEB$.

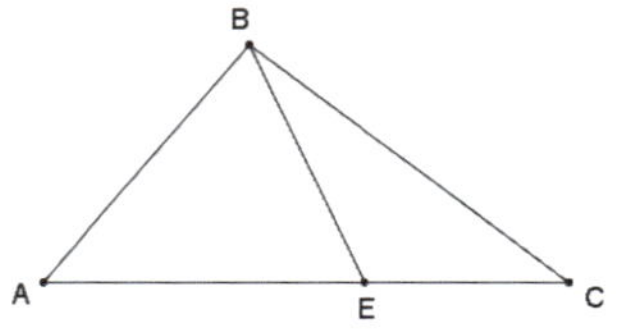

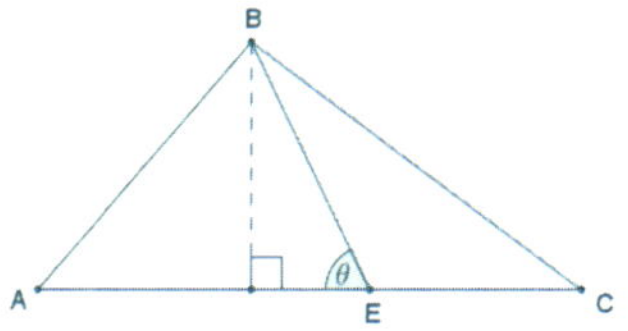

Let θ represent the degree measure of angle AEB, and let h represent the altitude of $\triangle ABC$ (and $\triangle ABE$).

$\text{Area}(\triangle ABC) = \frac{1}{2} \cdot AC \cdot h$

$\sin\theta = \frac{h}{BE}$*, which implies that* $h = BE \cdot \sin\theta$.

Therefore, by substitution:

$\text{Area}(\triangle ABC) = \frac{1}{2}AC \cdot BE \cdot \sin\angle AEB$.

11. Let $ABCD$ be a quadrilateral. Let w be the measure of the acute angle formed by diagonals $\overline{AC}$ and $\overline{BD}$. Show that $\text{Area}(ABCD) = \frac{1}{2}AC \cdot BD \cdot \sin w$.

(Hint: Apply the result from Problem 10 to $\triangle ABC$ and $\triangle ACD$.)

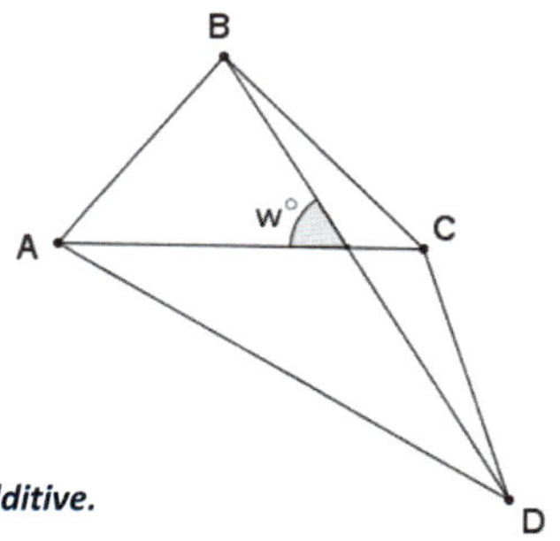

Let the intersection of $\overline{AC}$ and $\overline{BD}$ be called point P.

Using the results from Problem 10, solve the following:

$\text{Area}(\triangle ABC) = \frac{1}{2}AC \cdot BP \cdot \sin w$ *and*

$\text{Area}(\triangle ADC) = \frac{1}{2}AC \cdot PD \cdot \sin w$

$\text{Area}(ABCD) = \left(\frac{1}{2}AC \cdot BP \cdot \sin w\right) + \left(\frac{1}{2}AC \cdot PD \cdot \sin w\right)$ *Area is additive.*

$\text{Area}(ABCD) = \left(\frac{1}{2}AC \cdot \sin w\right) \cdot (BP + PD)$ *Distributive property*

$\text{Area}(ABCD) = \left(\frac{1}{2}AC \cdot \sin w\right) \cdot (BD)$ *Distance is additive*

And commutative addition gives us $\text{Area}(ABCD) = \frac{1}{2} \cdot AC \cdot BD \cdot \sin w$.

Lesson 32: Using Trigonometry to Find Side Lengths of an Acute Triangle

Student Outcomes

- Students find missing side lengths of an acute triangle given one side length and the measures of two angles.
- Students find the missing side length of an acute triangle given two side lengths and the measure of the included angle.

Lesson Notes

In Lesson 32, students learn how to determine unknown lengths in acute triangles. Once again, they drop an altitude in the given triangle to create right triangles and use trigonometric ratios and the Pythagorean theorem to solve triangle problems (**G-SRT.C.8**). This lesson and the next introduce the law of sines and cosines (**G-SRT.D.10** and **G-SRT.D.11**).

Based on availability of time in the module, teachers may want to divide the lesson into two parts by addressing everything until Exercises 1–2 on one day and the remaining content on the following day.

Classwork

Opening Exercise (3 minutes)

The objective for part (b) is that students realize that x and y cannot be found using the method they know with trigonometric ratios.

Opening Exercise

a. **Find the lengths of d and e.**

$\sin 60 = \frac{5}{e}; e = \frac{10}{\sqrt{3}}$

$\cos 60 = \frac{d}{\frac{10}{\sqrt{3}}}; d = \frac{5}{\sqrt{3}}$

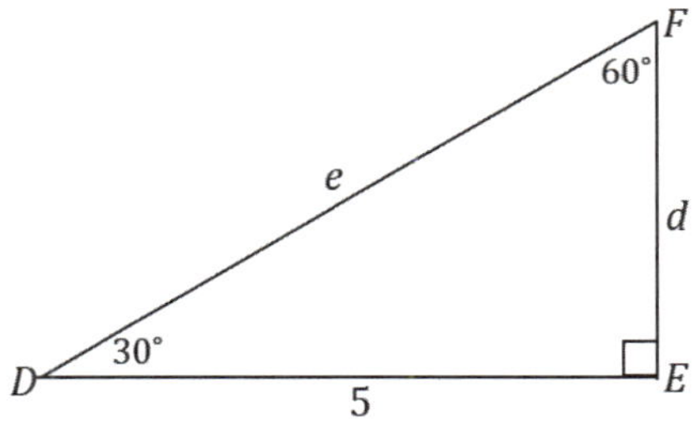

b. **Find the lengths of x and y. How is this different from part (a)?**

Accept any reasonable answer explaining that the triangle is not a right triangle; therefore, the trigonometric ratios used in part (a) are not applicable here.

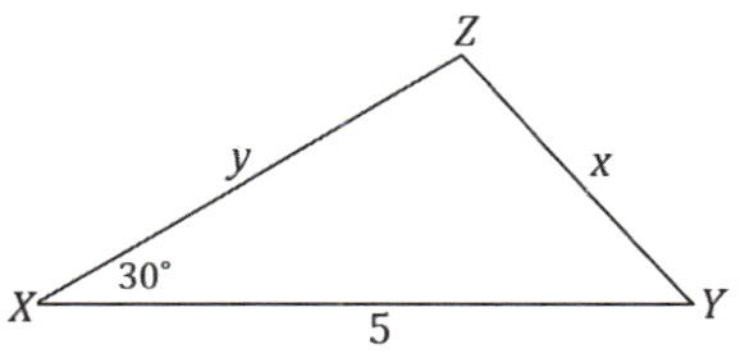

Discussion (10 minutes)

Lead students through an explanation of the law of sines for acute triangles.

> *Scaffolding:*
>
> Show students that the law of sines holds true by calculating the ratios using the following triangles:
>
> 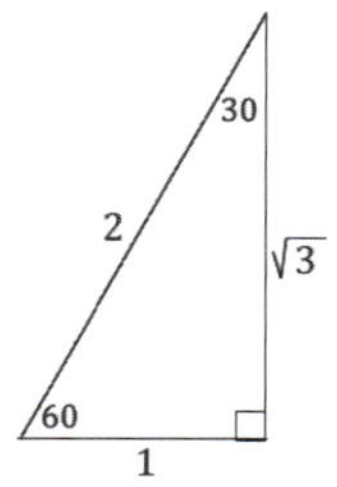
>
>
> 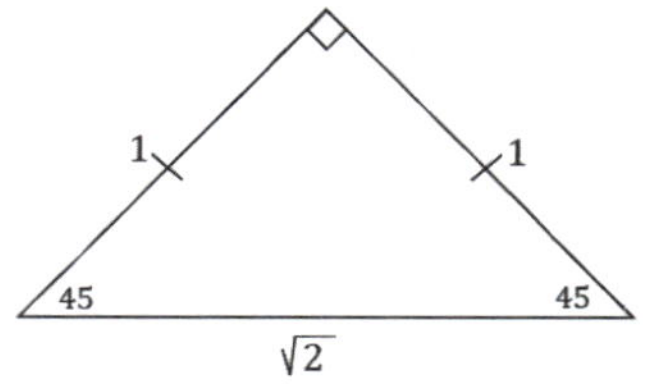
>

- Today we will show how two facts in trigonometry aid us to find unknown measurements in triangles that are not right triangles; we can use these facts for acute and obtuse triangles, but today we will specifically study acute triangles. The facts are called the law of sines and the law of cosines. We begin with the law of sines.
- **LAW OF SINES**: For an acute triangle $\triangle ABC$ with $\angle A$, $\angle B$, and $\angle C$ and the sides opposite them a, b, and c, the law of sines states:

$$\frac{\sin \angle A}{a} = \frac{\sin \angle B}{b} = \frac{\sin \angle C}{c}$$

- Restate the law of sines in your own words with a partner.

This may be difficult for students to articulate generally, without reference to a specific angle. State it for students if they are unable to state it generally.

  - *The ratio of the sine of an angle in a triangle to the side opposite the angle is the same for each angle in the triangle.*

- Consider $\triangle ABC$ with an altitude drawn from B to $\overline{AC}$. What is $\sin \angle C$?
  - $\sin \angle C = \frac{h}{a}$

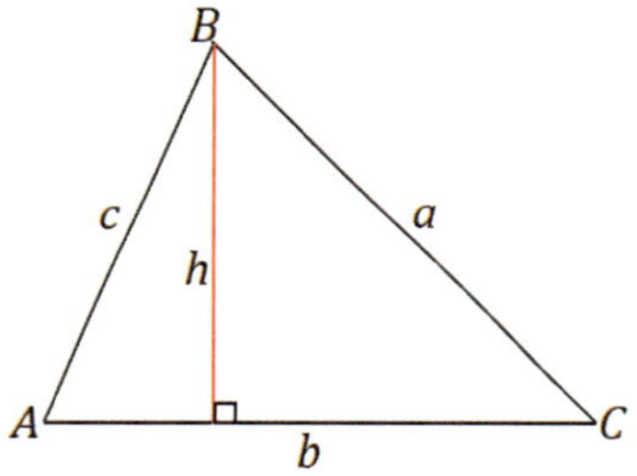

- Therefore, $h = a \sin \angle C$.
- What is $\sin A$?
  - $\sin \angle A = \frac{h}{c}$
- Therefore, $h = c \sin \angle A$.
- What can we conclude so far?
  - *Since $h = a \sin \angle C$ and $h = c \sin \angle A$, then $a \sin \angle C = c \sin \angle A$.*
- With a little algebraic manipulation, we can rewrite $a \sin \angle C = c \sin \angle A$ as $\frac{\sin \angle A}{a} = \frac{\sin \angle C}{c}$.
- MP.3 We have partially shown why the law of sines is true. What do we need to show in order to complete the proof, and how can we go about determining this?

Allow students a few moments to try and develop this argument independently.

- *We need to show that* $\frac{\sin \angle B}{b}$ *is equal to* $\frac{\sin \angle A}{a}$ *and to* $\frac{\sin \angle C}{c}$. *If we draw a different altitude, we can achieve this drawing:*

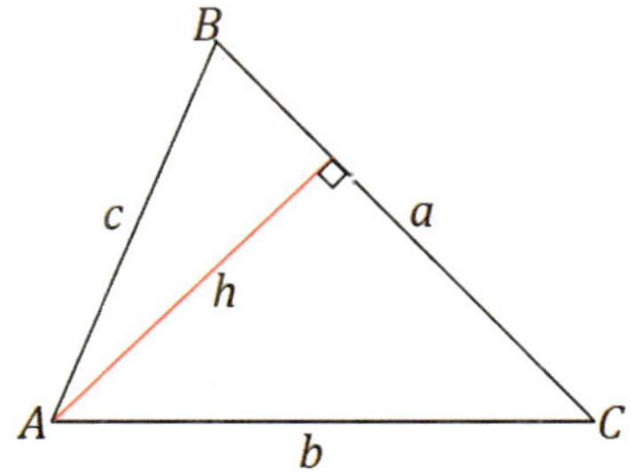

- *An altitude from A gives us* $\sin \angle B = \frac{h}{c}$ *or* $h = c \sin \angle B$.
- *Also,* $\sin \angle C = \frac{h}{b}$ *or* $h = b \sin \angle C$. *Therefore,* $c \sin \angle B = b \sin \angle C$ *or* $\frac{\sin \angle B}{b} = \frac{\sin \angle C}{c}$.
- *Since* $\frac{\sin \angle A}{a} = \frac{\sin \angle C}{c}$ *and* $\frac{\sin \angle B}{b} = \frac{\sin \angle C}{c}$, *then* $\frac{\sin \angle A}{a} = \frac{\sin \angle B}{b} = \frac{\sin \angle C}{c}$.

As soon as it has been established that $\frac{\sin \angle B}{b} = \frac{\sin \angle C}{c}$, the proof is really done, as the angles selected are arbitrary and, therefore, apply to any angle within the triangle. This can be explained for students if they are ready for the explanation.

Example 1 (4 minutes)

Students apply the law of sines to determine unknown measurements within a triangle.

Example 1

A surveyor needs to determine the distance between two points A and B that lie on opposite banks of a river. A point C is chosen 160 meters from point A, on the same side of the river as A. The measures of $\angle BAC$ and $\angle ACB$ are $41°$ and $55°$, respectively. Approximate the distance from A to B to the nearest meter.

Allow students a few moments to begin the problem before assisting them.

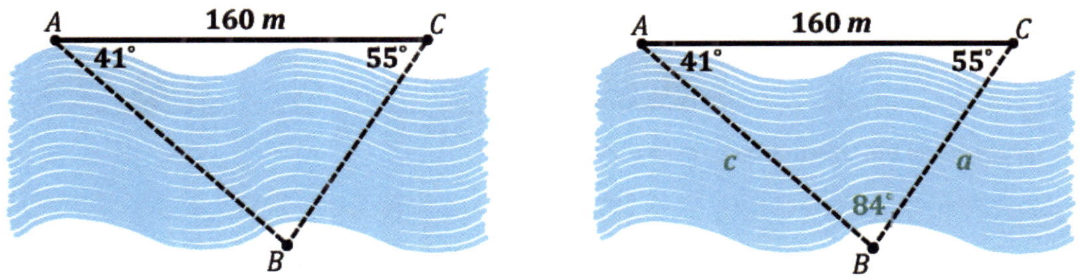

- What measurement can we add to the diagram based on the provided information?
  - *The measurement of $\angle B$ must be $84°$ by the triangle sum theorem.*
- Use the law of sines to set up all possible ratios applicable to the diagram.
  - $\frac{\sin 41}{a} = \frac{\sin 84}{160} = \frac{\sin 55}{c}$
- Which ratios will be relevant to determining the distance from A to B?
  - $\frac{\sin 84}{160} = \frac{\sin 55}{c}$
- Solve for c.
  - $c = \frac{160 \sin 55}{\sin 84}$

    $c = 132$

    The distance from A to B is 132 m.

Exercises 1–2 (6 minutes)

Depending on the time available, consider having students move directly to Exercise 2.

Exercises 1–2

1. **In $\triangle ABC$, $m\angle A = 30°$, $a = 12$, and $b = 10$. Find $\sin\angle B$. Include a diagram in your answer.**

$$\frac{\sin 30}{12} = \frac{\sin\angle B}{10}$$

$$\sin\angle B = \frac{5}{12}$$

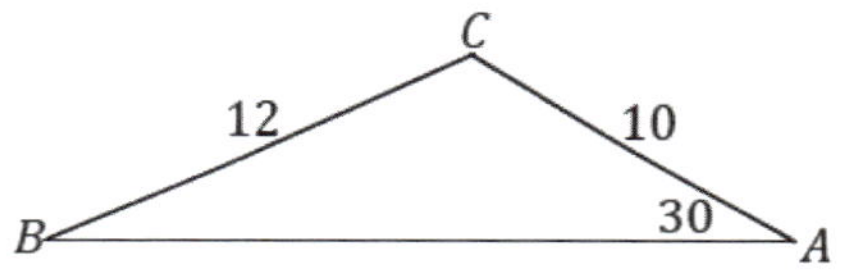

2. **A car is moving toward a tunnel carved out of the base of a hill. As the accompanying diagram shows, the top of the hill, H, is sighted from two locations, A and B. The distance between A and B is 250 ft. What is the height, h, of the hill to the nearest foot?**

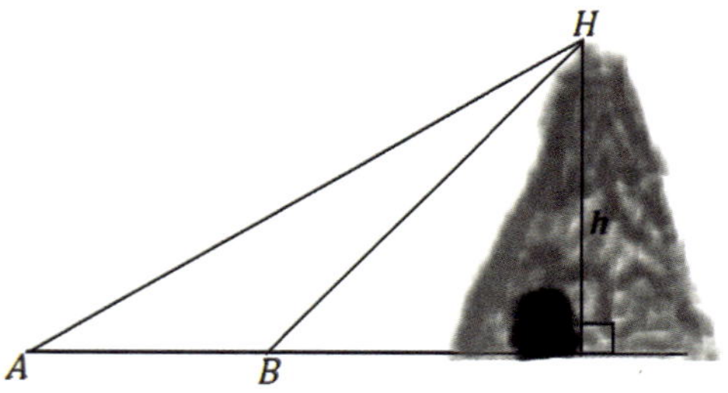

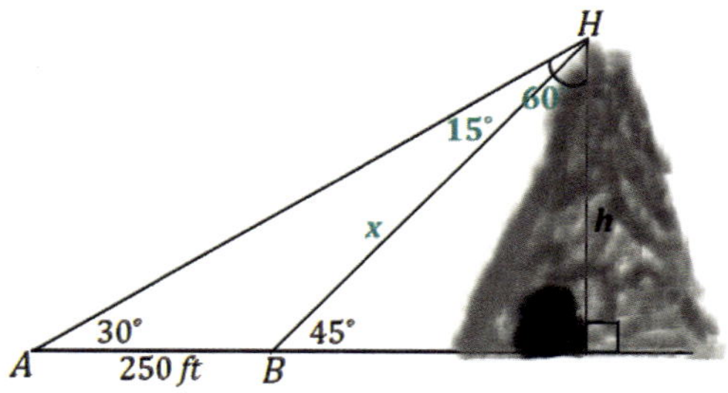

Let x represent BH, in feet. Applying the law of sines,

$$\frac{\sin 15}{250} = \frac{\sin 30}{x}$$

$$x = \frac{250 \sin 30}{\sin 15}$$

$$x = \frac{125}{\sin 15}$$

$$x \approx 482.96$$

> ***$\overline{BH}$ is the hypotenuse of a 45–45–90 triangle whose sides are in the ratio $1:1:\sqrt{2}$, or $h:h:h\sqrt{2}$.***
>
> $$h\sqrt{2} = x$$
> $$h\sqrt{2} = \frac{125}{\sin 15}$$
> $$h = \frac{125}{\sin 15 \cdot \sqrt{2}}$$
> $$h \approx 342$$
>
> ***The height of the hill is approximately 324 feet.***

Discussion (10 minutes)

Lead students through an explanation of the law of cosines for acute triangles. If more time is needed to cover the following Discussion, skip Exercise 1 as suggested earlier to allow for more time here.

- The next fact we will examine is the law of cosines.
- **LAW OF COSINES**: For an acute triangle $\triangle ABC$ with $\angle A$, $\angle B$, and $\angle C$ and the sides opposite them a, b, and c, the law of cosines states

$$c^2 = a^2 + b^2 - 2ab\cos\angle C.$$

- State the law of cosines in your own words.
  - *The square of one side of the triangle is equal to the sum of the squares of the other two sides minus twice the product of the other two sides and the cosine of the angle between them.*

The objective of being able to state the law in words is to move the focus away from specific letters and generalize the formula for any situation.

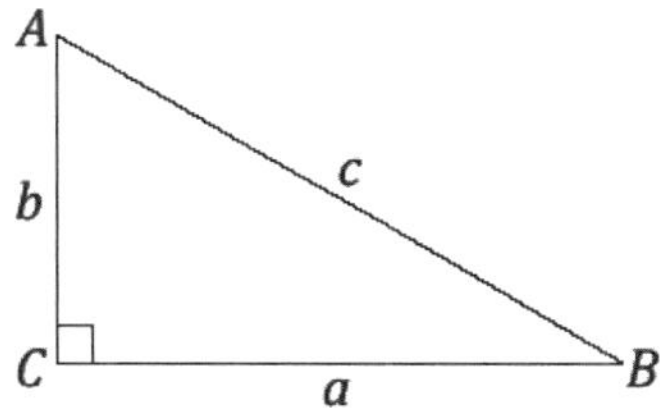

- The law of cosines is a generalization of the Pythagorean theorem, which can only be used with right triangles.
- Substitute $90°$ for $m\angle C$ into the law of cosines formula, and observe the result.
  - $c^2 = a^2 + b^2 - 2ab\cos 90$
- What is the value of $\cos 90$? What happens to the equation?
  - $\cos 90 = 0$; *the equation simplifies to* $c^2 = a^2 + b^2$.
- This explains why the law of cosines is a generalization of the Pythagorean theorem. We use the law of cosines for acute triangles in order to determine the side length of the third side of a triangle, provided two side lengths and the included angle measure.

> *Scaffolding:*
>
> For students ready for a challenge, instead of asking them to substitute $90°$ for $\angle C$, ask them what case results with the law of cosines being reduced to the Pythagorean theorem.

- We return to $\triangle ABC$ from Example 1, with an altitude drawn from B to $\overline{AC}$, but this time the point where the altitude meets $\overline{AC}$, point D, divides $\overline{AC}$ into lengths d and e.

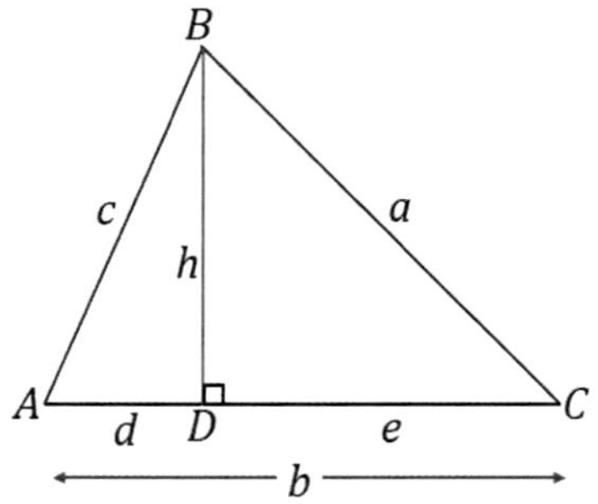

- Express e and h using trigonometry with respect to $\angle C$.
    - $h = a\ \sin\angle C$
    - $e = a\ \cos\angle C$
- Now we turn to right triangle $\triangle ABD$. What length relationship can be concluded between the sides of the triangle?
    - *By the Pythagorean theorem, the length relationship in* $\triangle ABD$ *is* $c^2 = d^2 + h^2$.
- Substitute the trigonometric expressions for d and h into this statement. Notice you will need length b.
    - *Then, the statement becomes* $c^2 = (b - a\ \cos\angle C)^2 + (a\ \sin\angle\ C)^2$.
- Simplify this statement as much as possible.
    - *The statement becomes:*

$$c^2 = b^2 - 2ab\cos\angle C + a^2\cos^2\angle C + \ a^2\sin^2\angle C$$
$$c^2 = b^2 - 2ab\cos\angle C + a^2((\cos\angle C)^2 + (\sin\angle C)^2)$$
$$c^2 = b^2 - 2ab\cos\angle C + a^2(1)$$
$$c^2 = a^2 + b^2 - 2ab\cos\angle C$$

- Notice that the right-hand side is composed of two side lengths and the cosine of the included angle. This will remain true if the labeling of the triangle is rearranged.
- What are all possible arrangements of the law of cosines for $\triangle ABC$?
    - $a^2 = b^2 + c^2 - 2(bc)\cos\angle A$
    - $\mathrm{b}^2 = \mathrm{a}^2 + \mathrm{c}^2 - 2(ac)\cos\angle B$

Example 2 (4 minutes)

Example 2

Our friend the surveyor from Example 1 is doing some further work. He has already found the distance between points A and B (from Example 1). Now he wants to locate a point D that is equidistant from both A and B and on the same side of the river as A. He has his assistant mark the point D so that $\angle ABD$ and $\angle BAD$ both measure $75°$. What is the distance between D and A to the nearest meter?

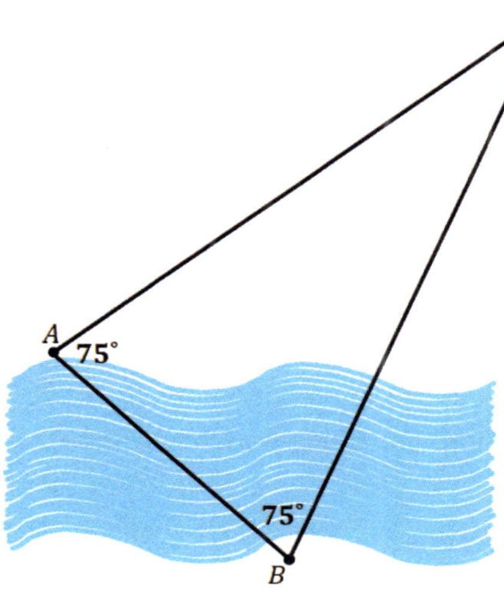

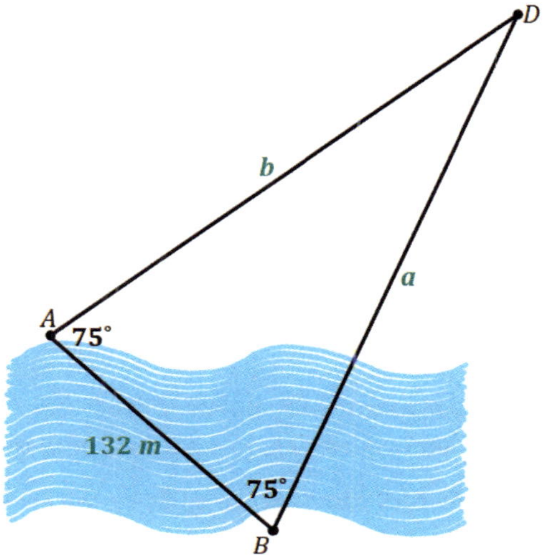

- What do you notice about $\triangle ABD$ right away?
  - *$\triangle ABD$ must be an isosceles triangle since it has two angles of equal measure.*
- We must keep this in mind going forward. Add all relevant labels to the diagram.

Students should add the distance of 132 m between A and B and add the label of a and b to the appropriate sides.

- Set up an equation using the law of cosines. Remember, we are trying to find the distance between D and A or, as we have labeled it, b.
  - $b^2 = 132^2 + a^2 - 2(132)(a)\cos 75$
- Recall that this is an isosceles triangle; we know that $a = b$. To reduce variables, we will substitute b for a. Rewrite the equation, and solve for b.
  - *Sample solution:*

$$b^2 = 132^2 + (b)^2 - 2(132)(b)\cos 75$$
$$b^2 = 132^2 + (b)^2 - 264(b)\cos 75$$
$$0 = 132^2 - 264(b)\cos 75$$
$$264(b)\cos 75 = 132^2$$
$$b = \frac{132^2}{264\cos 75}$$
$$b \approx 255 \text{ m}$$

Exercise 3 (2 minutes)

Exercise 3

3. **Parallelogram $ABCD$ has sides of lengths 44 mm and 26 mm, and one of the angles has a measure of $100°$. Approximate the length of diagonal $\overline{AC}$ to the nearest millimeter.**

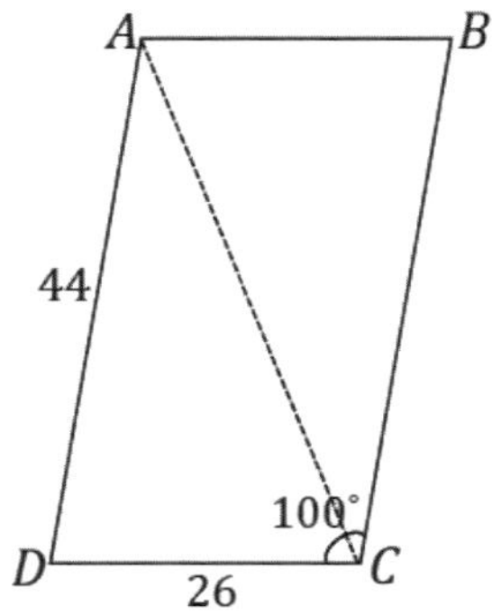

In parallelogram $ABCD$, $m\angle C = 100°$; therefore, $m\angle D = 80°$.

Let d represent the length of $\overline{AC}$.

$d^2 = 44^2 + 26^2 - 2(44)(26)\cos 80$

$d = 47$

The length of $\overline{AC}$ is 47 millimeters.

Closing (1 minute)

Ask students to summarize the key points of the lesson. Additionally, consider asking students to answer the following questions independently in writing, to a partner, or to the whole class.

- In what kinds of cases are we applying the laws of sines and cosines?
  - *We apply the laws of sines and cosines when we do not have right triangles to work with. We used the laws of sines and cosines for acute triangles.*
- State the law of sines. State the law of cosines.
  - For an acute triangle $\triangle ABC$ with $\angle A$, $\angle B$, and $\angle C$ and the sides opposite them a, b, and c,
    - The law of cosines states: $c^2 = a^2 + b^2 - 2ab\cos\angle C$.
    - The law of sines states: $\frac{\sin\angle A}{a} = \frac{\sin\angle B}{b} = \frac{\sin\angle C}{c}$.

Exit Ticket (5 minutes)

Name ______________________________ Date ______________

Lesson 32: Using Trigonometry to Find Side Lengths of an Acute Triangle

1. Use the law of sines to find lengths b and c in the triangle below. Round answers to the nearest tenth as necessary.

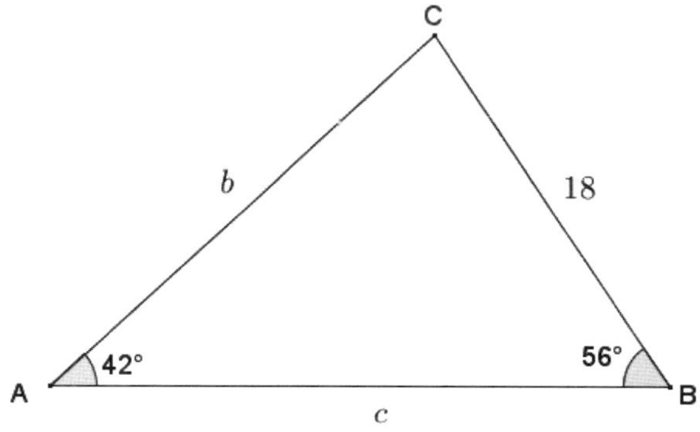

2. Given $\triangle DEF$, use the law of cosines to find the length of the side marked d to the nearest tenth.

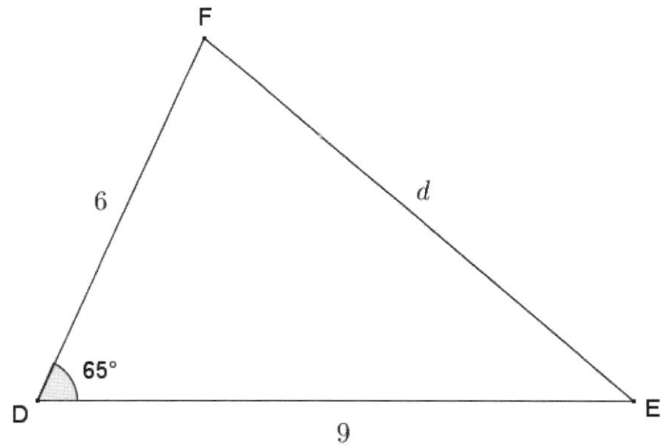

Exit Ticket Sample Solutions

1. Use the law of sines to find lengths b and c in the triangle below. Round answers to the nearest tenth as necessary.

$m\angle C = 82°$

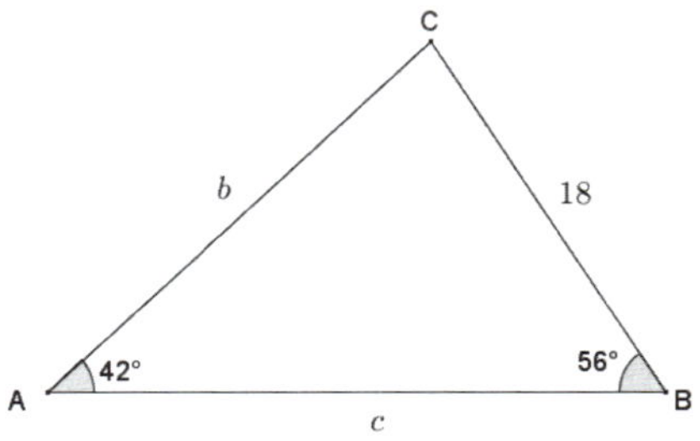

$$\frac{\sin A}{a} = \frac{\sin B}{b} = \frac{\sin C}{c}$$

$$\frac{\sin 42}{18} = \frac{\sin 56}{b} = \frac{\sin 82}{c}$$

$$b = \frac{18(\sin 56)}{\sin 42} \approx 22.3$$

$$c = \frac{18(\sin 82)}{\sin 42} \approx 26.6$$

2. Given $\triangle DEF$, use the law of cosines to find the length of the side marked d to the nearest tenth.

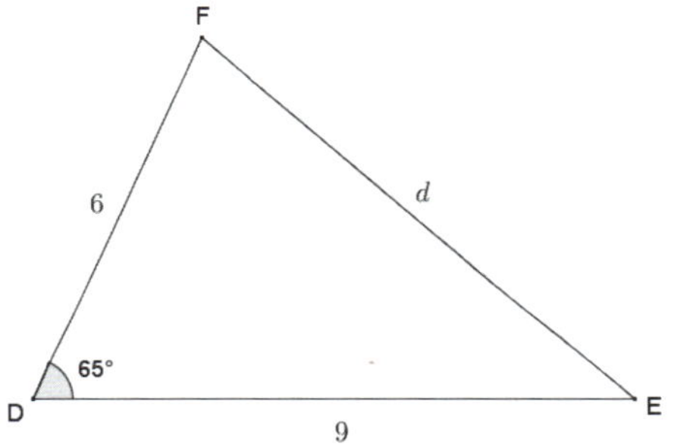

$$d^2 = 6^2 + 9^2 - 2(6)(9)(\cos 65)$$

$$d^2 = 36 + 81 - 108(\cos 65)$$

$$d^2 = 117 - 108(\cos 65)$$

$$d = \sqrt{117 - 108(\cos 65)}$$

$$d \approx 8.4$$

Problem Set Sample Solutions

1. Given $\triangle ABC$, $AB = 14$, $m\angle A = 57.2°$, and $m\angle C = 78.4°$, calculate the measure of angle B to the nearest tenth of a degree, and use the law of sines to find the lengths of $\overline{AC}$ and $\overline{BC}$ to the nearest tenth.

By the angle sum of a triangle, $m\angle B = 44.4°$.

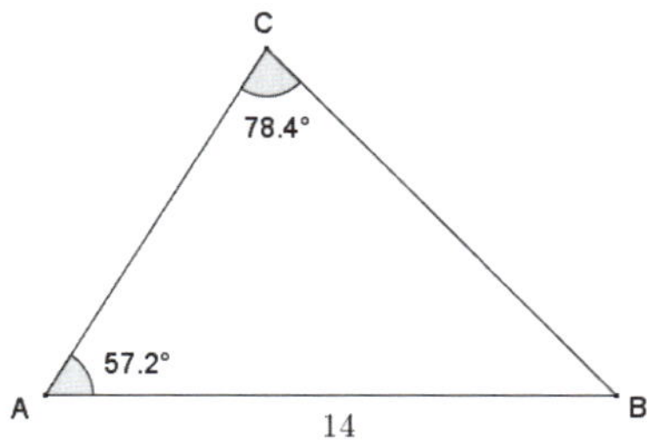

$$\frac{\sin A}{a} = \frac{\sin B}{b} = \frac{\sin C}{c}$$

$$\frac{\sin 57.2}{a} = \frac{\sin 44.4}{b} = \frac{\sin 78.4}{14}$$

$$a = \frac{14 \sin 57.2}{\sin 78.4} \approx 12.0$$

$$b = \frac{14 \sin 44.4}{\sin 78.4} \approx 10.0$$

Calculate the area of $\triangle ABC$ to the nearest square unit.

$$\text{Area} = \frac{1}{2}bc\sin A$$

$$\text{Area} = \frac{1}{2}(10)(14)\sin 57.2$$

$$\text{Area} = 70\sin 57.2$$

$$\text{Area} \approx 59$$

2. Given $\triangle DEF$, $m\angle F = 39°$, and $EF = 13$, calculate the measure of $\angle E$, and use the law of sines to find the lengths of $\overline{DF}$ and $\overline{DE}$ to the nearest hundredth.

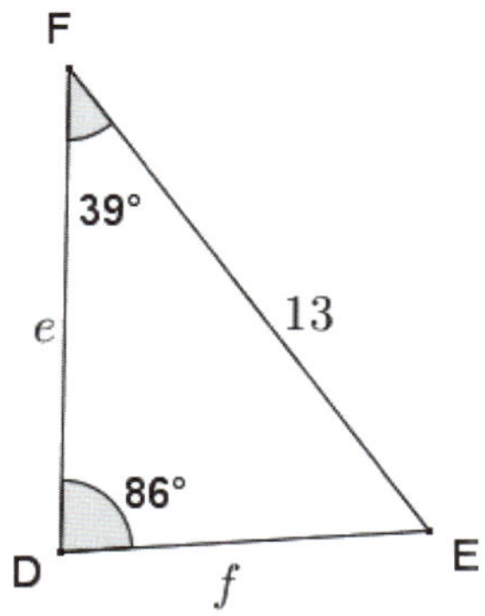

By the angle sum of a triangle, $m\angle E = 55°$.

$$\frac{\sin D}{d} = \frac{\sin E}{e} = \frac{\sin F}{f}$$

$$\frac{\sin 86}{13} = \frac{\sin 55}{e}$$

$$e = \frac{13\sin 55}{\sin 86}$$

$$e \approx 10.67$$

$$\frac{\sin 86}{13} = \frac{\sin 39}{f}$$

$$f = \frac{13\sin 39}{\sin 86}$$

$$f \approx 8.20$$

3. Does the law of sines apply to a right triangle? Based on $\triangle ABC$, the following ratios were set up according to the law of sines.

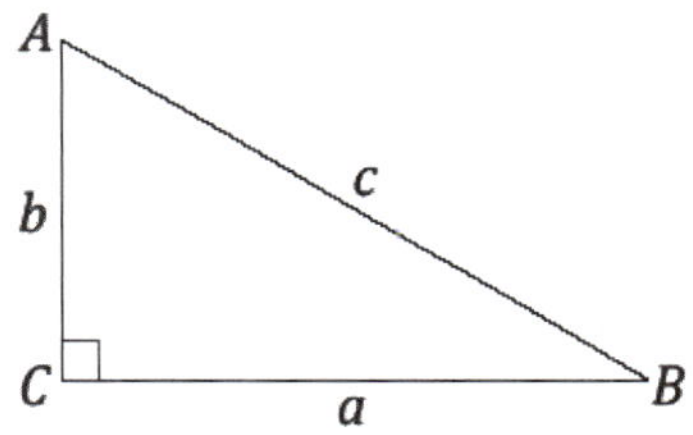

$$\frac{\sin\angle A}{a} = \frac{\sin\angle B}{b} = \frac{\sin 90}{c}$$

Fill in the partially completed work below:

$$\frac{\sin\angle A}{a} = \frac{\sin 90}{c} \qquad \frac{\sin\angle B}{b} = \frac{\sin 90}{c}$$

$$\frac{\sin\angle A}{a} = \frac{1}{c} \qquad \frac{\sin\angle B}{b} = \frac{1}{c}$$

$$\sin\angle A = \frac{\text{a}}{c} \qquad \sin\angle B = \frac{\text{b}}{c}$$

What conclusions can we draw?

The law of sines does apply to a right triangle. We get the formulas that are equivalent to $\sin\angle A = \frac{\text{opp}}{\text{hyp}}$ and $\sin\angle B = \frac{\text{opp}}{\text{hyp}}$, where A and B are the measures of the acute angles of the right triangle.

4. Given quadrilateral $GHKJ$, $m\angle H = 50°$, $m\angle HKG = 80°$, $m\angle KGJ = 50°$, $\angle J$ is a right angle, and $GH = 9$ in., use the law of sines to find the length of $\overline{GK}$, and then find the lengths of $\overline{GJ}$ and $\overline{JK}$ to the nearest tenth of an inch.

By the angle sum of a triangle, $m\angle HGK = 50°$; therefore, $\triangle GHK$ is an isosceles triangle since its base $\angle$'s have equal measure.

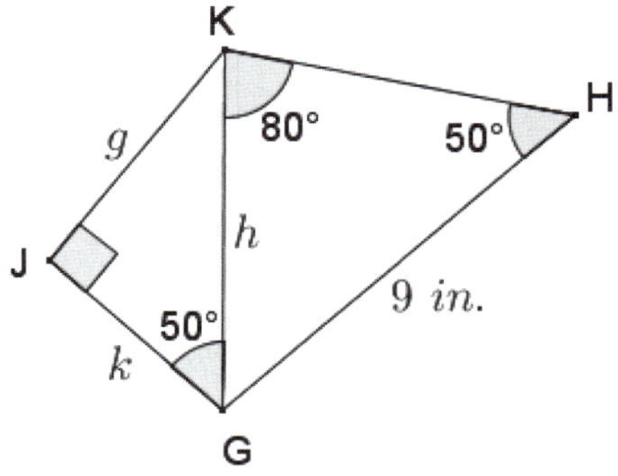

$$\frac{\sin 50}{h} = \frac{\sin 80}{9}$$

$$h = \frac{9 \sin 50}{\sin 80}$$

$$h \approx 7.0$$

$k = 7 \cos 50 \approx 4.5$

$g = 7 \sin 50 \approx 5.4$

5. Given triangle LMN, $LM = 10$, $LN = 15$, and $m\angle L = 38°$, use the law of cosines to find the length of $\overline{MN}$ to the nearest tenth.

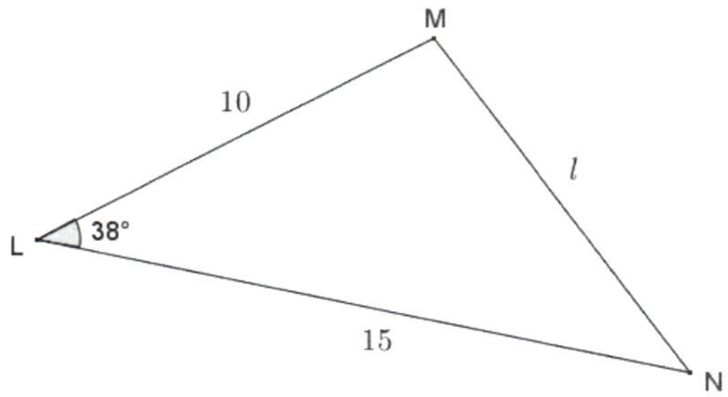

$l^2 = 10^2 + 15^2 - 2(10)(15)\cos 38$

$l^2 = 100 + 225 - 300\cos 38$

$l^2 = 325 - 300\cos 38$

$l = \sqrt{325 - 300\cos 38}$

$l \approx 9.4$

$MN = 9.4$

The length of $\overline{MN}$ is approximately 9.4 units.

6. Given triangle ABC, $AC = 6$, $AB = 8$, and $m\angle A = 78°$, draw a diagram of triangle ABC, and use the law of cosines to find the length of $\overline{BC}$.

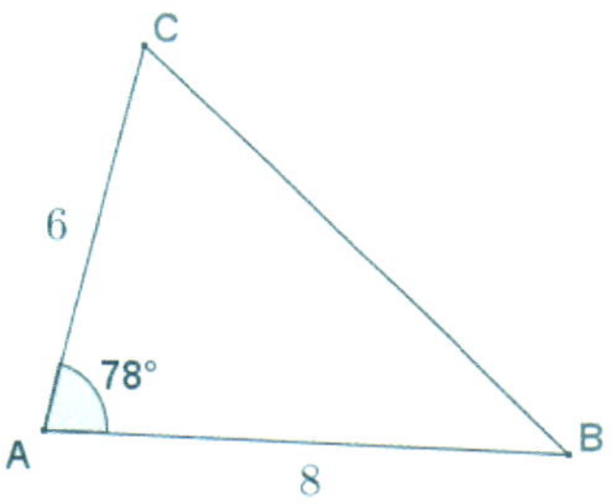

$a^2 = 6^2 + 8^2 - 2(6)(8)(\cos 78)$

$a^2 = 36 + 64 - 96(\cos 78)$

$a^2 = 100 - 96\cos 78$

$a = \sqrt{100 - 96\cos 78}$

$a \approx 8.9$

The length of $\overline{BC}$ is approximately 8.9 units.

Calculate the area of triangle ABC.

$\text{Area} = \frac{1}{2}bc(\sin A)$

$\text{Area} = \frac{1}{2}(6)(8)(\sin 78)$

$\text{Area} = 23.5(\sin 78)$

$\text{Area} \approx 23.5$

The area of triangle ABC is approximately 23.5 square units.

Lesson 33: Applying the Laws of Sines and Cosines

Student Outcomes

- Students understand that the law of sines can be used to find missing side lengths in a triangle when the measures of the angles and one side length are known.
- Students understand that the law of cosines can be used to find a missing side length in a triangle when the angle opposite the side and the other two side lengths are known.
- Students solve triangle problems using the laws of sines and cosines.

Lesson Notes

In this lesson, students apply the laws of sines and cosines learned in the previous lesson to find missing side lengths of triangles. The goal of this lesson is to clarify when students can apply the law of sines and when they can apply the law of cosines. Students are not prompted to use one law or the other; they must determine that on their own.

Classwork

Opening Exercise (10 minutes)

MP.3

Once students have made their decisions, have them turn to a partner and compare their choices. Pairs that disagree should discuss why and, if necessary, bring their arguments to the whole class so they may be critiqued. Ask students to provide justification for their choices.

Scaffolding:

- Consider having students make a graphic organizer to clearly distinguish between the law of sines and the law of cosines.
- Ask advanced students to write a letter to a younger student that explains the law of sines and the law of cosines and how to apply them to solve problems.

Opening Exercise

For each triangle shown below, identify the method (Pythagorean theorem, law of sines, law of cosines) you would use to find each length x.

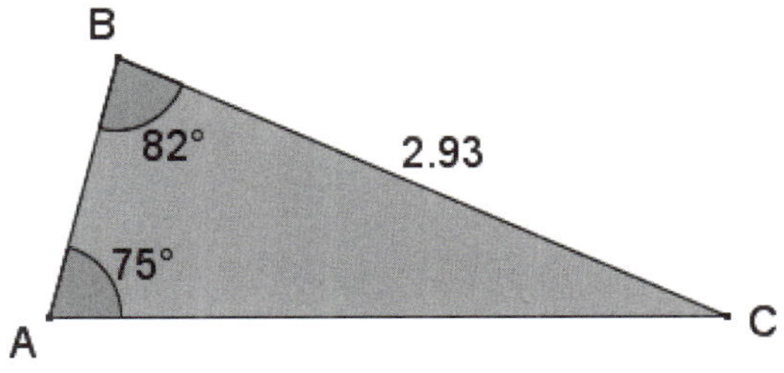	*law of sines*

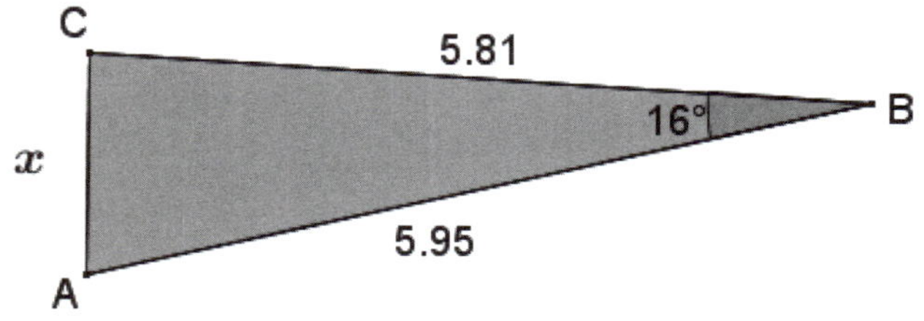	*law of cosines*
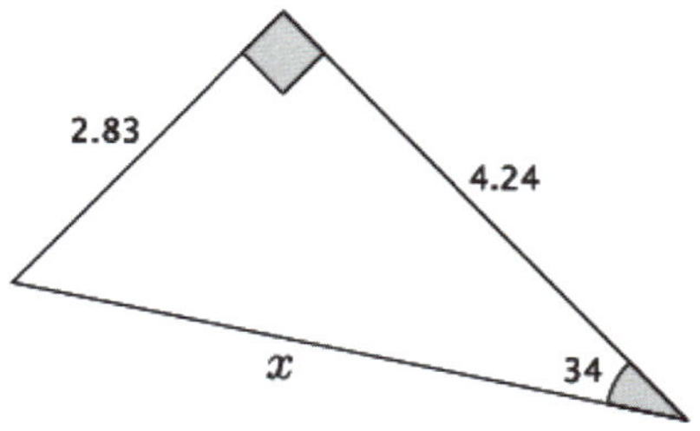	*Pythagorean theorem*
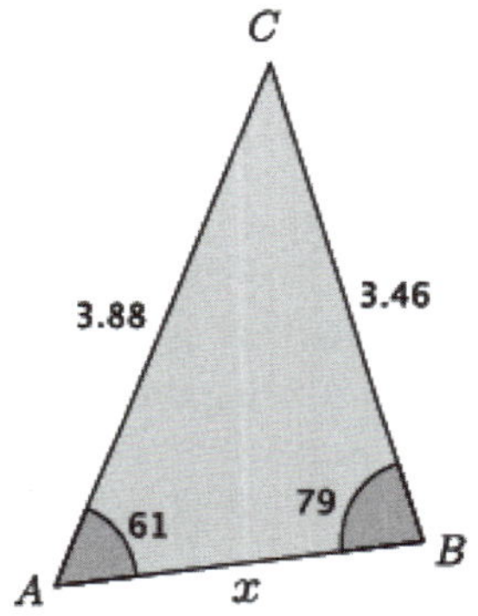	*law of sines* or *law of cosines*
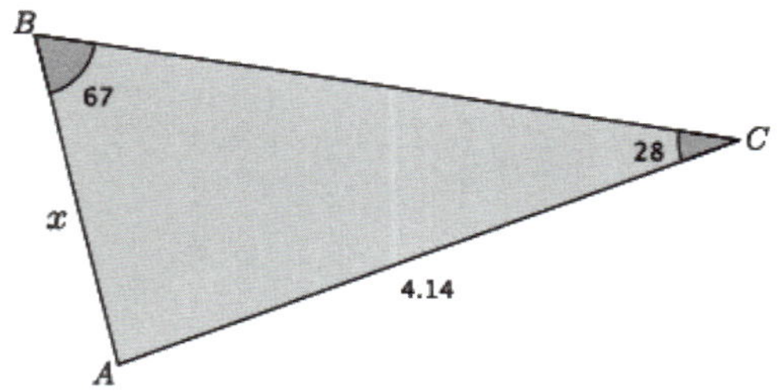	*law of sines*

Example 1 (5 minutes)

Students use the law of sines to find missing side lengths in a triangle.

Example 1

Find the missing side length in $\triangle ABC$.

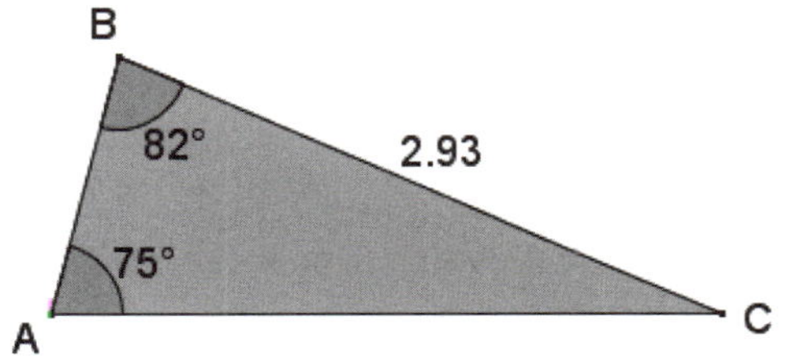

- Which method should we use to find length AB in the triangle shown below? Explain.

Provide a minute for students to discuss in pairs.

  - *Yes. The law of sines can be used because we are given information about two of the angles and one side. We can use the triangle sum theorem to find the measure of $\angle C$, and then we can use that information about the value of the pair of ratios $\frac{\sin A}{a} = \frac{\sin C}{c}$. Since the values are equivalent, we can solve to find the missing length.*
- Why can't we use the Pythagorean theorem with this problem?
  - *We can only use the Pythagorean theorem with right triangles. The triangle in this problem is not a right triangle.*
- Why can't we use the law of cosines with this problem?
  - *The law of cosines requires that we know the lengths of two sides of the triangle. We are only given information about the length of one side.*
- Write the equation that allows us to find the length of $\overline{AB}$.
  - *Let x represent the length of $\overline{AB}$.*

$$\frac{\sin 75}{2.93} = \frac{\sin 23}{x}$$

$$cx = \frac{2.93 \sin 23}{\sin 75}$$

- We want to perform one calculation to determine the answer so that it is most accurate and rounding errors are avoided. In other words, we do not want to make approximations at each step. Perform the calculation, and round the length to the tenths place.
  - *The length of $\overline{AB}$ is approximately* 1.2.

Example 2 (5 minutes)

Students use the law of cosines to find missing side lengths in a triangle.

Example 2

Find the missing side length in $\triangle ABC$.

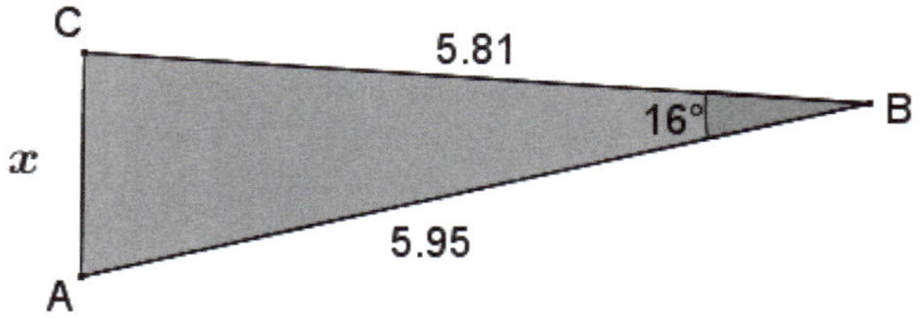

- Which method should we use to find side $\overline{AC}$ in the triangle shown below? Explain.

Provide a minute for students to discuss in pairs.

  - *We do not have enough information to use the law of sines because we do not have enough information to write any of the ratios related to the law of sines. However, we can use $b^2 = a^2 + c^2 - 2ac\cos B$ because we are given the lengths of sides a and c and we know the angle measure for $\angle B$.*

- Write the equation that can be used to find the length of $\overline{AC}$, and determine the length using one calculation. Round your answer to the tenths place.

  - $$x^2 = 5.81^2 + 5.95^2 - 2(5.81)(5.95)\cos 16$$
    $$x = \sqrt{5.81^2 + 5.95^2 - 2(5.81)(5.95)\cos 16}$$
    $$x \approx 1.6$$

Scaffolding:

It may be necessary to demonstrate to students how to use a calculator to determine the answer in one step.

Exercises 1–6 (16 minutes)

All students should be able to complete Exercises 1 and 2 independently. These exercises can be used to informally assess students' understanding in how to apply the laws of sines and cosines. Information gathered from these problems can inform how to present the next two exercises. Exercises 3–4 are challenging, and students should be allowed to work in pairs or small groups to discuss strategies to solve them. These problems can be simplified by having students remove the triangle from the context of the problem. For some groups of students, they may need to see a model of how to simplify the problem before being able to do it on their own. It may also be necessary to guide students to finding the necessary pieces of information, for example, angle measures, from the context or the diagram. Students who need a challenge should have the opportunity to make sense of the problems and persevere in solving them. The last two exercises are debriefed as part of the Closing.

Exercises 1–6

Use the laws of sines and cosines to find all missing side lengths for each of the triangles in the exercises below. Round your answers to the tenths place.

1. **Use the triangle to the right to complete this exercise.**

   a. **Identify the method (Pythagorean theorem, law of sines, law of cosines) you would use to find each of the missing lengths of the triangle. Explain why the other methods cannot be used.**

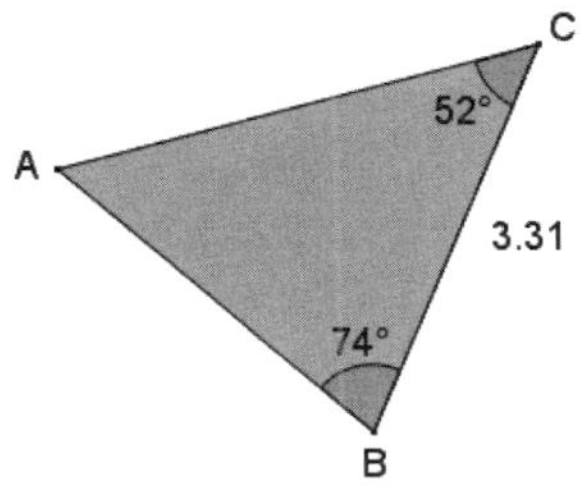

   Law of sines

   The Pythagorean theorem requires a right angle, which is not applicable to this problem. The law of cosines requires information about two side lengths, which is not given. Applying the law of sines requires knowing the measure of two angles and the length of one side.

   b. **Find the lengths of $\overline{AC}$ and $\overline{AB}$.**

   By the triangle sum theorem, $m\angle A = 54°$.
   Let b represent the length of side $\overline{AC}$.

$$\frac{\sin 54}{3.31} = \frac{\sin 74}{b}$$
$$b = \frac{3.31 \sin 74}{\sin 54}$$
$$b \approx 3.9$$

   Let c represent the length of side $\overline{AB}$.

$$\frac{\sin 54}{3.31} = \frac{\sin 52}{c}$$
$$c = \frac{3.31 \sin 52}{\sin 54}$$
$$c \approx 3.2$$

2. **Your school is challenging classes to compete in a triathlon. The race begins with a swim along the shore and then continues with a bike ride for 4 miles. School officials want the race to end at the place it began, so after the 4-mile bike ride, racers must turn $30°$ and run 3.5 miles directly back to the starting point. What is the total length of the race? Round your answer to the tenths place.**

   a. **Identify the method (Pythagorean theorem, law of sines, law of cosines) you would use to find the total length of the race. Explain why the other methods cannot be used.**

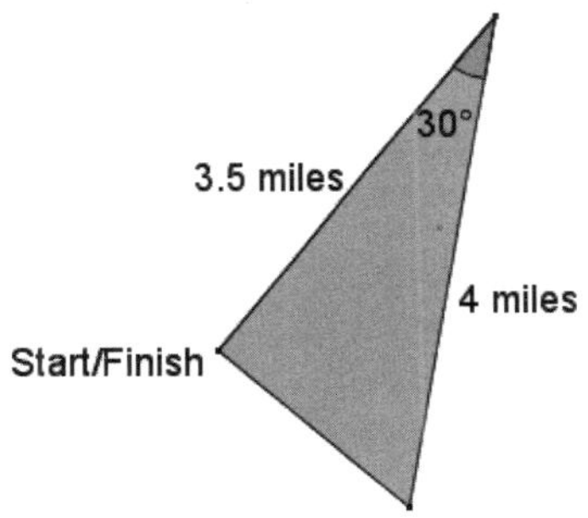

   Law of cosines

   The Pythagorean theorem requires a right angle, which is not applicable to this problem because we do not know if we have a right triangle. The law of sines requires information about two angle measures, which is not given. Applying the law of cosines requires knowing the measure of two sides and the included angle measure.

   b. **Determine the total length of the race. Round your answer to the tenths place.**

   Let a represent the length of the swim portion of the triathalon.

$$a^2 = 4^2 + 3.5^2 - 2(4)(3.5)\cos 30$$
$$a = \sqrt{4^2 + 3.5^2 - 2(4)(3.5)\cos 30}$$
$$a = 2.000322148\ldots$$
$$a \approx 2$$

   Total length of the race: $4 + 3.5 + (2) \approx 9.5$

   The total length of the race is approximately 9.5 ***miles.***

3. Two lighthouses are 30 miles apart on each side of shorelines running north and south, as shown. Each lighthouse keeper spots a boat in the distance. One lighthouse keeper notes the location of the boat as 40° east of south, and the other lighthouse keeper marks the boat as 32° west of south. What is the distance from the boat to each of the lighthouses at the time it was spotted? Round your answers to the nearest mile.

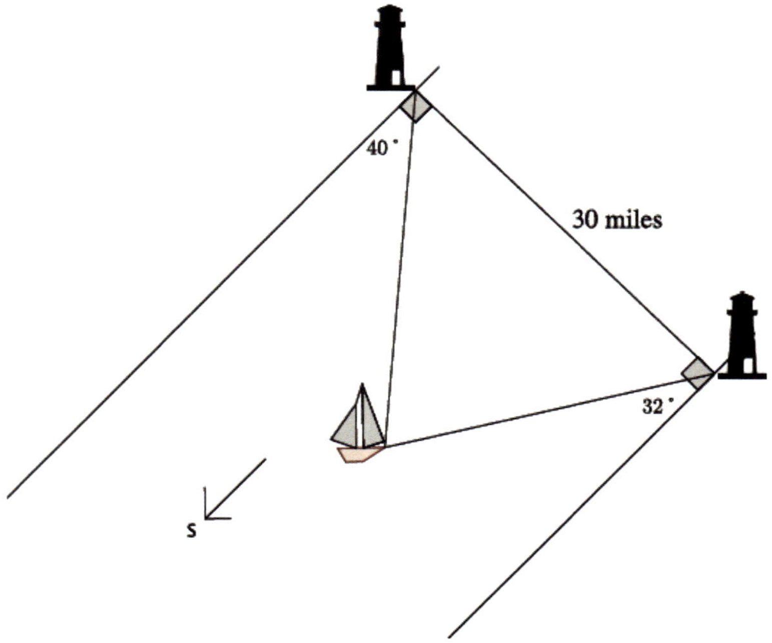

Students must begin by identifying the angle formed by one lighthouse, the boat, and the other lighthouse. This may be accomplished by drawing auxiliary lines and using facts about parallel lines cut by a transversal and the triangle sum theorem (or knowledge of exterior angles of a triangle). Once students know the angle is 72°, then the other angles in the triangle formed by the lighthouses and the boat can be found. The following calculations lead to the solution.

Let x be the distance from the southern lighthouse to the boat.

$$\frac{\sin 72}{30} = \frac{\sin 50}{x}$$
$$x = \frac{30 \sin 50}{\sin 72}$$
$$x = 24.16400382\ldots$$

The southern lighthouse is approximately 24 mi. from the boat.

With this information, students may choose to use the law of sines or the law of cosines to find the other distance. Shown below are both options.

Let y be the distance from the northern lighthouse to the boat.

$$\frac{\sin 72}{30} = \frac{\sin 58}{y}$$
$$y = \frac{30 \sin 58}{\sin 72}$$
$$y = 26.750\,716\,12\ldots$$

The northern lighthouse is approximately 27 mi. from the boat.

Let a be the distance from the northern lighthouse to the boat.

$$a^2 = 30^2 + 24^2 - 2(30)(24) \cdot \cos 58$$
$$a = \sqrt{30^2 + 24^2 - 2(30)(24) \cdot \cos 58}$$
$$a = 26.700\,491\,75\ldots$$

The northern lighthouse is approximately 27 mi. from the boat.

If groups of students work the problem both ways using the law of sines and the law of cosines, it may be a good place to have a discussion about why the answers were close but not exactly the same. When rounded to the nearest mile, the answers are the same, but if asked to round to the nearest tenths place, the results would be slightly different. The reason for the difference is that in the solution using the law of cosines, one of the values had already been approximated (24), leading to an even more approximated and less precise answer.

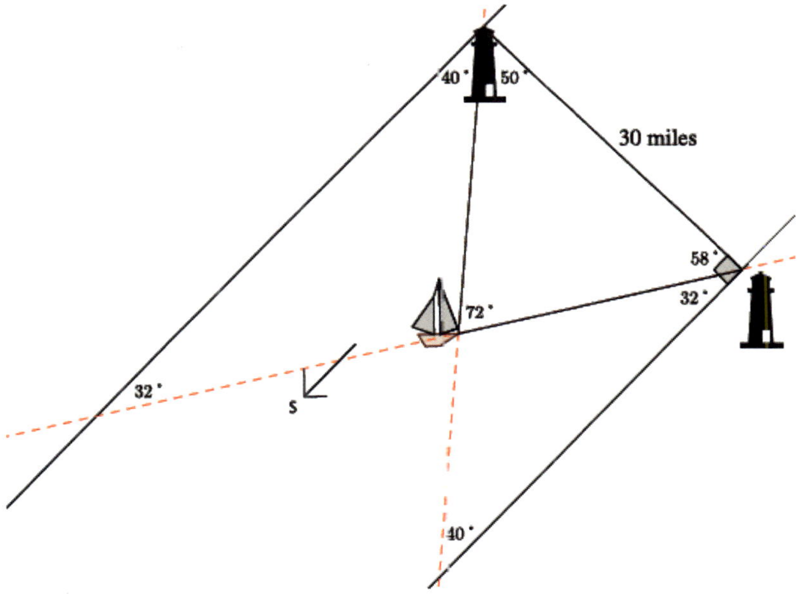

4. A pendulum 18 in. in length swings $72°$ from right to left. What is the difference between the highest and lowest point of the pendulum? Round your answer to the hundredths place, and explain how you found it.

Scaffolding:

Allow students time to struggle with the problem and, if necessary, guide them to drawing the horizontal and vertical dashed lines shown below.

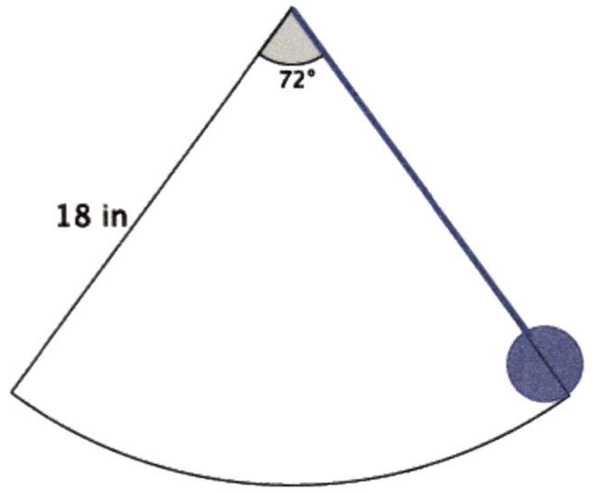

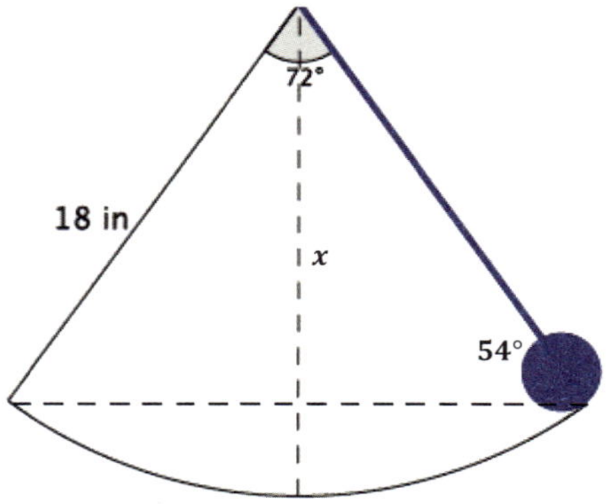

At the bottom of a swing, the pendulum is perpendicular to the ground, and it bisects the $72°$ angle; therefore, the pendulum currently forms an angle of $36°$ with the vertical. By the triangle sum theorem, the angle formed by the pendulum and the horizontal is $54°$. The $\sin 54$ will give us the length from the top of the pendulum to where the horizontal and vertical lines intersect, x, which happens to be the highest point of the pendulum.

$$\sin 54 = \frac{x}{18}$$
$$18 \sin 54 = x$$
$$14.562\,305\,9\ldots = x$$

The lowest point would be when the pendulum is perpendicular to the ground, which would be exactly 18 in. Then, the difference between those two points is approximately 3.44 in.

5. What appears to be the minimum amount of information about a triangle that must be given in order to use the law of sines to find an unknown length?

   To use the law of sines, you must know the measures of two angles and at least the length of one side.

6. What appears to be the minimum amount of information about a triangle that must be given in order to use the law of cosines to find an unknown length?

   To use the law of cosines, you must know at least the lengths of two sides and the angle measure of the included angle.

Closing (4 minutes)

Ask students to summarize the key points of the lesson. Additionally, consider asking students the following questions. Have students respond in writing, to a partner, or to the whole class.

- What is the minimum amount of information that must be given about a triangle in order to use the law of sines to find missing lengths? Explain.
- What is the minimum amount of information that must be given about a triangle in order to use the law of cosines to find missing lengths? Explain.

Exit Ticket (5 minutes)

Name ______________________________ Date ______________

Lesson 33: Applying the Laws of Sines and Cosines

Exit Ticket

1. Given triangle MLK, $KL = 8$, $KM = 7$, and $m\angle K = 75°$, find the length of the unknown side to the nearest tenth. Justify your method.

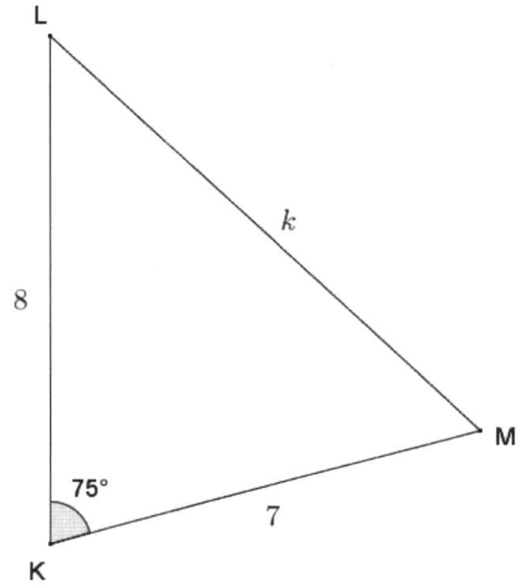

2. Given triangle ABC, $m\angle A = 36°$, $m\angle B = 79°$, and $AC = 9$, find the lengths of the unknown sides to the nearest tenth.

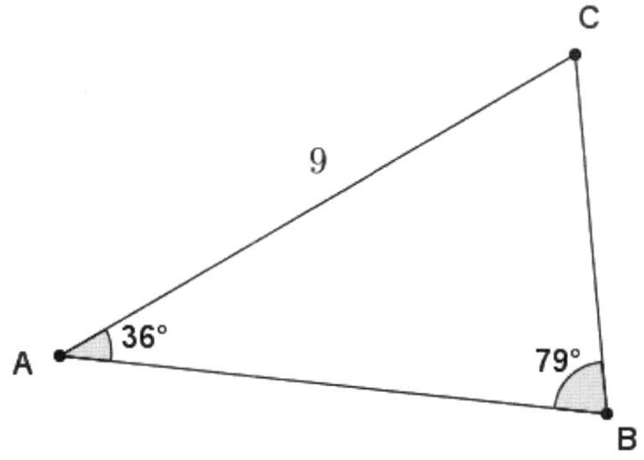

Exit Ticket Sample Solutions

1. Given triangle MLK, $KL = 8$, $KM = 7$, and $m\angle K = 75°$, find the length of the unknown side to the nearest tenth. Justify your method.

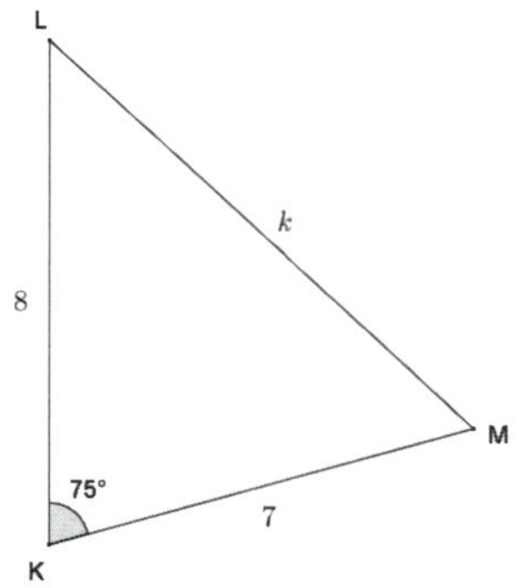

The triangle provides the lengths of two sides and their included angles, so using the law of cosines:

$k^2 = 8^2 + 7^2 - 2(8)(7)(\cos 75)$

$k^2 = 64 + 49 - 112(\cos 75)$

$k^2 = 113 - 112(\cos 75)$

$k = \sqrt{113 - 112(\cos 75)}$

$k \approx 9.2$

2. Given triangle ABC, $m\angle A = 36°$, $\angle B = 79°$, and $AC = 9$, find the lengths of the unknown sides to the nearest tenth.

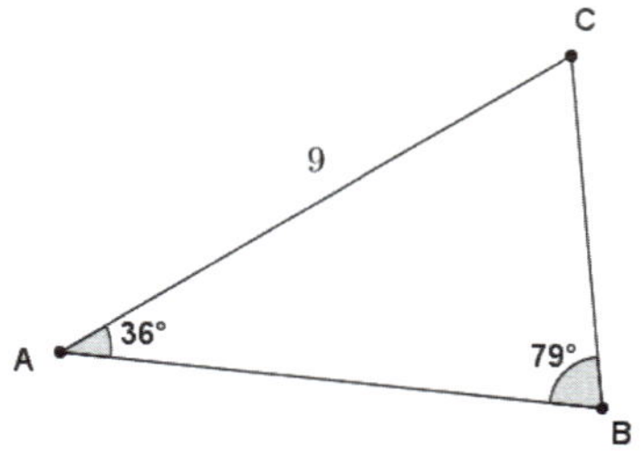

By the angle sum of a triangle, $m\angle C = 65°$.

Using the law of sines:

$$\frac{\sin A}{a} = \frac{\sin B}{b} = \frac{\sin C}{c}$$

$$\frac{\sin 36}{a} = \frac{\sin 79}{9} = \frac{\sin 65}{c}$$

$a = \dfrac{9 \sin 36}{\sin 79}$ $\qquad$ $c = \dfrac{9 \sin 65}{\sin 79}$

$a \approx 5.4$ $\qquad$ $c \approx 8.3$

$AB \approx 8.3$ *and* $BC \approx 5.4$.

Problem Set Sample Solutions

1. Given triangle EFG, $FG = 15$, angle E has a measure of $38°$, and angle F has a measure of $72°$, find the measures of the remaining sides and angle to the nearest tenth. Justify your method.

Using the angle sum of a triangle, the remaining angle G *has a measure of* $70°$.

The given triangle provides two angles and one side opposite a given angle, so it is appropriate to apply the law of sines.

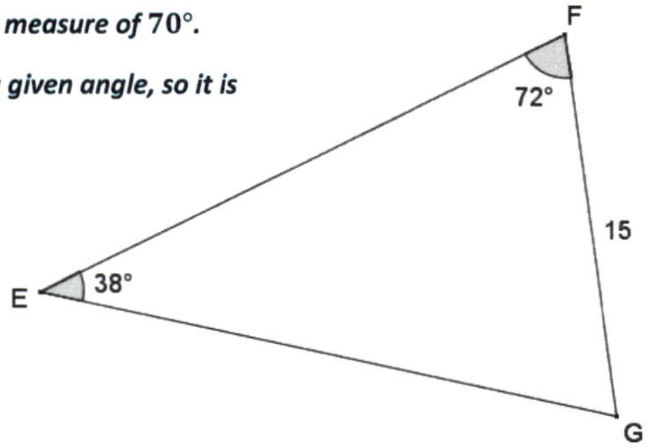

$$\frac{\sin 38}{15} = \frac{\sin 72}{EG} = \frac{\sin 70}{EF}$$

$\dfrac{\sin 38}{15} = \dfrac{\sin 72}{EG}$ $\qquad$ $\dfrac{\sin 38}{15} = \dfrac{\sin 70}{EF}$

$EG = \dfrac{15 \sin 72}{\sin 38}$ $\qquad$ $EF = \dfrac{15 \sin 70}{\sin 38}$

$EG \approx 23.2$ $\qquad$ $EF \approx 22.9$

2. Given triangle ABC, angle A has a measure of $75°$, $AC = 15.2$, and $AB = 24$, find BC to the nearest tenth. Justify your method.

The given information provides the lengths of the sides of the triangle and an included angle, so it is appropriate to use the law of cosines.

$BC^2 = 24^2 + 15.2^2 - 2(24)(15.2)(\cos 75)$

$BC^2 = 576 + 231.04 - 729.6 \cos 75$

$BC^2 = 807.04 - 729.6 \cos 75$

$BC = \sqrt{807.04 - 729.6 \cos 75}$

$BC \approx 24.9$

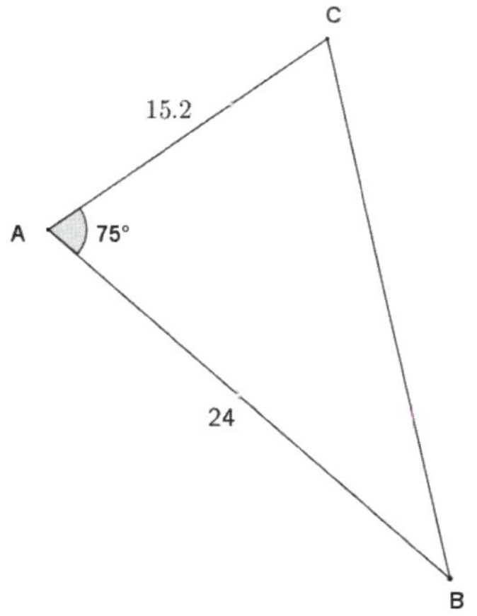

3. James flies his plane from point A at a bearing of $32°$ east of north, averaging a speed of 143 miles per hour for 3 hours, to get to an airfield at point B. He next flies $69°$ west of north at an average speed of 129 miles per hour for 4.5 hours to a different airfield at point C.

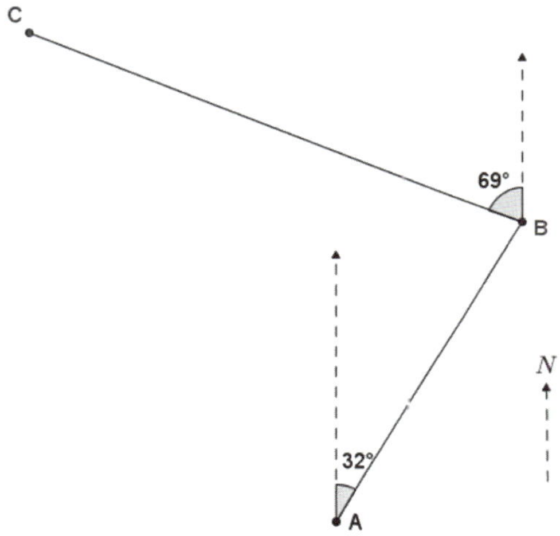

a. Find the distance from A to B.

distance = *rate* · *time*
$d = 143 \cdot 3$
$d = 429$

The distance from A to B is 429 mi.

b. Find the distance from B to C.

distance = *rate* · *time*
$d = 129 \cdot 4.5$
$d = 580.5$

The distance from B to C is 580.5 mi.

c. Find the measure of angle ABC.

All lines pointing to the north/south are parallel; therefore, by alternate interior ∠'s, the return path from B to A is $32°$ west of south. Using angles on a line, this mentioned angle, the measure of the angle formed by the path from B to C with north ($69°$) and $m\angle ABC$ sum to 180; thus, $m\angle ABC = 79°$.

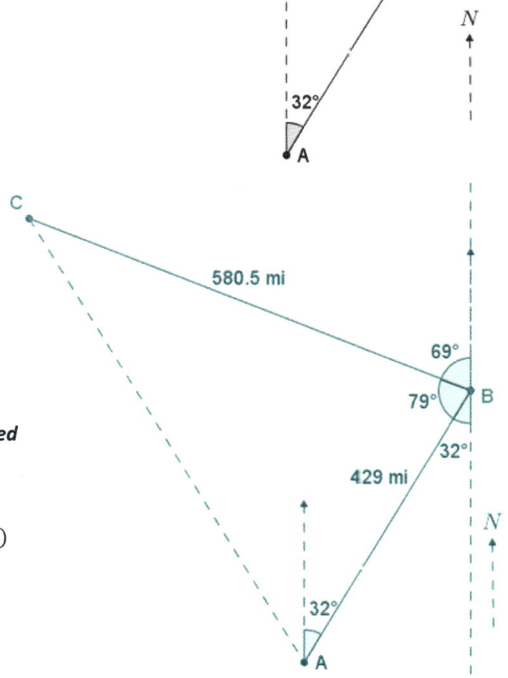

d. Find the distance from C to A.

The triangle shown provides two sides and the included angle, so using the law of cosines,

$b^2 = a^2 + c^2 - 2ac(\cos \mathrm{B})$

$b^2 = (580.5)^2 + (429)^2 - 2(580.5)(429)(\cos 79)$

$b^2 = 336\,980.25 + 184\,041 - 498\,069(\cos 79)$

$b^2 = 521\,021.25 - 498\,069(\cos 79)$

$b = \sqrt{521\,021.25 - 498\,069(\cos 79)}$

$b \approx 652.7$

The distance from C to A is approximately 652.7 mi.

e. What length of time can James expect the return trip from C to A to take?

distance = rate · time
$652.7 = 129 \cdot t_1$
$5.1 \approx t_1$

distance = rate · time
$652.7 = 143t_2$
$4.6 \approx t_2$

James can expect the trip from C to A to take between 4.6 and 5.1 hours.

4. Mark is deciding on the best way to get from point A to point B as shown on the map of Crooked Creek to go fishing. He sees that if he stays on the north side of the creek, he would have to walk around a triangular piece of private property (bounded by $\overline{AC}$ and $\overline{BC}$). His other option is to cross the creek at A and take a straight path to B, which he knows to be a distance of 1.2 mi. The second option requires crossing the water, which is too deep for his boots and very cold. Find the difference in distances to help Mark decide which path is his better choice.

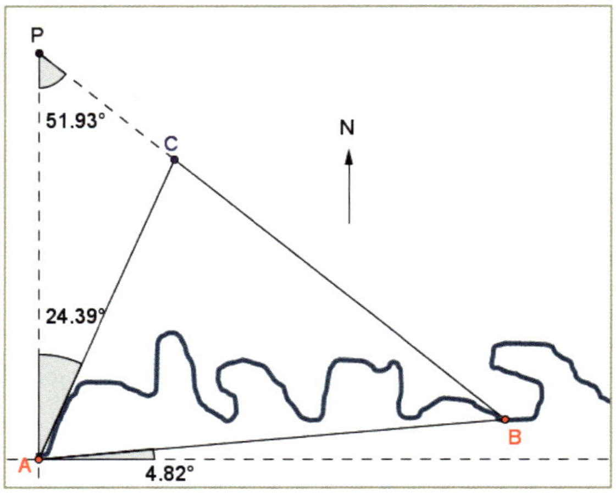

$\overline{AB}$ is $4.82°$ north of east, and $\overline{AC}$ is $24.39°$ east of north. The directions north and east are perpendicular, so the angles at point A form a right angle. Therefore, $m\angle CAB = 60.79°$.

By the angle sum of a triangle, $m\angle PCA = 103.68°$.

$\angle PCA$ and $\angle BCA$ are angles on a line with a sum of $180°$, so $m\angle BCA = 76.32°$.

Also, by the angle sum of a triangle, $m\angle ABC = 42.89°$.

Using the law of sines:

$$\frac{\sin A}{a} = \frac{\sin B}{b} = \frac{\sin C}{c}$$

$$\frac{\sin 60.79}{a} = \frac{\sin 42.89}{b} = \frac{\sin 76.32}{1.2}$$

$$b = \frac{1.2(\sin 42.89)}{\sin 76.32}$$
$$b \approx 0.8406$$

$$a = \frac{1.2(\sin 60.79)}{\sin 76.32}$$
$$a \approx 1.0780$$

Let d represent the distance from point A to B through point C in miles.

$d = AC + BC$

$d \approx 0.8406 + 1.0780$

$d \approx 1.9$

The distance from point A to B through point C is approximately 1.9 mi. This distance is approximately 0.7 mi. longer than walking a straight line from A to B.

5. **If you are given triangle ABC and the measures of two of its angles and two of its sides, would it be appropriate to apply the law of sines or the law of cosines to find the remaining side? Explain.**

Case 1:
Given two angles and two sides, one of the angles being an included angle, it is appropriate to use either the law of sines or the law of cosines.

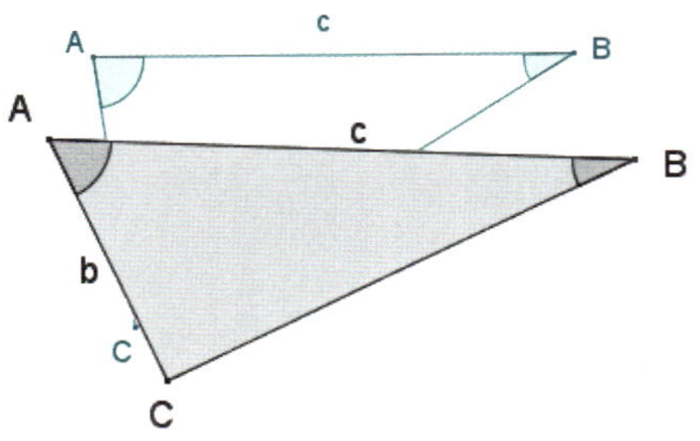

Case 2:
Given two angles and the sides opposite those angles, the law of sines can be applied as the remaining angle can be calculated using the angle sum of a triangle. The law of cosines then can also be applied as the previous calculation provides an included angle.

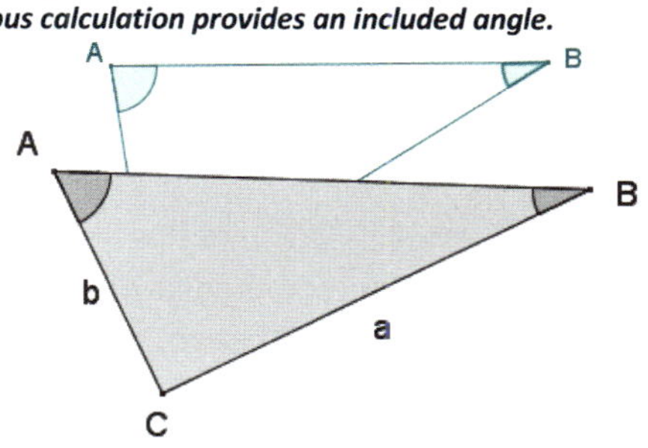

Lesson 34: Unknown Angles

Student Outcomes

- Students develop an understanding of how to determine a missing angle in a right triangle diagram and apply this to real-world situations.

Lesson Notes

This lesson introduces students to the use of reasoning based on trigonometric ratios to determine an unknown angle in a right triangle (**G-SRT.C.8**). At this stage, students are limited to understanding trigonometry in terms of ratios rather than functions. However, the concept of an inverse is not dealt with in this course; this is left until Algebra II (**F-BF.B.4**). Therefore, these ideas are introduced carefully and without the formalism of inverses, based on students' existing understanding of trigonometric ratios. It is important not to introduce the idea of inverses at this juncture, without care, which is why the lesson refers more generally to arcsin, arccos, and arctan as "words" that mathematicians have used to name, identify, or refer to the degree measures that give a certain trigonometric ratio.

Opening Exercise (12 minutes)

Ask students to complete this exercise independently or with a partner. Circulate and then discuss strategies.

Opening Exercise

a. **Dan was walking through a forest when he came upon a sizable tree. Dan estimated he was about 40 meters away from the tree when he measured the angle of elevation between the horizontal and the top of the tree to be 35 degrees. If Dan is about 2 meters tall, about how tall is the tree?**

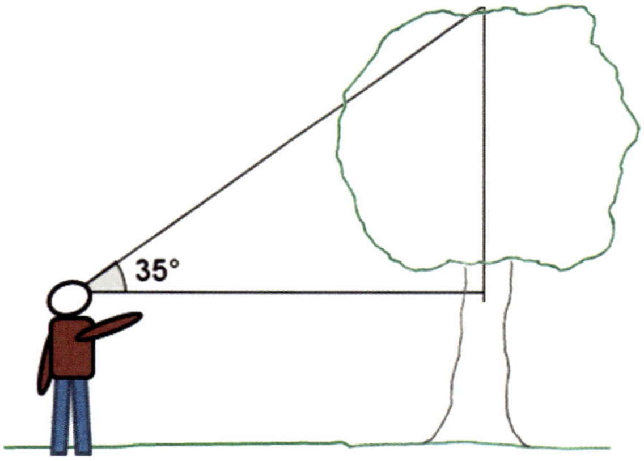

Let x represent the vertical distance from Dan's eye level to the top of the tree.

$$\tan 35 = \frac{x}{40}$$
$$40 \tan 35 = x$$
$$28 \approx x$$

The height of the tree is approximately 30 m.

b. **Dan was pretty impressed with this tree until he turned around and saw a bigger one, also 40 meters away in the other direction. "Wow," he said. "I bet that tree is *at least* 50 meters tall!" Then, he thought a moment. "Hmm … if it *is* 50 meters tall, I wonder what angle of elevation I would measure from my eye level to the top of the tree?" What angle will Dan find if the tree is 50 meters tall? Explain your reasoning.**

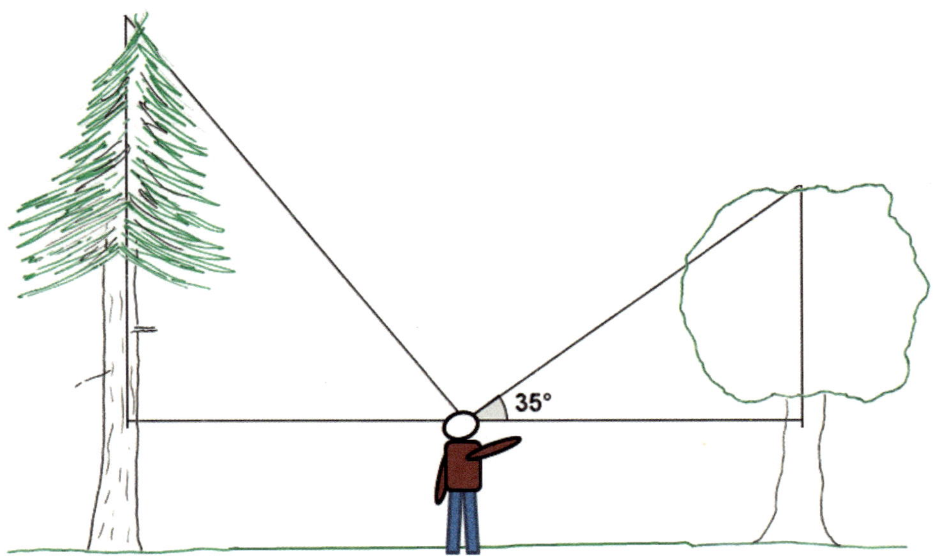

Let x represent the angle measure from the horizontal to the top of the tree.

$$\tan x = \frac{50}{40}$$
$$\tan x = \frac{5}{4}$$
$$\tan x = 1.25$$

Encourage students to develop conjectures about what the number of degrees are. Consider selecting some or all of the measures, placing them in a central visible location, and discussing which are the most reasonable and which are the least reasonable. Ideally, some students reason that measures less than $35°$ would be unreasonable; some may draw hypothetical angles and use these to compute the tangent ratios; still others may use the table of values for tangent given in Lesson 29 to see that the angle would be slightly greater than $50°$.

Discussion (15 minutes)

Just like in the second exercise, sometimes we are confronted with diagrams or problems where we are curious about what an angle measure might be.

In the same way that mathematicians have named certain ratios within right triangles, they have also developed terminology for identifying angles in a right triangle, given the ratio of the sides. Mathematicians often use the prefix *arc* to define these. The prefix *arc* is used because of how angles were measured, not just as an angle but also as the length of an *arc* on the unit circle. We will learn more about arc lengths in Module 5.

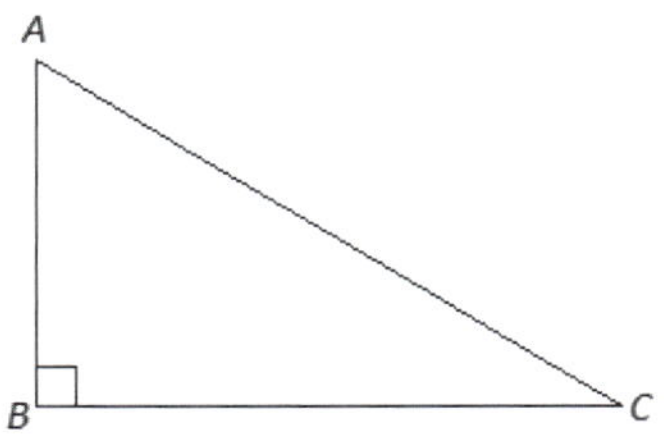

- Write ratios for sin, cos, and tan of angle C:
  - $\sin C = \frac{AB}{AC}, \cos C = \frac{BC}{AC}, \tan C = \frac{AB}{BC}$
- Write ratios for sin, cos, and tan of angle A:
  - $\sin A = \frac{BC}{AC}, \cos A = \frac{AB}{AC}, \tan A = \frac{BC}{AB}$
- Mathematicians have developed some additional terms to describe the angles and side ratios in right triangles. Examine the statements below, and see if you can determine the meaning of each one.

One by one, show each statement. Ask students to make and explain a guess about what these statements mean.

- $\arcsin\left(\frac{AB}{AC}\right) = m\angle C$
- $\arccos\left(\frac{BC}{AC}\right) = m\angle C$
- $\arctan\left(\frac{AB}{BC}\right) = m\angle C$

Once students have shared their guesses, formalize the ideas with a discussion:

- Mathematicians use *arcsin, arccos,* and *arctan* to refer to the angle measure that results in the given sin, cos, or tan ratio. For example, for this triangle, mathematicians would say, "$\arcsin\left(\frac{AB}{AC}\right) = m\angle C$." Explain the meaning of this in your own words.
  - *This means that the angle that has a sine ratio equal to* $\frac{AB}{AC}$ is $m\angle C$.
- Explain the meaning of $\arccos\left(\frac{BC}{AC}\right) = m\angle C$.
  - *This means that the angle that has a cosine ratio equal to* $\frac{BC}{AC}$ is $m\angle C$.
- Explain the meaning of $\arctan\left(\frac{AB}{BC}\right) = m\angle C$.
  - *This means that the angle that has a tangent ratio equal to* $\frac{AB}{BC}$ *is* $m\angle C$.
- We can use a calculator to help us determine the values of arcsin, arccos, and arctan. On most calculators, these are represented by buttons that look like "$\sin^{-1}$," "$\cos^{-1}$," and "$\tan^{-1}$."

- Let's revisit the example from the opening. How could we determine the angle of elevation that Dan would measure if he is 40 meters away and the tree is 50 meters tall?
  - *Let x represent the angle measure from the horizontal to the top of the tree.*

$$\tan x = \frac{48}{40}$$
$$\tan x = \frac{6}{5}$$
$$\tan x = 1.2$$
$$x \approx 50$$

Exercises 1–5 (10 minutes)

Students complete the exercises independently or in pairs.

Exercises 1–5

1. Find the measure of angles a through d to the nearest degree.

a.

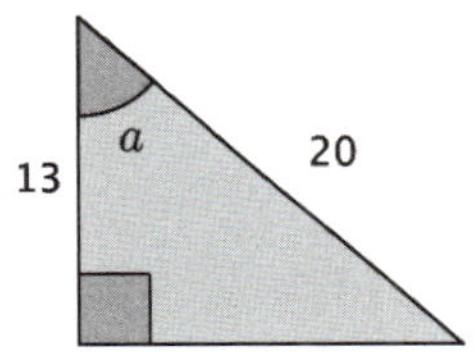

$$\mathbf{arccos}\left(\frac{13}{20}\right) \approx 49$$
$$m\angle a = 49^\circ$$

b.

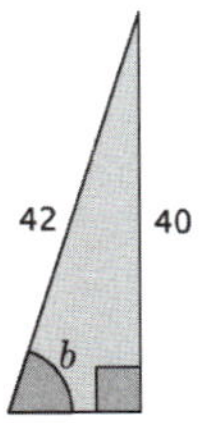

$$\mathbf{arcsin}\left(\frac{40}{42}\right) \approx 72$$
$$m\angle b = 72^\circ$$

c.

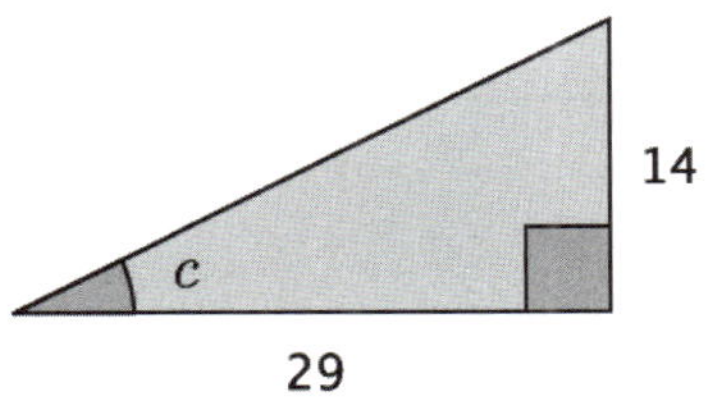

$$\mathbf{arctan}\left(\frac{14}{29}\right) \approx 26$$
$$m\angle c = 26^\circ$$

d.

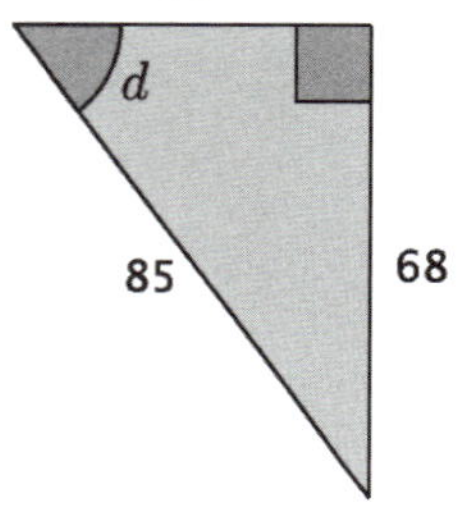

Several solutions are acceptable. One is shown below.

$$\arccos\left(\frac{51}{85}\right) \approx 53$$

$$m\angle d = 53^\circ$$

2. **Shelves are being built in a classroom to hold textbooks and other supplies. The shelves will extend 10 in. from the wall. Support braces will need to be installed to secure the shelves. The braces will be attached to the end of the shelf and secured 6 in. below the shelf on the wall. What angle measure will the brace and the shelf make?**

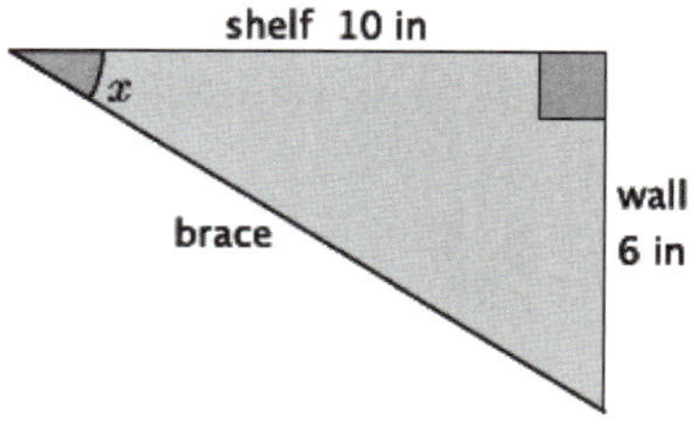

$$\arctan\left(\frac{6}{10}\right) \approx 31$$

The angle measure between the brace and the shelf is 31°.

3. **A 16 ft. ladder leans against a wall. The foot of the ladder is 7 ft. from the wall.**

   a. **Find the vertical distance from the ground to the point where the top of the ladder touches the wall.**

   Let x represent the distance from the ground to the point where the top of the ladder touches the wall.

$$16^2 = 7^2 + x^2$$
$$16^2 - 7^2 = x^2$$
$$207 = x^2$$
$$14 \approx x$$

   The top of the ladder is 14 ft. above the ground.

   b. **Determine the measure of the angle formed by the ladder and the ground.**

$$\arccos\left(\frac{7}{16}\right) \approx 64$$

   The angle formed by the ladder and the ground is approximately 64°.

4. A group of friends have hiked to the top of the Mile High Mountain. When they look down, they can see their campsite, which they know is approximately 3 miles from the base of the mountain.

   a. Sketch a drawing of the situation.

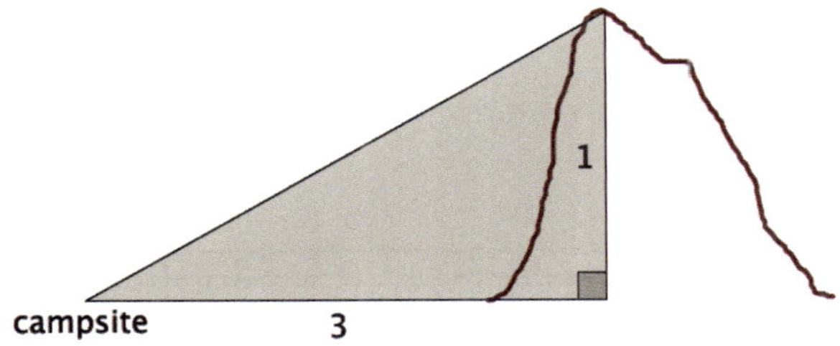

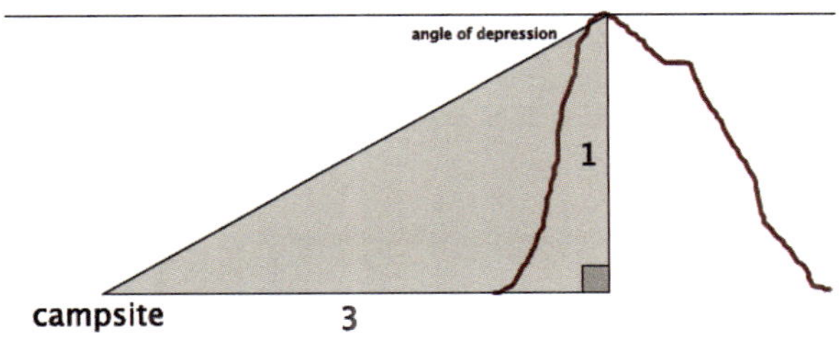

   b. What is the angle of depression?

$$\arctan\left(\frac{3}{1}\right) \approx 72$$

The angle of depression is approximately $18°$.

5. A roller coaster travels 80 ft. of track from the loading zone before reaching its peak. The horizontal distance between the loading zone and the base of the peak is 50 ft.

   a. Model the situation using a right triangle.

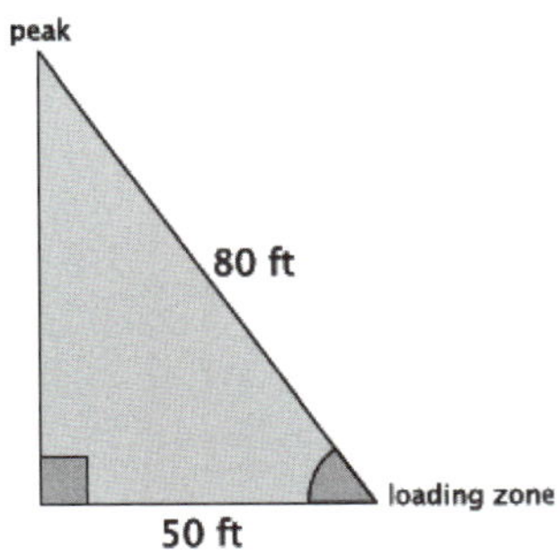

   b. At what angle is the roller coaster rising according to the model?

$$\arccos\left(\frac{50}{80}\right) \approx 51$$

The roller coaster is rising at approximately $51°$.

Closing (3 minutes)

Ask students to summarize the key points of the lesson, and consider asking them the following questions. Have students respond in writing, to a partner, or to the whole class.

- Explain the meaning of $\arccos\left(\frac{8}{9}\right) \approx 27$.
- Explain how to find the unknown measure of an angle given information about only two of the sides of a right triangle.

Lesson Summary

In the same way that mathematicians have named certain ratios within right triangles, they have also developed terminology for identifying angles in a right triangle, given the ratio of the sides. Mathematicians often use the prefix *arc* to define these because an angle is not just measured as an angle, but also as a length of an *arc* on the unit circle.

Given a right triangle ABC, the measure of angle C can be found in the following ways:

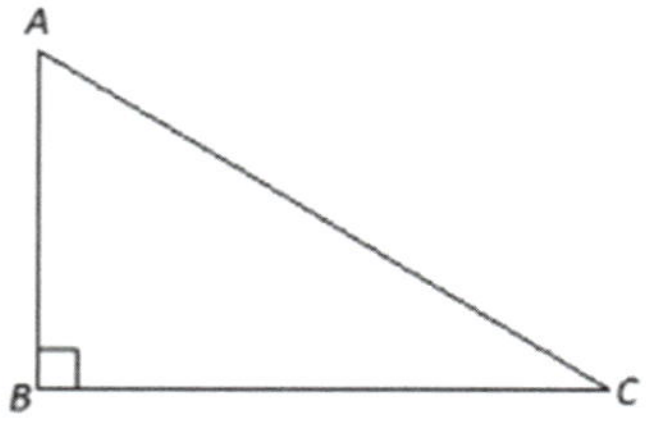

- $\arcsin\left(\frac{AB}{AC}\right) = m\angle C$
- $\arccos\left(\frac{BC}{AC}\right) = m\angle C$
- $\arctan\left(\frac{AB}{BC}\right) = m\angle C$

We can write similar statements to determine the measure of angle A.

We can use a calculator to help us determine the values of arcsin, arccos, and arctan. Most calculators show these buttons as "$\sin^{-1}$," "$\cos^{-1}$," and "$\tan^{-1}$." This subject is addressed again in future courses.

Exit Ticket (5 minutes)

Name ______________________________ Date ______________

Lesson 34: Unknown Angles

Exit Ticket

1. Explain the meaning of the statement "$\arcsin\left(\frac{1}{2}\right) = 30°$." Draw a diagram to support your explanation.

2. Gwen has built and raised a wall of her new house. To keep the wall standing upright while she builds the next wall, she supports the wall with a brace, as shown in the diagram below. What is the value of p, the measure of the angle formed by the brace and the wall?

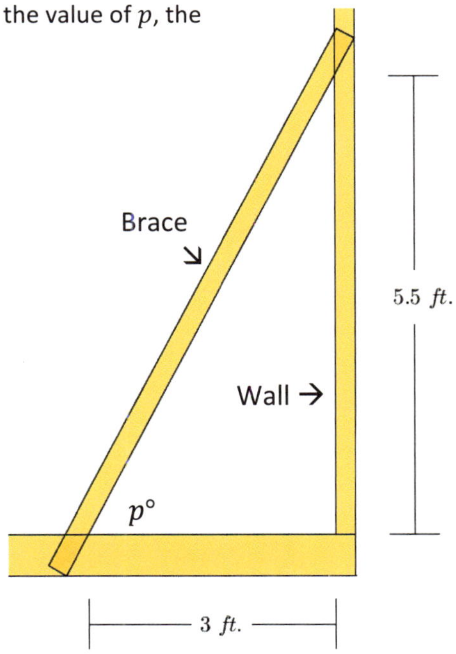

Exit Ticket Sample Solutions

1. Explain the meaning of the statement "$\arcsin\left(\frac{1}{2}\right) = 30°$." Draw a diagram to support your explanation.

   This means that the measure of the angle that has a sine ratio equal to $\frac{1}{2}$ *is* $30°$.

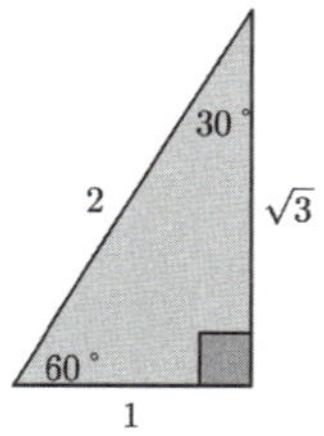

2. Gwen has built and raised a wall of her new house. To keep the wall standing upright while she builds the next wall, she supports the wall with a brace, as shown in the diagram below. What is the value of p, the measure of the angle formed by the brace and the wall?

$$\arctan \frac{5.5}{3} = p$$
$$61 \approx p$$

The measure of the angle formed by the brace and the wall is approximately $61°$.

Problem Set Sample Solutions

1. For each triangle shown, use the given information to find the indicated angle to the nearest degree.

   a.

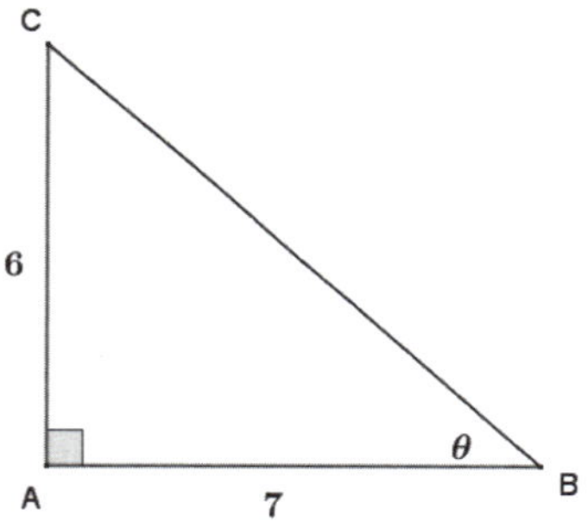

$$\tan\theta = \frac{6}{7}$$
$$\arctan\left(\frac{6}{7}\right) = \theta$$
$$\theta \approx 41°$$

   b.

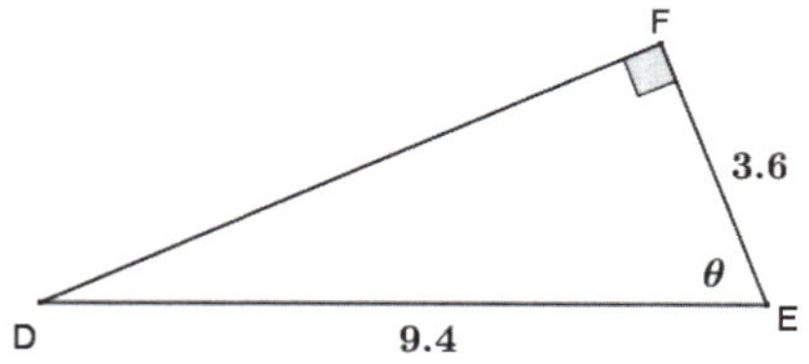

$$\cos\theta = \frac{3.6}{9.4}$$
$$\arccos\left(\frac{3.6}{9.4}\right) = \theta$$
$$\theta \approx 67°$$

c.

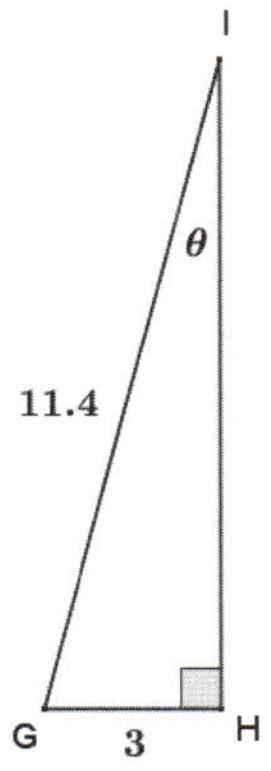

$\sin \theta = \frac{3}{11.4}$

$\arcsin\left(\frac{3}{11.4}\right) = \theta$

$\theta \approx 15°$

2. *Solving a right triangle* means using given information to find all the angles and side lengths of the triangle. Use $\arcsin$ and $\arccos$, along with the given information, to solve right triangle ABC if leg $AC = 12$ and hypotenuse $AB = 15$.

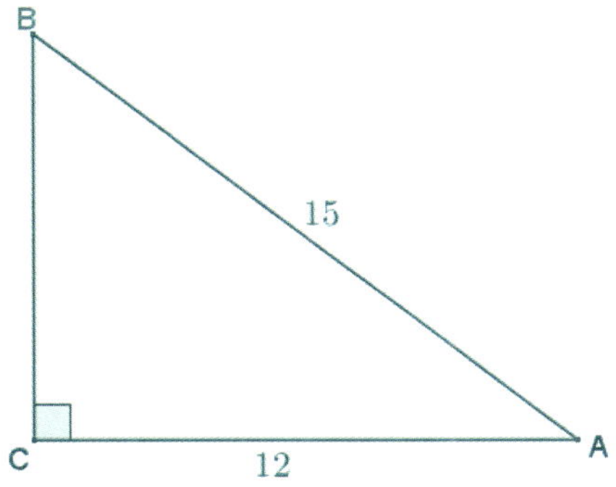

By the Pythagorean theorem, $BC = 9$.

$\sin B = \frac{12}{15}$

$\arcsin\left(\frac{12}{15}\right) = m\angle B$

$m\angle B \approx 53°$

$\cos A = \frac{12}{15}$

$\arccos\left(\frac{12}{15}\right) = m\angle A$

$m\angle A \approx 37°$

Once you have found the measure of one of the acute angles in the right triangle, can you find the measure of the other acute angle using a different method than those used in this lesson? Explain.

Yes. We could use the angle sum of a triangle after finding the measure of one acute angle.

3. A pendulum consists of a spherical weight suspended at the end of a string whose other end is anchored at a pivot point P. The distance from P to the center of the pendulum's sphere, B, is 6 inches. The weight is held so that the string is taut and horizontal, as shown to the right, and then dropped.

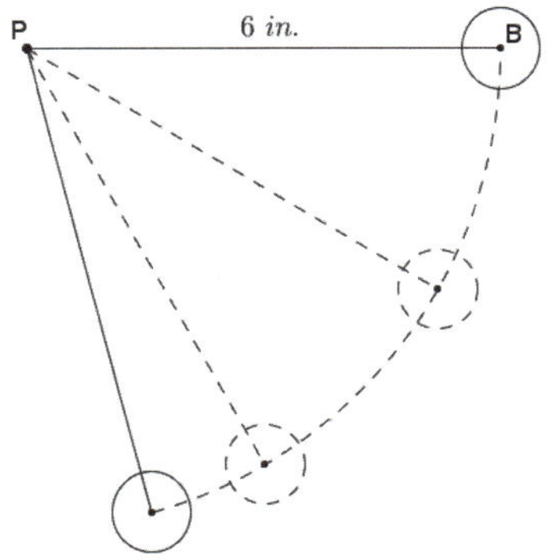

a. What type of path does the pendulum's weight take as it swings?

Since the string is a constant length, the path of the weight is circular.

b. **Danni thinks that for every vertical drop of 1 inch that the pendulum's weight makes, the degree of rotation is $15°$. Do you agree or disagree with Danni? As part of your explanation, calculate the degree of rotation for every vertical drop of 1 inch from 1 inch to 6 inches.**

Disagree A right triangle can model the pendulum and its vertical drops as shown in the diagrams.

The angle of rotation about P for a vertical drop of 1 inch is equal to the $\arcsin\left(\frac{1}{6}\right)$, which is approximately $10°$.

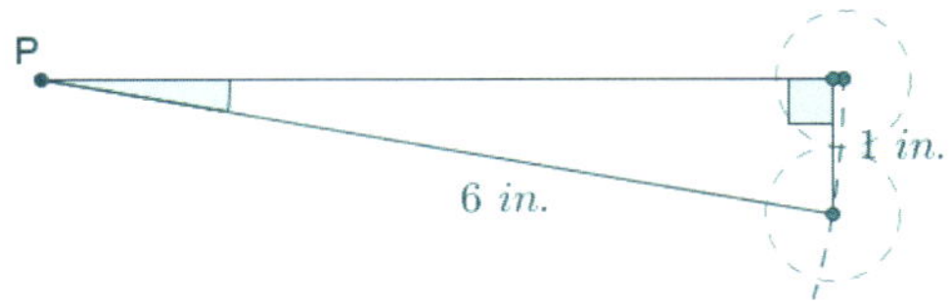

The angle of rotation about P for a vertical drop of 2 inches is equal to the $\arcsin\left(\frac{2}{6}\right)$, which is approximately $20°$.

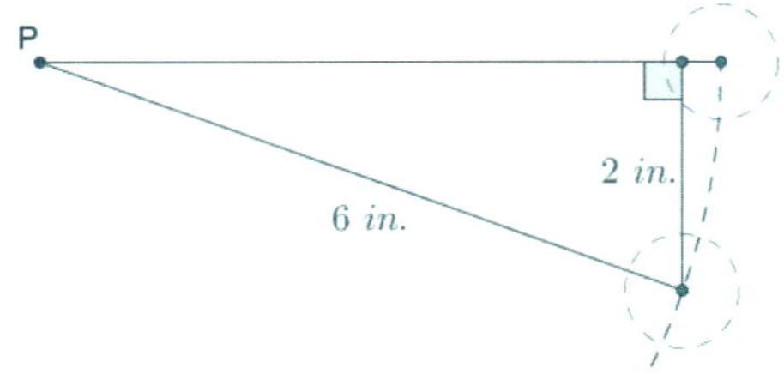

The angle of rotation about P for a vertical drop of 3 inches is equal to the $\arcsin\left(\frac{3}{6}\right)$, which is $30°$.

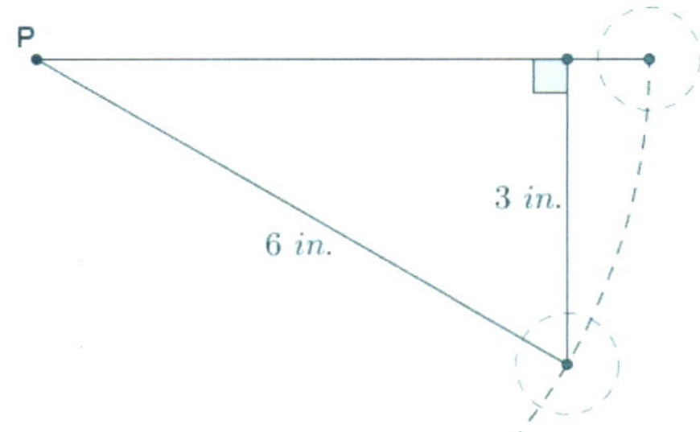

The angle of rotation about P for a vertical drop of 4 inches is equal to the $\arcsin\left(\frac{4}{6}\right)$, which is approximately $42°$.

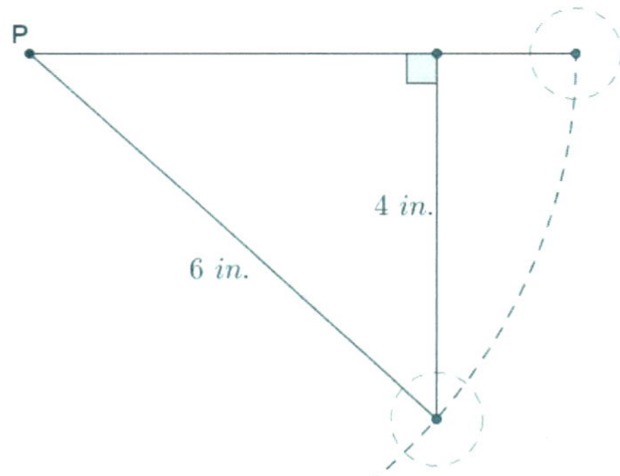

The angle of rotation about P for a vertical drop of 5 inches is equal to the $\arcsin\left(\frac{5}{6}\right)$, which is approximately $56°$.

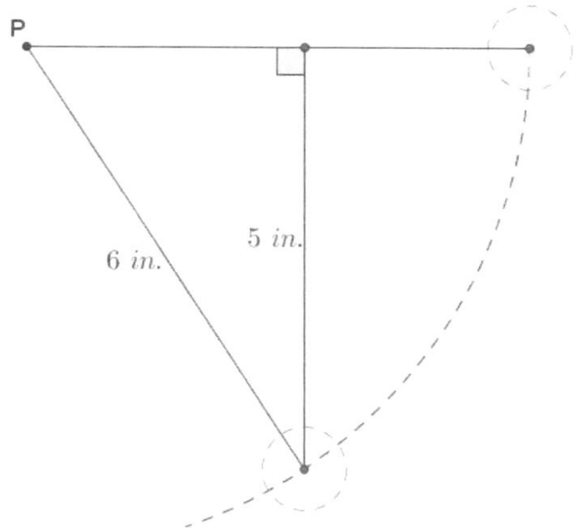

The pendulum's weight will be at its maximum distance below the pivot point, which means that the weight must be directly below the pivot point. This means that the string would be perpendicular to the horizontal starting position; therefore, the degree of rotation would be $90°$.

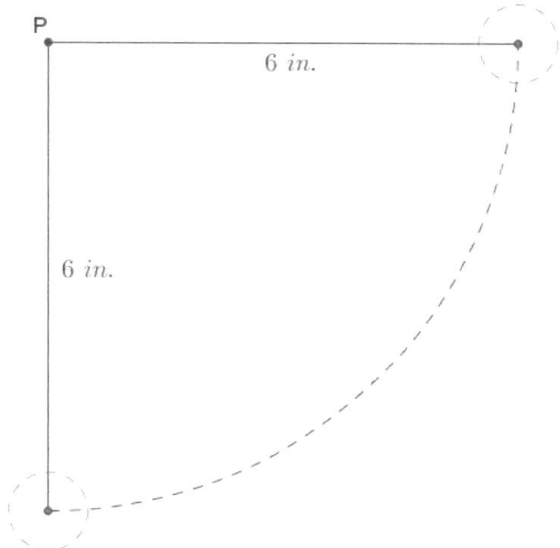

4. A stone tower was built on unstable ground, and the soil beneath it settled under its weight, causing the tower to lean. The cylindrical tower has a diameter of 17 meters. The height of the tower on the low side measured 46.3 meters and on the high side measured 47.1 meters. To the nearest tenth of a degree, find the angle that the tower has leaned from its original vertical position.

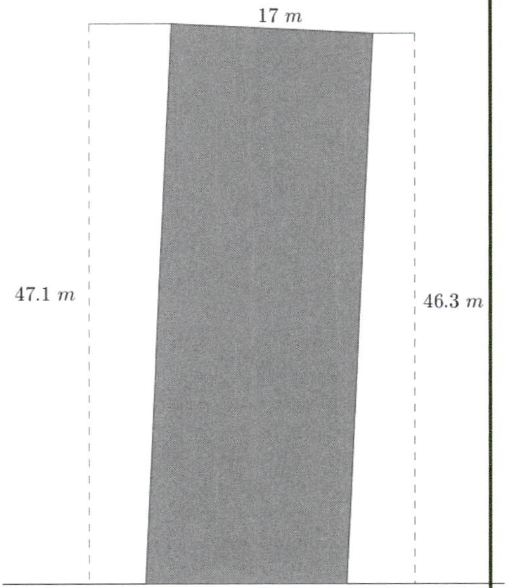

The difference in heights from one side of the tower to the other is $47.1\text{ m} - 46.3\text{ m} = 0.8\text{ m}$.

Model the difference in heights and the diameter of the tower using a right triangle. (The right triangle shown below is not drawn to scale).

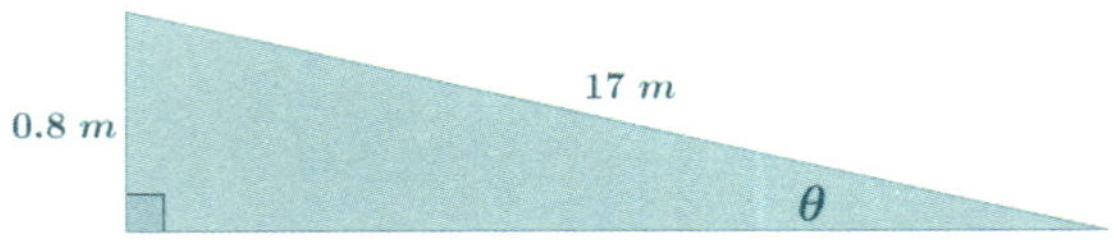

The unknown value θ *represents the degree measure that the tower has leaned.*

$$\sin\theta = \frac{0.8}{17}$$
$$\theta = \arcsin\left(\frac{0.8}{17}\right)$$
$$\theta \approx 3$$

The tower has leaned approximately $3°$ *from its vertical position.*

5. Doug is installing a surveillance camera inside a convenience store. He mounts the camera 8 ft. above the ground and 15 ft. horizontally from the store's entrance. The camera is meant to monitor every customer who enters and exits the store. At what angle of depression should Doug set the camera to capture the faces of all customers?

Note: This is a modeling problem and therefore will have various reasonable answers.

The solution below represents only one of many possible reasonable solutions.

Most adults are between $4\frac{1}{2}$ ft. *and* $6\frac{1}{2}$ ft. *tall, so the camera should be aimed to capture images within the range of* $1\frac{1}{2}$ ft. *to* $3\frac{1}{2}$ ft. *below its mounted height. Cameras capture a range of images, so Doug should mount the camera so that it points at a location in the doorway* $2\frac{1}{2}$ ft. *below its mounted height.*

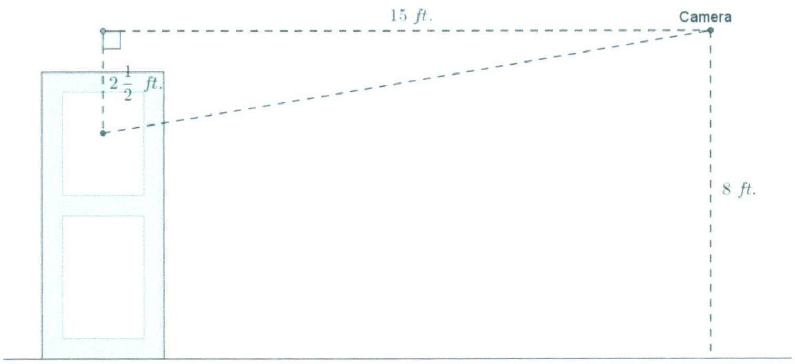

The angle of depression is equal to the $\arctan\left(\frac{2\frac{1}{2}}{15}\right)$*, which is approximately* $10°$.

Name ______________________________ Date ______________

1. In the figure below, rotate $\triangle EAB$ about E by $180°$ to get $\triangle EA'B'$. If $\overline{A'B'} \parallel \overline{CD}$, prove that $\triangle EAB \sim \triangle EDC$.

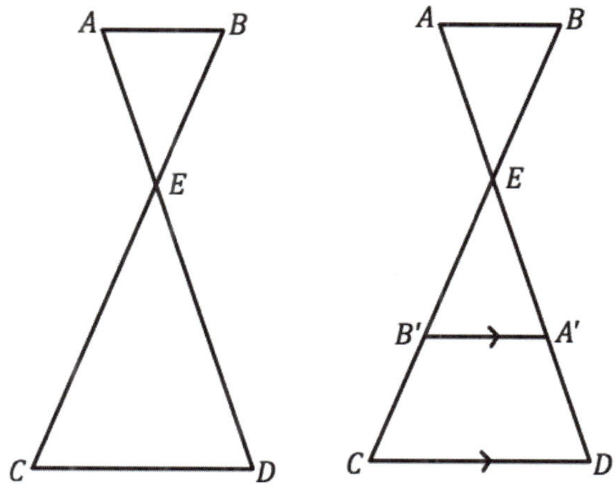

2. Answer the following questions based on the diagram below.

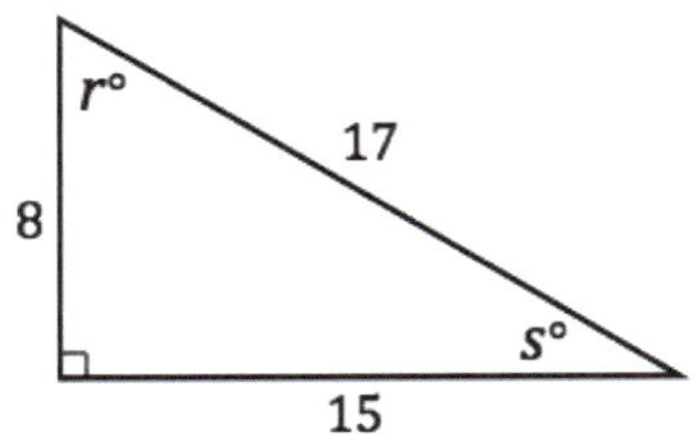

a. Find the sine and cosine values of angles r and s. Leave the answers as fractions.

$\sin r° =$ $\sin s° =$

$\cos r° =$ $\cos s° =$

$\tan r° =$ $\tan s° =$

b. Why is the sine of an acute angle the same value as the cosine of its complement?

c. Determine the measures of the angles to the nearest tenth of a degree in the right triangles below.

i. Determine the measure of $\angle a$.

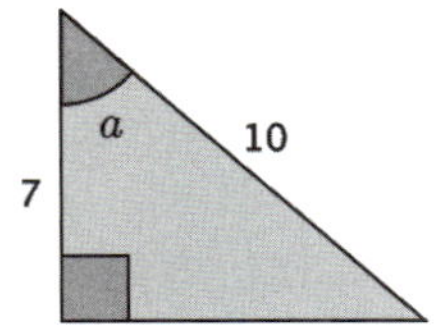

ii. Determine the measure of $\angle b$.

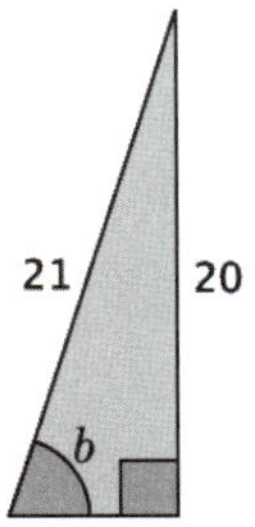

iii. Explain how you were able to determine the measure of the unknown angle in part (i) or part (ii).

d. A ball is dropped from the top of a 45 ft. building. Once the ball is released, a strong gust of wind blew the ball off course, and it dropped 4 ft. from the base of the building.

   i. Sketch a diagram of the situation.

   ii. By approximately how many degrees was the ball blown off course? Round your answer to the nearest whole degree.

3. A radio tower is anchored by long cables called guy wires, such as $\overline{AB}$ in the figure below. Point A is 250 m from the base of the tower, and $m\angle BAC = 59°$.

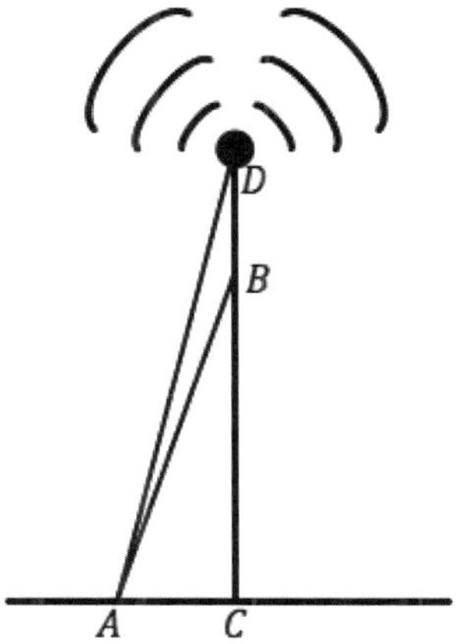

a. How long is the guy wire? Round to the nearest tenth.

b. How far above the ground is it fastened to the tower?

c. How tall is the tower, $\overline{DC}$, if $m\angle DAC = 71°$?

4. The following problem is modeled after a surveying question developed by a Chinese mathematician during the Tang dynasty in the seventh century C.E.

   A building sits on the edge of a river. A man views the building from the opposite side of the river. He measures the angle of elevation with a handheld tool and finds the angle measure to be $45°$. He moves 50 feet away from the river and remeasures the angle of elevation to be $30°$.

   What is the height of the building? From his original location, how far away is the viewer from the top of the building? Round to the nearest whole foot.

5. Prove the Pythagorean theorem using similar triangles. Provide a well-labeled diagram to support your justification.

6. In right triangle ABC with $\angle B$ a right angle, a line segment $B'C'$ connects side $\overline{AB}$ with the hypotenuse so that $\angle AB'C'$ is a right angle as shown. Use facts about similar triangles to show why $\cos C' = \cos C$.

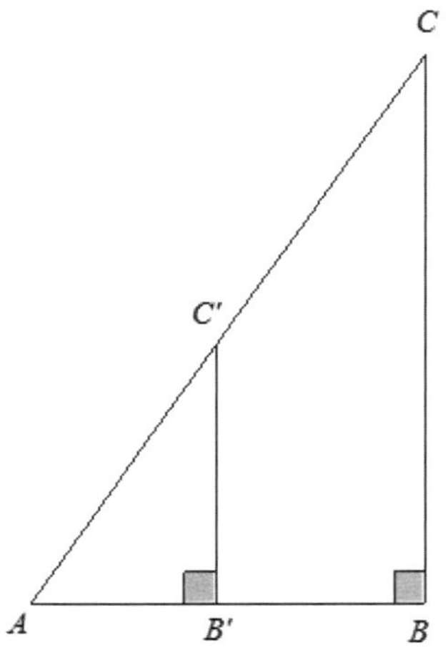

7. Terry said, "I will define the zine of an angle x as follows. Build an isosceles triangle in which the sides of equal length meet at angle x. The zine of x will be the ratio of the length of the base of that triangle to the length of one of the equal sides." Molly said, "Won't the zine of x depend on how you build the isosceles triangle?"

   a. What can Terry say to convince Molly that she need not worry about this? Explain your answer.

   b. Describe a relationship between zine and sin.

A Progression Toward Mastery					
Assessment Task Item		STEP 1 Missing or incorrect answer and little evidence of reasoning or application of mathematics to solve the problem.	STEP 2 Missing or incorrect answer but evidence of some reasoning or application of mathematics to solve the problem.	STEP 3 A correct answer with some evidence of reasoning or application of mathematics to solve the problem, OR an incorrect answer with substantial evidence of solid reasoning or application of mathematics to solve the problem.	STEP 4 A correct answer supported by substantial evidence of solid reasoning or application of mathematics to solve the problem.
1	G-SRT.B.5	Student shows little or no understanding of how to prove $\triangle EAB \sim \triangle EDC$.	Student makes an error in the rotation or it is not shown and makes an error in the justification.	Student makes an error in the rotation or it is not shown, but the rest of the justification is correct. OR Student correctly makes the rotation but makes an error in the justification.	Student correctly explains and proves $\triangle EAB \sim \triangle EDC$.
2	a–b G-SRT.C.7	Student makes one or more errors in part (a) and does not answer or provides poor reasoning for part (b).	Student makes one error in part (a) and makes an error in reasoning for part (b).	Student correctly answers part (a) but makes an error in reasoning for part (b).	Student correctly answers parts (a) and (b).
	c G-SRT.C.7	Student makes little or no attempt to find the angle measures.	Student correctly identifies the measure of each angle in part (i) and part (ii) but does not have an explanation in part (iii). OR Student correctly identifies the measure of one of the angles in part (i) or part (ii) and has a weak explanation for part (iii).	Student correctly identifies the measure of each angle in part (i) and part (ii) but has a weak explanation in part (iii). OR Student correctly identifies the measure of one of the angles in part (i) or part (ii) and has a complete explanation for part (iii).	Student correctly identifies the measure of each angle in part (i) and part (ii) and has a complete explanation for how the answer was determined.

	d G-SRT.C.7	Student makes little or no attempt to complete the problem. A sketch may be drawn, but it is not accurate for the situation.	Student draws a sketch and makes a major error in the calculation of the angle. For example, student may have used a function other than arctan.	Student draws an accurate sketch but may have made a mathematical error, leading to an incorrect answer.	Student draws an accurate sketch and has identified the correct angle measure.
3	a–c G-SRT.C.8	Student makes at least one computational or conceptual error in all three parts. OR Student makes at least one computational and one conceptual error in any two parts.	Student makes one computational or one conceptual error in any two parts. OR Student makes more than one computational or conceptual error in any one of the three parts.	Student makes one computational or conceptual error in any one of the three parts.	Student correctly answers all three parts.
4	G-SRT.C.8	Student makes three or more computational or conceptual errors.	Student makes two computational or conceptual errors.	Student makes one computational or conceptual error.	Student correctly finds the height of the building and the distance of the viewer to the top of the building.
5	G-SRT.B.4	Student shows little or no understanding of how similar triangles can be used to prove the Pythagorean theorem.	Student provides a correctly labeled diagram but offers an inaccurate justification toward proving the Pythagorean theorem.	Student provides a correctly labeled diagram and shows how each of the subtriangles has lengths in proportion to the corresponding lengths of the large triangle but does not tie these ideas together to show why the Pythagorean theorem must be true.	Student clearly demonstrates how to use similar triangles to prove the Pythagorean theorem and includes an accurately drawn and labeled diagram.
6	G-SRT.C.6	Student shows little or no understanding of similar triangles; student is unable to coherently show why $\triangle ABC \sim \triangle AB'C'$.	Student cites that $\triangle ABC \sim \triangle AB'C'$ but does not explicitly state facts about similar triangles to further explain why $\cos C' = \cos C$.	Student uses facts about similar triangles to show why $\frac{B'C'}{AC'} = \frac{BC}{AC}$ but does not explicitly conclude that $\cos C' = \cos C$.	Student correctly demonstrates that $\cos C' = \cos C$.
7	a–b G-SRT.C.6	Student shows little or no understanding of similar triangles.	Student shows evidence of understanding but lacks clarity in the reasoning for both parts (a) and (b).	Student shows evidence of understanding but lacks clarity in the reasoning for either part (a) or part (b).	Student provides accurate and well-reasoned responses for both parts (a) and (b).

Name ____ Date ____________________

1. In the figure below, rotate $\triangle EAB$ about E by $180°$ to get $\triangle EA'B'$. If $\overline{A'B'} \parallel \overline{CD}$, prove that $\triangle EAB \sim \triangle EDC$.

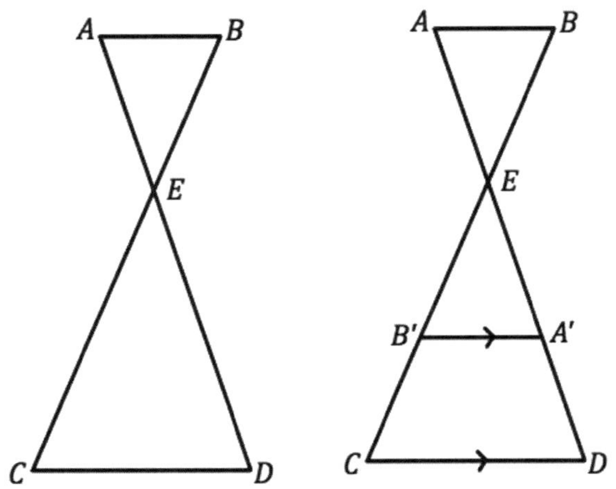

Triangles EAB and EA'B' are congruent since a 180° rotation is a rigid motion. Since $\overline{A'B'} \parallel \overline{CD}$, $m\angle EB'A' = m\angle ECD$, and $m\angle B'EA' = m\angle CED$. So, $\triangle EA'B' \sim \triangle EDC$ by AA similarity criteria, and $\triangle EAB \sim \triangle EDC$ by the transitive property of similarity.

2. Answer the following questions based on the diagram below.

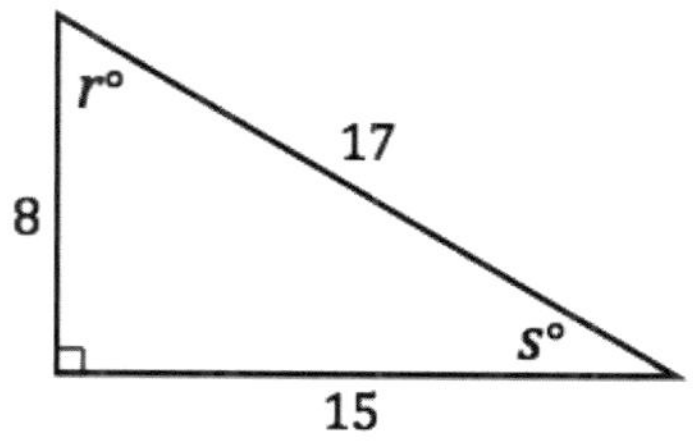

a. Find the sine and cosine values of angles r and s. Leave the answers as fractions.

$\sin r° = \frac{15}{17}$ $\sin s° = \frac{8}{17}$

$\cos r° = \frac{8}{17}$ $\cos s° = \frac{15}{17}$

$\tan r° = \frac{15}{8}$ $\tan s° = \frac{8}{15}$

b. Why is the sine of an acute angle the same value as the cosine of its complement?

By definition, sine is the ratio of the opposite side: hypotenuse, and cosine is the ratio of the adjacent side: hypotenuse; since the opposite side of an angle is the adjacent side of its complement, $\sin\theta = \cos(90 - \theta)$.

c. Determine the measures of the angles to the nearest tenth of a degree in the right triangles below.

i. Determine the measure of $\angle a$.

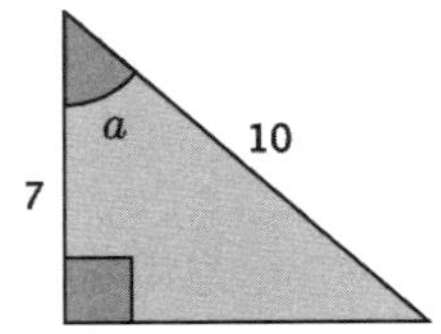

$m\angle a \approx 45.6°$

ii. Determine the measure of $\angle b$.

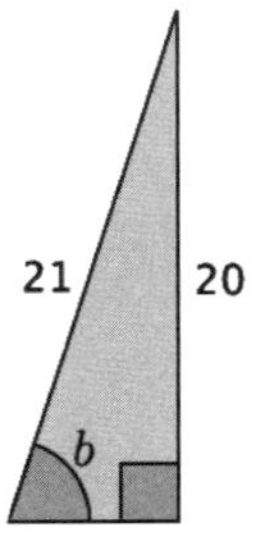

$m\angle b \approx 72.2°$

iii. Explain how you were able to determine the measure of the unknown angle in part (i) or part (ii).

For part (i), students should state that they had to use arccos to determine the unknown angle because the information given about the side lengths included the side adjacent to the unknown angle and the hypotenuse.

For part (ii), students should state that they had to use arcsin to determine the unknown angle because the information given about the side lengths included the side opposite to the unknown angle and the hypotenuse.

d. A ball is dropped from the top of a 45 ft. building. Once the ball is released, a strong gust of wind blew the ball off course, and it dropped 4 ft. from the base of the building.

i. Sketch a diagram of the situation.

Sample sketch shown below.

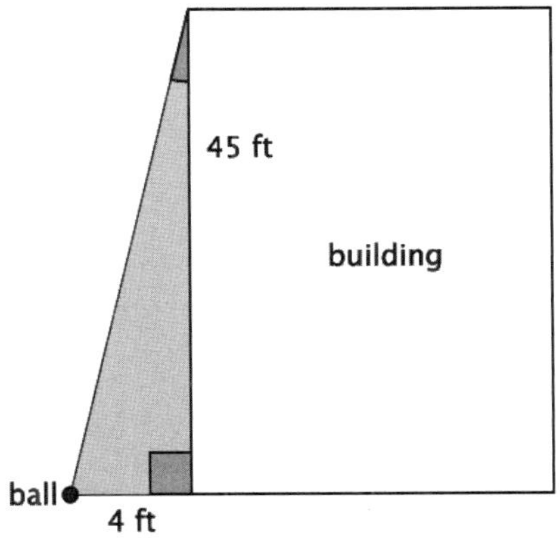

ii. By approximately how many degrees was the ball blown off course? Round your answer to the nearest whole degree.

The wind blew the ball about 5° off course.

3. A radio tower is anchored by long cables called guy wires, such as $\overline{AB}$ in the figure below. Point A is 250 m from the base of the tower, and $m\angle BAC = 59°$.

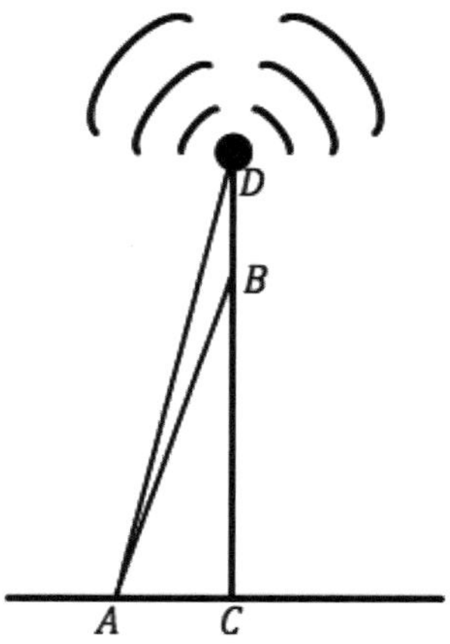

a. How long is the guy wire? Round to the nearest tenth.

$$\cos 59 = \frac{250}{AB}$$
$$AB = \frac{250}{\cos 59}$$
$$AB \approx 485.4$$

The guy wire is approximately 485.4 meters.

b. How far above the ground is it fastened to the tower?

$$\tan 59 = \frac{BC}{250}$$
$$BC = 250 \tan 59$$
$$BC \approx 416.1$$

It is fastened approximately 416.1 meters above the ground.

c. How tall is the tower, $\overline{DC}$, if $m\angle DAC = 71°$?

$$\tan 71 = \frac{DC}{250}$$
$$DC = 250 \tan 71$$
$$DC \approx 726.1$$

The tower is approximately 726.1 meters tall.

4. The following problem is modeled after a surveying question developed by a Chinese mathematician during the Tang dynasty in the seventh century C.E.

A building sits on the edge of a river. A man views the building from the opposite side of the river. He measures the angle of elevation with a handheld tool and finds the angle measure to be $45°$. He moves 50 feet away from the river and remeasures the angle of elevation to be $30°$.

What is the height of the building? From his original location, how far away is the viewer from the top of the building? Round to the nearest whole foot.

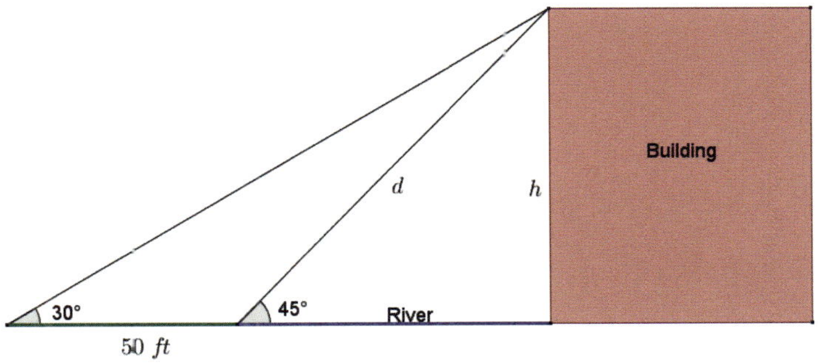

The angle of depression from the top of the building to the man's original spot is also $45°$, and the angle of depression to his final position is $60°$, so the difference of the angles is $15°$. Let d represent the distance from the man's position at the edge of the river to the top of the building, and let h represent the height of the building in feet.

Using the law of sines:

$$\frac{\sin 30}{d} = \frac{\sin 15}{50}$$

$$d = \frac{50 \sin 30}{\sin 15} \approx 96.6$$

The distance d is approximately 96.6 feet.

By the Pythagorean theorem, the distance $d = h\sqrt{2}$.

$$50\frac{\sin 30}{\sin 15} = h\sqrt{2}$$

$$\frac{25}{\sin 15(\sqrt{2})} = h$$

$$h \approx 68.3$$

The distance from the viewer to the top of the building is approximately 97 ft., and the height of the building is approximately 68 ft.

5. Prove the Pythagorean theorem using similar triangles. Provide a well-labeled diagram to support your justification.

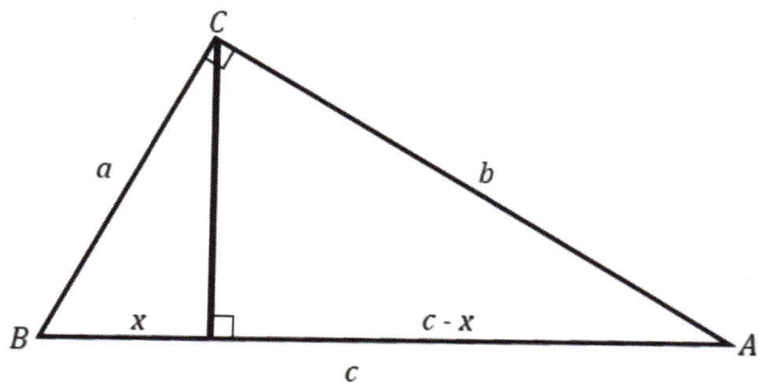

A right triangle ABC has side lengths a, b, and c. An altitude is drawn from C to the opposite side, dividing c into lengths x and $c-x$. Since the altitude from C divides the triangle into smaller similar right triangles by the AA criterion, then:

$$\frac{a}{x} = \frac{c}{a}$$
$$a^2 = cx$$
$$\frac{b}{c-x} = \frac{c}{b}$$
$$b^2 = c(c-x)$$
$$a^2 + b^2 = cx + c(c-x) = c^2$$
$$a^2 + b^2 = c^2$$

6. In right triangle ABC with $\angle B$ a right angle, a line segment $B'C'$ connects side $\overline{AB}$ with the hypotenuse so that $\angle AB'C'$ is a right angle as shown. Use facts about similar triangles to show why $\cos C' = \cos C$.

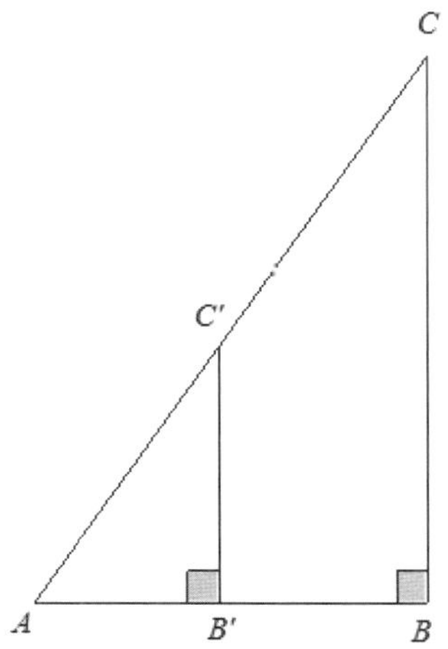

By the AA criterion, $\triangle ABC \sim \triangle AB'C'$. Let r be the scale factor of the similarity transformation. Then, $B'C' = r \cdot BC$, and $AC' = r \cdot AC$. Thus,

$$\cos C' = \frac{B'C'}{AC'} = \frac{r \cdot BC}{r \cdot AC} = \frac{BC}{AC} = \cos C$$

$$\cos C' = \cos C$$

7. Terry said, "I will define the zine of an angle x as follows. Build an isosceles triangle in which the sides of equal length meet at angle x. The zine of x will be the ratio of the length of the base of that triangle to the length of one of the equal sides." Molly said, "Won't the zine of x depend on how you build the isosceles triangle?"

   a. What can Terry say to convince Molly that she need not worry about this? Explain your answer.

      Isosceles triangles with vertex angle x are similar to each other; therefore, the value of zine x, the ratio of the length of the base of that triangle to the length of one of the equal sides, is the same for all such triangles.

   b. Describe a relationship between zine and sin.

      $\text{zine } x = 2 \sin \frac{1}{2}x$

EUREKA MATH™